Preliminary Edition

Differential Equations

Preliminary Edition

Differential Equations

Paul Blanchard

Robert L. Devaney

Glen R. Hall

Boston University

Brooks/Cole Publishing Company

I⟨T⟩P®

ITP An International Thomson Publishing Company

Pacific Grove • Albany • Belmont • Bonn • Boston • Cincinnati • Detroit
Johannesburg • London • Madrid • Melbourne • Mexico City • New York • Paris
Singapore • Tokyo • Toronto • Washington

Brooks/Cole Publishing Company

I(T)P® The ITP logo is a registered trademark under license.
International Thomson Publishing

For more information, contact:

Brooks/Cole Publishing Co.
511 Forest Lodge Rd.
Pacific Grove, CA 93950

International Thomson Publishing Europe
Berkshire House I68-I73
High Holborn
London WC1V 7AA
England

Thomas Nelson Australia
102 Dodds Street
South Melbourne, 3205
Victoria, Australia

Nelson Canada
1120 Birchmount Road
Scarborough, Ontario
Canada M1K 5G4

International Thomson Editores
Campos Eliseos 385, Piso 7
Col. Polanco
11560 Mexico D.F., Mexico

International Thomson Publishing GmbH
Konigswinterer Strasse 418
53227 Bonn, Germany

International Thomson Publishing Asia
221 Henderson Road
#05-10 Henderson Building
Singapore 0315

International Thomson Publishing Japan
Hirakawacho Kyowa Building, 31
2-2-1 Hirakawacho
Chiyoda-ku, Tokyo 102
Japan

Developmental Editor: Leslie Bondaryk
Production Editor and Interior Design: Pamela Rockwell
Cover Design: Julia Gecha
Manufacturing Coordinator: Wendy Kilborn
Marketing Manager: Marianne Rutter
Cover and Text Printer and Binder: Courier Westford, Inc.

Printed and bound in the United States of America.

96 97 98 99 -- 10 9 8 7 6 5 4 3 2

CONTENTS

PREFACE

The study of differential equations is a beautiful application of the ideas and techniques of calculus to our everyday lives. Indeed, it could be said that calculus was developed mainly so that the fundamental principles that govern many phenomena could be expressed in the language of differential equations. Unfortunately, it was difficult to convey the beauty of the subject in the traditional first course on differential equations because the number of equations that can be treated by analytic techniques is very limited. Consequently, the course tended to focus on technique rather than on concept.

This book is an outgrowth of our opinion that we are now able to effect a radical revision, and we approach our updated course with several goals in mind. First, the traditional emphasis on specialized tricks and techniques for solving differential equations is no longer appropriate given the technology that is readily available. Second, many of the most important differential equations are nonlinear, and numerical and qualitative techniques are more effective than analytic techniques in this setting. Finally, the differential equations course is one of the few undergraduate courses where it is possible to give students a glimpse of the nature of contemporary mathematical research.

The Qualitative, Numeric, and Analytic Approaches

Accordingly, this book is a radical departure from the typical "cookbook" differential equations text. We have eliminated most specialized techniques for deriving formulas for solutions, and we have replaced them with topics that focus on the formulation of differential equations and the interpretation of their solutions. To obtain an understanding of the solutions, we generally attack a given equation from three different points of view.

One major approach we adopt is qualitative. We expect students to be able to visualize differential equations and their solutions in many geometric ways. For example, we readily use slope fields, graphs of solutions, vector fields and solution curves in the phase portrait as tools to gain a better understanding of solutions. We also ask students to become adept at moving among these geometric representations and more traditional analytic representations.

Since differential equations are readily studied using the computer, we also emphasize numerical techniques. We assume that students have access to some sort of technology that approximates solutions and graphs these solutions easily. Even if we can find an explicit formula for a solution, we often work with the equation both numerically and qualitatively to understand the geometry and the long-term behavior of solutions. When we can find explicit solutions easily (such as in the case of separable first-order equations or constant-coefficient, linear systems), we do the calculations. But we never fail to examine the resulting formulas we obtain using qualitative and numerical points of view as well.

Specific Changes

There are several specific ways in which this book differs from other books at this level. First, we incorporate modeling throughout. We expect students to understand the meaning of the variables and parameters in a differential equation and to be able to interpret this meaning in terms of a particular model. Certain models reappear often as running themes, and they are drawn from a variety of disciplines so that students with various backgrounds will find something familiar.

We also adopt a dynamical systems point of view. Thus, we are always concerned with the long-term behavior of solutions of an equation, and using all of the appropriate approaches outlined above, we ask students to predict this long-term behavior of solutions. In addition, we emphasize the role of parameters in many of our examples, and we specifically address the manner in which the behavior of solutions changes as these parameters are varied.

Like other texts, we begin with first-order equations, but the only analytic technique we use to find closed-form solutions is separation of variables (and, at the end of the chapter, an integrating factor or two to handle certain linear equations). Instead, we emphasize the meaning of a differential equation and its solutions in terms of its slope field and the graphs of its solutions. If the differential equation is autonomous, we also discuss its phase line. This discussion of the phase line serves as an elementary introduction to the idea of a phase plane, which plays a fundamental role in subsequent chapters.

We then move directly to systems of differential equations. Rather than consider second-order equations separately, we immediately convert these equations to systems. When these equations are viewed as systems, we are able to use qualitative and numerical techniques more readily. Of course, we then use the information about these systems gleaned from these techniques to recover information about the solutions of the original equation.

We also begin the treatment of systems with a general approach. We do not immediately restrict our attention to linear systems. Qualitative and numerical techniques work just as easily when a system is nonlinear, and one can proceed a long way toward understanding systems without resorting to algebraic techniques. However, qualitative ideas do not tell the whole story, and we are led naturally to the idea of linearization. With this background in the fundamental geometric and qualitative concepts, we then discuss linear systems in detail. As always, we not only emphasize the formula for the general solution of a linear system but also the geometry of its solution curves and of the related eigenvectors and eigenvalues.

While our study of systems requires the minimal use of some linear algebra, it is definitely not a prerequisite. As we deal primarily with two-dimensional systems, we easily develop all of the necessary algebraic techniques as we proceed.

These topics form the core of our approach. However, there are many additional topics that one would like to cover in the course. Consequently, we have included discussions of nonlinear systems, forced second-order equations, discrete dynamical systems, numerical methods, and Laplace transforms. Although some of these topics are quite traditional, we always present them in a manner that is consistent with the philosophy developed in the first half of the text.

At the end of each chapter, we have included several "labs." Doing detailed numerical experimentation and writing reports has been our most successful modification of the traditional course at Boston University. Good labs are tough to write and to grade, but we feel that the benefit to students is extraordinary.

Pathways Through This Book

There are a number of possible tracks that instructors can follow in using this book. We feel that Chapters 1–3 form the core (with the possible exception of Sections 2.7 and 3.7, which cover systems in three dimensions). Most of the later chapters assume familiarity with this material. Certain sections such as Section 1.7 (bifurcations) and Section 1.9 (changing variables) may be skipped if some care is taken in choosing material from subsequent sections. However, the material on phase lines and phase planes, qualitative analysis, and solutions of linear systems is central.

A typical track for an engineering-oriented course would follow Chapters 1–3 (perhaps skipping Sections 1.9, 2.7, and 3.7). These chapters will take roughly two-thirds of a semester. The final third of the course might cover Section 4.1 (linearization of nonlinear systems), Sections 5.1 and 5.2 (forced, second-order linear equations and resonance), and Chapter 8 (Laplace transforms).

Incidentally, it is possible to skip to the Section 8.1 (Laplace transforms and first-order equations) immediately after covering Chapter 1. As we have learned from our colleagues in the College of Engineering at Boston University, some engineering programs have adopted a circuit theory course that uses the Laplace transform at an earlier point than is typically the case. Consequently, Section 8.1 is written so that the differential equations course and such a circuits course could proceed in parallel. However, if possible, we recommend waiting to cover Chapter 8 entirely until after the material in Sections 5.1 and 5.2 has been discussed.

Instructors may wish to substitute material on discrete dynamics (Chapter 6) for Laplace transforms. A course for students with a strong background in physics might involve more of Chapter 4, including a treatment of Hamiltonian (Section 4.2) and gradient systems (Section 4.3). A course geared toward applied mathematics might include a more detailed discussion of numerical methods (Chapter 7),

In What Way is This Edition Preliminary?

The easy answer is "in every way." In writing this book, we have had to make many choices about what to include, what to emphasize, and what to leave out. We have tried to make these choices based on the philosophy described above, but the choices we have made are definitely not the only ones possible. We encourage comments, suggestions, and even criticism. The best way to comment is to send email to odes@math.bu.edu. We'll do our best to acknowledge the email, but we will definitely read and consider every comment. The first edition of this text is due out late in 1997, so suggestions up until the summer of 1997 can be incorporated into that edition.

The Boston University Differential Equations Project

This book is one product of the NSF-sponsored Boston University Differential Equations Project (NSF Grant DUE-9352833). The goal of this project is to rethink completely the traditional, sophomore-level differential equations course. In addition to producing this text to support a revised course, the project will produce software to supplement the course, and this software will be freely distributed via the Internet. We also hold workshops on the teaching of differential equations. Additional information about our project is available from our Web site, http://math.bu.edu/odes, and our publisher has set up a related site, http://www.pws.com/diffeq.html, to distribute materials written by others. We also maintain an electronic discussion group related to the teaching of differential equations. To join this group, send email to odes@math.bu.edu.

Paul Blanchard
Robert L. Devaney
Glen R. Hall
Boston University
September 1995

ACKNOWLEDGMENTS

Tremendous thanks are due to **Sam Kaplan**. Sam has been involved with the project almost since its beginning, helping to teach early versions of the course and making valuable contributions to the exposition involving both content and style. As project manager, he has been crucial to the creation of the text and its pictures. It's hard to say too much good about Sam. Thanks, Sam.

With the exception of a few professionally drawn figures, this book was entirely produced at Boston University's Department of Mathematics using Alex Kasman's ASTEX macro package in concert with LATEX 2_ε. Alex is a true TEX wizard, and anyone who is writing a textbook should consider his package. In fact, Alex's graphics macros are extremely useful in many contexts (see Alex's Web page available from the site http://math.bu.edu).

Much of the production work, accuracy checking, and rendering of computer output was done by our team of graduate students and postdocs — Gareth Roberts, Adrian Vajiac, Kinya Ono, Adrian Iovita, and Nuria Fagella. They worked many long nights, way beyond the call of duty, to bring this work to completion on schedule. This would not have been possible without all of their hard work.

Many other individuals at Boston University have made important contributions. In particular, our teaching assistants Duff Campbell, Amnon Ben-Pazi, Bridget Neale, Nick Galbreath, and Clara Bodelon had to put up with the headaches associated to our experimentation. We tested our ideas with at least 10 different classes of various sizes here, and we are grateful that so many of the students gracefully endured our incomplete drafts and rough exposition. We would also like to thank two undergraduates, Beverly Steinhoff and Bill French, who made definite contributions during the final summer of production.

We received support from many of our colleagues at BU as well. Our chair, Marvin Freedman, encouraged us throughout. And it was a special pleasure for us to be able to receive assistance from colleagues in the College of Engineering — Michael Ruane (who coordinates the circuits course), Moe Wasserman (who permitted one of the authors to audit his course), and John Baillieul (a member of our advisory board).

As is mentioned in the Preface, this book would not exist if our project had not received support from the National Science Foundation's Division of Undergraduate Education, and we would like to thank the program directors at NSF/DUE for their enthusiastic support over the last two years. We would also like to thank the members of the advisory board — John Baillieul, Morton Brown, John Franks, Deborah Hughes Hallett, Philip Holmes, and Nancy Kopell. All have contributed their scarce time during workshops and trips to BU.

We have been pleased that so many of our colleagues outside of Boston University have been willing to help us with this project. Bill Krohn gave us valuable advice regarding our exposition of Laplace transforms, and Bruce Elenbogen did a thorough reading of early drafts of the beginning chapters. Throughout the editorial process, our publisher solicited independent evaluations, and we found those comments particularly informative as we did our revisions. Preliminary drafts of the notes were class tested in a number of different settings by Gregory Buck, Scott Sutherland, Kathleen Alligood, Diego Benedette, Jack Dockery, Mako Haruta, Jim Henle, Ed Packel, and Ben Pollina. Thanks, all.

Finally, as any author knows, writing a book requires significant sacrifices from one's family. Extra special thanks goes to Lori, Kathy, and Dottie.

G.R.H., R.L.D., P.B.

A NOTE TO THE STUDENT

This book is probably different from most of your previous mathematics texts. If you thumb through it, you will see that there are few "boxed" formulas, no margin notes, and very few n-step procedures. We've written the book this way because we think that you are now at a point in your education where you should be learning to identify and work effectively with the mathematics inherent in everyday life. As you pursue your careers, no one is going to ask you to do all of the odd exercises at the end of some employee manual. They are going to give you some problem whose mathematical component may be difficult to identify and ask you to do your best with it. One of our goals in this book is to start preparing you for this type of work by avoiding artificial algorthmic exercises.

Our intention is that you will read this book as you would any other text, then work on the exercises, rereading sections and examples as necessary. Even though there are no template examples, you will find the discussions full of examples. Since one of our main goals is to demonstrate how differential equations are used to model physical systems, we often start with the description of a physical system, build a model, and then study the model to make conclusions and predictions about the original system. Many of the exercises ask you to produce or modify a model of a physical system, analyze it, and explain your conclusions. This is hard stuff, and you will need to practice. Since the days when you could make a living plugging and chugging through computations are over (computers do that now), you will need to learn these skills, and we hope that this book helps you develop these skills.

Another way in which this book may differ from your previous texts is that we expect you to make judicious use of a graphing calculator or a computer as you work the exercises and labs. The computer won't do the thinking for you, but it will provide you with numerical evidence that is essentially impossible for you to get in any other way. One of our goals is to give you practice as a sophisticated consumer of computer cycles as well as a skeptic of computer results.

Finally, you should know that the authors take the study of differential equations very seriously. However, we don't take ourselves very seriously (and we certainly don't take the other two authors seriously). We have tried to express both the beauty of the mathematics and some of the fun we have doing mathematics. If you think (some of?) the jokes are old or stupid, you're probably right.

All of us who worked on this book have learned something about differential equations along the way, and we hope that we are able to communicate our appreciation for the subject's beauty and range of application.

We had fun writing this book. We hope you have fun reading it.

R.L.D., P.B., G.R.H.

REVIEWERS OF THE PRELIMINARY EDITION

FIRST-ORDER DIFFERENTIAL EQUATIONS

1

This book is about how to predict the future. All we have available to do this is a knowledge of how things are at the present time plus an understanding of the rules governing the changes that will occur. From calculus we know that rate of change is measured by the derivative. Using the derivative to express the rules governing how a quantity changes gives a "differential equation."

The process of turning the rules governing how a quantity changes into a differential equation is called modeling. In this chapter we study many examples of modeling using differential equations. Then the goal is to use the differential equation to predict future values of the quantity. There are three types of techniques for making these predictions. Analytical techniques involve finding formulas for the future values of the quantity. Qualitative techniques involve obtaining a rough sketches of the graph of the quantity versus time and a description of its long-term behavior. Numerical techniques involve doing arithmetic (or having a computer do arithmetic) giving numerical approximations of the future values of the quantity. We introduce and use all three types of these techniques in this chapter.

1.1 MODELING VIA DIFFERENTIAL EQUATIONS

The hardest part of using mathematics to study applications is translating the description of the "physical" situation into equations. This translation from real life into mathematical formalism is usually difficult because it involves the conversion of imprecise assumptions into very precise formulas. There is no way to avoid it. Modeling is difficult, and the best way to get good at it is the same way as you get to Carnegie Hall — practice, practice, practice.

What Is a Model?

It is important to remember that mathematical models are like other types of models — the goal is not to produce an exact copy of the "real" object, but rather to give a representation of some aspect of the real thing. For example, a portrait of a person, a store mannequin, and a pig can all be models of a human being. None is a perfect copy of a human, but each has certain aspects in common with a human: the painting gives a description of what a particular person looks like, the mannequin wears clothes like a person, and the pig is alive. Which of the three is the appropriate model depends on how you will use the model — to remember old friends, to buy clothes, or to study biology.

The mathematical models we study are systems that evolve in time. But these systems often depend on other variables as well. Real-world systems can be notoriously complicated — the population of rabbits in Wyoming depends on the number of coyotes, the number of bobcats, the number of mountain lions, the number of mice (alternative food for the predators), farming practices, the weather, any number of rabbit diseases, etc. We can make a model of the rabbit population simple enough to understand only by making simplifying assumptions and lumping together effects that may or may not belong together. Once we've built the model, we should compare predictions of the model with data from the system. If the model and the system agree, then we gain confidence that the assumptions we made in creating the model are reasonable, and we can use the model to make predictions. If the system and the model disagree, then we must study and improve our assumptions. In either case we learn more about the system by comparing it to the model.

The types of predictions that are reasonable depend on our assumptions. If our model is based on precise rules such as Newton's laws of motion or the rules of compound interest, then we can use the model to make very accurate quantitative predictions. If the assumptions are less precise or if the model is a simplified version of the system, then precise quantitative predictions would be silly. In this case we would use the model to make qualitative predictions such as "the population of rabbits in Wyoming will increase . . . ." The dividing line between qualitative and quantitative is itself imprecise, but we will see that it is frequently better and easier to make qualitative use of even the most precise models.

Some Hints for Model Building

The basic steps in creating the model are

Step 1 Clearly state the assumptions on which the model will be based. These assumptions should describe the relationships among the quantities to be studied.

Step 2 Completely describe the variables and parameters to be used in the model — "you can't tell the players without a program."

Step 3 Use the assumptions formulated in Step 1 to derive equations relating the quantities in Step 2.

Step 1 is the "science" step. In Step 1, we describe how we think the physical system works or, at least, what the most important aspects of the system are. In some cases these assumptions will be fairly speculative as, for example, "rabbits don't mind being overcrowded." In other cases the assumptions will be quite precise and well accepted, such as "force is equal to the product of mass and acceleration." The quality of the assumptions determines the validity of the model and the situations to which the model is relevant. For example, some population models apply only to small populations in large environments, whereas others consider limited space and resources. Most important, we must avoid "hidden assumptions" that make the model seem mysterious or magical.

Step 2 is where we name the quantities to be studied and, if necessary, describe the units and scales involved. Leaving this step out is like deciding you will speak your own language without telling anyone what the words mean.

The quantities in our models fall into three basic categories: the **independent variable**, the **dependent variables**, and the **parameters**. In this book the independent variable will (almost) always be time. Time is "independent" of any other quantity in the model. On the other hand, the dependent variables are quantities that are functions of the independent variable. For example, when we say that "position is a function of time," we mean that position is a variable that depends on time. We can vaguely state the goal of a model expressed in terms of a differential equation as "Describe the behavior of the dependent variable as the independent variable changes." For example, we may ask whether the dependent variable increases or decreases, or whether it oscillates or tends to a limit.

Parameters are quantities that don't change with time (or with the independent variable) but that can be adjusted (by a scientist running the experiment or by natural causes). For example, if we are studying the motion of a rocket, the initial mass of the rocket is a parameter. If we are studying the amount of ozone in the upper atmosphere, then the rate of release of fluorocarbons from refrigerators is a parameter. Determining how the behavior of the dependent variables changes when we adjust the parameters can be the most important aspect of the study of a model.

In Step 3 we create the equations. Most of the models we will consider are expressed in terms of differential equations. In other words, we expect to find derivatives in our equations. Look for phrases such as "rate of change of ..." or "rate of increase of ..." since rate of change is synonymous with derivative. Of course, also watch for "velocity" (derivative of position) and "acceleration" (derivative of velocity) in problems from physics. The word *is* means "equals" and indicates where the equality lies. The words *A is proportional to B* means $A = kB$, where k is the "constant of proportionality" (a parameter).

An important rule of thumb we use when formulating models is: *Always make the algebra as simple as possible.* For example, when modeling the velocity v of a cat falling from a tall building, we could assume:

● Air resistance increases as the cat's velocity increases.

This assumption says that air resistance provides a force that counteracts the force of gravity and that this force increases as the velocity v of the cat increases. We could choose kv or kv^2 for the air resistance term, where k is the friction coefficient, a parameter. Both expressions increase as v increases, so they satisfy the assumption. However, we would most likely try kv first because it is the simplest expression that satisfies the assumption. In fact, it turns out that kv yields a good model for falling bodies with low densities like snowflakes, but kv^2 is a more appropriate model for dense objects like raindrops.

Now we turn to a series of models of population growth based on various assumptions about the species involved. Our goal here is to study how to go from a set of assumptions to a model. These examples are not "state-of-the-art" models from population ecology, but they are good ones to consider initially. We also begin to describe the analytic, qualitative, and numerical techniques that we use to make predictions based on these models. Our approach is meant to be illustrative only; we discuss these mathematical techniques in much more detail throughout the entire book.

Unlimited Population Growth

An elementary model of population growth is based on the assumption that

• The rate of growth of the population is proportional to the size of the population.

Note that the rate of change of a population depends only on the size of the population and nothing else. In particular, limitations of space or resources have no effect. This assumption is reasonable for small populations in large environments, for example, the first few spots of mold on a piece of bread or the first European settlers in the United States.

Because the assumption is so simple, we expect the model to be simple as well. The quantities involved are

$$t = \text{time (independent variable),}$$
$$P = \text{population (dependent variable), and}$$
$$k = \text{proportionality constant (parameter) between the rate}$$
$$\text{of growth of the population and the size of the population.}$$

The parameter k is often called the "growth-rate coefficient."

The units for these quantities depend on the application. If we are modeling the growth of mold on bread, then t might be measured in days and $P(t)$ might be either the area of bread covered by the mold or the weight of the mold. If we are talking about the European population of the United States, then t probably should be measured in years and $P(t)$ in millions of people. In this case we could set $t = 0$ to be any time we wanted, but the year 1790 (first census) is a convenient choice.

Now let's express our assumption using this notation. The rate of growth of the population is the derivative dP/dt. Being proportional to the population is expressed as the product, kP, of the population P and the proportionality constant k. Hence our assumption is expressed as the differential equation

$$\frac{dP}{dt} = kP.$$

In other words, the rate of change of P is proportional to P.

This equation is our first example of a differential equation. Associated with it are a number of adjectives that describe the type of differential equation that we are considering. In particular, it is a **first-order** equation because it contains only first derivatives of the dependent variable, and it is an **ordinary differential equation** because it contains no partial derivatives. In this book we deal only with ordinary differential equations.

We have written this differential equation using the dP/dt Leibniz notation — the notation that we tend to use. However, there are many other ways to express the same differential equation. In particular, we could also write this equation as $P' = kP$ or $\dot{P} = kP$. The "dot" notation is often used when the independent variable is time t.

What does the model predict?

More important than the adjectives or how the equation is written is what the equation tells us about the situation being modeled. Since $dP/dt = kP$ for some constant k, $dP/dt = 0$ if $P = 0$. That is, if the population starts at zero, it stays at zero. Thus the constant function $P(t) = 0$ is a solution of the differential equation. This special type of solution is called an **equilibrium solution** because it is constant forever.

If $P(t_0) \neq 0$ at some time t_0, then

$$\frac{dP}{dt}(t_0) = kP(t_0) \neq 0.$$

As a consequence, the population is not constant. If $k > 0$ and $P(t_0) > 0$, we have

$$\frac{dP}{dt}(t_0) = kP(t_0) > 0,$$

and the population is increasing (as one would expect). As t increases, $P(t)$ becomes larger, so dP/dt becomes larger. In turn, $P(t)$ increases even faster. That is, the rate of growth increases as the population increases. We therefore expect that the graph of the function $P(t)$ might look like Figure 1.1.

The value of $P(t)$ at $t = 0$ is called an **initial condition**. If we start with a different initial condition we get a different function $P(t)$ as is indicated in Figure 1.2. If $P(0)$ is negative (remembering $k > 0$) we then have $P'(0) < 0$ so $P(t)$ is initially decreasing. As t increases, $P(t)$ becomes more negative. The picture below the t-axis is the flip of the picture above, although this isn't "physically meaningful" because a negative population doesn't make much sense.

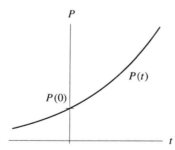

Figure 1.1
The graph of a function that satisfies the differential equation

$$\frac{dP}{dt} = kP.$$

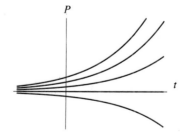

Figure 1.2
The graphs of several different functions that satisfy the differential equation $dP/dt = kP$. Each has a different initial condition at $t = 0$.

Our analysis of the way in which $P(t)$ increases as t increases is called a **qualitative analysis** of the differential equation. If all we care about is whether the model predicts "population explosions," then we can answer "yes, as long as $P(0) > 0$."

Analytic solutions of the differential equation

If, on the other hand, we know the exact value P_0 of $P(0)$ and we want to predict the value of $P(10)$ or $P(100)$, then we need more precise information about the function $P(t)$. The pair of equations

$$\frac{dP}{dt} = kP, \quad P(0) = P_0,$$

is called an **initial-value problem**. A **solution** to the initial-value problem is a function $P(t)$ that satisfies both equations. That is,

$$\frac{dP}{dt} = kP \text{ for all } t \quad \text{and} \quad P(0) = P_0.$$

Consequently, to find a solution to this differential equation we must find a function $P(t)$ whose derivative is k times itself. One (not very subtle) way to find such a function is to guess. In this case, it is relatively easy to guess the right form for $P(t)$ because we know that the derivative of an exponential function is essentially itself. (We can eliminate this guesswork by using the method of separation of variables, which we describe in the next section. But for now, let's just try the exponential and see where that leads us.) After a couple of tries with various forms of the exponential, we see that

$$P(t) = e^{kt}$$

is a function whose derivative, $P'(t) = ke^{kt}$, is k times $P(t)$. But there are other possible solutions, since $P(t) = ce^{kt}$ (where c is a constant) yields $P'(t) = c(ke^{kt}) = k(ce^{kt}) = kP(t)$. Thus $dP/dt = kP$ for all t no matter what the value of the constant c is.

We have infinitely many solutions to the differential equation, one for each value of c. To determine which of these solutions is the correct one for the situation at hand, we use the given initial condition. We have

$$P_0 = P(0) = c \cdot e^{k \cdot 0} = c \cdot e^0 = c \cdot 1 = c.$$

Consequently, we should choose $c = P_0$, so our solution for the initial-value problem is

$$P(t) = P_0 e^{kt}.$$

We have obtained an an actual formula for our solution, not just a qualitative picture of its graph.

The function $P(t)$ is called the solution to the initial-value problem as well as a **particular solution** of the differential equation. The collection of functions $P(t) = ce^{kt}$ is called the **general solution** of the differential equation because we can use it to find the particular solution corresponding to any initial-value problem. Figure 1.2 consists of the graphs of exponential functions of the form $P(t) = ce^{kt}$ with various values of the constant c, that is, with different initial values. In other words, it is a picture of the general solution to the differential equation.

The U. S. Population

As an example of how this model can be used, consider the U. S. census figures since 1790 given in Table 1.1.

Let's see how well the unlimited growth model fits this data. We measure time in years and the population $P(t)$ in millions of people. We also let $t = 0$ be the year 1790, so the initial condition is $P(0) = 3.9$. The corresponding initial-value problem

$$\frac{dP}{dt} = kP, \quad P(0) = 3.9,$$

has solution $P(t) = 3.9e^{kt}$. We cannot use this model to make predictions yet because we don't know the value of k. However, we do know that the population in the year 1800 was 5.3 million, and we can use this value to determine k. If we set

$$5.3 = P(10) = 3.9e^{k \cdot 10},$$

then we have

$$e^{k \cdot 10} = \frac{5.3}{3.9}$$

$$10k = \ln\left(\frac{5.3}{3.9}\right)$$

$$k \approx 0.03067.$$

Thus our model predicts that the United States population is given by

$$P(t) = 3.9e^{0.03067t}.$$

Table 1.1 U. S. census figures, in millions of people (see Funk and Wagnalls *1994 World Almanac*)

Year	t	Actual	$P(t) = 3.9e^{0.03067t}$	Year	t	Actual	$P(t) = 3.9e^{0.03067t}$
1790	0	3.9	3.9	1930	140	122	286
1800	10	5.3	5.3	1940	150	131	388
1810	20	7.2	7.2	1950	160	151	528
1820	30	9.6	9.8	1960	170	179	717
1830	40	12	13	1970	180	203	975
1840	50	17	18	1980	190	226	1,320
1850	60	23	25	1990	200	249	1,800
1860	70	31	33	2000	210		2,450
1870	80	38	45	2010	220		3,320
1880	90	50	62	2020	230		4,520
1890	100	62	84	2030	240		6,140
1900	110	75	114	2040	250		8,340
1910	120	91	155	2050	260		11,300
1920	130	105	210				

As we see from Figure 1.3, this model of $P(t)$ does a decent job of predicting the population until roughly 1860, but after 1860 the prediction is much too large. (Table 1.1 includes a comparison of the predicted values to the actual data.)

Our model is fairly good provided the population is relatively small. However, as time goes on, the model predicts that the population will continue to grow without any limits, and obviously, this cannot happen in the real world. Consequently, if we want a model that is accurate over a large time scale, we should account for the the fact that populations exist in a finite amount of space and with limited resources.

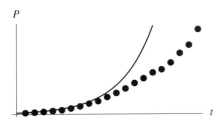

Figure 1.3
The dots represent actual census data and the solid line is the solution of the exponential growth model

$$\frac{dP}{dt} = 0.03067P.$$

Logistic Population Model

To adjust the exponential growth population model to account for a limited environment and limited resources, we add the assumptions:

- If the population is small, the rate of growth of the population is proportional to its size.
- If the population is too large to be supported by its environment and resources, the population will decrease. That is, the rate of growth is negative.

For this model, we again use

$$t = \text{time (independent variable),}$$
$$P = \text{population (dependent variable),}$$
$$k = \text{growth-rate coefficient for small}$$
$$\text{populations (parameter).}$$

However, our assumption about limited resources introduces another quantity, the size of the population that corresponds to being "too large." This quantity is a second parameter, denoted by N, that we call the "carrying capacity" of the environment. In terms of the carrying capacity, we are assuming that $P(t)$ is increasing if $P(t) < N$. However, if $P(t) > N$, we assume that $P(t)$ is decreasing.

Using this notation, we can restate our assumptions as:

- $\dfrac{dP}{dt} \approx kP$ if P is small (first assumption).
- If $P > N$, $\dfrac{dP}{dt} < 0$ (second assumption).

We also want the model to be "algebraically simple," or at least as simple as possible, so we try to modify the exponential model as little as possible. For instance, we might look for an expression of the form

$$\frac{dP}{dt} = k \cdot (\text{something}) \cdot P.$$

We want the "(something)" factor to be close to 1 if P is small, but if $P > N$ we want "(something)" to be negative. The simplest expression that has these properties is the function

$$(\text{something}) = \left(1 - \frac{P}{N}\right).$$

Note that this expression equals 1 if $P = 0$, and it is negative if $P > N$. Thus our model is

$$\frac{dP}{dt} = k\left(1 - \frac{P}{N}\right)P.$$

This is called the **logistic population model** with growth rate k and carrying capacity N. It is another first-order differential equation. This equation is said to be **nonlinear** because its right-hand side is not a linear function of P as it was in the exponential growth model.

Qualitative analysis of the logistic model

Although the logistic differential equation is just slightly more complicated than the exponential growth model, there is no way that we can just *guess* solutions. The method of separation of variables discussed in the next section produces a formula for the solution of this particular differential equation. But for now, we rely solely on qualitative methods to see what this model predicts over the long term.

First, let

$$f(P) = k\left(1 - \frac{P}{N}\right)P$$

denote the right-hand side of the differential equation. In other words,

$$\frac{dP}{dt} = f(P) = k\left(1 - \frac{P}{N}\right)P.$$

We can derive qualitative information about the solutions to the differential equation from a knowledge of where dP/dt is zero, where it is positive, and where it is negative.

If we sketch the graph of the quadratic function f (see Figure 1.4), we see that it crosses the P-axis at exactly two points, $P = 0$ and $P = N$. In either case we have $dP/dt = 0$. Since the derivative of P vanishes for all t, the population remains constant whenever $P = 0$ or $P = N$. That is, the constant functions $P(t) = 0$ and $P(t) = N$ are solutions of the differential equation. These two constant solutions make perfect sense: If the population is zero, the population remains zero indefinitely; if the population is exactly at the carrying capacity, it neither increases nor decreases. As

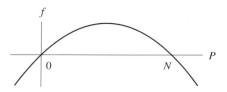

Figure 1.4
Graph of the right-hand side

$$f(P) = k\left(1 - \frac{P}{N}\right)P$$

of the logistic differential equation.

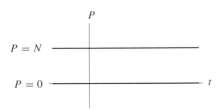

Figure 1.5
The equilibrium solutions of the logistic
differential equation

$$\frac{dP}{dt} = k\left(1 - \frac{P}{N}\right)P.$$

before, we say that $P = 0$ and $P = N$ are *equilibria*. The constant functions $P(t) = 0$ and $P(t) = N$ are called *equilibrium solutions* (see Figure 1.5).

The long-term behavior of the population is very different for other values of the population. If the initial population lies between 0 and N, then we have $f(P) > 0$. In this case the rate of growth $dP/dt = f(P)$ is positive, and consequently the population $P(t)$ is increasing. As long as $P(t)$ lies between 0 and N, the population continues to increase. However, as the population approaches the carrying capacity N, $dP/dt = f(P)$ approaches zero, so we expect that the population will level off as it approaches N (see Figure 1.6).

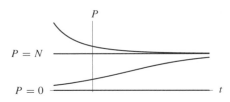

Figure 1.6
Solutions of the logistic differential equation

$$\frac{dP}{dt} = k\left(1 - \frac{P}{N}\right)P$$

approaching the equilibrium solution $P = N$.

If $P(0) > N$, then $dP/dt = f(P) < 0$, and the population is decreasing. As above, when the population approaches the carrying capacity N, dP/dt approaches zero, and we again expect the population to level off at N.

Finally, if $P(0) < 0$ (which does not make much sense in terms of populations), we also have $dP/dt = f(P) < 0$. Again we see that $P(t)$ decreases, but this time it does not level off at a particular value since dP/dt becomes more and more negative as $P(t)$ decreases.

Thus, from just a knowledge of the graph of f, we can sketch a number of different solutions with different initial conditions, all on the same axes. The only information that we need is the fact that $P = 0$ and $P = N$ are equilibrium solutions, $P(t)$ increases if $0 < P < N$ and $P(t)$ decreases if $P > N$ or $P < 0$. Of course the exact values of $P(t)$ at any given time t depend on the values of $P(0)$, k, and N (see Figure 1.7).

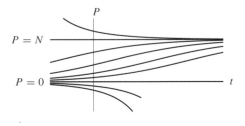

Figure 1.7
Solutions of the logistic differential
equation

$$\frac{dP}{dt} = k\left(1 - \frac{P}{N}\right)P$$

approaching the equilibrium solution
$P = N$ and moving away from the
equilibrium solution $P = 0$.

Predator-Prey Systems

No species lives in isolation, and the interactions among species give some of the most interesting models to study. We begin by studying some simple predator-prey systems of differential equations where one species "eats" another. The most obvious difference between the model here and previous models is that we have *two* quantities that depend on time. Thus our model has two dependent variables that are both functions of time. Since both predator and prey begin with "p," we call the prey "rabbits" and the predators "foxes," and we denote the prey by R and the predators by F. The assumptions for our model are:

- If no foxes are present, the rabbits reproduce at a rate proportional to their population, and they are not affected by overcrowding.
- The foxes eat the rabbits, and the rate at which the rabbits are eaten is proportional to the rate at which the foxes and rabbits interact.
- Without rabbits to eat, the fox population declines at a rate proportional to itself.
- The rate at which foxes are born is proportional to the number of rabbits eaten by foxes, which by the second assumption, is proportional to the rate at which the foxes and rabbits interact.

To formulate this model in mathematical terms, we need four parameters in addition to our independent variable t and our two dependent variables F and R. The parameters are

$$\alpha = \text{growth-rate coefficient of rabbits,}$$
$$\beta = \text{constant of proportionality that measures the number}$$
$$\text{of rabbit-fox interactions in which the rabbit is eaten,}$$
$$\gamma = \text{death-rate coefficient of foxes,}$$
$$\delta = \text{constant of proportionality that measures the}$$
$$\text{benefit to the fox population of an eaten rabbit.}$$

When we formulate our model, we follow the convention that α, β, γ, and δ are all positive.

Our first and third assumptions above are similar to the assumption in the unlimited growth model discussed earlier in this section. Consequently, they give terms of the form αR and $-\gamma F$ (since the fox population declines) in the equations for dR/dt and dF/dt, respectively.

The rate at which the rabbits are eaten is proportional to the rate at which the foxes and rabbits interact, so we need a term that models the rate of interaction of the two populations. We want a term that increases if either R or F increases, but it should vanish if either $R = 0$ or $F = 0$. A simple term that incorporates these assumptions is RF. Thus we model the effects of rabbit-fox interactions on dR/dt by a term of the form $-\beta RF$. The fourth assumption gives a similar term in the equation for dF/dt. In this case, eating rabbits helps the foxes, so we add a term of the form γRF.

Given these assumptions, we obtain the model

$$\frac{dR}{dt} = \alpha R - \beta RF$$
$$\frac{dF}{dt} = -\gamma F + \delta RF.$$

Considered together, this pair of equations is called a **first-order system** (only first derivatives, but more than one dependent variable) of ordinary differential equations. The system is said to be *coupled* because the rates of change of R and F depend on both R and F.

It is important to note the signs of the terms in this system. Because $\beta > 0$, the term "$-\beta RF$" is nonpositive, so an increase in the number of foxes causes a decline in the growth rate of the rabbit population. Also, since $\delta > 0$, the term "δRF" is nonnegative. Consequently, an increase in the number of rabbits increases the growth rate of the fox population.

Although this model may seem relatively simpleminded, it has been the basis of some interesting ecological studies. In particular, Volterra and D'Ancona successfully used the model to explain the increase in the population of sharks in the Mediterranean during World War I when the fishing of "prey" species decreased. The model can also be used as the basis for studying the effects of pesticides on the populations of predator and prey insects.

A *solution* to this system of equations is, unlike our previous models, a pair of functions, $R(t)$ and $F(t)$, that describe the populations of rabbits and foxes as functions of time. Since the system is coupled, we cannot simply determine one of these functions first and then the other. Rather, we must solve both differential equations simultaneously. Unfortunately, for most values of the parameters, it is impossible to determine explicit formulas for $R(t)$ and $F(t)$. These functions cannot be expressed in terms of known functions such as polynomials, sines, cosines, exponentials, and the like. However, as we will see in Chapter 2, these solutions do exist, although we have no hope of ever finding them exactly. Since analytic methods for solving this system are destined to fail, we must use either qualitative or numerical methods to "find" $R(t)$ and $F(t)$.

The Analytic, Qualitative, and Numerical Approaches

Our discussion of the three population models in this section illustrates three different approaches to the study of the solutions of differential equations. The **analytic** approach searches for explicit formulas that describe the behavior of the solutions. Here we saw that exponential functions give us explicit solutions to the exponential growth model. Unfortunately, a large number of important equations cannot be handled with the analytic approach; there simply is no way to find an exact formula that describes the situation. We are therefore forced to turn to alternative methods.

One particularly powerful method of predicting the behavior of solutions is the **qualitative** approach. This method involves using geometry to give an overview of the behavior of the model, just as we did with the logistic population growth model. We do not use this method to give precise values of the solution at specific times, but we will often be able to use this method to determine the long-term behavior of the solutions. Frequently, this is just the kind of information we need.

The third approach to solving differential equations is **numerical**. The computer approximates the solution we seek. Although we did not illustrate any numerical techniques in this section, we will soon see that numerical approximation techniques are a powerful tool for giving us intuition regarding the solutions we seek.

All three of the methods we use have certain advantages, and all have drawbacks. Sometimes certain methods are useful while others are not. One of our main tasks as we study the solutions to differential equations will be to determine which method or combination of methods works in each specific case. In the next three sections, we elaborate on these three techniques.

Exercises for Section 1.1

1. Consider the population model

$$\frac{dP}{dt} = 0.4P\left(1 - \frac{P}{230}\right),$$

where $P(t)$ is the population at time t.

(a) For what values of P is the population increasing?

(b) For what values of P is the population decreasing?

(c) For what values of P is the population in equilibrium?

2. Consider the population model

$$\frac{dP}{dt} = 0.3\left(1 - \frac{P}{200}\right)\left(\frac{P}{50} - 1\right)P,$$

where $P(t)$ is the population at time t.

(a) For what values of P is the population increasing?

(b) For what values of P is the population decreasing?

(c) For what values of P is the population in equilibrium?

3. Consider the differential equation

$$\frac{dy}{dt} = y^3 - y^2 - 12y.$$

(a) For what values of y is $y(t)$ increasing?

(b) For what values of y is $y(t)$ decreasing?

(c) For what values of y is $y(t)$ in equilibrium?

4. The following table provides the land area in Australia colonized by the American marine toad (*Bufo marinis*) every five years from 1939–1974. Model the migration of this toad using an exponential growth model

$$\frac{dP}{dt} = kP,$$

where, in this case $P(t) = $ land area occupied at time t. Make predictions about the land area occupied in the years 2010, 2050, and 2100. You should do this by

(a) solving the initial-value problem,

(b) determining the constant k,

(c) computing the predicted areas, and

(d) comparing your solution to the actual data. Do you believe your prediction?

Year	Cumulative area occupied (km^2)
1939	32,800
1944	55,800
1949	73,600
1954	138,000
1959	202,000
1964	257,000
1969	301,000
1974	584,000

(Note that there are many exponential growth models that you can form using this data. Is one a more reasonable model than the others? Note also that the area of Queensland is 1,728,000 km^2 and the area of Australia is 7,619,000 km^2.)*

Remark: The American marine toad was introduced to Australia to control sugar cane beetles and in the words of J. W. Hedgpath (see *Science*, July 1993 and *The New York Times*, July 6, 1993),

> Unfortunately the toads are nocturnal feeders and the beetles are abroad by day, while the toads sleep under rocks, boards and burrows. By night the toads flourish, reproduce phenomenally well and eat up everything they can find. The cane growers were warned by Walter W. Froggart, president of the New South Wales Naturalist Society, that the introduction was not a good idea and that the toads would eat the native ground fauna. He was immediately denounced as an ignorant meddlesome crank. He was also dead right.

Exercises 5–7 consider an elementary model of the learning process: Although human learning is an extremely complicated process, it is possible to build models of certain simple types of memorization. For example, consider a person presented with a list to be studied. The subject is given periodic quizzes to determine exactly how much of the list has been memorized. (The lists are usually things like nonsense syllables, randomly generated three-digit numbers, or techniques of integration.) If we let $L(t)$ be the fraction of the list learned at time t, where $L = 0$ corresponds to knowing nothing and $L = 1$ corresponds to knowing the entire list, then we can form a simple model of this type of learning based on the assumption:

- The rate of learning is proportional to the amount left to be learned.

Since $L = 1$ corresponds to knowing the entire list, the model is

$$\frac{dL}{dt} = k(1 - L),$$

where k is the constant of proportionality.

5. For what value of L, $0 \leq L \leq 1$, does learning occur most rapidly?

---

*All data taken from "Cumulative Geographical Range of *Bufo Marinis* in Queensland, Australia from 1935 to 1974," by Michael D. Sabath, Walter C. Boughton, and Simon Easteal, in Copeia, No. 3, 1981, pp. 676–680.

6. Suppose two students memorize lists according to the same model:

$$\frac{dL}{dt} = 2(1 - L).$$

(a) If one of the students knows one-half of the list at time $t = 0$ and the other knows none of the list, which student is learning most rapidly at this instant?

(b) Will the student who starts out knowing none of the list ever catch up to the student who starts out knowing one-half of the list?

7. Consider the following two differential equations that model two students' rate of memorizing a poem. Jillian's rate is proportional to the amount to be learned with proportionality constant $k = 2$. Beth's rate is proportional to the square of the amount to be learned with proportionality constant 3. The corresponding differential equations are

$$\frac{dL_J}{dt} = 2(1 - L_J) \quad \text{and} \quad \frac{dL_B}{dt} = 3(1 - L_B)^2,$$

where $L_J(t)$ and $L_B(t)$ are the fractions of the poem learned at time t by Jillian and Beth, respectively.

(a) Which student has a faster rate of learning at $t = 0$ if they both start memorizing together having never seen the poem before?

(b) Which student has a faster rate of learning at $t = 0$ if they both start memorizing together having already learned half the poem?

(c) Which student has a faster rate of learning at $t = 0$ if they both start memorizing together having already learned one-third of the poem?

In Exercises 8–12, we consider the phenomenon of radioactive decay which, from experimentation, we know behaves according to the law:

The rate at which a quantity of a radioactive isotope decays is proportional to the amount of the isotope present. The proportionality constant depends only on which radioactive isotope is used.

8. Model radioactive decay using the notation

$$t = \text{time (independent variable)},$$
$$r(t) = \text{amount of particular radioactive isotope}$$
$$\text{present at time } t \text{ (dependent variable)},$$
$$-\lambda = \text{decay rate (parameter)}.$$

Note that the minus sign is used so that $\lambda > 0$.

(a) Using this notation, write a model for the decay of a particular radioactive isotope.

(b) If the amount of the isotope present at $t = 0$ is r_0, state the corresponding initial-value problem for the model in part (a).

9. The **half-life** of a radioactive isotope is the amount of time it takes for a quantity of radioactive material to decay to one-half of its original amount.

(a) The half-life of Carbon 14 (C-14) is 5230 years. Determine the decay-rate parameter λ for C-14.

(b) The half-life of Iodine 131 (I-131) is 8 days. Determine the decay rate parameter for I-131.

(c) What are the units of the decay-rate parameters in parts (a) and (b)?

(d) To determine the half-life of an isotope, we could start with 1000 atoms of the isotope and measure the amount of time it takes 500 of them to decay, or we could start with 10,000 atoms of the isotope and measure the amount of time it takes 5000 of them to decay. Will we get the same answer? Why?

10. Carbon dating is a method of determining the time elapsed since the death of organic material. The assumptions implicit in carbon dating are that

   • Carbon 14 (C-14) makes up a constant proportion of the carbon that living matter ingests on a regular basis, and
   • once the matter dies, the C-14 present decays, but no new carbon is added to the matter.

   Hence, by measuring the amount of C-14 still in the organic matter and comparing it to the amount of C-14 typically found in living matter, a "time-since-death" can be approximated. Using the decay-rate parameter you computed in Exercise 9, determine the time-since-death if

   (a) 88% of the original C-14 is still in the material?

   (b) 12% of the original C-14 is still in the material?

   (c) 2% of the original C-14 is still in the material?

   (d) 98% of the original C-14 is still in the material?

   Remark: There has been speculation that the amount of C-14 available to living creatures has not been exactly constant over long periods (thousands of years). This makes accurate dates much trickier to determine.

11. In order to apply the carbon dating technique of Exercise 10, we must measure the amount of C-14 in a sample. Chemically, radioactive Carbon 14 (C-14) and regular carbon behave identically. How can we determine the amount of C-14 in a sample? [*Hint*: See Exercise 8.]

12. The radioactive isotope I-131 is used in the treatment of hyperthyroid. I-131 administered to a patient naturally collects in the thyroid gland where it decays and kills part of the gland.

   (a) Suppose that it takes 72 hours to ship I-131 from the hospital. What percentage of the original amount shipped actually arrives at the hospital? (See Exercise 9.)

   (b) If the I-131 is stored at the hospital for an additional 48 hours before it is used, how much of the original amount shipped from the producer is left when it is used?

   (c) How long will it take for the I-131 to *completely* decay so that the remnants can be thrown away without special precautions?

13. Suppose a species of fish in a particular lake has a population that is modeled by the logistic population model with growth rate k, carrying capacity N and time t measured in years. Adjust the model to account for each of the following situations.

   (a) 100 fish are harvested each year.

   (b) One-third of the fish population is harvested annually.

(c) The number of fish harvested each year is proportional to the square root of the number of fish in the lake.

14. Suppose that the growth rate parameter $k = 0.3$ and the carrying capacity $N = 2500$ in the logistic population model of Exercise 13. Suppose $P(0) = 2500$.

(a) If 100 fish are harvested each year, what does the model predict for the long-term behavior of fish population? In other words, what does a qualitative analysis of the model yield?

(b) If one-third of the fish are harvested each year, what does the model predict for the long-term behavior of fish population?

15. Consider the following assumptions concerning the fraction of a piece of bread covered by mold.

- Mold spores fall on the bread at a constant rate.
- When the proportion covered is small, the fraction of the bread covered by mold increases at a rate proportional to the amount of bread covered.
- When the fraction of bread covered by mold is large, the growth rate decreases.
- In order to survive, mold must be in contact with the bread.

Using these assumptions, write a differential equation that models the proportion of a piece of bread covered by mold.

16. The rhinoceros is now extremely rare. Suppose enough game preserve land is set aside so that there is sufficient room for many more rhinoceros territories than there are rhinoceroses. Consequently, there will be no danger of overcrowding. However, if the population is too small, fertile adults have difficulty finding each other when it is time to mate. Write a differential equation that models the rhinoceros population based on these assumptions. (Note that there is more than one reasonable model that fits these assumptions.)

17. The following table contains data for the population of tawny owls in Wyman Woods, Oxford England (collected by Southern). *

(a) What population model would you use to model this population?

(b) Can you approximate (or even make reasonable guesses for) the parameter values?

(c) What does your model predict for the population today?

Year	Population	Year	Population
1947	34	1954	52
1948	40	1955	60
1949	40	1956	64
1950	40	1957	64
1951	42	1958	62
1952	48	1959	64
1953	48		

---

*See J.P. Dempster, *Animal Population Ecology*, Academic Press, 1975, p. 99.

18. For the following predator-prey systems, identify which dependent variable, x or y, is the prey population and which is the predator population. Is the growth of the prey limited by any factors other than the number of predators? Do the predators have sources of food other than the prey? (The parameters α, β, γ, δ, and N are all positive.)

(a)
$$\frac{dx}{dt} = -\alpha x + \beta xy$$
$$\frac{dy}{dt} = \gamma y - \delta xy$$

(b)
$$\frac{dx}{dt} = \alpha x - \alpha \frac{x^2}{N} - \beta xy$$
$$\frac{dy}{dt} = \gamma y + \delta xy$$

19. In the following predator-prey population models, x represents the prey, and y represents the predators.

(i)
$$\frac{dx}{dt} = 5x - 3xy$$
$$\frac{dy}{dt} = -2y + \tfrac{1}{2}xy$$

(ii)
$$\frac{dx}{dt} = x - 8xy$$
$$\frac{dy}{dt} = -2y + 6xy$$

(a) In which system does the prey reproduce more quickly when there are no predators (when $y = 0$) and equal numbers of prey?

(b) In which system are the predators more successful at catching prey? In other words, if the number of predators and prey are equal for the two systems, in which system do the predators have a greater effect on the rate of change of the prey?

(c) Which system requires more prey for the predators to achieve a given growth rate (assuming identical numbers of predators in both cases?)

20. The system

$$\frac{dx}{dt} = ax - by\sqrt{x}$$
$$\frac{dy}{dt} = cy\sqrt{x}$$

has been proposed as a model for a predator-prey system of two particular species of microorganisms (where a, b, and c are positive parameters).

(a) Which variable, x or y, represents the predator population? Which variable represents the prey population?

(b) What happens to the predator population if the prey is extinct?

21. The following systems are models of the populations of pairs of species that either *compete* for resources (an increase in one species decreases the growth rate of the other) or *cooperate* (an increase in one species increases the growth rate of the other). For each system, identify the variables (independent and dependent) and the parameters (carrying capacity, measures of interaction between species, etc.) Do the species compete or cooperate? (Assume all parameters are positive.)

(a)
$$\frac{dx}{dt} = \alpha x - \alpha \frac{x^2}{N} + \beta xy$$
$$\frac{dy}{dt} = \gamma y + \delta xy$$

(b)
$$\frac{dx}{dt} = -\gamma x - \delta xy$$
$$\frac{dy}{dt} = \alpha y - \beta xy$$

1.2 ANALYTIC TECHNIQUE: SEPARATION OF VARIABLES

What Is a Differential Equation and What Is a Solution?

A first-order differential equation is an equation for an unknown function in terms of its derivative. As we saw in the previous section, there are three types of "variables" in differential equations — the independent variable (almost always t for time in our examples), one or more dependent variables (which are functions of the independent variable), and the parameters. This terminology is standard but a bit confusing. The dependent variable is actually a function, so technically it should be called the dependent function.

The standard form for a first-order differential equation is

$$\frac{dy}{dt} = f(t, y).$$

Here the right-hand side typically depends on both the dependent and independent variables, although we often encounter cases where either t or y is missing.

A **solution** of the differential equation is a function of the independent variable that, when substituted into the equation as the dependent variable, satisfies the equation for all values of the independent variable. That is, a function $y(t)$ is a solution if it satisfies $dy/dt = y'(t) = f(t, y(t))$. This terminology doesn't tell us how to find solutions, but it does tell us how to check whether a candidate function is or is not a solution. For example, consider the simple differential equation

$$\frac{dy}{dt} = y.$$

We can easily check that the function $y_1(t) = 3e^t$ is a solution, whereas $y_2(t) = \sin t$ is not a solution. The function $y_1(t)$ is a solution because

$$\frac{dy_1}{dt} = \frac{d(3e^t)}{dt} = 3e^t = y_1 \quad \text{for all } t.$$

On the other hand, $y_2(t)$ is not a solution since

$$\frac{dy_2}{dt} = \frac{d(\sin t)}{dt} = \cos t,$$

and certainly the function $\cos t$ is not the same function as $y_2(t) = \sin t$.

Checking that a given function is a solution to a given differential equation

If we look at a slightly more complicated equation such as

$$\frac{dy}{dt} = \frac{y^2 - 1}{t^2 + 2t},$$

then we have considerably more trouble finding a solution. On the other hand, if somebody hands us a function $y(t)$, then we know how to check whether or not it is a solution.

For example, suppose we meet three differential equations textbook authors — say Paul, Bob, and Glen — at our local espresso bar, and we ask them to find solutions of this differential equation. After a few minutes of furious calculation, Paul says that

$$y_1(t) = 1 + t$$

is a solution. Glen then says that

$$y_2(t) = 1 + 2t$$

is a solution. After several more minutes, Bob says that

$$y_3(t) = 1$$

is a solution. Which of these functions is a solution? Let's see who is right by substituting each function into the differential equation.

First we test Paul's function. We compute the left-hand side by differentiating $y_1(t)$

$$\frac{dy_1}{dt} = \frac{d(1+t)}{dt} = 1.$$

Substituting $y_1(t)$ into the right-hand side, we find

$$\frac{(y_1(t))^2 - 1}{t^2 + 2t} = \frac{(1+t)^2 - 1}{t^2 + 2t} = \frac{t^2 + 2t}{t^2 + 2t} = 1.$$

The left-hand side and the right-hand side of the differential equation are identical, so Paul is correct.

To check Glen's function, we again compute the derivative

$$\frac{dy_2}{dt} = \frac{d(1+2t)}{dt} = 2.$$

With y_2, the right-hand side of the differential equation is

$$\frac{(y_2(t))^2 - 1}{t^2 + 2t} = \frac{(1+2t)^2 - 1}{t^2 + 2t} = \frac{4t^2 + 4t}{t^2 + 2t} = \frac{4(t+1)}{t+2}.$$

The left-hand side of the differential equation does not equal the right-hand side for all t since the right-hand side is not the constant function 2. Glen's function is *not* a solution.

Finally, we check Bob's function the same way. The left-hand side is

$$\frac{dy_3}{dt} = \frac{d(1)}{dt} = 0$$

because $y_3(t) = 1$ is a constant. The right-hand side is

$$\frac{y_3(t)^2 - 1}{t^2 + t} = \frac{1 - 1}{t^2 + t} = 0.$$

Both the left-hand side and the right-hand side of the differential equation vanish for all t. Hence, Bob's function *is* a solution of the differential equation.

The lessons we learn from this example are that a differential equation may have solutions that look very different from each other algebraically and that (of course) not every function is a solution. Given a function, we can test to see whether it is a solution by just substituting it into the differential equation and checking to see whether the left-hand side is identical to the right-hand side. This is a very nice aspect of differential equations: *We can always check our answers.* So we should never be wrong.

Initial-Value Problems and The General Solution

When we encounter differential equations in practice, they often come with **initial conditions**. We seek a solution of the given equation that assumes a given value at a particular time. A differential equation along with an initial condition is called an **initial-value problem**. Thus the usual form of an initial-value problem is

$$\frac{dy}{dt} = f(t, y), \quad y(t_0) = y_0.$$

Here we are looking for a function $y(t)$ that is a solution of the differential equation *and* assumes the value y_0 at time t_0. Often, the particular time in question is $t = 0$ (hence the name *initial condition*), but any other time could be specified.

For example,

$$\frac{dy}{dt} = t^3 - 2\sin t, \quad y(0) = 3.$$

is an initial-value problem. To solve this problem, note that the right-hand side of the differential equation depends only on t, not on y. We are looking for a function whose derivative is $t^3 - 2\sin t$. This is a typical antidifferentiation problem from calculus, so all we need do is integrate this expression. We find

$$\int (t^3 - 2\sin t)\,dt = \frac{t^4}{4} + 2\cos t + c,$$

where c is a constant of integration. Thus the solution of the differential equation must be of the form

$$y(t) = \frac{t^4}{4} + 2\cos t + c.$$

We now use the initial condition $y(0) = 3$ to determine c by

$$3 = y(0) = \frac{0^4}{4} + 2\cos 0 + c = 0 + 2 \cdot 1 + c = 2 + c.$$

Thus $c = 1$, and the solution to this initial-value problem is

$$y(t) = \frac{t^4}{4} + 2\cos t + 1.$$

The expression

$$y(t) = \frac{t^4}{4} + 2\cos t + c$$

is called the **general solution** of the differential equation because we can use it to solve any initial-value problem whatsoever. For example, if the initial condition is $y(0) = \pi$, then we would choose $c = \pi - 2$.

Separable Equations

Now that we know how to check that a given function is a solution to a differential equation, the question is: How can we get our hands on a solution in the first place?

Unfortunately, it is rarely the case that we can find explicit solutions of a differential equation. Many differential equations have solutions that cannot be expressed in terms of known functions such as polynomials, exponentials, or trigonometric functions. However, there are a few special types of differential equations for which we can derive explicit solutions, and in this section we discuss one of these types of differential equations.

The typical first-order differential equation is given in the form

$$\frac{dy}{dt} = f(t, y).$$

The right-hand side of this equation generally involves both the independent variable t and the dependent variable y (although there are many important examples where either the t or the y is missing). A differential equation is called **separable** if the function $f(t, y)$ can be written as the product of two functions: one that depends on t alone and another that depends only on y. That is, a differential equation is separable if it assumes the form

$$\frac{dy}{dt} = f(t, y) = g(t)h(y).$$

For example, the differential equation

$$\frac{dy}{dt} = yt$$

is clearly separable, and the equation

$$\frac{dy}{dt} = y + t$$

is not. We might have to do a little work to see that an equation is separable. For instance,

$$\frac{dy}{dt} = \frac{t+1}{ty+t}$$

is separable since we can rewrite the equation as

$$\frac{dy}{dt} = \frac{(t+1)}{t(y+1)} = \left(\frac{t+1}{t}\right)\left(\frac{1}{y+1}\right).$$

Two important types of separable equations occur if either t or y is missing from the right-hand side of the equation. The differential equation

$$\frac{dy}{dt} = g(t)$$

is separable since we may regard the right-hand side as $g(t) \cdot 1$, where we consider 1 as a (very simple) function of y. Similarly,

$$\frac{dy}{dt} = h(y)$$

is also separable. This last type of differential equation is said to be **autonomous**. Many of the most important first-order differential equations that arise in applications (including all of our models in the previous section) are autonomous. For example, the right-hand side of the logistic equation

$$\frac{dP}{dt} = kP\left(1 - \frac{P}{N}\right)$$

depends on the dependent variable P alone, so this equation is autonomous.

How to solve separable differential equations

To find explicit solutions of separable differential equations, we use a technique familiar from calculus. To illustrate the method, consider the differential equation

$$\frac{dy}{dt} = \frac{t}{y^2}.$$

There is a temptation to solve this equation by simply integrating both sides of the equation with respect to t. This yields

$$\int \frac{dy}{dt}\, dt = \int \frac{t}{y^2}\, dt,$$

and, consequently,

$$y(t) = \int \frac{t}{y^2}\, dt.$$

Now we are stuck. We can't evaluate the integral on the right-hand side because we don't know the function $y(t)$. In fact, that is precisely the function we wish to find. We have simply replaced the differential equation with an *integral equation*.

We need to do something to this equation *before* we try to integrate. Returning to the original differential equation

$$\frac{dy}{dt} = \frac{t}{y^2},$$

we first do some "informal" algebra and rewrite this equation in the form

$$y^2\, dy = t\, dt.$$

That is, we multiply both sides by $y^2\, dt$. Of course, it makes no sense to split up dy/dt by multiplying by dt. However, this should remind you of the technique of integration known as u-substitution in calculus. We will soon see that substitution is exactly what we are doing here.

Continuing, we now integrate both sides, the left with respect to y and the right with respect to t. We have

$$\int y^2\, dy = \int t\, dt,$$

which yields

$$\frac{y^3}{3} = \frac{t^2}{2} + c.$$

Technically there is a constant of integration on both sides of this equation, but we can lump them together as a single constant c on the right. We may rewrite this expression as

$$y(t) = \left(\frac{3t^2}{2} + 3c \right)^{1/3};$$

or, since c is an arbitrary constant, we may write this even more compactly as

$$y(t) = \left(\frac{3t^2}{2} + k \right)^{1/3},$$

where k is an arbitrary constant. As usual, we can check that this expression really is a solution of the differential equation, so despite the questionable separation we performed above, we do obtain a solution in the end.

Note that this process yields many solutions of the differential equation. Each choice of the constant k gives a different solution.

What is really going on in the informal algebra above

If you read the previous example closely, you probably became nervous at one point. Treating dt as a variable is a tip-off that something a little more complicated is actually going on. Here is the real story.

We began with a separable equation

$$\frac{dy}{dt} = g(t)h(y),$$

and then rewrote it as

$$\frac{1}{h(y)}\frac{dy}{dt} = g(t).$$

This equation actually has a function of t on each side of the equals sign because y is a function of t. So we really should write it as

$$\frac{1}{h(y(t))}\frac{dy}{dt} = g(t).$$

In this form, we can integrate both sides with respect to t to get

$$\int \frac{1}{h(y(t))}\frac{dy}{dt}\,dt = \int g(t)\,dt.$$

Now for the important step: We make a "u-substitution" just as in calculus by replacing the function $y(t)$ by the new variable, say y. (In this case, the substitution is actually a y-substitution.) Of course, we must also replace the expression $(dy/dt)\,dt$ by dy. The method of substitution from calculus tells us that

$$\int \frac{1}{h(y(t))}\frac{dy}{dt}\,dt = \int \frac{1}{h(y)}\,dy,$$

and therefore we can combine the last two equations to obtain

$$\int \frac{1}{h(y)}\,dy = \int g(t)\,dt.$$

Hence, we can integrate the left-hand side with respect to y and the right-hand side with respect to t.

Separating variables and multiplying both sides of the differential equation by dt is simply a notational convention that helps us remember the method. It is justified by the argument above.

Missing Solutions

If it is possible to separate variables in a differential equation, it appears that solving the equation reduces to a matter of computing several integrals. This is true, but there are some hidden pitfalls, as the following example shows. Consider the differential equation

$$\frac{dy}{dt} = y^2.$$

This is an autonomous and hence separable equation, and its solution looks straightforward. If we separate and integrate as usual, we obtain

$$\int \frac{dy}{y^2} = \int dt$$

$$-\frac{1}{y} = t + c,$$

or

$$y(t) = -\frac{1}{t+c}.$$

We are tempted to say that this expression for $y(t)$ is the general solution. However, we cannot solve all initial-value problems with solutions of this form. In fact, we have $y(0) = -1/c$, so we cannot use this expression to solve the initial-value problem $y(0) = 0$.

What's wrong? Note that the right-hand side of the differential equation vanishes if $y = 0$. So the constant function $y(t) = 0$ is a solution to this differential equation. In other words, in addition to those solutions that we derived using the method of separation of variables, this differential equation possesses the equilibrium solution $y(t) = 0$ for all t, and it is this equilibrium solution that satisfies the initial-value problem $y(0) = 0$. Even though it is "missing" from the family of solutions that we obtain by separating variables, it is a solution that we need if we want to solve every initial-value problem for this differential equation. Thus the general solution consists of functions of the form $y(t) = -1/(t + c)$ together with the equilibrium solution $y(t) = 0$.

Getting Stuck

As another example, consider the differential equation

$$\frac{dy}{dt} = \frac{y}{1 + y^2}.$$

As before, this equation is autonomous. So we first separate to obtain

$$\left(\frac{1 + y^2}{y}\right) dy = dt.$$

Then we integrate

$$\int \frac{1}{y} + y \, dy = \int dt,$$

which yields

$$\ln|y| + \frac{y^2}{2} = t + c.$$

But now we are stuck; there is no way to solve the equation

$$\ln|y| + \frac{y^2}{2} = t + c$$

for y alone. Thus we cannot generate an explicit formula for y. We do, however, have an **implicit form** for the solution which, for many purposes, is perfectly acceptable.

Even though we don't obtain explicit solutions by separating variables for this equation, we can find one explicit solution. The right-hand side vanishes if $y = 0$. Thus the constant function $y(t) = 0$ is an equilibrium solution. Note that this equilibrium solution does not appear in the implicit solution we derived from the method of separation of variables.

There is another problem that arises with this method. It is often impossible to perform the required integrations. For example, the differential equation

$$\frac{dy}{dt} = \sec(y^2)$$

is autonomous. Separating variables and integrating we get

$$\int \frac{1}{\sec(y^2)} \, dy = \int dt,$$

or equivalently,

$$\int \cos(y^2) \, dy = \int dt.$$

The integral on the left-hand side is difficult, to say the least. (In fact, there is a special function that was defined just to give us a name for this integral.) The lesson is that, even for special equations of the form

$$\frac{dy}{dt} = f(y),$$

carrying out the required algebra or integration is frequently impossible. We will not be able to rely solely on analytic tools and explicit solutions when studying differential equations, even if we can separate variables.

A Savings Model

Suppose we deposit \$5000 in a savings account with interest accruing at the rate of 5% compounded continuously. If we let $A(t)$ denote the amount of money in the account at time t, then the differential equation for A is

$$\frac{dA}{dt} = 0.05A.$$

As we saw in the previous section, the general solution to this equation is the exponential function

$$A(t) = ce^{0.05t},$$

where $c = A(0)$. Thus

$$A(t) = 5000e^{0.05t}$$

is our particular solution.

Assuming interest rates never change, after 10 years we will have

$$A(10) = 5000e^{0.5} \approx 8244$$

dollars in this account. That is a nice little nest egg, so we decide we should have some fun in life. We decide to withdraw $1000 (mad money) from the account each year in a continuous way beginning in year 10. So the question is: How long will this money last? Will we ever go broke?

The differential equation for $A(t)$ must change, but only beginning in year 10. For $0 \leq t \leq 10$, our previous model works fine. However, for $t > 10$, the differential equation becomes

$$\frac{dA}{dt} = 0.05A - 1000.$$

Thus we really have a differential equation of the form

$$\frac{dA}{dt} = \begin{cases} 0.05A & \text{for } t < 10; \\ 0.05A - 1000 & \text{for } t > 10, \end{cases}$$

whose right-hand side consists of two pieces.

To solve this two-part equation, we solve the first part and determine $A(10)$. We just did that and obtained $A(10) \approx 8244$. Then we solve the second equation using $A(10) \approx 8244$ as the initial value. This equation is also separable, and we have

$$\int \frac{dA}{0.05A - 1000} = \int dt.$$

We calculate this integral using substitution and the natural logarithm function. Let $u = 0.05A - 1000$. Then $du = 0.05 \, dA$, or $20 \, du = dA$ since $0.05 = 1/20$. We obtain

$$\int \frac{20 \, du}{u} = t + c_1$$

$$20 \ln|u| = t + c_1$$

$$20 \ln|0.05A - 1000| = t + c_1,$$

for some constant c_1.

At $t = 10$, we know that $A \approx 8244$. Thus, $0.05A - 1000 \approx -587.8 < 0$. In other words, we are withdrawing at a rate that exceeds the rate at which we are earning interest. Since $dA/dt = 0.05A - 1000$ at $t = 10$ is negative, A will decrease and $0.05A - 1000$ will continue to be negative for all $t > 10$. If $0.05A - 1000 < 0$, then $|0.05A - 1000| = -(0.05A - 1000) = 1000 - 0.05A$. Consequently, we have

$$20 \ln(1000 - 0.05A) = t + c_1,$$

or

$$\frac{\ln(1000 - 0.05A)}{0.05} = t + c_1.$$

Multiplying both sides by 0.05 and exponentiating yields

$$1000 - 0.05A = e^{0.05(t+c_1)}$$
$$1000 - 0.05A = c_2 e^{0.05t},$$

where $c_2 = e^{0.05c_1}$. Solving for A, we obtain

$$A = \frac{1000 - c_2 e^{0.05t}}{0.05}$$
$$= 20\left(1000 - c_2 e^{0.05t}\right) = 20000 - c_3 e^{0.05t},$$

where $c_3 = 20c_2$. (Although we have been careful to spell out the relationships among the constants c_1, c_2, and c_3, we need only remember that c_3 is a constant that is determined from the initial condition.)

Now we use the initial condition to determine c_3. We know that

$$8244 \approx A(10) = 20000 - c_3 e^{0.05(10)} \approx 20000 - c_3(1.0513).$$

Solving for c_3, we obtain $c_3 \approx 7130$. Our solution for $t \geq 10$ is

$$A(t) \approx 20000 - 7130 e^{0.05t}.$$

We see that

$$A(11) \approx 7641$$
$$A(12) \approx 7008$$

and so forth. Our account is being depleted, but not by that much. In fact, we can find out just how long the good times will last by asking when our money will run out. That means we must solve the equation $A(t) = 0$ for t. We have

$$0 = 20000 - 7130 e^{0.05t},$$

which yields

$$t = 20\ln\left(\frac{20000}{7130}\right) \approx 20.63.$$

After letting the $5000 accumulate interest for ten years, we can withdraw $1000 per year for more than ten years.

A Mixing Problem

The name *mixing problem* refers to a large collection of different problems where two or more substances are mixed together. This can range from pollutants in a lake to chemicals in a vat to cigar smoke in the air in a room to spices in a serving of curry.

Mixing in a vat

Consider a large vat containing sugar-water that is to be made into soft drinks (see Figure 1.8). Suppose:

- The vat contains 100 gallons of liquid. Moreover, the amount flowing in equals the amount flowing out, so there are always 100 gallons in the vat.
- The vat is kept well mixed, so the concentration of sugar is uniform throughout the vat.
- Sugar-water containing 5 tablespoons of sugar per gallon enters the vat through Pipe A at a rate of 2 gallons per minute.
- Sugar-water containing 10 tablespoons of sugar per gallon enters the vat through Pipe B at a rate of 1 gallon per minute.
- Sugar-water leaves the vat through Pipe C at a rate of 3 gallons per minute.

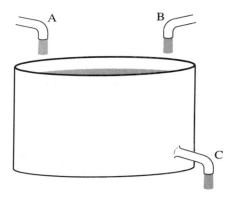

Figure 1.8
Mixing Vat

To make the model, we let t be time measured in minutes (the independent variable). For the dependent variable, we have two choices. We could choose either the total amount of sugar, $S(t)$, in the vat at time t measured in tablespoons, or $C(t)$, the concentration of sugar in the vat at time t measured in tablespoons per gallon. We will develop the model for S, leaving the model for C as an exercise.

Using the total sugar $S(t)$ in the vat as the dependent variable, the rate of change of S is the difference between the amount of sugar being added and the amount of sugar being removed. The sugar entering the vat comes from pipes A and B and can be easily computed by multiplying the number of gallons per minute of sugar mixture entering the vat by the amount of sugar per gallon. The amount of sugar leaving the vat through pipe C at any given moment depends on the concentration of sugar in the vat at that moment. The concentration is given by $S/100$, so the sugar leaving the vat is the product of the number of gallons leaving per minute (3 gallons per minute) and the concentration ($S/100$). The model is

$$\frac{dS}{dt} = \underbrace{2 \cdot 5}_{\substack{\text{sugar in} \\ \text{from pipe A}}} + \underbrace{1 \cdot 10}_{\substack{\text{sugar in} \\ \text{from pipe B}}} - \underbrace{3 \cdot \frac{S}{100}}_{\substack{\text{sugar out} \\ \text{from pipe C}}}.$$

That is,

$$\frac{dS}{dt} = 20 - \frac{3S}{100} = \frac{2000 - 3S}{100}.$$

To solve this equation analytically, we separate and integrate. We find

$$\frac{dS}{2000 - 3S} = \frac{dt}{100}$$

$$\frac{\ln|2000 - 3S|}{-3} = \frac{t}{100} + c_1$$

$$\ln|2000 - 3S| = -\frac{3t}{100} - 3c_1$$

$$\ln|2000 - 3S| = -0.03t + c_2,$$

where $c_2 = -3c_1$. Exponentiating we obtain

$$|2000 - 3S| = e^{(-0.03t + c_2)} = c_3 e^{-0.03t},$$

where $c_3 = e^{c_2}$. Note that this means that c_3 is a positive constant. Now we must be careful. Removing the absolute value signs yields

$$2000 - 3S = \pm c_3 e^{-0.03t},$$

where we choose $+$ if $S(t) < 2000/3$ and $-$ if $S(t) > 2000/3$. Therefore we may write this equation more simply as

$$2000 - 3S = c_4 e^{-0.03t},$$

where c_4 is an arbitrary constant (positive, negative, or zero). Solving for S yields the general solution

$$S(t) = ce^{-0.03t} + \frac{2000}{3},$$

where $c = -c_4/3$ is an arbitrary constant. We can determine the precise value of c if we know the exact amount of sugar that is present in the vat initially. Note that, if $c = 0$, the solution is simply $S(t) = 2000/3$, an equilibrium solution.

Exercises for Section 1.2

1. Bob, Glen and Paul are once again sitting around having a nice, cold glass of iced cappuccino when one of their students asks them to come up with solutions to the differential equation

$$\frac{dy}{dt} = \frac{y + 1}{t + 1}.$$

After much discussion, Bob says $y(t) = t$, Glen says $y(t) = 2t + 1$, and Paul says $y(t) = t^2 - 2$.

(a) Who is right?

(b) What solution should they have seen right away?

2. Make up a differential equation of the form

$$\frac{dy}{dt} = 2y - t + g(y)$$

that has the function $y(t) = e^{2t}$ as a solution.

3. Make up a differential equation of the form

$$\frac{dy}{dt} = f(t, y)$$

that has $y(t) = e^{t^3}$ as a solution. (Try to come up with one whose right-hand side $f(t, y)$ depends explicitly on both t and y.)

4. Use some relatively simple function to produce a differential equation that has that function as a solution. Then ask a classmate to guess a solution to the resulting differential equation without telling them the solution you have in mind. (Be kind. Your classmate will ask the same thing of you.)

(a) Did your classmate come up with a solution?

(b) Was it the same one you had in mind?

(c) Why is it easier to write differential equations problems than it is to solve them?

In Exercises 5–24, find the general solution of the differential equation specified. (You may not be able to reach the ideal answer of an equation with only the dependent variable on the left and only independent variable on the right, but get as far as you can.)

5. $\dfrac{dy}{dt} = ty$

6. $\dfrac{dy}{dt} = t^4 y$

7. $\dfrac{dy}{dt} = 2y + 1$

8. $\dfrac{dy}{dt} = 2 - y$

9. $\dfrac{dy}{dt} = e^{-y}$

10. $\dfrac{dy}{dt} = (ty)^2$

11. $\dfrac{dy}{dt} = \dfrac{t}{t^2 y + y}$

12. $\dfrac{dy}{dt} = t\sqrt[3]{y}$

13. $\dfrac{dy}{dt} = \dfrac{1}{2y + 1}$

14. $\dfrac{dy}{dt} = \dfrac{t}{1 + y^2}$

15. $\dfrac{dy}{dt} = y(1 - y)$

16. $\dfrac{dy}{dt} = \dfrac{t}{y}$

17. $\dfrac{dy}{dt} = t^2 y + 1 + y + t^2$

18. $\dfrac{dy}{dt} = \dfrac{1}{ty + t + y + 1}$

19. $\dfrac{dy}{dt} = \dfrac{e^t y}{1 + y^2}$

20. $\dfrac{dy}{ds} = \sec y$

21. $\dfrac{dw}{dt} = \dfrac{w}{t}$

22. $\dfrac{dy}{dt} = \dfrac{2y + 1}{t}$

23. $\dfrac{dy}{dt} = \dfrac{t^2 + 1}{y^4 + 3y}$

24. $\dfrac{dy}{dt} = 1 + \dfrac{1}{y^2}$

In Exercises 25–34, solve the given initial-value problem.

25. $\dfrac{dy}{dt} = 2y + 1, \quad y(0) = 3$

26. $\dfrac{dy}{dt} = ty^2 + 2y^2, \quad y(0) = 1$

27. $\dfrac{dy}{dt} = -y^2, \quad y(0) = 1/2$

28. $\dfrac{dx}{dt} = -xt, \quad x(0) = 1/\sqrt{\pi}$

29. $\dfrac{dy}{dt} = -y^2, \quad y(0) = 0$

30. $\dfrac{dy}{dt} = \dfrac{t}{y - t^2 y}, \quad y(0) = 4$

31. $\dfrac{dy}{dt} = \dfrac{t^2}{y + t^3 y}, \quad y(0) = -2$

32. $\dfrac{dy}{dt} = 2ty^2 + 3t^2 y^2, \quad y(1) = -1$

33. $\dfrac{dy}{dt} = (y^2 + 1)t, \quad y(0) = 1$

34. $\dfrac{dy}{dt} = \dfrac{1}{2y + 3}, \quad y(0) = 1$

35. A 5-gallon bucket is full of pure water. Suppose we begin dumping salt into the bucket at a rate of 1/4 pounds per minute. Also, we open the spigot so that 1/2 gallons per minute leaves the bucket, and we add pure water to keep the bucket full. If the water is always well mixed, what is the amount of salt in the bucket after

(a) 1 minute?   (b) 10 minutes?   (c) 60 minutes?
(d) 1000 minutes?   (e) a very very long time?

36. High levels of cholesterol in the blood are known to be a risk factor for heart disease. Cholesterol is manufactured by the body for use in the construction of cell walls and is absorbed from foods containing cholesterol. The following is a very simple model of blood cholesterol levels. Let $C(t)$ be the amount of cholesterol in the blood of a particular person at time t (in milligrams per deciliter). Then

$$\frac{dC}{dt} = k_1(C_0 - C) + k_2 E,$$

where

$C_0 = $ the person's "natural" cholesterol level,
$k_1 = $ production" parameter,
$E = $ amount of cholesterol eaten (per day), and
$k_2 = $ "absorption" parameter.

(a) Suppose $C_0 = 200$, $k_1 = 0.1$, $k_2 = 0.1$, $E = 400$, and $C(0) = 150$. What will the person's cholesterol level be after 2 days on this diet?

(b) With the initial conditions as above, what will the person's cholesterol level be after 5 days on this diet?

(c) What will the person's cholesterol level be after a long time on this diet?

(d) Suppose that, after a long time on the high cholesterol diet described above, the person goes on a very low cholesterol diet, so E changes to $E = 100$. (The initial cholesterol level at the starting time of this diet is the result of part (c).) What will the person's cholesterol level be after 1 day on the new diet, after 5 days on the new diet, and after a very long time on the new diet?

(e) Suppose the person stays on the high cholesterol diet but takes drugs that block some of the uptake of cholesterol from food, so k_2 changes to $k_2 = 0.075$. With the cholesterol level from part (c), what will the person's cholesterol level be after 1 day, after 5 days, and after a very long time?

37. A cup of hot chocolate is initially $170°$ and is left in a room with an ambient temperature of $70°$. Suppose that at time $t = 0$ it is cooling at a rate of $20°$ per minute.

    (a) Assume that Newton's law of cooling applies: The rate of cooling is proportional to the <u>difference</u> between the current temperature and the ambient temperature. Write an initial-value problem that models the temperature of the hot chocolate.

    (b) How long does it take the hot chocolate to cool to a temperature of $110°$?

38. In the mixing problem in this section, we had to make a choice of dependent variable. We used the amount of sugar as the dependent variable, but we could have used the concentration of sugar as the dependent variable. If $S(t)$ is the total amount of sugar in the vat at time t, then the concentration at time t is given by $C(t) = S(t)/100$ and is measured in tablespoons per gallon. Write a differential equation modeling the assumptions in the section using $C(t)$ as the dependent variable.

39. Use the techniques of this section to solve the differential equation in Exercise 38. Are there any equilibrium solutions for this differential equation?

40. Suppose you are having a dinner party for a large group of people, and you decide to make 2 gallons of chili. The recipe calls for 2 teaspoons of hot sauce per gallon, but you misread the instructions and put in 2 tablespoons of hot sauce per gallon. (Since each tablespoon is 3 teaspoons, you have put in 12 teaspoons of hot sauce in the chili.) You don't want to throw the chili out because there isn't much else to eat (and some people like hot chili), so you serve the chili anyway. However, as each person takes some chili, you fill up the pot with beans and tomatoes without hot sauce until the concentration of hot sauce agrees with the recipe. Suppose the guests take 1 cup of chili per minute from the pot (there are 16 cups in a gallon), how long will it take to get the chili back to the recipe's concentration of hot sauce? How many cups of chili will have been taken from the pot?

41. Suppose Ms. Lee is buying a new house and must borrow $150,000. She wants a 20-year mortgage and she has two choices. She can either borrow money at 7% per year with no "points," or she can borrow the money at 6.85% per year with a charge of 3 points. (A "point" is a fee of 1% of the loan amount that the borrower must pay the lender at the outset, so a mortgage of 3 points would require Ms. Lee to pay $4,500 extra to get the loan.) As an approximation, we assume that interest is compounded and payments are made continuously. Let

$$M(t) = \text{amount owed at time } t,$$
$$i = (\text{annual interest rate})/100,$$
$$p = \text{annual payment}.$$

Then the model for the amount owed is

$$\frac{dM}{dt} = iM - p.$$

    (a) How much does Ms. Lee pay in each case?

    (b) Which is a better deal over the entire time of the loan (assuming Ms. Lee does not invest the money she would have paid in points)?

    (c) If Ms. Lee can invest the $4,500 she would pay in points for the second mortgage at 5% compounded continuously, which is the better deal?

1.3 QUALITATIVE TECHNIQUE: SLOPE FIELDS

Whenever possible, it is useful to have a visual representation for a mathematical theory or problem. This is especially true for the study of ordinary differential equations, and in this section we develop a method for visualizing the graphs of the solutions to the differential equation

$$\frac{dy}{dt} = f(t, y).$$

If the function $y(t)$ is a solution of the equation $dy/dt = f(t, y)$, and if its graph passes through the point (t_1, y_1) where $y_1 = y(t_1)$, then the differential equation says that the derivative dy/dt at $t = t_1$ is given by the value $f(t_1, y_1)$. Geometrically, this equality of dy/dt at $t = t_1$ with $f(t_1, y_1)$ means that the slope of the tangent line to the graph of $y(t)$ at the point (t_1, y_1) is the number $f(t_1, y_1)$ (see Figure 1.9). Note that there is nothing special about the point (t_1, y_1) other than the fact that it is a point on the graph of the solution $y(t)$. The equality of dy/dt and $f(t, y)$ must hold for all t for which $y(t)$ satisfies the differential equation. In other words, the values of the right-hand side of the differential equation yield the slopes of the tangents at all points on the graph of $y(t)$ (see Figure 1.10).

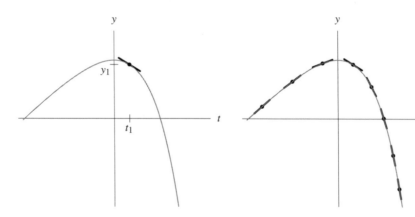

Figure 1.9
Slope of the tangent at (t_1, y_1) is given by the value of $f(t_1, y_1)$.

Figure 1.10
If $y = y(t)$ is a solution, then the slope of any tangent must equal $f(t, y)$.

Sketching Slope Fields

This simple geometric observation leads to our main device for the visualization of the solutions to a first-order differential equation

$$\frac{dy}{dt} = f(t, y).$$

If we are given the function $f(t, y)$, we obtain a rough idea of the graphs of the solutions to the differential equation by sketching its corresponding **slope field**. We make

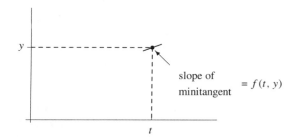

Figure 1.11
The slope field is made up of (infinitely)
many minitangent segments. The slope of
the minitangent at the point (t, y) is
determined by the right-hand side $f(t, y)$
of the differential equation.

this sketch by selecting points in the ty-plane and computing the numbers $f(t, y)$ at these points. At each point (t, y) selected, we use $f(t, y)$ to draw a minitangent line whose slope is $f(t, y)$ (see Figure 1.11). Once we have many such line segments, it is easy to visualize the graphs of the solutions.

For example, consider the differential equation

$$\frac{dy}{dt} = f(t, y) = y - t.$$

To get some practice with the idea of a slope field, we sketch its slope field by hand at a small number of points. Then we discuss a computer-generated version of this slope field.

Generating slope fields by hand is tedious, so we consider only the nine points in the ty-plane. For example, at the point $(t, y) = (1, -1)$, we have $f(t, y) = f(1, -1) = -1 - 1 = -2$. Therefore we sketch a "small" line segment with slope -2 centered at the point $(1, -1)$ (see Figure 1.12). To sketch the slope field for all nine points, we use the function $f(t, y)$ to compute the appropriate slopes. The results are summarized in Table 1.2. Once we have these values, we use them to give a sparse sketch of the slope field for this equation (see Figure 1.12).

Sketching slope fields is best done using a computer. Figure 1.13 is a sketch of the slope field for this equation over the region $-3 \leq t \leq 3$ and $-3 \leq y \leq 3$ in the ty-plane. We calculated values of the function $f(t, y)$ over 25×25 points (625 points) in that region.

A glance at this slope field suggests that the graph of one solution is a diagonal line passing through the points $(-1, 0)$ and $(0, 1)$. Solutions corresponding to initial conditions that are below this line seem to increase until they reach an absolute maximum.

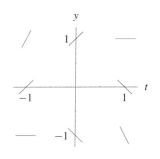

Figure 1.12
A "sparse" slope field
generated from Table 1.2.

Table 1.2 Selected slopes corresponding to the differential equation
$dy/dt = y - t$

(t, y)	$f(t, y)$	(t, y)	$f(t, y)$	(t, y)	$f(t, y)$
$(-1, 1)$	2	$(0, 1)$	1	$(1, 1)$	0
$(-1, 0)$	1	$(0, 0)$	0	$(1, 0)$	-1
$(-1, -1)$	0	$(0, -1)$	-1	$(1, -1)$	-2

Solutions corresponding to initial conditions that are above the line seem to increase more and more rapidly.

In fact, in Section 1.8, we will learn an analytic technique for finding solutions of this equation. We will see that the general solution consists of the family of functions

$$y(t) = t + 1 + ce^t,$$

where c is an arbitrary constant. (At this point it is important to emphasize that, even though we have not studied the technique that gives us these solutions, we can still check to see whether these functions are indeed solutions. If $y(t) = t + 1 + ce^t$, then $dy/dt = 1 + ce^t$. Also $f(t, y) = y - t = (t + 1 + ce^t) - t = 1 + ce^t$. Hence all these functions are solutions.)

In Figure 1.14, we sketch the graphs of these functions with $c = -2, -1, 0, 1, 2, 3$. Note that each of these graphs is tangent to the slope field. Also note that, if $c = 0$, the graph is a straight line with slope 1 through the points $(-1, 0)$ and $(0, 1)$.

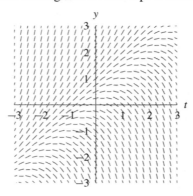

Figure 1.13
A computer-generated version of the slope field for $dy/dt = y - t$.

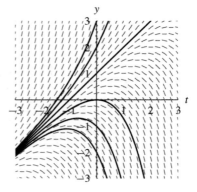

Figure 1.14
The graphs of six solutions to $dy/dt = y - t$ superimposed on its slope field.

Important Special Cases

From an analytic point of view, differential equations of the forms

$$\frac{dy}{dt} = f(t) \quad \text{and} \quad \frac{dy}{dt} = f(y)$$

are somewhat easier to consider than more complicated equations because they are separable. The geometry of their slope fields is equally special.

Slope fields for $dy/dt = f(t)$

If the right-hand side of the differential equation in question is solely a function of t, or in other words, if

$$\frac{dy}{dt} = f(t),$$

the slope at any point is the same as the slope of any other point with the same t-coordinate (see Figure 1.15).

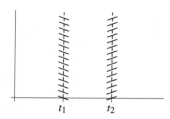

Figure 1.15
If the right-hand side of the differential equation is a function of t alone, then the slopes in the slope field are determined by their t-coordinate.

Geometrically, this implies that all of the tangents in the slope field on any vertical line are parallel. Whenever this holds for all vertical lines throughout the domain in question, we know that the corresponding differential equation is really an equation of the form

$$\frac{dy}{dt} = f(t).$$

(Note that finding solutions to this type of differential equation is the same thing as finding an antiderivative of $f(t)$ in calculus.)

For example, consider the slope field shown in Figure 1.16. We generated this slope field from the equation

$$\frac{dy}{dt} = 2t,$$

and from calculus we know that

$$y(t) = \int 2t \, dt = t^2 + c,$$

where c is the constant of integration. Hence the general solution of the differential equation consists of functions of the form

$$y(t) = t^2 + c.$$

In Figure 1.17, we have superimposed graphs of such solutions on this field. Note that all of these graphs simply differ by a vertical translation. If one graph is tangent to the slope field, we can get infinitely many graphs, each of which is tangent to the slope field, by translating the original graph either up or down.

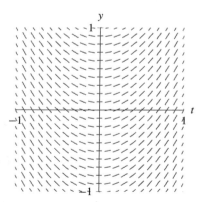

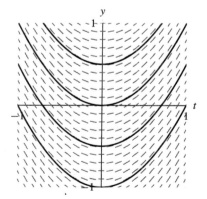

Figure 1.16
A slope field with parallel slopes along vertical lines.

Figure 1.17
Graphs of solutions to $dy/dt = 2t$ superimposed on its slope field.

Slope fields for autonomous equations

In the case of an autonomous differential equation

$$\frac{dy}{dt} = f(y),$$

the right-hand side of the equation does not depend on the independent variable t. The slope field in this case is also somewhat special. Here, the slopes that correspond to two different points with the same y-coordinate are equal. That is, $f(t_1, y) = f(t_2, y) = f(y)$, since the right-hand side of the differential equation depends only on y. In other words, the slope field of an autonomous equation is parallel along each horizontal line (see Figure 1.18).

Figure 1.18
If the right-hand side of the differential equation is a function of y alone, then the slopes in the slope field are determined by their y-coordinate.

For example, the slope field for the autonomous equation

$$\frac{dy}{dt} = 4y(1 - y)$$

is given in Figure 1.19. Note that, along each horizontal line, the tangents are parallel. In fact, if $0 < y < 1$, then dy/dt is positive, and the tangents suggest that a solution with $0 < y < 1$ is increasing. On the other hand, if $y < -1$ or if $y > 1$, then dy/dt is negative and any solution with either $y < -1$ or $y > 1$ is decreasing.

We have equilibrium solutions at $y = 0$ and at $y = 1$ since the right-hand side of the differential equation vanishes along these lines. The slope field is horizontal all along these two horizontal lines, and therefore we know that these lines are the graphs of solutions. Solutions whose graphs are between these two lines are increasing. Solutions that are above the line $y = 1$ or are below the line $y = 0$ are decreasing (see Figure 1.20).

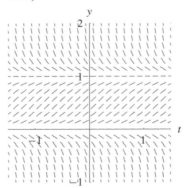

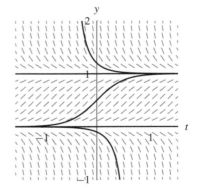

Figure 1.19
The slope field for
$dy/dt = 4y(1 - y)$.

Figure 1.20
The graphs of five solutions superimposed on the slope field for
$dy/dt = 4y(1 - y)$.

The fact that autonomous equations produce slope fields that are parallel along horizontal lines indicates that we can get infinitely many solutions from one solution simply by translating the graph of the given solution left or right (see Figure 1.21). We will make extensive use of this simple geometric observation about the solutions to autonomous equations in Section 1.6.

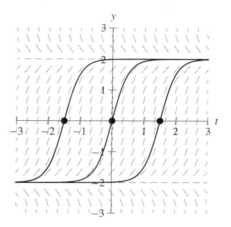

Figure 1.21
The graphs of three solutions to an autonomous equation. Note that each is a horizontal translate of the other.

Analytic versus Qualitative Analysis

For the autonomous equation

$$\frac{dy}{dt} = 4y(1 - y),$$

we could have used the analytic techniques of the previous section to find explicit formulas for the solutions. In fact, we can perform all of the required integrations to determine the general solution (see Exercise 15 in that section). However, these integrations are quite involved, and the formulas that result are by no means easy to interpret. This points out the power of geometric and qualitative methods for solving differential equations. With very little work, we gain a lot of insight into the behavior of solutions. Although we cannot use qualitative methods to answer specific questions, such as what the exact value of the solution is at any given time, we can use these methods to understand the long-term behavior of a solution.

These ideas are especially important if the differential equation in question cannot be handled by analytic techniques. As an example, consider the differential equation

$$\frac{dy}{dt} = e^{y^2/10} \sin^2 y.$$

This equation is autonomous and hence separable. To solve this equation analytically, we must evaluate the integrals

$$\int \frac{dy}{e^{y^2/10} \sin^2 y} = \int dt.$$

However, the integral on the left-hand side cannot be evaluated so easily (if at all). Thus we resort to qualitative methods. The right-hand side of this differential equation is positive except where $y = n\pi$ for any integer n. These special lines correspond to equilibrium solutions of the equation. Between these equilibria, solutions must always increase. From the slope field, we expect that their graphs either lie on one of the horizontal lines of the form $y = n\pi$ or increase from one of these lines to the next higher as $t \to \infty$ (see Figure 1.22). Hence we can predict the long-term behavior of the solutions even though we cannot explicitly solve the equation.

Although the computer pictures of solutions of this differential equation are convincing, some subtle questions remain. For example, how do we *really* know that these pictures are correct? In particular, for $dy/dt = e^{y^2/10} \sin^2 y$, how do we know that solutions do not cross the horizontal line solutions given by the equilibria (see Figure 1.22)? Such a solution could not cross these lines at a nonzero angle since we know that the tangent line to the solution must be horizontal. But what prevents certain solutions from crossing these lines tangentially and then continuing to increase?

In the differential equation

$$\frac{dy}{dt} = 4y(1 - y)$$

we can eliminate these questions because we can evaluate all of the integrals and check the accuracy of the pictures using analytic techniques. But using analytic techniques to check our qualitative analysis does not work if we cannot find explicit solutions. Besides, having to resort to analytic techniques to check the qualitative results defeats the purpose of using these methods in the first place. In Section 1.5 we discuss powerful theorems that answer many of these questions without undue effort.

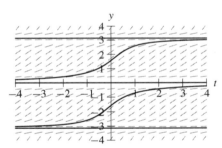

Figure 1.22
Slope field and graphs of solutions for

$$\frac{dy}{dt} = e^{y^2/10} \sin^2 y.$$

The Mixing Problem Revisited

Recall that in the previous section we found precise analytic solutions for the differential equation

$$\frac{dS}{dt} = \frac{20000 - 3S}{1000},$$

where S describes the amount of sugar in a vat at time t. We found that the general solution of this equation was

$$S(t) = ce^{-0.03t} + \frac{2000}{3},$$

where c is an arbitrary constant.

Using the slope field of this equation, we can easily derive a qualitative description of these solutions. In Figure 1.23, we display the slope field and graphs of selected solutions. Note that, as expected, the slope field is horizontal when $S = 20000/3$, the equilibrium solution. Slopes are positive when $S < 20000/3$ and negative when $S > 20000/3$. So we expect solutions to tend toward the equilibrium solution as t increases. This qualitative analysis indicates that, no matter what the initial amount of sugar, the amount of sugar in the vat tends to $20000/3$ as $t \to \infty$. Of course, we obtain the same information by taking the limit of the general solution as $t \to \infty$, but it is nice to see the same result in a geometric setting. Furthermore, in other examples, taking such a limit may not be as easy as in this case, but qualitative methods may still be used to determine the long-term behavior of the solutions.

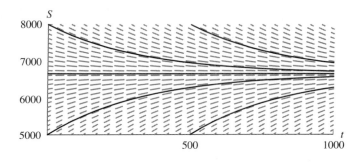

Figure 1.23
The slope field and graphs of a few solutions of

$$\frac{dS}{dt} = \frac{20000 - 3S}{1000}.$$

An RC Circuit

The simple electric circuit pictured in Figure 1.24 contains a capacitor, a resistor, and a voltage source. The behavior of the resistor is specified by a positive parameter R (the "resistance"), and the behavior of the capacitor is specified by a positive parameter C (the "capacitance"). The input voltage across the voltage source at time t is denoted by $V(t)$. This voltage source could be a constant source such as a battery, or it could be a source that varies with time such as alternating current. In any case, we consider $V(t)$ to be a function that is specified by the circuit designer. In other words, it's given as part of the design of the circuit.

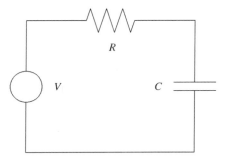

Figure 1.24
Circuit diagram with resistor, capacitor, and voltage source.

The quantities that specify the behavior of the circuit at a particular time t are the current $i(t)$ and the voltage across the capacitor $v_c(t)$. In this example we are interested in the voltage $v_c(t)$ across the capacitor. From the theory of electric circuits, we know that $v_c(t)$ satisfies the differential equation

$$RC\frac{dv_c}{dt} + v_c = V(t).$$

If we rewrite this in our standard form $dv_c/dt = f(t, v_c)$, we have

$$\frac{dv_c}{dt} = \frac{V(t) - v_c}{RC}.$$

We use slope fields to visualize solutions for four different types of voltage sources $V(t)$. (If you don't know anything about electric circuits, don't worry; Paul, Bob, and Glen don't either. In examples like this, all we need to do is accept the differential equation and "go with it.")

Zero input

If $V(t) = 0$ for all t, the equation becomes

$$\frac{dv_c}{dt} = \frac{-v_c}{RC}.$$

A sample slope field for a particular choice of R and C is given in Figure 1.25. We see clearly that all solutions "decay" toward $v_c = 0$ as t increases. If there is no voltage source, the voltage across the capacitor $v_c(t)$ decays to zero. This prediction for the voltage agrees with what we obtain analytically since the general solution of this equation is $v_c(t) = v_0 e^{-t/RC}$, where v_0 is the initial voltage across the capacitor. (Note that this equation is almost exactly the same as the exponential growth model that we studied in Section 1.1, and consequently, we can solve it analytically by either guessing the correct form of a solution or by separating variables.)

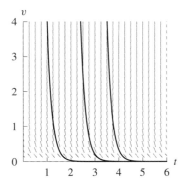

Figure 1.25
Slope field for

$$\frac{dv_c}{dt} = -\frac{v_c}{RC}$$

with $R = 0.2$ and $C = 1$, and the graph of a solution.

Constant nonzero voltage source

Suppose $V(t)$ is a nonzero constant K for all t. The equation for voltage across the capacitor becomes

$$\frac{dv_c}{dt} = \frac{K - v_c}{RC}.$$

This equation is autonomous with one equilibrium solution at $v_c = K$. The slope field for this equation shows that all solutions tend toward this equilibrium as t increases (see Figure 1.26). Given any initial voltage $v_c(0)$ across the capacitor, the voltage $v_c(t)$ tends to the value $v = K$ as time increases.

We could find a formula for the general solution by separating variables and integrating, but we leave this as an exercise.

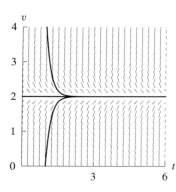

Figure 1.26
Slope field for
$$\frac{dv_c}{dt} = \frac{K - v_c}{RC}$$
for $R = 0.2$, $C = 1$, and $K = 2$, and the graphs of three solutions with different initial conditions.

On-off voltage source

Suppose $V(t) = K > 0$ for $t < 3$, but at $t = 3$, this voltage is "turned off." Then $V(t) = 0$ for $t > 3$. Our differential equation is

$$\frac{dv_c}{dt} = \frac{V(t) - v_c}{RC} = \begin{cases} \dfrac{K - v_c}{RC} & \text{for } t < 3; \\[2ex] \dfrac{-v_c}{RC} & \text{for } t > 3. \end{cases}$$

The right-hand side is given by two different formulas depending on the value of t. We can see this in the slope field for this equation (see Figure 1.27). It resembles Figures 1.25 and 1.26 pasted together along $t = 3$. Since the differential equation is not defined at $t = 3$, we must add an additional assumption to our model. We assume that the voltage $v_c(t)$ is a continuous function at $t = 3$.

The particular solution with the initial condition $v_c(0) = K$ is constant for $t < 3$, but for $t > 3$ it decays exponentially. Solutions with $v_c(0) \neq K$ move toward K for $t < 3$, but then decay toward zero for $t > 3$. We could find formulas for the solutions by first finding the solution for $t < 3$, then starting over for $t > 3$ (see Section 1.2), but we again leave this as an exercise.

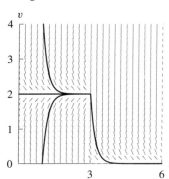

Figure 1.27
Slope field for
$$\frac{dv_c}{dt} = \frac{V(t) - v_c}{RC}$$
for $V(t)$, which "turns off" at $t = 3$ for $R = 0.2$, $C = 1$ and $K = 2$, along with graphs of three solutions with different initial conditions.

Periodic on-off voltage source

Suppose $V(t)$ alternates periodically between the values K and zero every three seconds. That is,

$$
V(t) = \begin{cases}
K & \text{for } 0 \le t < 3; \\
0 & \text{for } 3 < t < 6; \\
K & \text{for } 6 < t < 9; \\
\quad\vdots
\end{cases}
$$

This corresponds to someone switching the source voltage off every three seconds and back on three seconds later. The slope field for the differential equation

$$
\frac{dv_c}{dt} = \frac{V(t) - v_c}{RC}
$$

is given in Figure 1.28. Parts of the slope fields in Figures 1.25 and 1.26 are patched together every three seconds. The solutions are also patched together from these two equations. When $V(t) = K$, the solution approaches the equilibrium value $v_c = K$, and when $V(t) = 0$, the solution decays toward zero.

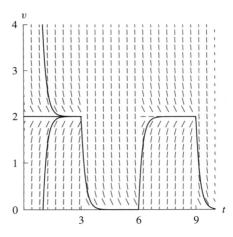

Figure 1.28
Slope field for

$$
\frac{dv_c}{dt} = \frac{V(t) - v_c}{RC}
$$

where $V(t)$ alternates between K and zero every three seconds for $R = 0.2$, $C = 1$, and $K = 1$, along with the graphs of three solutions with different initial conditions.

Combining Qualitative with Quantitative Results

When only qualitative behavior of the solution is required, sketches of solutions obtained from slope fields can sometimes suffice. In some applications it is necessary to know the exact value (or almost exact value) of the solution with a given initial condition. In these situations analytic and/or numerical methods can't be avoided. But even then, it is nice to have a mental picture of what solutions look like.

Exercises for Section 1.3

In Exercises 1–6, sketch the slope fields for the given differential equation. (You may use a computer or compute the slopes by hand. If you compute by hand, use a grid of points (t, y) with $t = 0, 1/2, 1, 3/2, 2$ and $y = 0, 1/2, 1, 3/2, 2$. If you use a computer, you should compute the slope field at many more points to get a more accurate picture.)

1. $\dfrac{dy}{dt} = t^2 - t$

2. $\dfrac{dy}{dt} = 1 - y$

3. $\dfrac{dy}{dt} = y + t + 1$

4. $\dfrac{dy}{dt} = t^2 + 4$

5. $\dfrac{dy}{dt} = 2y(1 - y)$

6. $\dfrac{dy}{dt} = 4y^2$

In Exercises 7–10, a differential equation and its associated slope field are given. For each equation,

(a) sketch a number of different solutions on the slope field, and

(b) describe briefly the behavior of the solution with $y(0) = 1/2$ as t increases.

7. $\dfrac{dy}{dt} = 3y(1 - y)$

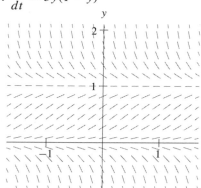

8. $\dfrac{dy}{dt} = 2y - t$

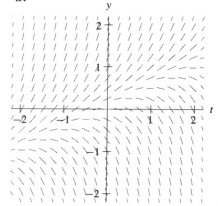

9. $\dfrac{dy}{dt} = (y + 1/2)(y + t)$

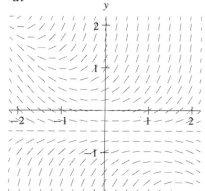

10. $\dfrac{dy}{dt} = (t + 1)y$

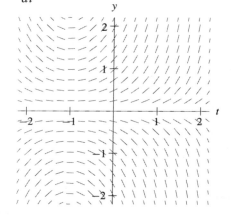

11. Eight differential equations and four slope fields are given below. Determine the equation that corresponds to each slope field and state briefly how you know your choice is correct. You should do this exercise without using technology.

(i) $\dfrac{dy}{dt} = t - 1$ (ii) $\dfrac{dy}{dt} = 1 - y^2$ (iii) $\dfrac{dy}{dt} = y^2 - t^2$ (iv) $\dfrac{dy}{dt} = 1 - t$

(v) $\dfrac{dy}{dt} = 1 - y$ (vi) $\dfrac{dy}{dt} = t^2 - y^2$ (vii) $\dfrac{dy}{dt} = 1 + y$ (viii) $\dfrac{dy}{dt} = y^2 - 1$

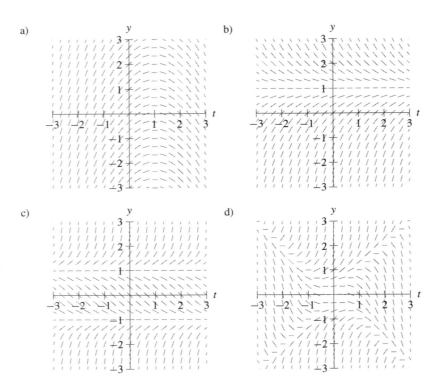

12. Consider the autonomous differential equation

$$\frac{dS}{dt} = S^3 - 2S^2 + S.$$

(a) By hand, give a rough sketch of the slope field.

(b) Using this sketch, sketch graphs of solutions $S(t)$ with initial conditions $S(0) = 0$, $S(0) = 1/2$, $S(1) = 1/2$, $S(0) = 3/2$, $S(0) = -1/2$.

13. Suppose we know that the function $f(t, y)$ is continuous and that $f(t, 3) = -1$ for all t.

(a) What does this information tell us about the slope field for the differential equation $dy/dt = f(t, y)$?

(b) What can we conclude about solutions $y(t)$ of $dy/dt = f(t, y)$? For example, if $y(0) < 3$, can $y(t) \to \infty$ as t increases?

14. Suppose we know that the graph below is the graph of the right-hand side $f(t)$ of the differential equation $dy/dt = f(t)$. Give a rough sketch of the slope field that corresponds to this differential equation.

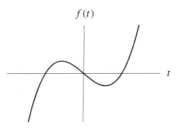

15. Suppose we know that the graph below is the graph of the right-hand side $f(y)$ of the differential equation $dy/dt = f(y)$. Give a rough sketch of the slope field that corresponds to this differential equation.

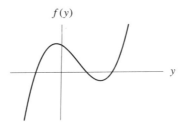

16. Suppose we know that the graph below is the graph of a solution to $dy/dt = f(t)$.

(a) How much of the slope field can you sketch from this information? [*Hint*: Note that the differential equation depends only on t.]

(b) What can you say about the solution with $y(0) = 2$? (For example, can you sketch the graph of this solution?)

17. Suppose we know that the graph below is the graph of a solution to $dy/dt = f(y)$.

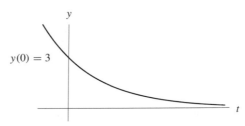

(a) How much of the slope field can you sketch from this information? [*Hint*: Note that the equation is autonomous.]

(b) What can you say about the solution with $y(0) = 2$? Sketch this solution.

18. Suppose the constant function $y(t) = 2$ for all t is a solution of the differential equation

$$\frac{dy}{dt} = f(t, y).$$

   (a) What does this tell you about the slope field? In other words, how much of the slope field can you sketch using this information?

   (b) What does this tell you about solutions?

Exercises 19–23 refer to the RC circuit discussed in this section. The differential equation for the voltage v across the capacitor is

$$\frac{dv}{dt} = \frac{V(t) - v}{RC}.$$

19. Find the formula for the general solution of the RC circuit equation above if the voltage source is constant for all time. In other words, $V(t) = K$ for all t. (Your solution will contain the three parameters R, C, and K, along with a constant that depends on the initial condition.)

20.   **(a)** Find the solution for the voltage $v(t)$ with initial value $v(0) = 1$ in the RC circuit equation given above if the voltage source $V(t)$ is the step function given by

$$V(t) = \begin{cases} K & \text{for } t < 3; \\ 0 & \text{for } t > 3. \end{cases}$$

   Your answer should contain the three parameters R, C, and K.

   (b) Find the general solution for this equation.

21. Given the source voltage $V(t) = 2t$ and the parameter values $R = 0.2$ and $C = 1$,

   (a) sketch the slope field,

   (b) sketch the solution curve with initial condition $v(0) = 0$, and

   (c) sketch the solution curve with initial condition $v(0) = 3$.

22. Given the source voltage

$$V(t) = \begin{cases} 0 & \text{for } t < 1; \\ 2 & \text{for } t \geq 1; \end{cases}$$

   and the parameter values $R = 0.2$ and $C = 1$,

   (a) sketch the slope field,

   (b) sketch the solution curve with initial condition $v(0) = 0$, and

   (c) sketch the solution curve with initial condition $v(0) = 3$.

23. Given the source voltage

$$V(t) = \begin{cases} 2t & \text{for } t < 1; \\ 2 & \text{for } t \geq 1; \end{cases}$$

and parameter values $R = 0.2$ and $C = 1$,

(a) sketch the slope field,

(b) sketch the solution curve with initial condition $v(0) = 0$,

(c) sketch the solution curve with initial condition $v(0) = 3$, and

(d) discuss in a few sentences the differences between the solutions for this differential equation and the solutions for the differential equations in Exercises 21 and 22.

24. Suppose that a population can be accurately modeled by the logistic equation

$$\frac{dp}{dt} = 0.4p\left(1 - \frac{p}{30}\right).$$

(Note that the growth rate parameter is 0.4 and the carrying capacity is 30.) Suppose that, at time $t = 5$, a disease is introduced into the population that kills 25% of the population per year. To adjust the model. we change the differential equation to

$$\frac{dp}{dt} = \begin{cases} 0.4p\left(1 - \dfrac{p}{30}\right) & \text{for } t < 5; \\[2ex] 0.4p\left(1 - \dfrac{p}{30}\right) - 0.25p & \text{for } t > 5. \end{cases}$$

(a) Sketch the slope field for this equation.

(b) Using the slope field, sketch some representative solution curves for this equation.

(c) Find formulas for the solutions of this equation for initial conditions $y(0) = 30$ and $y(0) = 20$.

(d) In a few sentences, describe the behavior of the solutions with initial conditions $y(0) = 30$ and $y(0) = 20$. (You can use either the sketches from the slope field or the formulas, but give a qualitative description of the solutions.)

1.4 NUMERICAL TECHNIQUE: EULER'S METHOD

The geometric concept of a slope field as discussed in the previous section is closely related to a fundamental numerical method for approximating solutions to a differential equation. Given an initial-value problem

$$\frac{dy}{dt} = f(t, y), \quad y(t_0) = y_0,$$

we can get a rough idea of the graph of its solution by first sketching the slope field in the ty-plane and then, starting at the initial value (t_0, y_0), sketching the solution by drawing a graph that is tangent to the slope field at each point along the graph. In this section, we describe a numerical procedure that automates this idea. Using a computer or a calculator, we obtain numbers and graphs that approximate solutions to initial-value problems.

Numerical methods provide quantitative information about solutions even if we cannot find their formulas. There is also the advantage that most of the work can be done by machine. The disadvantage is that we only obtain approximations, not precise solutions. If we remain aware of this fact and are prudent, numerical methods become powerful tools for the study of differential equations. It is not uncommon to turn to numerical methods, even when it is possible to find formulas for solutions. (Most of the graphs of solutions of differential equations in this text were drawn using numerical approximations, even when formulas were available.)

The numerical technique that we discuss in this section is called Euler's method. A more detailed discussion of the accuracy of Euler's method as well as other numerical methods is given in Chapter 7.

Stepping Along The Slope Field

To describe Euler's method, we begin with the initial-value problem

$$\frac{dy}{dt} = f(t, y), \quad y(t_0) = y_0.$$

Since we are given $f(t, y)$, we can plot slope field in the ty-plane. The idea of the method is to start at the point (t_0, y_0) in the slope field and take tiny steps dictated by the the tangents in the slope field.

We begin by choosing a (small) **step size** Δt. The slope of the approximate solution is updated every Δt units of t. In other words, for each step, we move Δt units along the t-axis. The size of Δt determines the accuracy of the approximate solution as well as the number of computations that are necessary to obtain the approximation.

Starting at (t_0, y_0), our first step is to the point (t_1, y_1), where $t_1 = t_0 + \Delta t$ and (t_1, y_1) is the point on the line through (t_0, y_0) with slope given by the slope field at (t_0, y_0) (see Figure 1.29). At (t_1, y_1) we repeat the procedure. Taking a step whose size along the t-axis is Δt and whose direction is determined by the slope field at (t_1, y_1), we reach the new point (t_2, y_2). The new time is given by $t_2 = t_1 + \Delta t$ and (t_2, y_2) is on the line segment that starts at (t_1, y_1) and has slope $f(t_1, y_1)$. Continuing, we use the slope field at the point (t_k, y_k) to determine the next point (t_{k+1}, y_{k+1}). The sequence of values $y_0, y_1, y_2, \ldots$ serves as an approximation to the solution at the times $t_0, t_1,$ $t_2, \ldots$. Geometrically it is best to think of the method as producing a sequence of tiny line segments connecting (t_k, y_k) to (t_{k+1}, y_{k+1}) for each k (see Figure 1.30). Basically, we are stitching together little pieces of the slope field to form a graph that approximates our solution curve.

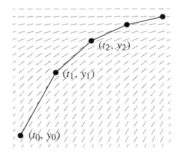

Figure 1.29
Stepping along the slope field.

This method uses tangent line segments, given by the slope field, to approximate the graph of the solution. Consequently, at each stage we make a slight error (see Figure 1.30). Hopefully, if the step size is sufficiently small, these errors do not get out of hand as we continue to step, and the resulting graph is close to the desired solution.

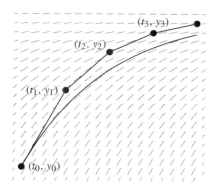

Figure 1.30
The graph of a solution and its approximation obtained using Euler's method.

Euler's Method

To put Euler's method into practice, we need a formula for determining (t_{k+1}, y_{k+1}) from (t_k, y_k). Finding t_{k+1} is easy. We specify the step size Δt at the outset, so

$$t_{k+1} = t_k + \Delta t.$$

To obtain y_{k+1} from (t_k, y_k), we use the differential equation. We know that the slope of the solution to the equation $dy/dt = f(t, y)$ at the point (t_k, y_k) is $f(t_k, y_k)$, and Euler's method uses this slope to determine y_{k+1}. In fact, the method determines the point (t_{k+1}, y_{k+1}) by assuming that it lies on the line through (t_k, y_k) with slope $f(t_k, y_k)$.

Now we can use our basic knowledge of slopes to determine y_{k+1}. The formula for the slope of a line gives

$$\frac{y_{k+1} - y_k}{t_{k+1} - t_k} = f(t_k, y_k).$$

Since $t_{k+1} = t_k + \Delta t$, the denominator $t_{k+1} - t_k$ is just Δt, and therefore we have

$$\frac{y_{k+1} - y_k}{\Delta t} = f(t_k, y_k)$$

$$y_{k+1} - y_k = f(t_k, y_k)\, \Delta t$$

$$y_{k+1} = y_k + f(t_k, y_k)\, \Delta t.$$

This is the formula for Euler's method (see Figures 1.31 and 1.32).

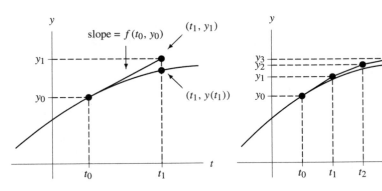

Figure 1.31
Euler's method uses the slope at one point to approximate the graph over an interval

Figure 1.32
Euler's method after several steps and the actual solution $y(t)$

| **Euler's Method for** $\dfrac{dy}{dt} = f(t, y)$ | *Given the initial condition $y(t_0) = y_0$ and the step size Δt, compute the point (t_{k+1}, y_{k+1}) from the preceding point (t_k, y_k) as follows:* |

1. *Use the differential equation to compute the slope $f(t_k, y_k)$.*

2. *Calculate the next point (t_{k+1}, y_{k+1}) by $t_{k+1} = t_k + \Delta t$ and*

$$y_{k+1} = y_k + f(t_k, y_k)\Delta t.$$

Approximating an Autonomous Equation

To illustrate Euler's method, we first use it to approximate the solution to a differential equation whose solution we already know. In this way, we are able to compare the approximation we obtain to the known solution. Consequently, we are able to gain some insight into the effectiveness of the method in addition to seeing how it is implemented.

Consider the initial-value problem

$$\frac{dy}{dt} = 2y - 1, \quad y(0) = 1.$$

This equation is separable, and when we separate variables and integrate, we obtain the solution

$$y(t) = \frac{e^{2t} + 1}{2}.$$

In this example, $f(t, y) = 2y - 1$, so Euler's method is given by

$$y_{k+1} = y_k + (2y_k - 1)\Delta t.$$

To illustrate the method, we start with a relatively large step size of $\Delta t = 0.1$ and approximate the solution over the interval $0 \leq t \leq 1$. In order to approximate the solution over an interval whose length is 1 with a step size of 0.1, we must compute ten iterations of the method. The initial condition $y(0) = 1$ provides the initial value $y_0 = 1$. Given $\Delta t = 0.1$, we have $t_1 = t_0 + 0.1 = 0 + 0.1 = 0.1$. We compute the y-coordinate for the first step by

$$y_1 = y_0 + (2y_0 - 1)\Delta t = 1 + (1)\,0.1 = 1.1.$$

Thus the first point (t_1, y_1) on the graph of the approximate solution is $(0.1, 1.1)$.

To compute the y-coordinate y_2 for the second step, we now use y_1 rather than y_0. That is,

$$y_2 = y_1 + (2y_1 - 1)\Delta t = 1.1 + (1.2)\,0.1 = 1.22,$$

and the second point for our approximate solution is $(t_2, y_2) = (0.2, 1.22)$.

Continuing this procedure, we obtain the results given in Table 1.3. After 10 steps, we obtain the approximation of $y(1)$ by $y_{10} = 3.59$. Since we know that

$$y(1) = \frac{e^2 + 1}{2} \approx 4.194,$$

the approximation y_{10} is off by slightly more than 0.6. This is not a very good approximation, but we'll soon see how to avoid this (usually). The reason for the error can be seen by looking at the graph of the solution and its approximation. The slope field for this differential equation always lies below the graph (see Figure 1.33), so we expect our approximation to come up short.

Using a smaller step size usually reduces the error, but more computations must be done to approximate the solution over the same interval. For example, if we halve the step size in this example ($\Delta t = 0.05$), then we must calculate twice as many steps since $t_1 = 0.05$, $t_2 = 0.1, \ldots, t_{20} = 1.0$. Again we start with $(t_0, y_0) = (0, 1)$ as specified by the initial condition. However, with $\Delta t = 0.05$, we obtain

$$y_1 = y_0 + (2y_0 - 1)\Delta t = 1 + (1)\,0.05 = 1.05.$$

This step yields the point $(t_1, y_1) = (0.05, 1.05)$ on the graph of our approximate solution. For the next step, we compute

$$y_2 = y_1 + (2y_1 - 1)\Delta t = 1.05 + (1.1)\,0.05 = 1.105.$$

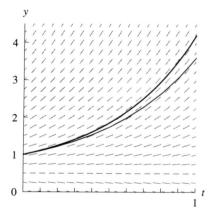

Figure 1.33
The graph of the solution to

$$\frac{dy}{dt} = 2y - 1$$

with $y(0) = 1$ and the approximation produced by Euler's method with $\Delta t = 0.1$.

Table 1.3 Euler's method for $dy/dt = 2y - 1$, $y(0) = 1$ with $\Delta t = 0.1$.

k	t_k	y_k	$f(t_k, y_k)$
0	0	1	1
1	0.1	1.1	1.2
2	0.2	1.22	1.44
3	0.3	1.36	1.73
4	0.4	1.54	2.07
5	0.5	1.74	2.49
6	0.6	1.99	2.98
7	0.7	2.29	3.58
8	0.8	2.65	4.30
9	0.9	3.08	5.16
10	1.0	3.59	

Now we have the point $(t_2, y_2) = (1.1, 1.105)$. This type of calculation gets tedious fairly quickly, but luckily, calculations such as these are perfect for a computer or a calculator. For $\Delta t = 0.05$, the results of Euler's method are given in Table 1.4.

If we carefully compare the results of our two computations, we see that, with $\Delta t = 0.1$, we approximate $y(1) \approx 4.194$ with $y_{10} = 3.59$. With $\Delta t = 0.05$, we approximate $y(1)$ with $y_{20} = 3.864$. The error in the first approximation is slightly more than 0.6, whereas the error in the second approximation is 0.33. Roughly speaking we halve the error by halving the step size. This type of improvement is typical of Euler's method. (We will be much more precise about how the error in Euler's method is related to the step size in Chapter 7.)

With the even smaller step size of $\Delta t = 0.01$, we must do much more work since we need 100 steps to go from $t = 0$ to $t = 1$. However, in the end, we obtain a much better approximation to the solution (see Table 1.5).

This example illustrates the typical trade-off that occurs with numerical methods. There are always decisions to be made such as the choice of the step size Δt. Lowering Δt often results in a better approximation — at the expense of more computation.

Table 1.4 Euler's method for $dy/dt = 2y - 1$, $y(0) = 1$ with $\Delta t = 0.05$.

k	t_k	y_k	$f(t_k, y_k)$
0	0	1	1
1	0.05	1.05	1.1
2	0.1	1.105	1.21
3	0.15	1.166	1.331
⋮	⋮	⋮	⋮
19	0.95	3.558	6.116
20	1.0	3.864	

Table 1.5 Euler's method for $dy/dt = 2y - 1$, $y(0) = 1$ with $\Delta t = 0.01$.

k	t_k	y_k	$f(t_k, y_k)$
0	0	1	1
1	0.01	1.01	1.02
2	0.02	1.0202	1.0404
3	0.03	1.0306	1.0612
$\vdots$	$\vdots$	$\vdots$	$\vdots$
98	0.98	3.9817	6.9633
99	0.99	4.0513	7.1026
100	1.0	4.1223	

A Nonautonomous Example

Note that it is the value $f(t_k, y_k)$ of the right-hand side of the differential equation at (t_k, y_k) that determines the next point (t_{k+1}, y_{k+1}). The last example was an autonomous differential equation, so $f(t_k, y_k)$ depended only on y_k. However, if the differential equation is nonautonomous, the value of t_k also plays an important role in the computations.

To illustrate Euler's method applied to a nonautonomous equation, we consider the initial-value problem

$$\frac{dy}{dt} = -2ty^2, \quad y(0) = 1.$$

This differential equation is also separable, and we can separate variables to obtain the solution

$$y(t) = \frac{1}{1 + t^2}.$$

We use Euler's method to approximate this solution over the interval $0 \leq t \leq 2$. The value of the solution at $t = 2$ is $y(2) = 1/5$. Again, it is interesting to see how close we come to this value with various choices of Δt. The formula for Euler's method is

$$y_{k+1} = y_k + f(t_k, y_k)\,\Delta t = y_k - (2t_k y_k^2)\Delta t$$

with $t_0 = 0$ and $y_0 = 1$. We begin by approximating the solution from $t = 0$ to $t = 2$ using just four steps. This involves so few computations that we can perform the arithmetic "by hand." To cover an interval of length 2 in four steps, we must use $\Delta t = 2/4 = 1/2$. The entire calculation is displayed in Table 1.6. Note that we end up approximating the exact value $y(2) = 1/5 = 0.2$ by $y_4 = 5/32 = 0.15625$. Figure 1.34 shows the graph of the solution as compared to the results of Euler's method over this interval.

Table 1.6 Euler's method for $dy/dt = -2ty^2$, $y(0) = 1$ with $\Delta t = 1/2$.

k	t_k	y_k	$f(t_k, y_k)$
0	0	1	0
1	1/2	1	-1
2	1	1/2	-1/2
3	3/2	1/4	-3/16
4	2	5/32	

As before, choosing smaller values of Δt yields better approximations. For example, if $\Delta t = 0.1$, the Euler approximation of the exact value $y(2) = 0.2$ is $y_{20} = 0.1933\ldots$. If $\Delta t = 0.001$, we need to compute 2000 steps, but the approximation improves to $y_{2000} = 0.199937\ldots$. (See Tables 1.7 and 1.8. You may get slightly different results on your computer or calculator since different machines use different algorithms for rounding numbers.)

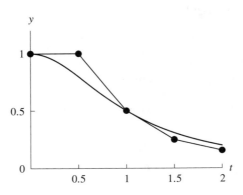

Figure 1.34
The graph of the solution to the initial-value problem

$$\frac{dy}{dt} = -2ty^2, \quad y(0) = 1,$$

and the approximation produced by Euler's method with $\Delta t = 1/2$.

Note that the convergence of the approximation to the actual value is slow. We computed 2000 steps and obtained an answer that is only accurate to three decimal places. In Chapter 7, we present more complicated algorithms for numerical approximation of solutions. Although the algorithms are more complicated from a conceptual point of view, they obtain better accuracy with less computation.

Table 1.7 Euler's method for $dy/dt = -2ty^2$, $y(0) = 1$ with $\Delta t = 0.1$.

k	t_k	y_k
0	0	1
1	0.1	1
2	0.2	0.98
3	0.3	$0.9416\ldots$
⋮	⋮	⋮
19	1.9	$0.2101\ldots$
20	2	$0.1933\ldots$

Table 1.8 Euler's method for $dy/dt = -2ty^2$, $y(0) = 1$ with $\Delta t = 0.001$.

k	t_k	y_k
0	0	1
1	0.001	$1.000000\ldots$
2	0.002	$0.999998\ldots$
3	0.003	$0.999994\ldots$
⋮	⋮	⋮
1999	1.999	$0.200097\ldots$
2000	2	$0.199937\ldots$

An RC Circuit with Periodic Input

Recall from Section 1.3 that the voltage v_c across the capacitor in the simple circuit drawn in Figure 1.35 is given by the differential equation

$$\frac{dv_c}{dt} = \frac{V(t) - v_c}{RC}$$

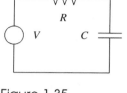
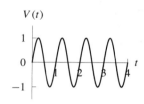

Figure 1.35
Circuit diagram with resistor, capacitor and voltage source.

where R is the resistance, C is the capacitance and $V(t)$ is the source or input voltage. We have seen how we can use slope fields to give a qualitative sketch of solutions. Using Euler's method we can also obtain numerical approximations of the solutions.

Suppose we consider a circuit where $R = 0.5$ and $C = 1$. (The usual units for resistance is "ohms" and for capacitance is "farads". We choose these numbers so that the numbers in the solution work out nice. A 1 farad capacitor would be extremely large.) Then the differential equation is

$$\frac{dv_c}{dt} = \frac{V(t) - v_c}{0.5} = 2(V(t) - v_c).$$

To understand how the voltage v_c varies if the voltage source $V(t)$ is periodic in time, we consider the case where $V(t) = \sin(2\pi t)$. Consequently, the voltage oscillates between -1 and 1 once each unit of time (see Figure 1.36). The differential equation is now

$$\frac{dv_c}{dt} = -2v_c + 2\sin(2\pi t).$$

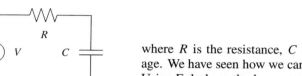

Figure 1.36
Graph of $V(t) = \sin(2\pi t)$, the input voltage.

From the slope field for this equation (see Figure 1.37), we might predict that the solutions oscillate. Using Euler's method applied to this equation for several different initial conditions, we see that the solutions do indeed oscillate. In addition, we see that they also approach each other and collect around a single solution (see Figure 1.38). This uniformity of long-term behavior is not so easily predicted from the slope field alone.

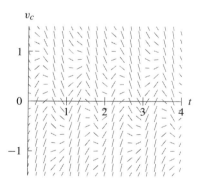

Figure 1.37
Slope field for
$dv_c/dt = -2v_c + 2\sin(2\pi t)$.

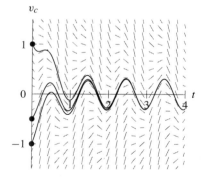

Figure 1.38
Graphs of approximate solutions to
$dv_c/dt = -2v_c + 2\sin(2\pi t)$ obtained
using Euler's method.

Problems with Numerical Methods

By their very nature, any numerical approximation scheme is inaccurate. For instance, in each step of Euler's method, we almost always make an error of some sort. These errors can accumulate and sometimes lead to disastrously wrong approximations. As an example, consider the differential equation

$$\frac{dy}{dt} = e^t \sin y.$$

There are equilibrium solutions for this equation if $\sin y = 0$. In other words, any constant function of the form $y(t) = n\pi$ for $n = \ldots -1, 0, 1, 2, \ldots$ is a solution.

Using the initial value $y(0) = 5$ and a step size $\Delta t = 0.1$, Euler's method yields the approximation graphed in Figure 1.39. It seems like something must be wrong. At first, the solution tends toward the equilibrium solution $y(t) = \pi$, but then just before $t = 5$ something strange happens. The graph of the approximation jumps dramatically. If we lower Δt to 0.05, we still find erratic behavior although t is slightly greater than 5 before this happens (see Figure 1.40).

The difficulty arises in Euler's method for this equation because of the term e^t on the right-hand side. It becomes very large as t increases, and consequently slopes in the slope field are quite large for large t. Even a very small step in the t-direction throws us far from the actual solution.

This is typical of the use of numerics in the study of differential equations. Numerical methods, when they work, work beautifully. But they sometimes fail. We always have to be aware of this possibility and be ready with an alternate approach. In the next section we present theoretical results which help identify when numerical approximations have gone awry.

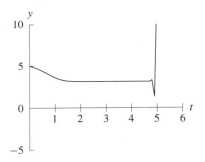

Figure 1.39
Euler's method applied to
$dy/dt = e^t \sin y$ with $\Delta t = 0.1$

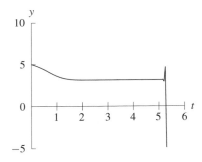

Figure 1.40
Euler's method applied to
$dy/dt = e^t \sin y$ with $\Delta t = 0.05$.

The Big Three

We have now introduced examples of all three of the fundamental methods for attacking differential equations — the analytic, the numeric and the qualitative approaches. Which method is the best depends both on the differential equation in question and on what we want to know about the solutions. Often all three methods "work", but a great deal of labor can be saved if we think first about which method gives the most direct route to the information we need.

Exercises for Section 1.4

In Exercises 1–8, use Euler's method with the given step size Δt to approximate the solution to the given initial-value problem over the time interval specified. Your answer should include a table of the approximate values of the dependent variable. It should also include a sketch of the graph of the approximate solution.

1. $\dfrac{dy}{dt} = 2y + 1, \quad y(0) = 3, \quad 0 \le t \le 2, \quad \Delta t = 0.5$

2. $\dfrac{dy}{dt} = t - y^2, \quad y(0) = 1, \quad 0 \le t \le 1, \quad \Delta t = 0.25$

3. $\dfrac{dy}{dt} = y^2 - 2y + 1, \quad y(0) = 2, \quad 0 \le t \le 2, \quad \Delta t = 0.5$

4. $\dfrac{dy}{dt} = \sin y, \quad y(0) = 1, \quad 0 \le t \le 3, \quad \Delta t = 0.5$

5. $\dfrac{dw}{dt} = (3 - w)(w + 1), \quad w(0) = 4, \quad 0 \le t \le 5, \quad \Delta t = 1.0$

6. $\dfrac{dw}{dt} = (3 - w)(w + 1), \quad w(0) = 0, \quad 0 \le t \le 5, \quad \Delta t = 0.5$

7. $\dfrac{dy}{dt} = e^{(2/y)}, \quad y(0) = 2, \quad 0 \le t \le 2, \quad \Delta t = 0.5$

8. $\dfrac{dy}{dt} = e^{(2/y)}, \quad y(1) = 2, \quad 1 \le t \le 3, \quad \Delta t = 0.5$

9. Compare your answers to Exercises 7 and 8 and explain your observations.

10. Compare your answers to Exercises 5 and 6. Is Euler's method doing a good job in this case? What would you do to avoid the difficulties that arise in this case?

11. Do a qualitative analysis of the solution of the initial-value problem in Exercise 6 and compare your conclusions with your results in Exercise 6. What's wrong with the approximate solutions given by Euler's method?

12. Consider the initial-value problem

$$\frac{dy}{dt} = \sqrt{y}, \quad y(0) = 1.$$

Using Euler's method, compute three different approximate solutions corresponding to $\Delta t = 1.0$, 0.5, and 0.25 over the interval $0 \le t \le 4$. Graph all three solutions. What predictions do you make about the actual solution to the initial-value problem?

13. Consider the initial-value problem

$$\frac{dy}{dt} = 2 - y, \quad y(0) = 1.$$

Using Euler's method, compute three different approximate solutions corresponding to $\Delta t = 1.0$, 0.5, and 0.25 over the interval $0 \le t \le 4$. Graph all three solutions.

What predictions do you make about the actual solution to the initial-value problem? How do the graphs of these approximate solutions relate to the graph of the actual solution? Why?

In Exercises 14–17, we consider the RC circuit model equation from the text

$$\frac{dv}{dt} = \frac{V(t) - v}{RC}.$$

Suppose $V(t) = e^{-0.1t}$ (the voltage source $V(t)$ is decaying exponentially). If $R = 0.2$ and $C = 1$, use Euler's method to compute values of the solutions with the given initial conditions over the interval $0 \leq t \leq 10$:

14. $v(0) = 0$ **15.** $v(0) = 2$

16. $v(0) = -2$ **17.** $v(0) = 4$

18. Consider the polynomial

$$p(y) = -y^3 - 2y + 2.$$

Using appropriate technology,
   (a) sketch the slope field for $dy/dt = p(y)$;
   (b) sketch the solution curves of $dy/dt = p(y)$ using the slope field; and
   (c) using Euler's method, find the real root(s) of $p(y)$ to three decimal places.

19. Consider the polynomial

$$p(y) = -y^3 + 4y + 1.$$

Using appropriate technology,
   (a) sketch the slope field for $dy/dt = p(y)$;
   (b) sketch the solution curves of $dy/dt = p(y)$ using the slope field; and
   (c) using Euler's method, find the real root(s) of $p(y)$ to three decimal places.

1.5 EXISTENCE AND UNIQUENESS OF SOLUTIONS

What Does It Mean to Say Solutions Exist?

We have seen analytic, qualitative and numerical techniques for studying solutions of differential equations. One problem we have not considered is: How do we know there are solutions? Although this may seem to be a subtle and abstract question, it is also a question of great importance. If solutions to the differential do not exist, then there is no use trying to find or approximate them. More important, if a differential equation is supposed to model a physical system but the solutions of the differential equation do not exist, then we should have serious doubts about the validity of the model.

To get an idea of what is meant by existence of solutions, consider the algebraic equation

$$x^5 - 5x + \tfrac{5}{2} = 0.$$

A solution to this equation is a value of x for which the left-hand side equals zero, that is, a root of the fifth-degree polynomial. We can easily compute that the left-hand side is equal to -1.5 if $x = 1$ and 6.5 if $x = -1$. Since polynomials are continuous, there must therefore be a value of x between -1 and 1 for which the left-hand side is zero.

So we have established the existence of at least one solution of this equation between -1 and 1. We did not construct the value of x or approximate it (other than to say it was between -1 and 1). It is possible that there is more than one solution between -1 and 1; that is, our solution might not be unique. Nonetheless, we do know that there is at least one solution in this interval.

Unfortunately, there is no "quadratic equation" for finding roots of fifth-degree polynomials, so there is no way to write down the exact values of the solutions of this equation. But this does not make us any less sure of the existence of this solution. The point here is that we can discuss existence of solutions without having to compute them.

In the same way, if we are given an initial-value problem

$$\frac{dy}{dt} = f(t, y), \quad y(0) = y_0,$$

we can ask if there is a solution. This is a different question than asking what the solution is or what its graph looks like. We can say there is a solution without having any knowledge of a formula for the solution, just as we can say that the algebraic equation above has a solution between -1 and 1 without knowing its exact or even approximate value.

Existence

Luckily, the question of existence of solutions for differential equations has been extensively studied and some very good results have been established. For our purposes, we will use the standard existence theorem.

Existence Theorem

Suppose $f(t, y)$ is a continuously differentiable function; that is, the partial derivatives of $f(t, y)$ with respect to t and y exist and are continuous for all points (t, y) in a rectangle in the ty-plane of the form $\{(t, y) \mid a < t < b, c < y < d\}$. If (t_0, y_0) is a point in this rectangle, then there exists an $\epsilon > 0$ and a function $y(t)$ defined for $t_0 - \epsilon < t < t_0 + \epsilon$ that solves the initial-value problem

$$\frac{dy}{dt} = f(t, y), \quad y(t_0) = y_0.$$

This theorem says that as long as the function on the right-hand side of the differential equation is reasonable, solutions exist. (It does not rule out the possibility that solutions exist even if $f(t, y)$ is not so nice a function, but it doesn't guarantee it either.) This is reassuring. When we are studying the solutions of a reasonable initial-value problem, there is something there to study.

Extendability

Given an initial-value problem

$$\frac{dy}{dt} = f(t, y), \quad y(t_0) = y_0,$$

the Existence Theorem guarantees that there is a solution. If you read the theorem very closely (with a lawyer's eye for loopholes), you will see that the solution may have a very small domain of definition. The theorem says that there exists an $\epsilon > 0$ and that the solution has domain $(t_0 - \epsilon, t_0 + \epsilon)$. The ϵ may be very, very small, so although the theorem guarantees that a solution exists, it may be defined for only a very short amount of time.

Unfortunately, this is a serious but necessary restriction. Consider the initial-value problem

$$\frac{dy}{dt} = 1 + y^2, \quad y(0) = 0.$$

The slopes in the slope field for this equation increase in steepness very rapidly as y increases (see Figure 1.41). Hence, the rate of increase of solutions $y(t)$ increases more and more rapidly as $y(t)$ increases. There is a danger that solutions will "blow up" (tend to infinity very quickly) as t increases. By looking at solutions sketched by the slope field, we can't really tell if the solutions blow up in finite time or if they stay finite for all time, so we try analytic methods.

This is an autonomous equation, so we may separate variables and integrate as usual:

$$\int \frac{1}{1 + y^2} \, dy = \int \, dt.$$

Integration yields

$$\arctan y = t + c,$$

where c is an arbitrary constant. Therefore

$$y(t) = \tan(t + c),$$

which is the general solution of the differential equation. Using the initial-value

$$0 = y(0) = \tan(0 + c),$$

we find $c = 0$ (or $c = n\pi$ for any integer n). Thus, the particular solution is $y(t) = \tan t$, and the domain of definition for this particular solution is $-\pi/2 < t < \pi/2$.

The graph of this solution shows that our fears were well founded (see Figure 1.42). This particular solution has vertical asymptotes at $t = \pm\pi/2$. As t approaches $\pi/2$ from the left and $-\pi/2$ from the right, the solution blows up. If this were a model of a physical system, then we would expect that as t approaches $\pi/2$ the system would break.

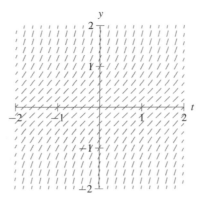

Figure 1.41
The slope field for the equation $dy/dt = 1 + y^2$.

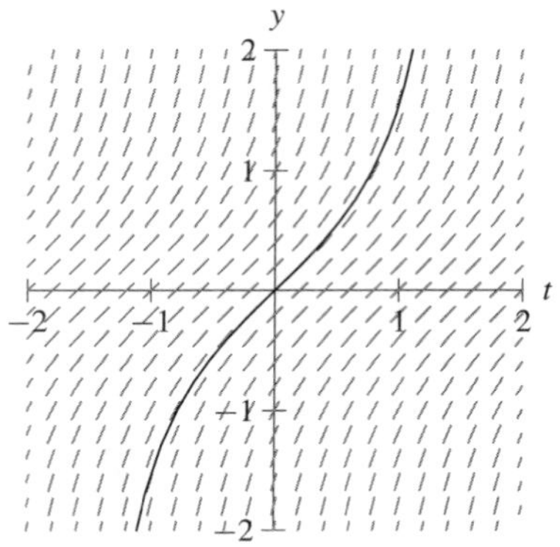

Figure 1.42
The graph of the solution $y(t) = \tan t$ and the slope field for $dy/dt = 1 + y^2$.

Uniqueness

When dealing with initial-value problems of the form

$$\frac{dy}{dt} = f(t, y), \quad y(t_0) = y_0,$$

we have always said "consider *the* solution." By the Existence Theorem we know there is a solution, but how do we know there is only one? Why don't we have to say "consider *a* solution" instead of "consider *the* solution?" In other words, how do we know the solution is unique?

Knowing that the solution to an initial-value problem is unique is very valuable from both theoretical and practical standpoints. If solutions weren't unique, then we would have to worry about all the possible solutions, even when we were doing numerical or qualitative work. Following different solutions could give completely different predictions concerning how the system works. Fortunately, there is a very strong theorem available concerning uniqueness of solutions. We combine the Existence Theorem with a uniqueness statement to give us our fundamental theoretical result.

Existence and Uniqueness Theorem

Suppose $f(t, y)$ is a continuously differentiable function; that is, the partial derivatives of $f(t, y)$ with respect to t and y exist and are continuous for all points (t, y) in a rectangle in the ty-plane of the form $\{(t, y) \mid a < t < b, \ c < y < d\}$. If (t_0, y_0) is a point in this rectangle, then there exists an $\epsilon > 0$ and a function $y(t)$ defined for $t_0 - \epsilon < t < t_0 + \epsilon$ that solves the initial-value problem

$$\frac{dy}{dt} = f(t, y), \quad y(t_0) = y_0.$$

This function is defined for $t_0 - \epsilon < t < t_0 + \epsilon$. Moreover, if $y_1(t)$ is also a solution to the same initial-value problem that is defined in the same interval, then

$$y_1(t) = y(t)$$

for $t_0 - \epsilon < t < t_0 + \epsilon$. That is, the solution to the initial-value problem is unique.

Before giving applications of the Existence and Uniqueness Theorem we should emphasize that both the existence and uniqueness parts of the theorem have *hypotheses*, conditions that must hold before we can use the theorem. Before we say that the solution of an initial-value problem

$$\frac{dy}{dt} = f(t, y), \quad y(t_0) = y_0$$

exists and is unique, we must check that $f(t, y)$ is a nice function. It turns out that it is pretty hard to construct an example whose solutions fail to exist, but we often encounter difficulties when the right-hand side of the equation is undefined at certain points.

For example, consider the equation

$$\frac{dy}{dt} = \frac{y}{t}.$$

This equation is not defined at $t = 0$. Nevertheless, we can separate the variables and then write down the general solution, which is $y(t) = ct$. Note that *all* of these solutions pass through $y = 0$ when $t = 0$, so solutions are by no means unique there. However, $y(t) = ct$ is not a solution at $t = 0$ because at $t = 0$ the differential equation is not defined.

It is not so hard to find examples where the uniqueness fails and $f(t, y)$ is a reasonable function (but doesn't have continuous partial derivatives in both t and y). For example, consider the differential equation

$$\frac{dy}{dt} = \tfrac{3}{2}y^{1/3}.$$

The right-hand side is a continuous function of both y and t. However, the derivative of $y^{1/3}$ with respect to y fails to exist when $y = 0$, so we cannot apply the Existence and Uniqueness Theorem. We do have one solution that satisfies the initial condition $y(0) = 0$, namely the equilibrium solution $y_0(t) = 0$. We proceed further and use the method of separation of variables to find that any function of the form

$$y(t) = \pm(t + c)^{3/2}$$

is also a solution. In particular, when $c = 0$, we see that both $y_\pm(t) = \pm t^{3/2}$ are solutions. Now both these functions are only defined for $t \geq 0$, but we can extend them for all values of t by setting

$$y_\pm(t) = \begin{cases} 0 & \text{if } t < 0; \\ \pm t^{3/2} & \text{if } t \geq 0. \end{cases}$$

Then $y_0(t)$ and $y_\pm(t)$ are all solutions of the differential equation that satisfy the initial condition $y(0) = 0$. So we have existence but not uniqueness of solutions in this case.

Applications of the Uniqueness Theorem

The uniqueness part of the Existence and Uniqueness Theorem says that two solutions to the same initial-value problem are identical. This fact is reassuring, but it may not sound useful in a practical sense. Here we discuss a few examples to illustrate why this theorem is, in fact, very useful.

Suppose $y_1(t)$ and $y_2(t)$ are both solutions of a differential equation

$$\frac{dy}{dt} = f(t, y),$$

where $f(t, y)$ satisfies the hypotheses of the Existence and Uniqueness Theorem. If for some t_0, we have $y_1(t_0) = y_2(t_0)$, then both of these functions are solutions of the same initial-value problem

$$\frac{dy}{dt} = f(t, y), \quad y(t_0) = y_1(t_0) = y_2(t_0).$$

The uniqueness part of the Existence and Uniqueness Theorem guarantees that $y_1(t) = y_2(t)$, at least for all t for which both solutions are defined. We can phrase the uniqueness part of the Existence and Uniqueness Theorem as:

"If two solutions are ever in the same place at the same time, then they are the same function."

This form of uniqueness part of the Existence and Uniqueness Theorem is very valuable, as the following examples show.

Role of equilibrium solutions

Consider the initial-value problem

$$\frac{dy}{dt} = \frac{(y^2 - 4)(\sin^2 y^3 + \cos y - 2)}{2}, \quad y(0) = \frac{1}{2}.$$

Finding the explicit solution to this equation is not easy because, even though the equation is autonomous and hence separable, the integrals involved are very difficult (try them). On the other hand, we know that if $y = 2$ the right-hand side of the equation vanishes. Thus the constant function $y_1(t) = 2$ is an equilibrium solution for this equation.

Suppose $y_2(t)$ is another solution to the differential equation that satisfies the initial condition $y_2(0) = 1/2$. The uniqueness half of the Existence and Uniqueness Theorem implies that $y_2(t) < 2$ for all t since the graph of y_2 cannot touch the line $y = 2$, the graph of the constant solution (see Figure 1.43).

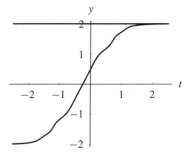

Figure 1.43
The graphs of two solutions of

$$\frac{dy}{dt} = \frac{(y^2 - 4)(\sin^2 y^3 + \cos y - 2)}{2}.$$

Although it looks like these two graphs agree for $t > 2$, we know that there is always a little space between them.

This is not a lot of information about the solution of the initial-value problem with $y(0) = 1/2$. On the other hand, we didn't have to do a lot of work to get this information. Identifying $y_1(t) = 2$ as a solution is pretty easy, and the rest follows from the uniqueness statement. By doing a little bit of work, we get some information. If all we care about is how large the solution of the original initial-value problem can possibly become, then the fact that it is bounded above by $y = 2$ may suffice. If we need more detailed information, we must look more carefully at the equation.

Comparing solutions

We can use this technique to obtain information about solutions of more complicated equations. For example, consider

$$\frac{dy}{dt} = \frac{(1+t)^2}{(1+y)^2}.$$

It is easy to check that $y_1(t) = t$ is a solution to the differential equation with initial condition $y_1(0) = 0$. If $y_2(t)$ is the solution satisfying the initial condition $y(0) = -0.1$, then $y_2(0) < y_1(0)$, so $y_2(t) < y_1(t)$ for all t. Thus, $y_2(t) < t$ for all t (see Figure 1.44). Again, this is only a little bit of information about the solution of the initial-value problem, but then we only did a little work.

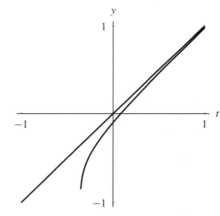

Figure 1.44
The graphs of two solutions $y_1(t)$ and $y_2(t)$ of

$$\frac{dy}{dt} = \frac{(1+t)^2}{(1+y)^2}.$$

The graph of the solution $y_1(t)$ that satisfies the initial condition $y_1(0) = 0$ is a line, and the graph of the solution that satisfies the initial condition $y_2(0) = -0.1$ lies below the line.

Uniqueness and qualitative analysis

In some cases we can use the uniqueness statement and some qualitative information to give more exact information about solutions. For example, consider the differential equation

$$\frac{dy}{dt} = f(y) = (y - 2)(y + 1).$$

Note that $f(2) = f(-1) = 0$. Thus $y = 2$ and $y = -1$ are equilibrium solutions (see the slope field in Figure 1.45). By the Existence and Uniqueness Theorem, any solution $y(t)$ with initial condition $y(0)$ between 2 and -1 must satisfy $-1 < y(t) < 2$ for all t.

In this case we can say even more about solutions. For example, consider a solution to the initial-value problem

$$\frac{dy}{dt} = f(y) = (y - 2)(y + 1), \quad y(0) = 0.5.$$

As above, we know that the solution $y(t)$ to this initial-value problem remains between $y = -1$ and $y = 2$ for all t. Because this equation is autonomous, the sign of the right-hand side of the differential equation depends only on the value of y, and for $-1 < y < 2$, we have that $f(y) < 0$. Hence, the solution $y(t)$ for our initial-value problem satisfies $dy/dt = f(y(t)) < 0$ for all t. Consequently, this solution is decreasing for all t.

Since the solution is decreasing for all t and since it always remains above $y = -1$, it is reasonable to guess that, as $t \to \infty$, $y(t) \to 1$. In fact, this is the case. If $y(t)$ were to limit to any value y_0 larger than -1 as t gets large, then when t is very large, $y(t)$ must be close to y_0. But $f(y_0)$ is negative because $-1 < y_0 < 2$. So when $y(t)$ is close to y_0, we have dy/dt close to $f(y_0)$, which is negative, so the solution must continue to decrease past y_0. That is, solutions of this differential equation can be asymptotic only to equilibrium solutions.

We can sketch the solution of this initial-value problem. For all t the graph is between the lines $y = -1$ and $y = 2$, and for all t it decreases (see Figure 1.46).

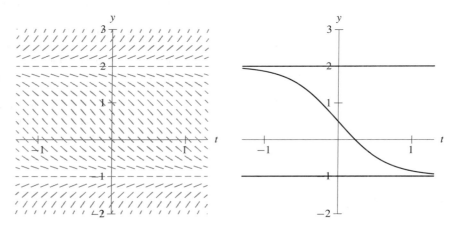

Figure 1.45
The slope field for
$dy/dt = (y - 2)(y + 1)$.

Figure 1.46
Graphs of the equilibrium solutions and the solution with initial condition $y(0) = 0.5$ for $dy/dt = (y - 2)(y + 1)$.

Uniqueness and Numerical Approximation

As the preceding examples show, the Uniqueness Theorem gives us qualitative information concerning the behavior of solutions. We can use this information to check the behavior of numerical approximations of solutions. If numerical approximations of solutions violate the Uniqueness Theorem, then we are certain that something is wrong.

The graph of the Euler approximation to the solution of the initial-value problem

$$\frac{dy}{dt} = e^t \sin y, \quad y(0) = 5,$$

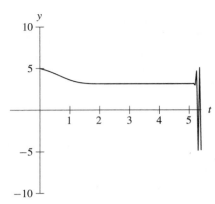

Figure 1.47
Euler's method applied to

$$\frac{dy}{dt} = e^t \sin y$$

with $\Delta t = 0.05$. The graph of the approximation behaves as expected for $t < 5$, but for t slightly larger than 5, the approximation is no longer valid.

with $\Delta t = 0.05$ is shown in Figure 1.47. As noted in Section 1.4, the behavior seems erratic and hence we are suspicious.

We can easily check that the constant function $y(t) = n\pi$ is a solution for any integer n and hence by the Uniqueness Theorem, each solution is trapped between $y = n\pi$ and $y = (n + 1)\pi$ for some integer n. The approximations in Figure 1.47 violate this requirement. This confirms our suspicions that the numerical results in this case are not to be believed.

This equation is unusual because of the e^t term on the right-hand side. When t is large, the slopes of solutions become gigantic and hence Euler's method overshoots the true solution for even a very small step size.

Exercises for Section 1.5

In Exercises 1–4, the function f is continuously differentiable in the entire ty-plane. However, we do not provide its formula. We do provide various solutions to the given differential equation. Finally, we specify an initial condition. Using the Uniqueness Theorem, what can you conclude about the solution to the equation with the given initial condition?

1. $\dfrac{dy}{dt} = f(t, y)$

$y_1(t) = 3$ is a solution for all t,
initial condition $y(0) = 1$

2. $\dfrac{dy}{dt} = f(y)$

$y_1(t) = 4$ is a solution for all t,
$y_2(t) = 2$ is a solution for all t,
$y_3(t) = 0$ is a solution for all t,
initial condition $y(0) = 1$

3. $\dfrac{dy}{dt} = f(t, y)$

$y_1(t) = t + 2$ is a solution for all t,
$y_2(t) = -t^2$ is a solution for all t,
initial condition $y(0) = 1$

4. $\dfrac{dy}{dt} = f(t, y)$

$y_1(t) = -1$ is a solution for all t,
$y_2(t) = 1 + t^2$ is a solution for all t,
initial condition $y(0) = 0$

In Exercises 5–8, an initial condition for the differential equation

$$\frac{dy}{dt} = (y - 2)(y - 3)y$$

is given. What does the Existence and Uniqueness Theorem say about the corresponding solution?

5. $y(0) = 4$ **6.** $y(0) = 3$ **7.** $y(0) = 1$ **8.** $y(0) = -1$

9.   (a) Show that $y_1(t) = 1 + t^2$ is a solution to

$$\frac{dy}{dt} = -y^2 + y + 2yt^2 + 2t - t^2 - t^4.$$

   (b) Show that, if $y(t)$ is a solution to the equation in part (a) and if $y(0) < 1$, then $y(t) < 1 + t^2$ for all t.

10. Consider the differential equation

$$\frac{dy}{dt} = y^{2/3}.$$

   (a) Show that $y_1(t) = 0$ for all t is a solution.

   (b) Show that $y_2(t) = t^3/27$ is a solution.

   (c) Verify that $y_1(0) = y_2(0)$ but that $y_1(t) \neq y_2(t)$ for all t. Why doesn't this contradict the Existence and Uniqueness Theorem?

11. Consider a differential equation of the form $dy/dt = f(y)$ — an autonomous equation — and assume that the function $f(y)$ is continuously differentiable.

   (a) Suppose $y_1(t)$ is a solution and $y_1(t)$ has a local maximum at $t = t_0$. Let $y_0 = y_1(t_0)$. Show that $f(y_0) = 0$.

   (b) Use the information of part (a) to sketch the slope field along the line $y = y_0$ in the ty-plane.

   (c) Show that the constant function $y_2(t) = y_0$ is a solution (in other words, $y_2(t)$ is an equilibrium solution).

   (d) Show that $y_1(t) = y_0$ for all t.

   (e) Show that, if a solution of $dy/dt = f(y)$ has a local minimum, then it is a constant function; that is, it also corresponds to an equilibrium solution.

12.   (a) Show that

$$y_1(t) = \frac{1}{t - 1} \quad \text{and} \quad y_2(t) = \frac{1}{t - 2}$$

   are solutions of $dy/dt = -y^2$.

   (b) What can you say about solutions of $dy/dt = -y^2$ for which the initial condition $y(0)$ satisfies the inequality $-1 < y(0) < -1/2$? [*Hint*: You could find the general solution, but what information can you get from your answer to part (a) alone?]

13. Consider the differential equation

$$\frac{dy}{dt} = \frac{y}{t^2}.$$

   (a) Show that the constant function $y_1(t) = 0$ is a solution.

(b) Show that there are infinitely many other functions which are solutions that agree with this solution when $t \leq 0$ but that are nonzero when $t > 0$.

(c) Why doesn't this contradict the Existence and Uniqueness Theorem?

In Exercises 14–17, an initial-value problem is given.

(a) Find a formula for the solution.

(b) State the domain of definition of the solution.

(c) Describe what happens to the solution as it approachs the limits of its domain of definition; that is, why can't the solution be extended for more time?

14. $\dfrac{dy}{dt} = y^3$, $y(0) = 1$

15. $\dfrac{dy}{dt} = \dfrac{1}{(y+1)(t-2)}$, $y(0) = 0$

16. $\dfrac{dy}{dt} = \dfrac{1}{(y+2)^2}$, $y(0) = 1$

17. $\dfrac{dy}{dt} = \dfrac{t}{(y-2)}$, $y(-1) = 0$

18. We have emphasized that the Existence and Uniqueness Theorem does not apply to every differential equation. There are hypotheses that must be verified before we can apply the theorem. There is a temptation to think that, since models of "real-world" problems must have solutions, we don't need to worry about the hypotheses of the Existence and Uniqueness Theorem when we are working with differential equations modeling the physical world. The following exercise illustrates the flaw in this assumption.

Suppose we wish to study the formation of raindrops in the atmosphere. We make the reasonable hypothesis that raindrops are approximately spherical. We also assume that the rate of growth of the volume of a raindrop is proportional to its surface area.

Let $v(t)$ equal the volume of the raindrop at time t, and let $r(t)$ equal its radius. We have

$$v = \tfrac{4}{3}\pi r^3$$

by the usual formula for the volume of a sphere. Therefore

$$r = \left(\frac{3v}{4\pi}\right)^{1/3}.$$

The surface area of the drop is given by $4\pi r^2$, which is therefore $3^{2/3}(4\pi)^{1/3}v^{2/3}$. Hence the differential equation for the volume of the drop is

$$\frac{dv}{dt} = kv^{2/3},$$

where k is the product of the proportionality constant with $3^{2/3}(4\pi)^{1/3}$ (see Exercise 10 in this section).

(a) Why doesn't this equation satisfy the hypotheses of the Existence and Uniqueness Theorem?

(b) Can you give a "physical" interpretation of the fact that solutions to this equation with initial condition $v(0) = 0$ are not unique? That is, does this model say anything about the way raindrops begin to form?

1.6 EQUILIBRIA AND THE PHASE LINE

Given a differential equation

$$\frac{dy}{dt} = f(t, y),$$

we can get an idea of what solutions "look like" as graphs in the ty-plane by drawing slope fields and sketching their graphs or by using Euler's method and computing approximate solutions. Sometimes we can even derive explicit formulas for solutions and plot the results. Each of these techniques requires quite a bit of work, either numerical (computation of slopes or Euler's method) or analytic (integration).

In this section we consider differential equations where the right-hand side is independent of t — that is, **autonomous** equations. For these differential equations, there are qualitative techniques that allow us to sketch the solutions with less arithmetic than required for slope fields. These techniques also help us more carefully study the behavior of solutions of autonomous equations, and they will be invaluable later when we study higher dimensional systems.

Autonomous Equations

Autonomous equations are differential equations of the form $dy/dt = f(y)$. For autonomous equations, the rate of change of the dependent variable depends on a function of the dependent variable only and not on the time. Autonomous equations appear frequently as models for two reasons. First, many physical systems work the same way at any time. For example, a spring compressed the same amount at 10:00AM and at 3:00PM provides the same force. Second, for many systems, the time dependence "averages out" over the time scales being considered. If, for example, we are studying how wolves and field mice interact, we might find that wolves eat many more field mice during the day than they do at night. However, if we are interested in how the wolf and mouse populations behave over a period of years or decades, then we can average the number of mice eaten by each wolf per week. The daily fluctuations are irrelevant.

We have already noticed that autonomous equations have slope fields that have a special form (see Section 1.3). Because the right-hand side of the equation does not depend on t, the slope field is constant along horizontal lines on the ty-plane. That is, two points with the same y-coordinate but different t-coordinates have the same slope line for an autonomous equation (see Figure 1.48).

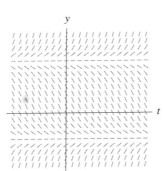

Figure 1.48
Slope field for the autonomous differential equation

$$\frac{dy}{dt} = (y - 2)(y + 1).$$

The slopes are parallel along horizontal lines.

Hence there is a great deal of redundancy in the slope field of an autonomous equation. If we know the slope field along one vertical line $t = t_0$ for any time t_0, then we know the slope field in the entire ty-plane. So instead of drawing the entire slope field, we should be able to draw just one line containing the same information. This line is called the **phase line** for the autonomous equation. The remainder of this section will be devoted to drawing and using phase lines.

Metaphor of the rope

Suppose you are given an autonomous differential equation

$$\frac{dy}{dt} = f(y).$$

Think of a rope hanging vertically and stretching infinitely far up and infinitely far down. The dependent variable y tells you a position on the rope (the rope is the y-axis). The function $f(y)$ gives a number for each position on the rope. Suppose the number $f(y)$ is actually printed on the rope at height y for every value of y. For example, at the height $y = 2.17$, the number $f(2.17)$ is printed on the rope.

Suppose that you are placed on the rope at height y_0 at time $t = 0$ and given the following instructions: Read the number that is printed on the rope and climb up or down the rope with velocity equal to that number. Climb up the rope if the number is positive or down the rope if the number is negative. (So a large positive number means you climb up very quickly, whereas a negative number near zero means you climb down slowly.) As you move, continue to read the numbers off the rope and adjust your velocity so that it is always equal to the number printed on the rope.

If you follow this rather bizarre set of instructions, you will generate a function $y(t)$ that gives your position on the rope at time t. Your position at time $t = 0$ is $y(0) = y_0$ because that is where you were placed initially. The velocity of your motion dy/dt at time t will be given by the number on the rope, so $dy/dt = f(y(t))$ for all t. Hence, your position function $y(t)$ is a solution to the initial-value problem

$$\frac{dy}{dt} = f(y), \quad y(0) = y_0.$$

The phase line is a picture of this rope. Because it is tedious to record the numerical values of all the velocities, we record on the phase line only the places where the velocity is zero and whether the velocity is positive or negative. The phase line provides qualitative sketches of solutions.

Phase Line of a Logistic Equation

For example, consider the differential equation

$$\frac{dy}{dt} = f(y) = (1 - y)y.$$

Then $f(y) = 0$ precisely when $y = 0$ and $y = 1$. Therefore $y_1(t) = 0$ for all t and $y_2(t) = 1$ for all t are equilibrium solutions for this equation. We call the points $y = 0$ and $y = 1$ on the y-axis **equilibrium points**. Also note that $f(y)$ is positive if $0 < y < 1$, whereas $f(y)$ is negative if $y < 0$ or $y > 1$. We can draw the phase line (or "rope") by placing dots at the equilibrium points $y = 0$ and $y = 1$. For $0 < y < 1$, we put arrows pointing up because $f(y) > 0$ means you climb up, and for $y < 0$ or $y > 1$, we put arrows pointing down because $f(y) < 0$ means you climb down (see Figure 1.49).

If we compare the phase line to the slope field, we see that the phase line contains all the information about equilibrium points and whether solutions are increasing or decreasing. Information about the *speed* of increase or decrease of solutions is lost (see Figure 1.50). We can give rough sketches of solutions using the phase line alone. These will not be quite as accurate as the sketches from the slope field, but they will contain all the information about the behavior of solutions as t gets large (see Figure 1.51).

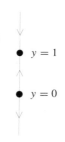

Figure 1.49
Phase line for
$dy/dt = (1 - y)y$.

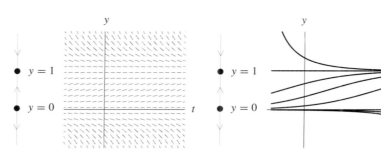

Figure 1.50
Phase line and slope field of
$dy/dt = (1 - y)y$.

Figure 1.51
Phase line and sketches of the graphs of solutions
for $dy/dt = (1 - y)y$.

How to Draw Phase Lines

We can give a more precise definition of the phase line by giving the steps required to draw it. For the autonomous equation

$$\frac{dy}{dt} = f(y) :$$

- Draw the y-line.
- Find the **equilibrium points** (the numbers such that $f(y) = 0$), and mark them on the line.
- Find the intervals of y-values for which $f(y) > 0$, and draw arrows pointing up in these intervals.
- Find the intervals of y-values for which $f(y) < 0$, and draw arrows pointing down in these intervals.

We sketch several examples of phase lines in Figure 1.52. When looking at the phase line, you should remember the metaphor of the rope and think of solutions of the differential equation "dynamically," that is, as people climbing up and down the rope as time increases.

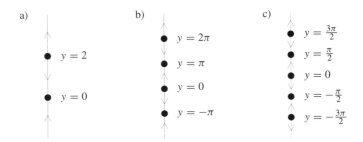

Figure 1.52
Phase lines for a) $dy/dt = (y - 2)(y + 3)$, b) $dy/dt = \sin y$, and c) $dy/dt = y \cos y$.

How to Use Phase Lines to Sketch Solutions

Now that we can draw phase lines, we would like to use them to sketch solutions of autonomous equations. We can obtain rough sketches of solutions directly from the phase lines, provided we are careful in interpreting these sketches. The sort of information that phase lines are very good at predicting is the limiting behavior of solutions as t increases or decreases.

Consider the equation

$$\frac{dw}{dt} = (2 - w) \sin w.$$

The phase line for this differential equation is given in Figure 1.53. Note that the equilibrium points are $w = 2$ and $w = k\pi$ for $k = \ldots, -2, -1, 0, 1, 2, \ldots$. Suppose we want to sketch the solution $w(t)$ with initial-value $w(0) = 0.4$. Because $w = 0$ and $w = 2$ are equilibrium points of this equation and $0 < 0.4 < 2$, we know from the Existence and Uniqueness Theorem that $0 < w(t) < 2$ for all t. Moreover, because $(2 - w) \sin w > 0$ for $0 < w < 2$, the solution is always increasing. Because the velocity of the solution is small only when $(2 - w) \sin w$ is close to zero, and this happens only near equilibrium points, we know that the solution $w(t)$ increases toward $w = 2$ as $t \to \infty$ (see Section 1.5).

Similarly, if we run the clock backwards, the solution $w(t)$ decreases. It will always remain above $w = 0$ and cannot stop, since $0 < w < 2$. Thus, as $t \to -\infty$, the solution tends toward $w = 0$. We can draw a qualitative picture of the solution with initial condition $w(0) = 0.4$ (see Figure 1.54).

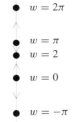

Figure 1.53
Phase line for $dw/dt = (2-w) \sin w$.

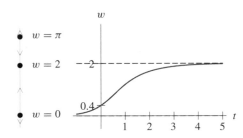

Figure 1.54
Graph of the solution to the initial-value problem

$$\frac{dw}{dt} = (2 - w) \sin w, \quad w(0) = 0.4.$$

Likewise, we can sketch other solutions in the tw-plane from the information on the phase line. The equilibrium solutions are easy to find and draw because they are marked on the phase line. The intervals on the phase line with upward-pointing arrows correspond to increasing solutions, and those with downward-pointing arrows correspond to decreasing solutions. Graphs of the solutions do not cross by the Uniqueness Theorem. In particular, they cannot cross the graphs of the equilibrium solutions. Also, solutions must continue to increase or decrease until they come close to an equilibrium solution. Hence we can sketch many solutions with different initial conditions quite easily. The only information that we do not have is how quickly the solutions increase or decrease (see Figure 1.55).

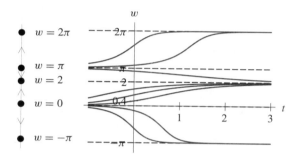

Figure 1.55
Graphs of many solutions to $dw/dt = (2 - w)\sin w$.

These observations lead to some general statements that can be made for all solutions of autonomous equations. Suppose $y(t)$ is a solution to an autonomous equation

$$\frac{dy}{dt} = f(y).$$

- If $f(y(0)) = 0$, then $y(0)$ is an equilibrium point and $y(t) = y(0)$ for all t.
- If $f(y(0)) > 0$, then $y(t)$ is increasing for all t and either $y(t) \to \infty$ as t increases or $y(t)$ tends to the first equilibrium point larger than $y(0)$.
- If $f(y(0)) < 0$, then $y(t)$ is decreasing for all t and either $y(t) \to -\infty$ as t increases or $y(t)$ tends to the first equilibrium point smaller than $y(0)$.

As t decreases (as time runs backwards), similar results hold except that if $f(y(0)) > 0$, then, as t runs backwards, $y(t)$ decreases and either tends to $-\infty$ or the next smaller equilibrium point. If $f(y(0)) < 0$, then, as t runs backwards, $y(t)$ increases and either tends to $+\infty$ or the next larger equilibrium point.

An example with three equilibrium points

For example, consider the differential equation

$$\frac{dP}{dt} = \left(1 - \frac{P}{20}\right)^3 \left(\frac{P}{5} - 1\right) P^7.$$

If the initial condition is given by $P(0) = 8$, what happens as t becomes very large? First we draw the phase line for this equation. Let

$$f(P) = \left(1 - \frac{P}{20}\right)^3 \left(\frac{P}{5} - 1\right) P^7.$$

We find the equilibrium points by solving $f(P) = 0$. Thus $P = 0$, $P = 5$, and $P = 20$ are the equilibrium points.

If $0 < P < 5$, $f(P)$ is negative; if $P < 0$ or $5 < P < 20$, $f(P)$ is positive; and if $P > 20$, $f(P)$ is negative. We can place the arrows on the phase line appropriately (see Figure 1.56). Note that we only have to check the value of $f(P)$ at one point in each of these intervals to determine the sign of $f(P)$ in the entire interval.

The solution with initial condition $P(0) = 8$ is in the region between the equilibrium points $P = 5$ and $P = 20$, so $5 < P(t) < 20$ for all t. The arrows point up in this interval, so the value of $P(t)$ is increasing for all t. As $t \to \infty$, $P(t)$ tends toward the equilibrium value $P = 20$.

If we run time backwards, the solution with initial condition $P(0) = 8$ decreases toward the next smaller equilibrium point, which is $P = 5$. Hence $P(t)$ is always greater than $P = 5$. If we compute the solution $P(t)$ numerically, we see that it increases from $P(0) = 8$ to close to $P = 20$ very quickly (see Figure 1.57). From the phase line alone, we cannot tell how quickly the solution increases.

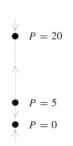

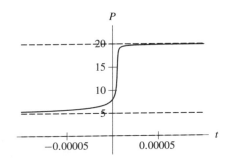

Figure 1.56
Phase line for $dP/dt = f(P) = (1 - P/20)^3 ((P/5) - 1) P^7$.

Figure 1.57
Graph of the solution to the initial-value problem
$dP/dt = (1 - P/20)^3 ((P/5) - 1) P^7$,
$P(0) = 8$.

Warning: Not All Solutions Exist for All Time

Suppose y_0 is an equilibrium point for the equation $dy/dt = f(y)$. Then $f(y_0) = 0$. We are assuming $f(y)$ is continuous, so if solutions are close to y_0, the value of f is small. Thus solutions move slowly when they are close to equilibrium points. A solution that approaches an equilibrium point as t increases (or decreases) moves more and more slowly as it approaches the equilibrium point. By the Existence and Uniqueness Theorem, a solution that approaches an equilibrium point never actually gets there. It is *asymptotic* to the equilibrium point, and the graph of the solution in the ty-plane has a horizontal asymptote.

On the other hand, unbounded solutions often speed up as they move. For example, the equation

$$\frac{dy}{dt} = f(y) = (1 + y)^2$$

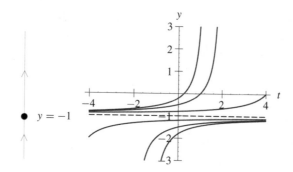

Figure 1.58
Phase line for

$$\frac{dy}{dt} = (1+y)^2$$

and graphs of solutions that tend to infinity in finite time.

has one equilibrium point at $y = -1$ and $f(y) > 0$ everywhere else (see Figure 1.58).

The phase line indicates that solutions with initial condition $y(0) > -1$ increase for all t and tend to $+\infty$ as t increases. In fact, these solutions blow up in a finite amount of time. To see this, we use separation of variables and compute the explicit form of the solution. This computation shows that any nonequilibrium solution is given by

$$y(t) = -1 - \frac{1}{t+c}$$

for some constant c. Since we are assuming that $y(0) > -1$, we must have

$$y(0) = -1 - \frac{1}{c} > -1$$

or $c < 0$. Therefore these solutions are defined only for $t < -c$; they tend to ∞ as t approaches $-c$ from the left. We cannot tell whether solutions "blow up" in finite time like this by looking at the phase line.

The solutions with initial conditions $y(0) < -1$ are asymptotic to the equilibrium point $y = -1$ as t increases, so they are defined for all $t > 0$. However, these solutions tend to $-\infty$ in a finite time as t decreases. So they are not defined for all $t < 0$.

Another dangerous example is

$$\frac{dy}{dt} = g(y) = \frac{1}{1-y}.$$

If $y > 1$, $g(y)$ is negative, and if $y < 1$, $g(y)$ is positive. If $y = 1$, $g(y)$ does not exist. The phase line has a hole in it. There is no standard way to denote such points on the phase line. We will use a small empty circle to mark them (see Figure 1.59).

All solutions tend toward $y = 1$ as t increases. Because the value of $g(y)$ is large if y is close to 1, solutions speed up as they get close to $y = 1$ and solutions reach

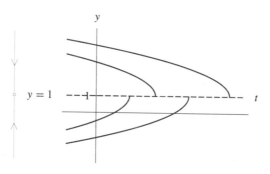

Figure 1.59
Phase line for

$$\frac{dy}{dt} = \frac{1}{1-y}$$

(not defined at $y = 1$), and graphs of solutions reaching the hole $y = 1$ in finite time.

$y = 1$ in a finite amount of time. Once a solution reaches $y = 1$, it cannot be continued because it has left the domain of definition of the differential equation. It has fallen into a hole in the phase line.

Drawing Phase Lines from Qualitative Information Alone

To draw the phase line for the differential equation $dy/dt = f(y)$, we need to know the location of the equilibrium points and the intervals over which the solutions are increasing or decreasing. That is, we need to know the points where $f(y) = 0$, the intervals where $f(y) > 0$, and the intervals where $f(y) < 0$. Consequently, we can draw the phase line for the differential equation with only qualitative information about the function $f(y)$.

For example, suppose we do not know a formula for $f(y)$, but we do have its graph (see Figure 1.60). From the graph we can determine the values of y for which $f(y) = 0$ and decide on which intervals $f(y) > 0$ and $f(y) < 0$. With this information we can draw the phase line (see Figure 1.61). From the phase line we can then get qualitative sketches of solutions (see Figure 1.62). Thus we can go from qualitative information about $f(y)$ to graphs of solutions of the differential equation $dy/dt = f(y)$ without ever writing down a formula. For models where the information available is completely qualitative, this approach is very appropriate.

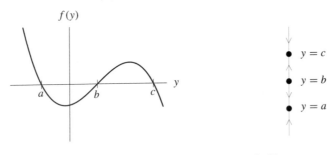

Figure 1.60
Graph of $f(y)$.

Figure 1.61
Phase line for $dy/dt = f(y)$ for $f(y)$ graphed in Figure 1.60.

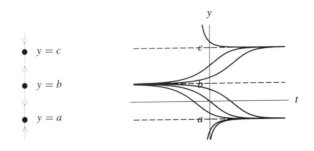

Figure 1.62
Sketch of solutions for $dy/dt = f(y)$ for $f(y)$ graphed in Figure 1.60.

The Role of Equilibrium Points

We have already determined that every solution to an autonomous differential equation $dy/dt = f(y)$ either tends to $+\infty$ or $-\infty$ as t increases (perhaps becoming infinite in finite time) or tends asymptotically to an equilibrium point as t increases. Hence the equilibrium points are extremely important in understanding the long-term behavior of solutions.

Also we have seen that, when drawing a phase line, we need to find the equilibrium points, the intervals on which $f(y)$ is positive, and the intervals on which $f(y)$ is negative. If f is continuous, it can switch from positive to negative only at points y_0 when $f(y_0) = 0$, that is, at equilibrium points. Hence, the equilibrium points also play a crucial role in sketching the phase line.

In fact, the equilibrium points are the key to understanding the entire phase line. For example, suppose we have an autonomous differential equation $dy/dt = g(y)$. Suppose all we know about this differential equation is that it has exactly two equilibrium points, at $y = 2$ and $y = 7$, and that the phase line near $y = 2$ and $y = 7$ is as shown in Figure 1.63. We can use this information to sketch the entire phase line. We know that the sign of $g(y)$ can change only at an equilibrium point. Hence the sign of $g(y)$ does not change for $2 < y < 7$ or for $y < 2$ or $y > 7$. Thus if we know the direction of the arrows anywhere in these intervals (say near the equilibrium points), then we know the directions on the entire phase line (see Figure 1.64). Consequently, if we understand the equilibrium points for an autonomous differential equation, we should be able to understand (at least qualitatively) any solution of the equation.

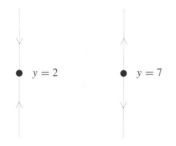

Figure 1.63
Pieces of the phase line for
$dy/dt = g(y)$ near $y = 2$ and $y = 7$.

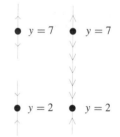

Figure 1.64
Entire phase line of $dy/dt = g(y)$
constructed from the phase line near
the equilibrium points.

Classification of Equilibrium Points

Given their significance, it is useful to give some names to the different types of equilibrium points and to classify them according to the behavior of nearby solutions. Consider an equilibrium point $y = y_0$, as shown in Figure 1.65. For y slightly less than y_0, the arrows point up, and for y slightly larger than y_0, the arrows point down. A solution with initial condition close to y_0 is asymptotic to y_0.

We say an equilibrium point y_0 is a **sink** if any solution with initial condition sufficiently close to y_0 is asymptotic to y_0 as t increases. (The name *sink* is supposed to

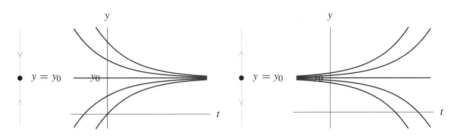

Figure 1.65
Phase line at a sink, and graphs of solutions near a sink.

Figure 1.66
Phase line at a source, and a sketch of solutions near a source.

bring to mind a kitchen sink with the equilibrium point as the drain. If water starts close enough to the drain, it will run toward it.)

Another possible phase line near an equilibrium point y_0 is shown in Figure 1.66. Here, the arrows point up at points just above y_0 and down for points just below y_0. A solution that has an initial-value near y_0 tends away from y_0 as t increases. If time is run backwards, solutions that start near y_0 tend toward y_0.

We say an equilibrium point y_0 is a **source** if all solutions that start sufficiently close to y_0 tend toward y_0 as t decreases. This means that all solutions that start close to y_0 (but not at y_0) will tend away from y_0 as t increases. So a source is a sink if time is run backwards. (The name *source* is supposed to help you picture solutions flowing out of or away from a point.)

Sinks and sources are the two major types of equilibrium points. Every equilibrium point that is neither a source nor a sink is called a **node**. Two possible phase line pictures near nodes are shown in Figure 1.67.

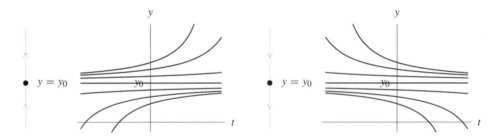

Figure 1.67
Examples of node equilibrium points and sketches of nearby solutions.

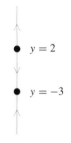

Figure 1.68
Phase line for $dy/dt = y^2 + y - 6$.

Given a differential equation, we can classify the equilibrium points as sinks, sources, or nodes from the phase line. For example, consider

$$\frac{dy}{dt} = f(y) = y^2 + y - 6 = (y + 3)(y - 2).$$

The equilibrium points are $y = -3$ and $y = 2$. Also $f(y) < 0$ for $-3 < y < 2$, and $f(y) > 0$ for $y < -3$ and $y > 2$. This is all the information we need to draw the phase line (see Figure 1.68). From the phase line we see that $y = -3$ is a sink and $y = 2$ is a source.

Suppose we are given the differential equation $dw/dt = g(w)$, where the graph of g is provided rather than its formula (see Figure 1.69). This differential equation has three equilibrium points, $w = -0.5$, $w = 2$, and $w = 2.5$; and $g(w) > 0$ if $w < -0.5$, $2 < w < 2.5$, and $w > 2.5$. For $-0.5 < w < 2$, $g(w) < 0$. From this information we can draw the phase line (see Figure 1.70) and classify the equilibrium points. The point $w = -0.5$ is a sink, $w = 2$ is a source, and $w = 2.5$ is a node.

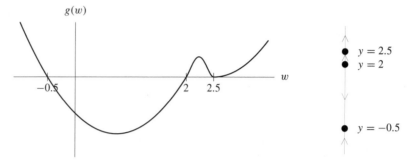

Figure 1.69
Graph of $g(w)$.

Figure 1.70
Phase line for
$dw/dt = g(w)$ for $g(w)$, as
displayed in Figure 1.69.

Identifying the type of an equilibrium point and "linearization"

From the examples above we know that we can determine the phase line and classify the equilibrium points for an autonomous differential equation $dy/dt = f(y)$ from the graph of $f(y)$ alone. The classification of an equilibrium point depends only on the phase line near the equilibrium point. If $y = y_0$ is an equilibrium point for the equation above — that is, $f(y_0) = 0$ — then we should be able to determine its type from the graph of $f(y)$ near y_0.

If y_0 is a sink, then the arrows on the phase line just below y_0 point up and the arrows just above y_0 point down. Hence $f(y)$ must be positive for y just smaller than y_0 and negative for y just larger than y_0 (see Figure 1.71). So f must be decreasing for y near y_0. Conversely, if $f(y_0) = 0$ and f is decreasing for all y near y_0, then $f(y)$ is positive just to the left of y_0 and negative just to the right of y_0. Hence, y_0 is a sink. Similarly, the equilibrium point y_0 is a source if and only if f is increasing for all y near y_0 (see Figure 1.72).

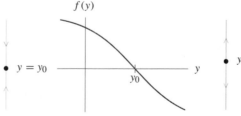

Figure 1.71
Phase line near a sink at $y = y_0$ for
$dy/dt = f(y)$, and graph of $f(y)$ near
$y = y_0$.

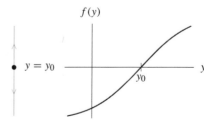

Figure 1.72
Phase line near a source at $y = y_0$ for
$dy/dt = f(y)$, and graph of $f(y)$ near
$y = y_0$.

From calculus we have a powerful tool for telling if a function is increasing or decreasing at a particular point — the derivative. Using the derivative of $f(y)$ combined with the geometric observations above, we can give criteria that specify the type of the equilibrium point.

Linearization Theorem

Suppose y_0 is an equilibrium point of the differential equation

$$\frac{dy}{dt} = f(y).$$

Then,

- *if $f'(y_0) < 0$, then y_0 is a sink;*
- *if $f'(y_0) > 0$, then y_0 is a source; or*
- *if $f'(y_0) = 0$ or if $f'(y_0)$ does not exist, then we need additional information to determine the type of y_0 (y_0 may be a source or a sink or a node).*

This theorem follows immediately from the discussion above once we recall that if $f'(y_0) < 0$, then f is decreasing near y_0, and if $f'(y_0) > 0$, then f is increasing near y_0. This is an example of **linearization**, a technique that we will often find useful. The derivative $f'(y_0)$ tells us the behavior of the best linear approximation to f near y_0. If we replace f with its best linear approximation, then the differential equation we obtain is very close to the original differential equation for y near y_0.

We cannot make any conclusion about the classification of y_0 if $f'(y_0) = 0$, because all three possibilities can occur (see Figure 1.73).

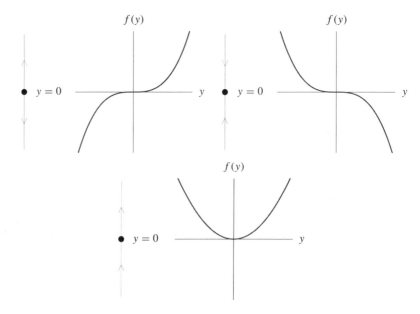

Figure 1.73
Graphs of various functions f along with the corresponding phase lines for the differential equation $dy/dt = f(y)$. In all cases, $y_0 = 0$ is an equilibrium point and $f'(y_0) = 0$.

As another example, consider the differential equation

$$\frac{dy}{dt} = h(y) = y(\cos(y^5 + 2y) - 27\pi y^4).$$

What does the phase line look like near $y = 0$? Drawing the phase line for this equation would be a very complicated affair. We would have to find the equilibrium points and determine the sign of $h(y)$. On the other hand, it is easy to see that $y = 0$ is an equilibrium point because $h(0) = 0$. We compute

$$h'(y) = (\cos(y^5 + 2y) - 27\pi y^4) + y \frac{d}{dy}(\cos(y^5 + 2y) - 27\pi y^4).$$

Thus $h'(0) = (\cos(0) - 0) + 0 = 1$. By the Linearization Theorem, we conclude that $y = 0$ is a source. Solutions that start sufficiently close to $y = 0$ will move away from $y = 0$ as t increases. Of course, there is the dangerous loophole clause "sufficiently close." Initial conditions might have to be very, very close to $y = 0$ for the above to apply. Again we did a little work and got a little information. To get more information, we would need to study the function $h(y)$ more carefully.

Modified Logistic Model

As an application of these ideas, we use the techniques of this section to discuss a modification of the logistic population model we introduced in Section 1.1.

The fox squirrel is a small mammal native to the Rocky Mountains. These squirrels are very territorial, so if their population is large, their rate of growth decreases or even becomes negative. On the other hand, if the population is too small, fertile adults run the risk of not being able to find suitable mates, so again the rate of growth is negative.

The model

We can restate these assumptions succinctly:

- If the population is too big, the rate of growth is negative.
- If the population is too small, the rate of growth is negative.

So the population only grows if it is between "too big" and "too small." Also, it is reasonable to assume that, if the population is zero, it will stay zero. Thus we also assume:

- If the population is zero, the growth rate is zero. (The reader should refer back to the logistic population model of Section 1.1 and compare the assumptions there with these assumptions.)

We let

$$t = \text{time (independent variable)},$$
$$S(t) = \text{population of squirrels at time } t \text{ (dependent variable)},$$
$$k = \text{growth-rate coefficient (parameter)}$$
$$N = \text{carrying capacity (parameter)}$$
$$M = \text{"sparsity" constant (parameter)}.$$

The carrying capacity N indicates what population is "too big," and the sparsity parameter M indicates when the population is "too small."

Now we want a model of the form

$$\frac{dS}{dt} = g(S)$$

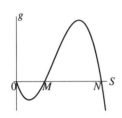

Figure 1.74
Graph of $g(S)$.

that conforms to the assumptions. We can think of the assumptions as determining the shape of the graph of $g(S)$, in particular where $g(S)$ is positive and where it is negative. Note that $g(S) < 0$ if $S > N$ because the population decreases if it is too big; that is, $dS/dt = g(S) < 0$. Also $g(S) < 0$ when $S < M$ because the population decreases if it is too small. Finally, $g(S) > 0$ when $M < S < N$ and $g(0) = 0$. That is, we want $g(S)$ to have a graph shaped like Figure 1.74. The graph of g for $S < 0$ does not matter because a negative number of squirrels (anti-squirrels?) is meaningless.

The logistic model would give "correct" behavior for populations near the carrying capacity, but for small populations (below the "sparsity" level M) the solutions of the logistic model do not agree with the assumptions. Hence we will need to modify the logistic model to include the behavior of small populations and to include the parameter M. We make a model of the form

$$\frac{dS}{dt} = g(S) = kS\left(1 - \frac{S}{N}\right)\text{(something)}.$$

The "something" term must be positive when $S > M$ and negative when $S < M$. The simplest choice that satisfies these conditions is

$$\text{(something)} = \left(\frac{S}{M} - 1\right).$$

Hence our the model is

$$\frac{dS}{dt} = kS\left(1 - \frac{S}{M}\right)\left(\frac{S}{M} - 1\right).$$

This is the logistic model with the extra term

$$\left(\frac{S}{M} - 1\right).$$

We call it the modified logistic population model. (Other models might also be called the modified logistic, but modified in a different way.)

Analysis of the model

To analyze solutions of this differential equation, we could use analytic techniques, since the equation is separable. However, qualitative techniques provide a lot of information about the solutions with a lot less work. The differential equation is

$$\frac{dS}{dt} = g(S) = kS\left(1 - \frac{S}{N}\right)\left(\frac{S}{M} - 1\right),$$

with $0 < M < N$ and $k > 0$. There are three equilibrium points — $S = 0$, $S = M$, and $S = N$. If $0 < S < M$, we have $g(S) < 0$, so solutions with initial conditions between 0 and M decrease. Similarly, if $S > N$, $g(S) < 0$, solutions with initial conditions larger than N also decrease. For $M < S < N$, we have $g(S) > 0$. Consequently, solutions with initial conditions between M and N increase. Thus we conclude that the equilibria at 0 and N are sinks, and the equilibrium point at M is a source. The phase line and graphs of typical solutions are shown in Figure 1.75.

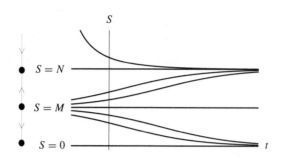

Figure 1.75
Solutions of the modified logistic
equation

$$\frac{dS}{dt} = k\left(1 - \frac{S}{N}\right)\left(\frac{S}{N} - 1\right)S,$$

with various initial conditions.

Exercises for Section 1.6

In Exercises 1–8, sketch the phase lines for the given differential equation. Identify the equilibrium points as sinks, sources, or nodes.

1. $\dfrac{dy}{dt} = 3y(1 - y)$ **2.** $\dfrac{dy}{dt} = y^2 - 6y - 16$ **3.** $\dfrac{dy}{dt} = \cos y$

4. $\dfrac{dw}{dt} = w \cos w$ **5.** $\dfrac{dw}{dt} = (w - 2) \sin w$ **6.** $\dfrac{dy}{dt} = \dfrac{1}{y - 2}$

7. $\dfrac{dw}{dt} = w^2 + 2w + 10$ **8.** $\dfrac{dy}{dt} = \tan y$

In Exercises 9–15, a differential equation and various initial conditions are specified. Sketch the graphs of the solutions satisfying these initial conditions. For each exercise, put all your graphs on one pair of axes.

9. Equation from Exercise 1; $y(0) = 1$, $y(2) = -1$, $y(0) = 1/2$, $y(0) = 2$.

10. Equation from Exercise 2; $y(0) = 1$, $y(1) = 0$, $y(0) = -10$, $y(0) = 5$.

11. Equation from Exercise 3; $y(0) = 0$, $y(-1) = 1$, $y(0) = -\pi/2$, $y(0) = \pi$.

12. Equation from Exercise 4; $w(0) = 0$, $w(3) = 1$, $w(0) = 2$, $w(0) = -1$.

13. Equation from Exercise 5; $w(0) = 1$, $w(0) = 7/4$, $w(0) = -1$, $w(0) = 3$.

14. Equation from Exercise 6; $y(0) = 0$, $y(1) = 3$, $y(0) = 2$ (trick question).

15. Equation from Exercise 7; $w(0) = 0$, $w(1/2) = 1$, $w(0) = 2$.

In Exercises 16–21, describe the long-term behavior of the solution to the differential equation

$$\frac{dy}{dt} = y^2 - 4y + 2$$

with the given initial condition.

16. $y(0) = 0$ **17.** $y(0) = 1$ **18.** $y(0) = -1$

19. $y(0) = -10$ **20.** $y(0) = 10$ **21.** $y(3) = 1$

22. Consider the autonomous equation $dy/dt = f(y)$. Suppose we know that $f(-1) = f(2) = 0$.

    (a) Describe all the possible behaviors of the solution $y(t)$ that has the initial value $y(0) = 1$.

    (b) Suppose also that $f(y) > 0$ for $-1 < y < 2$. Describe all the possible behaviors of the solution $y(t)$ that has initial value $y(0) = 1$.

In Exercises 23–26, the graph of a function $f(y)$ is given. Sketch the phase line for the autonomous differential equation $dy/dt = f(y)$.

23.

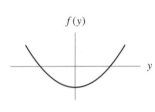

24.

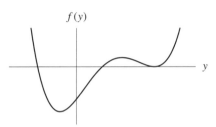

25.

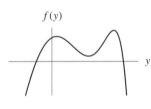

26.

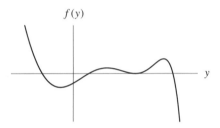

In Exercises 27–30, a phase lines for the equation $dy/dt = f(y)$ is shown. Make a rough sketch of the graph of the corresponding function $f(y)$. (Assume $y = 0$ is in the middle of the segment shown in each case.)

27.   **28.**   **29.**   **30.**

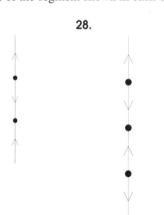
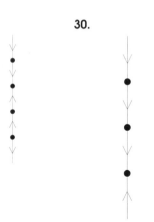

31. Suppose you wish to model a population with a differential equation of the form $dP/dt = f(P)$, where $P(t)$ is the population at time t. Experiments have been performed on the population that give the following information:

- The only equilibrium points in the population are $P = 0$, $P = 10$, and $P = 50$.
- If the population is 100, the population decreases.
- If the population is 25, the population increases.

(a) Sketch the possible phase lines for this system for $P > 0$ (there are two!).

(b) Give a rough sketch of the corresponding functions $f(P)$ for each of your phase lines.

(c) Give a formula for functions $f(P)$ whose graph agrees (qualitatively) with the rough sketches in part (b) for each of your phase lines.

32. Suppose the experimental information in Exercise 31 is changed as follows:

- The population $P = 0$ is fixed.
- A population close to 0 will decrease.
- A population of $P = 20$ will increase.
- A population of $P > 100$ will decrease.

(a) Sketch the simplest possible phase line that agrees with the experimental information above.

(b) Give a rough sketch of the function $f(P)$ for the phase line of part (a).

(c) Discuss other possible phase lines. What other phase lines are possible?

33. Given a continuous function $f(y)$:

(a) Suppose we know that $dy/dt = f(y)$ has $f(-10) > 0$ and $f(10) < 0$. Show that there is an equilibrium point for $dy/dt = f(y)$ between $y = -10$ and $y = 10$.

(b) Suppose we know that $dy/dt = f(y)$ has $f(-10) > 0$ and $f(10) < 0$ and that there are finitely many equilibrium points between $y = -10$ and $y = 10$. If $y = 1$ is a source, show that $dy/dt = f(y)$ must have at least two sinks between $y = -10$ and $y = 10$. (Can you say where they are located?)

34. Consider the differential equation $dy/dt = f(y)$. Suppose y_0 is an isolated equilibrium point; that is, y_0 is an equilibrium point and there is an open interval containing y_0 that has no other equilibrium points. We define the **index** of y_0 to be

$$\text{index}(y_0) = \begin{cases} +1 & \text{if } y_0 \text{ is a source;} \\ -1 & \text{if } y_0 \text{ is a sink;} \\ 0 & \text{if } y_0 \text{ is a node.} \end{cases}$$

(a) Suppose that there are finitely many equilibrium points in the interval $-100 < y < 100$ and that $f(-100) < 0$ and $f(100) > 0$. Show that, if y_1, $y_2, \ldots, y_n$ are the equilibrium points with $-100 < y_1 < y_2 < \cdots < y_n < 100$, then

$$\text{index}(y_1) + \text{index}(y_2) + \cdots + \text{index}(y_n) = 1.$$

(b) Suppose the conditions of part (a) hold, except that $f(-100) > 0$ and $f(100) < 0$. Show that

$$\text{index}(y_1) + \text{index}(y_2) + \cdots + \text{index}(y_n) = -1.$$

(c) Suppose the conditions of part (a) hold, except that $f(-100) < 0$ and $f(100) < 0$. Show that

$$\text{index}(y_1) + \text{index}(y_2) + \cdots + \text{index}(y_n) = 0.$$

35. Suppose $dy/dt = f(y)$ has an equilibrium point at $y = y_0$ and

(a) $f'(y_0) = 0$, $f''(y_0) = 0$, and $f'''(y_0) > 0$; then is y_0 a source, a sink, or a node?

(b) $f'(y_0) = 0$, $f''(y_0) = 0$, and $f'''(y_0) < 0$; then is y_0 a source, a sink, or a node?

(c) $f'(y_0) = 0$ and $f''(y_0) > 0$; then is y_0 a source, a sink, or a node?

36.   (a) Sketch the phase line for the differential equation

$$\frac{dy}{dt} = \frac{1}{(y-2)(y+1)},$$

and discuss the behavior of the solution with initial condition $y(0) = 1/2$.

(b) Apply analytic techniques to the initial-value problem

$$\frac{dy}{dt} = \frac{1}{(y-2)(y+1)}, \quad y(0) = \frac{1}{2},$$

and compare your results with your discussion in part (a).

The proper scheduling of city bus and train systems is a difficult and delicate problem, which the City of Boston seems to ignore. It is not uncommon in Boston to wait a long time for the trolley, only to have several trolleys arrive simultaneously. In Exercises 37–40 we study a very simple model of the behavior of trolley cars.

Consider two trolley cars on the same track moving toward downtown Boston. Let $x(t)$ denote the amount of time between the two cars at time t. That is, if the first car arrives at a particular stop at time t, then the other car will arrive at the stop $x(t)$ time units later. We assume that the first car runs at a constant average speed (not a bad assumption for a car running before rush hour). We wish to model how $x(t)$ changes at t increases.

We first assume that, if no passengers are waiting for the second train, then it has an average speed greater than the first train and hence will catch up to the first train. Thus the time between trains $x(t)$ will decrease at a constant rate if no people are waiting for the second train. However, the speed of the second train decreases if there are passengers to pick up. We assume that the speed of the second train decreases at a rate proportional to the number of passengers it picks up and that the passengers arrive at the stops at a constant rate. Hence the number of passengers waiting for the second train is proportional to the time between trains.

37. We claim that a reasonable model for $x(t)$ is

$$\frac{dx}{dt} = \beta x - \alpha.$$

Which term represents the rate of decrease of the time between the trains if no people are waiting, and which term represents the effect of the people waiting for the second train? (Justify your answer.) Should the parameters α and β be positive or negative?

38. For the model from the preceding problem:

(a) Find the equilibrium points.

(b) Classify the equilibrium points (source, sink, or node).

(c) Sketch the phase line.

(d) Sketch the graphs of solutions.

(e) Find the formula for the general solution, if you can.

39. Use the model to predict what happens to $x(t)$ as t increases. Include the effect of the initial value $x(0)$. Is it possible for the trains to run at regular intervals? Given that there are always slight variations in the number of passengers waiting at each stop, is it likely that a regular interval can be maintained? Write two brief reports (of one or two paragraphs):

(a) The first report is addressed to other students in the class (hence you may use technical language we use in class).

(b) The second report is addressed to the Mayor of Boston.

40. What happens if trolley cars leave the station at fixed intervals? Can you use the model to predict what will happen for a whole sequence of trains? Will it help to increase the number of trains so that they leave the station more frequently?

1.7 BIFURCATIONS

Equations with Parameters

In many of our models, a common feature is the presence of **parameters** along with the other variables involved. Parameters are quantities that do not depend on time (the independent variable) but that assume different values depending on the specifics of the application at hand. For instance, the exponential growth model for population

$$\frac{dP}{dt} = kP$$

contains the parameter k, the constant of proportionality for the growth rate dP/dt versus the total population P. One of the underlying assumptions of this model is that the growth rate dP/dt is a constant multiple of the total population. However, when we apply this model to different species, we expect to use different values for the constant of proportionality. In particular, the value of k that we would use for rabbits would be significantly larger than the value for humans.

How the behavior of solutions changes as the parameters vary is a particularly important aspect of the study of differential equations. For some models, we must study the behavior of solutions for all parameter values in a certain range. For example, the number of cars on a bridge may affect how the bridge reacts to wind. A model of the bridge must contain a parameter for the total mass of cars on the bridge, and we must know the behavior of solutions of the model equations for a variety of different values of the mass.

In many models we know only approximate values for the parameters. To be able to use the model to make predictions, we must know the effect of possible errors in our knowledge of the exact values of the parameters on the behavior of solutions. Also there may be effects that we have not included in our model that make the parameters vary in unexpected ways. In many complicated physical systems, the long-term effect of these intentional or unintentional adjustments in the parameters can be very dramatic.

In this section we study how solutions of a differential equation change as a parameter is varied. We study autonomous equations with one parameter. We find that a small change in the parameter usually results in only a small or gentle change in the nature of the solutions. However, occasionally a small change in the parameter can lead to a drastic change in the long-term behavior of solutions. Such a change is called a **bifurcation**. We say that a differential equation that depends on a parameter "bifurcates" if there is a qualitative change in the behavior of solutions as the parameter changes.

Notation for differential equations depending on a parameter

An example of an autonomous differential equation that depends on a parameter is

$$\frac{dy}{dt} = y^2 - 2y + \mu.$$

The parameter is μ. The independent variable is t and the dependent variable is y, as usual. Note that this equation really represents infinitely many different equations, one for each value of μ. We think of the value of μ as a constant in each equation, but different values of μ yield different differential equations, each with a different set of solutions. Because of their different roles in the differential equation, we use a notation that distinguishes the dependence of the right-hand side on y and μ. We let

$$f_\mu(y) = y^2 - 2y + \mu.$$

The parameter μ appears in the subscript, and the dependent variable y is in the parentheses. If we want to specify a particular value of μ, say $\mu = 3$, then we write

$$f_3(y) = y^2 - 2y + 3,$$

and we have a function that depends only on y. With $\mu = 3$, we obtain the corresponding differential equation

$$\frac{dy}{dt} = f_3(y) = y^2 - 2y + 3.$$

We use this notation in general. A function of the dependent variable y, which also depends on a parameter μ, is denoted by $f_\mu(y)$. The corresponding differential equation with dependent variable y and parameter μ is

$$\frac{dy}{dt} = f_\mu(y).$$

Since such a differential equation really refers to a collection of different equations, one for each value of μ, we call such an equation a **one-parameter family** of differential equations.

A One-Parameter Family with One Bifurcation

Let's consider the one-parameter family

$$\frac{dy}{dt} = f_\mu(y) = y^2 - 2y + \mu$$

more closely. For each value of μ we have an autonomous differential equation, and we can draw its phase line and analyze it using the techniques of the previous section. We begin our study of this family by studying the differential equations obtained from particular choices of μ. Since we do not yet know the most interesting values of μ, we just pick integer values, say, $\mu = -4$, $\mu = -2$, $\mu = 0$, $\mu = 2$, and $\mu = 4$, for starters. This gives us five different autonomous differential equations with five different phase lines. One equation is

$$\frac{dy}{dt} = f_{-2}(y) = y^2 - 2y - 2.$$

This differential equation has equilibrium points at values of y for which

$$f_{-2}(y) = y^2 - 2y - 2 = 0.$$

The equilibrium points are $y = 1 - \sqrt{3}$ and $y = 1 + \sqrt{3}$. Between the equilibrium points, the function f_{-2} is negative, and above and below the equilibrium points, f_{-2} is positive. Hence $y = 1 - \sqrt{3}$ is a sink and $y = 1 + \sqrt{3}$ is a source. With this information we can draw the phase line. For the other values of μ we follow a similar procedure and draw the phase lines. All these phase lines are shown in Figure 1.76.

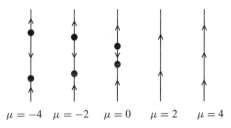

$\mu = -4 \quad \mu = -2 \quad \mu = 0 \quad \mu = 2 \quad \mu = 4$

Figure 1.76
Phase lines for

$$\frac{dy}{dt} = f_\mu(y) = y^2 - 2y + \mu$$

for $\mu = -4, -2, 0, 2,$ and 4.

Each of the phase lines is somewhat different from the others. However, the basic description of the phase lines for $\mu = -4$, $\mu = -2$, and $\mu = 0$ is the same: There are exactly two equilibrium points; the smaller one is a sink and the larger one is a source. Although the exact position of these equilibrium points changes as μ increases, their relative position and type do not change. Solutions of these equations with large initial values tend to ∞ as t increases (blowing up in finite time) and tend to an equilibrium point as t decreases. Solutions with very negative initial conditions tend to an equilibrium point as t increases and to $-\infty$ as t decreases. Solutions with initial values between the equilibrium points tend to the smaller equilibrium point as t increases and to the larger equilibrium point as t decreases (see Figure 1.77).

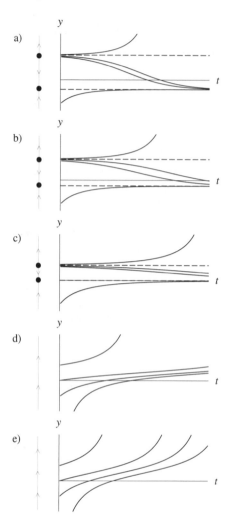

Figure 1.77
Phase lines and sketches of solutions for $dy/dt = f_\mu(y) = y^2 - 2y + \mu$ for a)
$\mu = -4$,   b) $\mu = -2$,   c) $\mu = 0$,   d) $\mu = 2$,   and   e) $\mu = 4$.

At $\mu = 2$ and $\mu = 4$, we see something very different. These equations do not have any equilibrium points. All solutions tend to $+\infty$ as t increases and to $-\infty$ as t decreases. Because there is a significant change in the nature of the solutions, we say that a bifurcation has occurred somewhere between $\mu = 0$ and $\mu = 2$. To investigate the nature of this bifurcation, we draw the graphs of f_μ for the μ-values above. For $\mu = -4, -2$, and 0, $f_\mu(y)$ has 2 roots, but for $\mu = 2$ and 4, the graph of $f_\mu(y)$ does not cross the axis. Somewhere between $\mu = 0$ and $\mu = 2$ the graph of $f_\mu(y)$ must be tangent to the axis (see Figure 1.78).

The roots of the quadratic equation

$$y^2 - 2y + \mu = 0$$

are $y = 1 \pm \sqrt{1 - \mu}$. If $\mu < 1$, the quadratic has two real roots; if $\mu = 1$, it has only one root; and if $\mu > 1$, it has no real roots. The corresponding differential equations

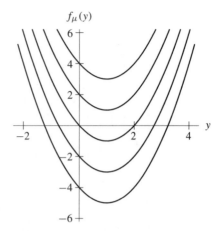

Figure 1.78
Graphs of $f_\mu(y) = y^2 - 2y + \mu$ for
$\mu = -4, -2, 0, 2,$ and 4.

have two equilibrium points if $\mu < 1$, one equilibrium point if $\mu = 1$, and no equilibrium points if $\mu > 1$. Hence the phase lines change when $\mu = 1$. This is called the **bifurcation value** of the parameter μ. We say that a bifurcation occurs when $\mu = 1$.

The graph of $f_1(y)$ and the phase line for $dy/dt = f_1(y)$ are shown in Figures 1.79 and 1.80. The phase line has one equilibrium point (which is a node), and everywhere else solutions increase. The fact that the bifurcation occurs at the parameter value for which the equilibrium point is a node is not a coincidence. In fact, this entire bifurcation scenario is quite common.

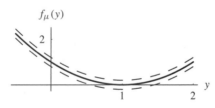

Figure 1.79
Graphs of $f_\mu(y) = y^2 - 2y + \mu$ for μ
slightly less than 1, equal to 1, and slightly
greater than 1.

Figure 1.80
Corresponding phase lines
for $dy/dt = f_\mu(y) =$
$y^2 - 2y + \mu$.

The Bifurcation Diagram

An extremely helpful way to view bifurcations is through the **bifurcation diagram**. This is a picture of the phase lines near a bifurcation value, highlighting the changes that the phase lines undergo as the parameter passes through this value. To plot the bifurcation diagram, we plot the parameter values along the horizontal axis. For each μ-value (not just integers), we draw the phase line corresponding to μ on the vertical line through $(\mu, 0)$. We think of the bifurcation diagram as a movie: As our eye scans the picture from left to right, we see the phase lines evolve through the bifurcation. Figure 1.81 shows the bifurcation diagram for $f_\mu(y) = y^2 - 2y + \mu = 0$.

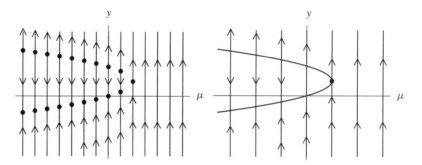

Figure 1.81
Bifurcation diagram for the differential equation $dy/dt = f_\mu(y) = y^2 - 2y + \mu$.
The horizontal axis is the μ-value and the vertical lines are the phase lines for the
differential equations with the corresponding μ-values.

A bifurcation from one to three equilibria

Let's look now at another one-parameter family of differential equations

$$\frac{dy}{dt} = g_\alpha(y) = y^3 - \alpha y = y(y^2 - \alpha).$$

In this equation, α is the parameter. There are three equilibria if $\alpha > 0$ (at $y = 0, \pm\sqrt{\alpha}$),
but there is only one equilibrium point (at $y = 0$) if $\alpha \leq 0$. Therefore a bifurcation occurs when $\alpha = 0$. To understand this bifurcation, we plot the bifurcation diagram.

First, if $\alpha \leq 0$, the term $y^2 - \alpha$ is always positive. Thus $f_\alpha(y) = y(y^2 - \alpha)$ has
the same sign as y. Solutions tend to ∞ if $y(0) > 0$ and to $-\infty$ if $y(0) < 0$. If $\alpha > 0$,
the situation is different. The graph of $g_\alpha(y)$ shows that $g_\alpha(y) > 0$ in the intervals
$\sqrt{\alpha} < y < \infty$ and $-\sqrt{\alpha} < y < 0$ (see Figure 1.82). Thus solutions increase in
these intervals. In the other intervals, $g_\alpha(y) < 0$, so solutions decrease. The bifurcation
diagram is depicted in Figure 1.83.

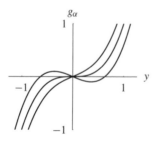

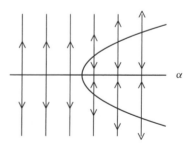

Figure 1.82
Graphs of $g_\alpha(y)$ for $\alpha > 0$,
$\alpha = 0$, and $\alpha < 0$. Note that
for $\alpha \geq 0$ the graph crosses the
y-axis once, whereas if $\alpha < 0$,
the graph crosses the y-axis
three times.

Figure 1.83
Bifurcation diagram for the
one-parameter family
$dy/dt = g_\alpha(y) = y^3 - \alpha y$.

Bifurcations of Equilibrium Points

Throughout the rest of this section, we assume that all the one-parameter families of differential equations that we consider depend "smoothly" on the parameter. That is, for a one-parameter family

$$\frac{dy}{dt} = f_\mu(y),$$

the partial derivatives of $f_\mu(y)$ with respect to y and μ exist and are continuous. So changing μ a little changes the graph of $f_\mu(y)$ only slightly.

When bifurcations do not happen

The most important fact about bifurcations is that they usually do not happen. A small change in the parameter usually leads to only a small change in the behavior of solutions. This is very reassuring. For example, suppose we have a one-parameter family

$$\frac{dy}{dt} = f_\mu(y),$$

and the differential equation for $\mu = \mu_0$ has an equilibrium point at $y = y_0$. Also suppose that $f'_{\mu_0}(y_0) < 0$, so the equilibrium point is a sink. We sketch the phase line and the graph of $f_{\mu_0}(y)$ near $y = y_0$ in Figure 1.84.

Now if we change μ just a little bit, say from μ_0 to μ_1, then the graph of $f_{\mu_1}(y)$ is very close to the graph of $f_{\mu_0}(y)$ (see Figure 1.85). So the graph of $f_{\mu_1}(y)$ is strictly decreasing near y_0, passing through the horizontal axis near $y = y_0$. The corresponding differential equation

$$\frac{dy}{dt} = f_{\mu_1}(y)$$

has a sink at some point $y = y_1$ very near y_0.

We can make this more precise: If y_0 is a sink for a differential equation

$$\frac{dy}{dt} = f_{\mu_0}(y)$$

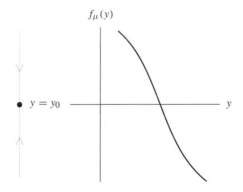

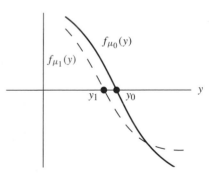

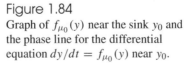

Figure 1.84
Graph of $f_{\mu_0}(y)$ near the sink y_0 and the phase line for the differential equation $dy/dt = f_{\mu_0}(y)$ near y_0.

Figure 1.85
Graphs of $f_{\mu_0}(y)$ and $f_{\mu_1}(y)$ for μ_1 close to μ_0. Note that $f_{\mu_1}(y)$ decreases across the y-axis at $y = y_1$ near y_0, so $dy/dt = f_{\mu_1}(y)$ has a sink at $y = y_1$.

with $f'_{\mu_0}(y_0) < 0$, then for all μ_1 sufficiently close to μ_0, the differential equation

$$\frac{dy}{dt} = f_{\mu_1}(y)$$

has a sink at a point $y = y_1$ very near y_0 (and no other equilibrium points near y_0). A similar statement holds if y_0 is a source and $f'_{\mu_0}(y_0) > 0$. These are the situations in which we can say for sure that no bifurcation occurs, at least not near y_0.

With these observations in mind, we see that bifurcations occur only if the above conditions do not hold. Consequently, given a one-parameter family of differential equations

$$\frac{dy}{dt} = f_\mu(y),$$

we look for values $\mu = \mu_0$ and $y = y_0$ for which $f_{\mu_0}(y_0) = 0$ and $f'_{\mu_0}(y_0) = 0$.

Determining bifurcation values

Consider the one-parameter family of differential equations given by

$$\frac{dy}{dt} = f_\mu(y) = y(1-y)^2 + \mu.$$

If $\mu = 0$, the equilibrium points are $y = 0$ and $y = 1$. Also $f'_0(0) = 1$. Hence $y = 0$ is a source for the differential equation $dy/dt = f_0(y)$. Thus for all μ sufficiently close to zero, the differential equation $dy/dt = f_\mu(y)$ has a source near $y = 0$.

On the other hand, for the equilibrium point $y = 1$, $f'_0(1) = 0$. The Linearization Theorem from Section 1.6 says nothing about what happens in this case. To see what is going on, we sketch the graph of $f_\mu(y)$ for several μ-values near $\mu = 0$ (see Figure 1.86). When $\mu = 0$, the graph of f_μ is tangent to the horizontal axis at $y = 1$. Since $f_0(y) > 0$ for all $y > 0$ except $y = 1$, it follows that the equilibrium point at $y = 1$ is a node for this parameter value. Changing μ moves the graph of $f_\mu(y)$ up (if μ is positive) or down (if μ is negative). If we make μ slightly positive, $f_\mu(y)$ does not touch the horizontal axis near $y = 1$. So the equilibrium point at $y = 1$ for $\mu = 0$ disappears. A bifurcation occurs at $\mu = 0$. For μ slightly negative, the corresponding differential equation has two equilibrium points near $y = 1$. Since f_μ is decreasing at one of these equilibria and increasing at the other, one of these equilibria is a source and the other is a sink.

There is a second bifurcation in this one-parameter family. To see this, note what happens as μ decreases. There is a value of μ for which the graph of $f_\mu(y)$ again has a tangency with the horizontal axis (see Figure 1.87). For larger μ-values, the graph crosses the horizontal axis three times, but for lower μ-values, the graph crosses only once. Thus a second bifurcation occurs at this μ-value.

To find this bifurcation value exactly, we must find the μ-values for which the graph of f_μ is tangent to the horizontal axis. That is, we must find the μ-values for which, at some equilibrium point y, we have $f'_\mu(y) = 0$. Since

$$f'_\mu(y) = (1-y)^2 - 2y(1-y) = (1-y)(1-3y),$$

it follows that the graph of $f_\mu(y)$ is horizontal at the two points $y = 1$ and $y = 1/3$. We

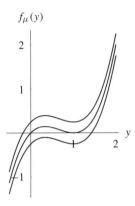

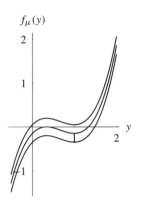

Figure 1.86
Graphs of
$f_\mu(y) = y(1 - y)^2 + \mu$ for
μ slightly greater than
zero, μ equal to zero, and
μ slightly less than zero.

Figure 1.87
Graphs of
$f_\mu(y) = y(1 - y)^2 + \mu$ for
μ slightly greater than
$-4/27$, for μ equal to
$-4/27$, and for μ slightly
less than $-4/27$.

know that the graph of $f_0(y)$ is tangent to the horizontal axis $y = 1$, so let's look at $y = 1/3$. We have $f_\mu(1/3) = \mu + 4/27$, so the graph is also tangent to the horizontal axis when $\mu = -4/27$. This is our second bifurcation value. Using analogous arguments to those above, we find that f_μ has three equilibria for $-4/27 < \mu < 0$ and only one equilibrium point when $\mu < -4/27$.

The bifurcation diagram summarizes all this information in one picture (see Figure 1.88).

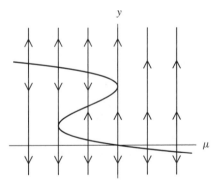

Figure 1.88
Bifurcation diagram for

$$\frac{dy}{dt} = f_\mu(y) = y(1 - y)^2 + \mu.$$

Note the two bifurcation values of μ,
$\mu = -4/27$, and $\mu = 0$.

Harvesting of Natural Resources

When harvesting a natural resource, it is important to control the amount harvested so that the resource is not completely depleted. To accomplish this, we must study the particular species involved and pay close attention to the possible changes that may occur when the harvesting level is increased.

Suppose we model the population $P(t)$ of a particular species of fish with a logistic model

$$\frac{dP}{dt} = kP\left(1 - \frac{P}{N}\right),$$

where k is the growth rate parameter and N is the carrying capacity of the area. Suppose that fishing removes a certain constant number C (for catch) of fish per season from the population. Then a modification of the model that takes fishing into account is

$$\frac{dP}{dt} = k\left(1 - \frac{P}{N}\right)P - C.$$

How does the population of fish vary as C is increased?

This model has three parameters, k, N, and C; but we are only concerned with what happens if C is varied. Therefore we think of k and N as fixed constants determined by the type of fish and their habitat. Our predictions involve the values of k and N. For example, if $C = 0$, we know from Section 1.1 that all positive initial conditions give solutions that tend toward the equilibrium point $P = N$. So if fishing is prohibited, we expect the population to be close to $P = N$.

Let

$$f_C(P) = k\left(1 - \frac{P}{N}\right)P - C.$$

As C increases, the graph of $f_C(P)$ slides down (see Figure 1.89). The points where $f_C(P)$ crosses the horizontal axis tend toward each other. This implies that the equilibrium points for the corresponding differential equations slide together.

We can compute the equilibrium points by solving $f_C(P) = 0$, which yields

$$k\left(1 - \frac{P}{N}\right)P - C = 0$$

or, equivalently,

$$-kP^2 + kNP - CN = 0.$$

This quadratic equation has solutions

$$P = \frac{N}{2} \pm \sqrt{\frac{N^2}{4} - \frac{CN}{k}}.$$

As long as the term under the square root (the "discriminant" of the quadratic) is positive, the function crosses the horizontal axis twice and the corresponding differential

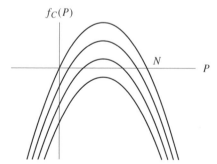

$f_C(P)$

N

P

Figure 1.89
Graphs of

$$f_C(P) = k\left(1 - \frac{P}{N}\right)P - C$$

for several values of C. Note that, as C increases, the graph of $f_C(P)$ slides down the vertical axis.

equation has two equilibrium points, a source and a sink. This is the case for small values of C in Figure 1.89.

If

$$\frac{N^2}{4} - \frac{CN}{k} < 0,$$

then the graph of $f_C(P)$ does not cross the horizontal axis and the corresponding differential equation has no equilibrium points. This occurs if

$$\frac{N^2}{4} < \frac{CN}{k}$$

or

$$C > \frac{kN}{4}.$$

When $C > kN/4$, the function $f_C(P)$ is negative for all values of P and the solutions of the corresponding differential equation decrease toward $-\infty$. Since negative populations do not make any sense, we say that the species has become extinct when the population reaches zero.

This is precisely the information we need to sketch the bifurcation diagram for this system (see Figure 1.90). A bifurcation occurs as we increase C. The bifurcation value for the parameter C is $kN/4$ because, at this value, the graph of $f_C(P)$ is tangent to the horizontal axis. The corresponding differential equation has a node at $P = N/2$. When C is just less than $kN/4$, the corresponding differential equation has two equilibrium points near $P = N/2$, a source and a sink. When C is just greater than $kN/4$, the corresponding differential equation has no equilibrium points (see Figure 1.90).

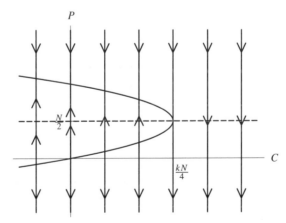

Figure 1.90
Bifurcation diagram for

$$\frac{dP}{dt} = f_C(P) = k\left(1 - \frac{P}{N}\right)P - C.$$

Note that when $C < kN/4$, the phase line has two equilibrium points, whereas when $C > kN/4$, the phase line has no equilibrium points and all solutions decrease.

It is interesting to consider what would happen to the fish population as the parameter C is slowly increased. When $C = 0$, the population tends to the sink at $P = N$, so we can assume that when fishing begins, the population is close to $P = N$. When C is slightly increased, the sink at $P = N$ moves to the slightly smaller value

$$P = \frac{N}{2} + \sqrt{\frac{N^2}{4} - \frac{CN}{k}}.$$

When C is very close to zero, this new sink is very close to $P = N$. Solutions starting near $P = N$ tend toward this new sink. As C slowly increases, the sink decreases and the fish population adjusts to stay close to this sink. We observe a gradual decrease in the fish population. When C is close to $kN/4$, the fish population is close to the sink for the corresponding differential equation, which is close to $P = N/2$. If C is increased just a little more so that $C > kN/4$, then the corresponding differential equation has no equilibrium points and all solutions decrease. When C is slightly larger than $kN/4$, $f_C(P)$ is slightly negative near $P = N/2$, so the population decreases slowly at first. As P decreases, $f_C(P)$ becomes more negative and the rate of decrease of P accelerates. The population reaches zero in a finite amount of time and the fish species is extinct.

So as the number of fish removed by fishing increases gradually, we expect at first to observe a gradual decline in the fish population. This decline continues until the fishing parameter C reaches the bifurcation value $C = kN/4$. At this point, if we allow even slightly more fishing, the fish population decreases slowly at first and then collapses, and the fish become extinct in the area. This is a pretty frightening scenario. The fact that a little fishing causes only a small population decline over the long term does not necessarily imply that a little more fishing causes only a little more population decline. Once the bifurcation value is passed, the fish population tends to zero.

Of course, this is a very simple population model. The lesson to be learned is that, if this sort of behavior can be observed in simple models, we would expect that the same (and even more surprising behavior) occurs in more complicated models and in actual populations. To properly manage resources, we need to have accurate models and to be aware of possible bifurcations.

Exercises for Section 1.7

In Exercises 1–4, locate the bifurcation values for the one-parameter family and draw the phase lines for values of the parameter slightly smaller than, slightly larger than, and at the bifurcation value.

1. $\dfrac{dy}{dt} = y^2 + a$

2. $\dfrac{dy}{dt} = y^2 + 3y + a$

3. $\dfrac{dy}{dt} = y^2 - ay + 1$

4. $\dfrac{dy}{dt} = \cos y + a$

5. For the one-parameter family

$$\frac{dy}{dt} = y^6 - 2y^3 + \alpha,$$

identify the bifurcation values of α and describe the bifurcations that take place as α increases. [*Hint*: Rewrite y^6 as $(y^3)^2$ and use the quadratic equation to find the equilibrium points.]

6. For the one-parameter family

$$\frac{dy}{dt} = y^6 - 2y^4 + \alpha,$$

identify the bifurcation values of α and describe the bifurcations that take place as α increases. [*Hint*: It might be useful to look at the graph of the right-hand side of the equation for various of α.]

7. Consider the population model

$$\frac{dP}{dt} = -\frac{P^2}{50} + 2P$$

for a species of fish in a lake. Suppose it is decided that fishing will be allowed in the lake, but it is unclear how many fishing licenses should be issued. Suppose the average catch of a fisherman with a license is 3 fish per year (these are hard fish to catch).

(a) What is the largest number of licenses that can be issued if the fish are to have a chance to survive in the lake?

(b) Suppose the number of fishing licenses in part (a) is issued. What will happen to the fish population — that is, how does the behavior of the population depend on the initial population?

(c) The simple population model above can be thought of as a model of an ideal fish population that is not subject to many of the environmental problems of an actual lake. For the actual fish population, there will be occasional changes in the population that were not considered in the building of the model. If the water level were high because of a heavy rainstorm, a few extra fish might be able to swim down a usually dry streambed to reach the lake; or the extra water might wash toxic waste into the lake, killing a few fish. Given the possibility of unexpected perturbation of the population not included in the model, what do you think will happen to the actual fish population if we fix the fishing level at the one determined in part (b)?

8. Consider our model

$$\frac{dS}{dt} = f(S) = kS\left(1 - \frac{S}{N}\right)\left(\frac{S}{M} - 1\right).$$

of a fox squirrel population from the previous section. The parameters M and k remain relatively constant over the long term, but as more people move into the area, the parameter N (the carrying capacity) decreases.

(a) Sketch the graph of the function $f(S)$ for fixed values of k and M and several values of N. (Recall that $M < N$.)

(b) At what value of N does a bifurcation occur?

(c) What bifurcation occurs at this value of N? How does the population of fox squirrels behave if the parameter N is slowly and continuously decreased through the bifurcation value?

9. Suppose we observe that an increase in the human population of an area makes the area less desirable to fox squirrels, so they emigrate from the region. We can adjust the fox squirrel population model to take account of this emigration by subtracting a fixed rate E of emigration. We have

$$\frac{dS}{dt} = kS\left(1 - \frac{S}{N}\right)\left(\frac{S}{M} - 1\right) - E,$$

where S, k, M, and N are as in Exercise 8. Suppose that the parameters k, M, and N are fixed.

(a) Sketch the graph of the function

$$f(S) = kS\left(1 - \frac{S}{N}\right)\left(\frac{S}{M} - 1\right) - E$$

for several values of E.

(b) What bifurcation takes place when E varies? How does the population of fox squirrels behave if the parameter E is slowly and continuously increased past the bifurcation value?

(c) What is the bifurcation value of E? (Your answer will be in terms of the other parameters, k, M, and N).

10. For the differential equation that models fish populations with harvesting,

$$\frac{dP}{dt} = f_C(P) = kP\left(1 - \frac{P}{N}\right) - C,$$

we saw that if $C > kN/4$ the fish population will decrease toward zero. If the fish population falls to near zero because the fishing level C is slightly greater than $kN/4$, why must fishing be banned completely in order for the population to recover? That is, if a level of fishing just above $C = kN/4$ caused the collapse of the population, why can't the population be restored by reducing the fishing level to just below
$C = kN/4$?

The differential equations in Exercises 11 and 12 depend on two parameters, α and β. The nature of the phase line depends on the values of both parameters. For each of these equations, determine all of the possible qualitatively distinct phase lines and the regions in the $\alpha\beta$-plane where they occur.

11. $\dfrac{dy}{dt} = y^2 + \alpha y + \beta$ **12.** $\dfrac{dy}{dt} = y^4 + \alpha y^2 + \beta$

In Exercise 34 in Section 1.6, we defined the notion of the index of an equilibrium point for an autonomous differential equation. A source has index $+1$, a sink has index -1, and nodes have index 0. Consider the one-parameter family of differential equations $dy/dt = f_\alpha(y)$ where f depends continuously on α and y. In Exercises 13–15 we suppose that this differential equation has finitely many equilibrium points in the interval $0 \le y \le 1$ for all α.

13. Suppose that for $\alpha = 0$, the sum of the indices of the equilibrium points in the interval $0 \leq y \leq 1$ for $dy/dt = f_0(y)$ is -1. If $f_\alpha(0) \neq 0$ and $f_\alpha(1) \neq 0$ for all α, show that $dy/dt = f_\alpha(y)$ must have at least one sink in $0 \leq y \leq 1$ for all α.

14. Suppose that for $\alpha = 0$, the sum of the indices of the equilibrium points in the interval $0 \leq y \leq 1$ for $dy/dt = f_0(y)$ is $+1$, but when $\alpha = 1$, the sum of the indices of the equilibrium points in the interval $0 \leq y \leq 1$ for $dy/dt = f_1(y)$ is 0. Show that $f_\alpha(0) = 0$ or $f_\alpha(1) = 0$ for some α between 0 and 1.

15. Suppose that for $\alpha = 0$, the sum of the indices of the equilibrium points in the interval $0 \leq y \leq 1$ for $dy/dt = f_0(y)$ is 0. If for each α, $f_\alpha(0) > 0$ and $f_\alpha(1) \neq 0$, show that for each α, there is a positive number M_α such that $dy/dt = f_\alpha(y) + M_\alpha$ has no equilibrium points in the interval $0 \leq y \leq 1$.

The differential equation

$$\frac{dy}{dt} = y^n,$$

where n is an even integer greater than or equal to 2, has a node at $y = 0$ and no other equilibrium points. One might be tempted to try to "classify" all the different bifurcations that can occur. In Exercises 16–19, we consider one-parameter families that include this differential equation.

16. For the one-parameter family

$$\frac{dy}{dt} = y^3 + \alpha y^2,$$

how many equilibrium points are near $y = 0$ if α is just slightly greater than zero? How many are there if α is just slightly less than zero?

17. For the one-parameter family

$$\frac{dy}{dt} = y^4 + \alpha y^2,$$

how many equilibrium points are near $y = 0$ if α is just slightly greater than zero? How many equilibrium points are near zero if α is just slightly less than zero?

18. For the one-parameter family

$$\frac{dy}{dt} = y^6 + \alpha(y^2 - 1000y^4),$$

how many equilibrium points are near $y = 0$ if α is just slightly greater than zero? How many equilibrium points are near $y = 0$ if α is just slightly less than zero?

19. Write a one-parameter family of differential equations with parameter α that has exactly one equilibrium point if $\alpha = 0$ and exactly six equilibrium points if $\alpha < 0$.

20. For the one-parameter family of differential equations

$$\frac{dy}{dt} = \frac{y}{y^2 + 1} + \alpha,$$

locate the equilibrium points if α is slightly greater than zero, if $\alpha = 0$, and if α is slightly less than zero. Discuss what bifurcation occurs at $\alpha = 0$.

1.8 LINEAR DIFFERENTIAL EQUATIONS

In Section 1.2 we developed an analytic method for finding the explicit solutions to separable differential equations. Although many interesting problems lead to separable differential equations, most differential equations are not separable. The qualitative and numerical techniques we developed in Sections 1.5 and 1.6 apply to a much wider range of problems. It would be nice if we could also extend the usefulness of the analytical methods by developing ways of finding explicit solutions of more general equations.

Unfortunately, there is no general way to compute explicit solutions of differential equations that works for all equations. In fact, the problem is even worse. We know from the Existence Theorem that every reasonable differential equation has solutions, but this is no guarantee that those solutions are at all familiar. Over the centuries, solution techniques have been found for many different special types of differential equations. Mathematical "cookbooks" are available giving recipes for these special types, and many of these techniques are available to us as one-line commands in sophisticated computer packages such as Maple and *Mathematica*. Nevertheless, we should be able to "explicitly solve" a few important types of equations. In this section, we develop a technique for solving *linear* differential equations.

Linear Differential Equations

A first-order differential equation is **linear** if it can be written in the form

$$\frac{dy}{dt} = g(t)y + r(t),$$

where $g(t)$ and $r(t)$ are arbitrary functions of t. Examples of linear equations include

$$\frac{dy}{dt} = t^2 y + \cos t,$$

where $g(t) = t^2$ and $r(t) = \cos t$, and

$$\frac{dy}{dt} = \frac{e^{4\sin t}}{t^3 + 7t} y + 23t^3 - 7t^2 + 3,$$

where $g(t) = e^{4\sin t}/(t^3 + 7t)$ and $r(t) = 23t^3 - 7t^2 + 3$. Sometimes it is necessary to do a little algebra in order to see that an equation is linear. For example, the differential equation

$$ty + 2 = \frac{dy}{dt} - 3y$$

can be rewritten as

$$\frac{dy}{dt} = (t + 3)y + 2.$$

In this form we see that the equation is linear — $g(t) = t + 3$ and $r(t) = 2$. Some differential equations fit into several categories. For example, the equation

$$\frac{dy}{dt} = -2y + 8$$

is linear with $g(t) = -2$ and $r(t) = 8$. (Both g and r are very simple functions of t.) It is also separable because it is autonomous.

The word *linear* in the name of these equations refers to the fact that the dependent variable y appears in the equation only to the first power. The equation

$$\frac{dy}{dt} = y^2$$

is not linear because the y^2 term cannot be rewritten in the form $g(t)y + r(t)$, no matter how $g(t)$ and $r(t)$, are chosen. Of course, there is nothing magical about the names of the variables. The equation

$$\frac{dP}{dt} = e^{2t}P - \sin t$$

is linear ($g(t) = e^{2t}$ and $r(t) = -\sin t$). Also,

$$\frac{dw}{dt} = (\sin t)w$$

is both linear (where $g(t) = \sin t$ and $r(t) = 0$) and separable. But

$$\frac{dz}{dt} = t \sin z$$

is not linear.

Solving Linear Differential Equations

Given a linear differential equation

$$\frac{dy}{dt} = g(t)y + r(t),$$

how can we go about finding the general solution? There is a clever trick that turns an equation of this form into a differential equation that can be solved by integration. As with many techniques in mathematics, the cleverness of this trick might leave you feeling a little depressed — the "how could I ever think of something like that" feeling. The thing to remember is that differential equations have been around for more than 300 years. If you were given 300 years to work on it, you would certainly come up with a method for solving these equations.

To see how to solve a linear differential equation, we first rewrite it as

$$\frac{dy}{dt} - g(t)y = r(t).$$

After staring at this equation for a while (say a couple of decades), we notice that the left-hand side looks somewhat like the result of a differentiation using the Product Rule. The Product Rule says that the derivative of the product of $y(t)$ with a function $\mu(t)$ is

$$\frac{d(\mu(t) \cdot y(t))}{dt} = \mu(t)\frac{dy}{dt} + \frac{d\mu}{dt}y(t).$$

Now here's the clever part. Let's multiply both sides of the differential equation by an as yet unspecified function $\mu(t)$. We obtain

$$\mu(t)\frac{dy}{dt} - \mu(t)g(t)y = \mu(t)r(t).$$

The left-hand side looks even more like the derivative of a product of two functions. For the moment, let's *assume* that we have found a function $\mu(t)$ so that the left-hand side is actually the derivative of $\mu(t) \cdot y(t)$. That is, we have found a function $\mu(t)$ that satisfies

$$\frac{d(\mu(t) \cdot y(t))}{dt} = \mu(t)\frac{dy}{dt} - \mu(t)g(t)y.$$

Then the differential equation becomes

$$\frac{d(\mu(t) \cdot y(t))}{dt} = \mu(t)r(t).$$

How does this help? We can now integrate both sides of this equation with respect to t to obtain

$$\mu(t)\,y(t) = \int \mu(t)\,r(t)\,dt$$

so that

$$y(t) = \frac{1}{\mu(t)} \int \mu(t)\,r(t)\,dt.$$

That is, assuming we have $\mu(t)$ and can evaluate this integral, we can compute our solution $y(t)$.

Therefore, the question is: How do we find such a function $\mu(t)$ in the first place? The Product Rule says that

$$\frac{d(\mu(t) \cdot y(t))}{dt} = \mu(t)\frac{dy}{dt} + \frac{d\mu}{dt}y(t),$$

but given the previous assumptions on $\mu(t)$, we know that we also want

$$\frac{d(\mu(t) \cdot y(t))}{dt} = \mu(t)\frac{dy}{dt} - \mu(t)g(t)y(t).$$

Because we want $\mu(t)$ to satisfy both of these equations, we have

$$\mu(t)\frac{dy}{dt} + \frac{d\mu}{dt}y(t) = \mu(t)\frac{dy}{dt} - \mu(t)g(t)y(t).$$

Canceling the first term on both sides yields

$$\frac{d\mu}{dt}y(t) = -\mu(t)\,g(t)\,y(t).$$

Dividing by $y(t)$, we see that we need to find a function $\mu(t)$ that satisfies the equation

$$\frac{d\mu}{dt} = -\mu(t)\,g(t).$$

This equation is another differential equation — this time for $\mu(t)$. Notice that this differential equation is separable. So we can separate variables and obtain

$$\frac{1}{\mu}\frac{d\mu}{dt} = -g(t),$$

which leads to the solution

$$\mu(t) = e^{\int -g(t)\,dt}.$$

Given this formula for $\mu(t)$, we now know that this strategy is going to work and we can solve a first-order linear differential equation. The function $\mu(t)$ is called an **integrating factor** for the original differential equation because, if we multiply the original equation by $\mu(t)$, we can then solve the new equation by integration. (When we calculate $\mu(t)$, there is an arbitrary constant of integration in the exponent. Since we only need one function, $\mu(t)$, we choose the constant to be whatever is most convenient. That choice is usually zero.)

To summarize, whenever we want to determine an explicit solution to a linear differential equation of the form

$$\frac{dy}{dt} = g(t)y + r(t),$$

we first compute the integrating factor

$$\mu(t) = e^{\int -g(t)\,dt}.$$

Then we can determine the solutions by multiplying both sides of the linear differential equation by $\mu(t)$ and integrating. To see this method at work, let's look at some examples. The method looks very general. However, because there are two integrals to calculate, it is quite easy to get stuck before we obtain an explicit solution.

Complete success

Consider the differential equation

$$\frac{dy}{dt} = -\frac{2}{t}y + (t-1).$$

This equation is linear with $g(t) = -2/t$ and $r(t) = t - 1$. We first compute the integrating factor

$$\mu(t) = e^{\int -g(t)\,dt} = e^{\int -(-2/t)\,dt} = e^{2\ln t} = e^{\ln(t^2)} = t^2.$$

Remember that the idea behind this method is to rewrite the equation in the form

$$\frac{dy}{dt} + \frac{2}{t}y = t - 1,$$

and then multiply both sides by $\mu(t)$ so that the left-hand side of the new equation is the result of the Product Rule. In this case, multiplying by $\mu(t) = t^2$ yields

$$t^2\frac{dy}{dt} + 2ty = t^2(t-1).$$

Note that the left-hand side is the derivative of the product of t^2 and $y(t)$. Thus this equation is the same as

$$\frac{d}{dt}(t^2 y) = t^2(t-1) = t^3 - t^2.$$

Integrating both sides with respect to t yields

$$t^2 y = \frac{t^4}{4} - \frac{t^3}{3} + c,$$

where c is an arbitrary constant. The general solution is

$$y(t) = \frac{t^2}{4} - \frac{t}{3} + \frac{c}{t^2}.$$

Of course, we can check that these functions satisfy the differential equation by substituting them back into the equation.

Problems with the integration

The previous example was rather carefully chosen. Another example, which doesn't look any more difficult, is

$$\frac{dy}{dt} = t^2 y + (t - 1),$$

which is a linear differential equation with $g(t) = t^2$ and $r(t) = t - 1$. Again we begin by computing the integrating factor

$$\mu(t) = e^{\int -g(t)\,dt} = e^{\int -t^2\,dt} = e^{-t^3/3}.$$

We rewrite the differential equation as

$$\frac{dy}{dt} - t^2 y = (t - 1)$$

and multiply both sides by $\mu(t)$. We obtain

$$e^{-t^3/3}\frac{dy}{dt} - e^{-t^3/3}t^2\,y = e^{-t^3/3}(t - 1).$$

Note that the left-hand side is the derivative of the product of $e^{-t^3/3}$ and $y(t)$, so we really have

$$\frac{d}{dt}\left(e^{-t^3/3}\,y\right) = e^{-t^3/3}(t - 1).$$

Integrating both sides yields

$$e^{-t^3/3}\,y = \int e^{-t^3/3}(t - 1)\,dt,$$

and we are stuck. It turns out that the integral on the right-hand side of this equation is not expressible in terms of the familiar functions (sin, cos, ln, and so on), so we can't obtain explicit formulas for the solutions.

This example indicates what can go wrong with techniques that involve the calculation of explicit integrals. Even reasonable-looking functions can quickly lead to complicated integrating factors and very complicated integrals. On the other hand, we can express the solution in terms of integrals with respect to t (which is some progress), and although many integrals are impossible to calculate explicitly, many others are possible. Indeed, as we mentioned at the beginning of this section, there are now a number of computer programs that are quite good at calculating indefinite integrals.

Mixing Problems Revisited

In Section 1.2, we considered a model of the concentration of a substance in solution. Typically in these problems we have a container in which there is a certain amount of fluid (such as water or air) to which a contaminant is added at some rate. The fluid is kept well mixed at all times. If the total volume of fluid is kept fixed, then the resulting differential equation for the amount of contaminant is an autonomous differential equation. If the total volume of fluid changes with time, then the differential equation is nonautonomous. If the problem requires finding an exact value for the amount or concentration of the contaminant at a particular time, then we need to use techniques like those of this section.

A polluted pond

Consider a pond that has an initial volume of $10,000$ cubic meters. Suppose that at time $t = 0$, the water in the pond is clean and that the pond has two streams flowing into it, stream A and stream B, and one stream flowing out, stream C (see Figure 1.91). Suppose 500 cubic meters per day of water flow into the pond from stream A, 750 cubic meters per day flow into the pond from stream B, and 1250 cubic meters flow out of the pond via stream C.

At time $t = 0$, the water flowing into the pond from stream A becomes contaminated with road salt at a concentration of 5 kilograms per 1000 cubic meters. Suppose the water in the pond is well mixed so the concentration of salt at any given time is constant. To make matters worse, suppose also that at time $t = 0$ someone begins dumping trash into the pond at a rate of 50 cubic meters a day. The trash settles to the bottom of the pond, reducing the volume by 50 cubic meters per day. To adjust for the incoming trash, the rate that water flows out via stream C increases to 1300 cubic meters per day and the banks of the pond do not overflow.

The description looks very much like the mixing problems we have already considered (where "pond" replaces "vat" and "stream" replaces "pipe"). The new element here is that the total volume is not constant. Because of the dumping of trash, the volume decreases by 50 cubic meters per day.

If we let $S(t)$ be the amount of salt (in kilograms) in the pond at time t, then the rate of change of S is given by the difference between the rate that salt enters the pond and the rate that salt leaves the pond. Salt enters the pond from stream A only, and the rate at

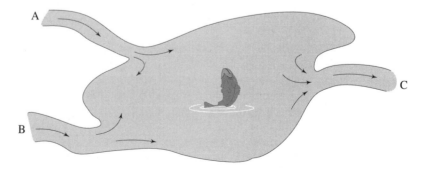

Figure 1.91
Schematic of the pond with three streams.

which it enters is the product of its concentration in the water and the rate at which the water flows in through stream A. Since the concentration is 5 kilograms per 1000 cubic meters and the rate that water flows into the pond from stream A is 500 cubic meters per day, the rate at which salt enters the pond is $(500)(5/1000) = 5/2$ kilograms per day. The rate at which the salt leaves via stream C is the product of its concentration *in the pond* and the rate at which water flows out of the pond. The rate at which water flows out is 1300 cubic meters per day. To determine the concentration, we note that it is the quotient of the total amount S of salt in the pond by the total volume V. Because the volume is initially 10,000 cubic meters and it decreases by 50 cubic meters per day, we know that $V(t) = 10,000 - 50t$. Hence, the concentration is $S/(10,000 - 50t)$, and the rate at which salt flows out of the pond is

$$1300 \left(\frac{S}{10,000 - 50t} \right) = \frac{26S}{200 - t}.$$

The differential equation that models the amount of salt in the pond is therefore

$$\frac{dS}{dt} = \frac{5}{2} - \frac{26S}{200 - t}.$$

This model is valid only as long as there is water in the pond, that is, as long as the volume $V(t) = 10,000 - 50t$ is positive. So the differential equation is valid for $0 \le t < 200$. Because the water is clean at time $t = 0$, the initial condition is $S(0) = 0$.

The differential equation for salt in the pond is nonautonomous. Its slope field is given in Figure 1.92. From the slope field, or by using Euler's method, we could approximate the solution with initial value $S(0) = 0$. Because the equation is linear, we can also find a formula for the solution.

The integrating factor is

$$\mu(t) = e^{\int \frac{26}{200-t} \, dt} = e^{-26 \ln(200-t)} = e^{\ln((200-t)^{-26})} = (200 - t)^{-26}.$$

Rewriting the differential equation as

$$\frac{dS}{dt} + \frac{26S}{200 - t} = \frac{5}{2}$$

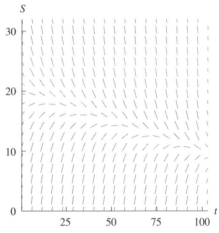

Figure 1.92
Slope field for the equation
$dS/dt = 5/2 - 26S/(200 - t)$.

and multiplying both sides by $\mu(t)$ gives

$$(200 - t)^{-26}\frac{dS}{dt} + 26(200 - t)^{-27}S = \frac{5}{2}(200 - t)^{-26}.$$

By the Product Rule, this equation is the same as the differential equation

$$\frac{d}{dt}\left((200 - t)^{-26}S\right) = \frac{5}{2}(200 - t)^{-26}.$$

Integrating both sides yields

$$(200 - t)^{-26}S = \frac{5}{2}\int (200 - t)^{-26}\,dt$$

$$= \frac{5}{2}\frac{(200 - t)^{-25}}{25} + c,$$

where c is an arbitrary constant. Solving for S, we obtain the general solution

$$S = \frac{200 - t}{10} + c(200 - t)^{26}.$$

Using the initial condition $S(0) = 0$, we find that $c = -20/200^{26}$ and the particular solution for the initial-value problem is

$$S = \frac{200 - t}{10} - 20\left(\frac{200 - t}{200}\right)^{26}.$$

This is an unusual-looking expression because of the large number 200^{26}. However, the graph reveals that its behavior is not at all unusual (see Figure 1.93). The amount of salt in the pond rises fairly quickly, reaching a maximum close to $S = 20$ at $t \approx 25$. After that time, the amount of salt decreases almost linearly, reaching zero at $t = 200$.

The behavior of this solution is quite reasonable if we recall that the pond starts out containing no salt and that eventually it is completely filled with trash. (It contains no salt or water at time $t = 200$.) As we mentioned above, the concentration of salt in the pond water is given by $C(t) = S(t)/V(t) = S(t)/(10,000 - 50t)$. Graphing $C(t)$, we see that it increases asymptotically toward 20 kilograms per cubic meter even as the water level decreases (see Figure 1.94).

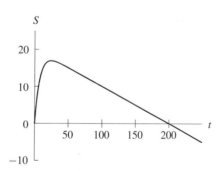

Figure 1.93
Graph of solution of
$dS/dt = 5/2 - 26S/(200 - t)$ with
$S(0) = 0$.

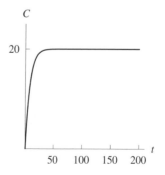

Figure 1.94
Graph of concentration of salt
versus time for the solution
graphed in Figure 1.93.

Exercises for Section 1.8

In Exercises 1–8, find the general solution of the differential equation specified.

1. $\dfrac{dy}{dt} = -\dfrac{y}{t} + 2$

2. $\dfrac{dy}{dt} = \dfrac{3}{t}y + t^5$

3. $\dfrac{dy}{dt} = y - 3e^{-t}$

4. $\dfrac{dy}{dt} = -2ty + 4e^{-t^2}$

5. $\dfrac{dy}{dt} = \dfrac{2t}{1+t^2}y + \dfrac{2}{1+t^2}$

6. $\dfrac{dy}{dt} = -2y - 5t$

7. $\dfrac{dy}{dt} = -y + t^2$

8. $\dfrac{dy}{dt} = -2y + \sin t$

In Exercises 9–14, find the particular solution with the given initial condition.

9. $\dfrac{dy}{dt} = -\dfrac{y}{1+t} + 2, \quad y(0) = 3$

10. $\dfrac{dy}{dt} = -y + e^t, \quad y(0) = 0.4$

11. $\dfrac{dy}{dt} = -\dfrac{y}{t} + 2, \quad y(1) = 3$

12. $\dfrac{dy}{dt} = \dfrac{2t}{1+t^2}y + \dfrac{2}{1+t^2}, \quad y(0) = -2$

13. $\dfrac{dy}{dt} = \dfrac{2y}{t} + 2t^2, \quad y(-2) = 4$

14. $\dfrac{dy}{dt} = -5y + \sin t, \quad y(0) = 1$

Exercises 15–18 are linear equations, so, theoretically, we can find their general solutions. Since finding the general solution of a linear equation involves computing at least two integrals, in practice it is frequently impossible to reach a formula for the solution that does not involve any intergrals. In Exercises 15–18, get as far as you can toward finding the general solution.

15. $\dfrac{dy}{dt} = (\sin t)y + 4$

16. $\dfrac{dy}{dt} = t^2 y + 4$

17. $\dfrac{dy}{dt} = \dfrac{y}{t^2} + 4\cos t$

18. $\dfrac{dy}{dt} = y + 4\cos t^2$

19. For what value(s) of the parameter a is it possible to find explicit formulas (without integrals) for the solutions to

$$\frac{dy}{dt} = aty + 4e^{-t^2}?$$

[*Hint:* First show that the integrating factor is $\mu(t) = e^{-at^2/2}$.]

20. For what value(s) of the parameter r is it possible to find explicit formulas (without integrals) for the solutions to

$$\frac{dy}{dt} = t^r y + 4?$$

21. A person initially places $500 in a savings account that pays interest at the rate of 4% per year compounded continuously. Suppose the person arranges for $10 per week to be deposited automatically into the savings account.

(a) Write a differential equation for $P(t)$, the amount on deposit after t years (assume that "weekly deposits" is close enough to "continuous deposits" so that we may use a differential equation model.)

(b) Find the amount on deposit after 5 years.

22. A student has saved $30,000$ for her college tuition. When she starts college, she invests the money in a savings account that pays 6% interest per year, compounded continuously. Suppose her college tuition is $10,000$ per year and she arranges with the college that the money will be deducted from her savings account in small payments (we assume she is paying continuously). How long will she be able to stay in school before she runs out of money?

23. A college professor contributes $1200 per year into her retirement fund by making many small deposits throughout the year. The fund grows at a rate of 8% per year compounded continuously. After 30 years, she retires and begins withdrawing from her fund at a rate of $3000 per month. If she does not make any deposits after retirement, how long will the money last? [*Hint*: Solve this in two steps, before retirement and after retirement.]

24. A 30-gallon tank initially contains 15 gallons of saltwater containing 6 pounds of salt. Suppose saltwater containing 1 pound of salt per gallon is pumped into the top of the tank at the rate of 2 gallons per minute, while a well-mixed solution leaves the bottom of the tank at a rate of 1 gallon per minute. How much salt is in the tank when the tank is full?

25. A 400-gallon tank initially contains 200 gallons of water containing 2 parts per billion by weight of dioxin, an extremely potent carcinogen. Suppose water containing 5 parts per billion of dioxin flows into the top of the tank at a rate of 4 gallons per minute. The water in the tank is kept well mixed and 2 gallons per minute are removed from the bottom of the tank. How much dioxin is in the tank when the tank is full?

26. A 100-gallon tank initially contains 100 gallons of sugar-water at a concentration of 0.25 pounds of sugar per gallon. Suppose sugar is added to the tank at a rate of p pounds per minute, sugar-water is removed at a rate of 1 gallon per minute, and the water in the tank is kept well mixed.

(a) What value of p should we pick so that, when 5 gallons of sugar solution is left in the tank, the concentration is 0.5 pounds of sugar per gallon?

(b) Is it possible to choose p so that the last drop of water out of the bucket has a concentration of 0.75 pounds of sugar per gallon?

27. Suppose a 50-gallon tank contains a volume V_0 of clean water at time $t = 0$. At time $t = 0$, we begin dumping 2 gallons per minute of salt solution containing 0.25 pounds of salt per gallon into the tank. Also at time $t = 0$, we begin removing 1 gallon per minute of saltwater from the tank. As usual, suppose the water in the tank is well mixed so that the salt concentration at any given time is constant throughout the tank.

(a) Set up the initial-value problem for the amount of salt in the tank. [*Hint*: The parameter V_0 will appear in the differential equation.]

(b) What is your model equation when $V_0 = 0$ (when the tank is initially empty)? Comment on the validity of the model in this situation. What will be the amount of salt in the tank at time t for this situation?

1.9 CHANGING VARIABLES

Many of the techniques we have dealt with in this chapter concern particular types of equations. For example, the analytic techniques we have studied apply only to separable and linear equations, and the concept of phase lines applies only to autonomous equations.

Given a differential equation, it is likely that it will not be in the form appropriate for the particular technique we would like to use. It would be nice if we could transform it into a new equation that has the proper form. This is the idea behind changing variables. By rewriting the equation in terms of a new variable, we may be able to put the equation into a form that is appropriate for a particular technique. In this section we give a method for transforming differential equations by changing variables into (hopefully) simpler forms.

u-Substitution

The idea behind changing variables is not new. You have seen it before in calculus when you did the "*u*-substitution" method for computing antiderivatives. For example, given the integral

$$\int t \sin t^2 \, dt,$$

we define a new variable u by $u = t^2$ and rewrite the integral in terms of u instead of t. Because $du/dt = 2t$ (or, informally, $du = 2t \, dt$), we obtain

$$\int t \sin t^2 \, dt = \int \frac{1}{2} \sin u \, du.$$

This is the same integral as before but written using the new variable u. The variable u was chosen so that the new integral is easy to evaluate. We have

$$\frac{1}{2} \int \sin u \, du = -\frac{\cos u}{2} + c.$$

We now find the solution to the original problem by replacing the new variable u with the original variable t, and we obtain

$$\int t \sin t^2 \, dt = \int \frac{\sin u}{2} \, du$$

$$= -\frac{\cos u}{2} + c$$

$$= -\frac{\cos t^2}{2} + c.$$

Changing variables from t to u changes the integral from a difficult problem to an easy problem. After computing the antiderivative in the new variable, we can recover the solution of the original problem by replacing the new variable with the old variable.

Changing Variables in Differential Equations

A similar u-substitution method works for differential equations. The idea is the same as for integrals: Define a new variable in terms of the old variable(s), then rewrite the equation in terms of the new variable. This gives us the same differential equation expressed in terms of a new variable. From an algebraic point of view, the new equation can be very different from the original equation. If we make a wise (or clever) choice of the new variable, then the new equation may be much easier to study than the old equation.

An example

Consider the differential equation

$$\frac{dy}{dt} = -\frac{y}{t} + \frac{t-1}{2y}.$$

This looks like a pretty complicated equation. It is neither autonomous, separable, nor linear, so if we had to find the general solution we would be at a loss. But of course we don't give up. We might try a little algebraic simplification first. The most annoying term in this equation is the second fraction on the right. The fact that the dependent variable y appears in the denominator is what makes the equation nonlinear. If the y were not there, then the equation would be linear. So let's try to simplify by multiplying both sides by y to obtain

$$y\frac{dy}{dt} = -\frac{y^2}{t} + \frac{t-1}{2}.$$

This equation is still not linear, and there is a y on the left-hand side. However, the right-hand side of the equation would be the right-hand side of a linear equation if the y^2 term were only y. This motivates our substitution. We define a new dependent variable u by

$$u = y^2,$$

and we rewrite the equation in terms of u. This means eliminating all the y's from the equation, including dy/dt, and replacing them with u's (and du/dt). To express dy/dt in terms of u, we differentiate the equation $u = y^2$ to obtain

$$\frac{du}{dt} = 2y\frac{dy}{dt},$$

so

$$\frac{1}{2}\frac{du}{dt} = y\frac{dy}{dt}.$$

Substituting this into the left-hand side and replacing the y^2 with u on the right-hand side, we obtain the new equation

$$\frac{1}{2}\frac{du}{dt} = -\frac{u}{t} + \frac{t-1}{2}.$$

Multiplying through by 2 gives

$$\frac{du}{dt} = -\frac{2}{t}u + (t-1).$$

This is a linear equation. In fact, it is one of the examples that we worked out in complete detail in Section 1.8. The general solution is

$$u(t) = \frac{t^2}{4} - \frac{t}{3} + \frac{c}{t^2}.$$

To find the general solution for $y(t)$ in the original equation, we just recall that $u = y^2$. Thus $y = \pm\sqrt{u}$ and

$$y(t) = \pm\sqrt{\frac{t^2}{4} - \frac{t}{3} + \frac{c}{t^2}},$$

where the choice of sign and the value of c depend on the initial condition.

This is pretty remarkable. By defining a new dependent variable and rewriting the differential equation in terms of that variable, we obtained a linear differential equation that we could solve. Of course, we had to figure out (or guess) that the "right" dependent variable or the right **change of variables** is $u = y^2$. As with u-substitution, it sometimes takes several attempts to find the best substitution, and sometimes there just isn't a good substitution.

Changing Variables and Qualitative Analysis

The technique of changing variables is not a "magic bullet" with which all differential equations can be solved. Finding a change of variables that makes the differential equation manageable can depend both on the form of the equation and on the goal of the analysis. In the following examples we consider various different types of changes of variables.

An example

We can use this method of changing variables in concert with qualitative and numerical methods. For example, consider the very complicated-looking equation

$$\frac{dy}{dt} = y^2 - 4ty + 4t^2 - 4y + 8t - 3.$$

We can draw the slope field for this equation and use Euler's method to sketch solutions (see Figure 1.95).

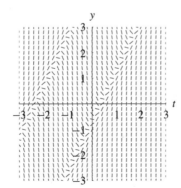

Figure 1.95
Slope field for

$$\frac{dy}{dt} = y^2 - 4ty + 4t^2 - 4y + 8t - 3.$$

Solutions seem to be increasing in two regions and decreasing in a third region.

This equation is neither linear nor separable, so we might try to look for a change of variables that simplifies the equation algebraically. To make an intelligent guess of a new dependent variable, we first rewrite the equation using some algebra. After staring at the right-hand side for a while, we see that we can make it look a little simpler by collecting terms and factoring, that is,

$$\frac{dy}{dt} = y^2 - 4ty + 4t^2 - 4y + 8t - 3 = (y - 2t)^2 - 4(y - 2t) - 3.$$

Looking at this equation, a possible choice of a new dependent variable jumps out at us. Let

$$u = y - 2t.$$

This new dependent variable u is a combination of the old dependent variable y and the independent variable t. To replace all the occurrences of the old variable y with the new dependent variable u, we compute dy/dt in terms of u by differentiating $u = y - 2t$. We get

$$\frac{du}{dt} = \frac{dy}{dt} - 2 \quad \text{or} \quad \frac{dy}{dt} = \frac{du}{dt} + 2.$$

Thus the equation

$$\frac{dy}{dt} = (y - 2t)^2 - 4(y - 2t) - 3$$

becomes

$$\frac{du}{dt} + 2 = u^2 - 4u - 3 \quad \text{or} \quad \frac{du}{dt} = u^2 - 4u - 5.$$

This is an autonomous equation. We can study it by drawing its phase line. Because $u^2 - 4u - 5 = (u - 5)(u + 1)$, the equilibrium points are $u = 5$ and $u = -1$. The phase line and a sketch of several solutions of this equation are given in Figure 1.96. The equilibrium point $u = -1$ is a sink and $u = 5$ is a source.

What does this tell us about the original differential equation? Because the new equation is separable, we can find explicit solutions by integration. However, we can also obtain information about the solutions $y(t)$ even more directly. We know that

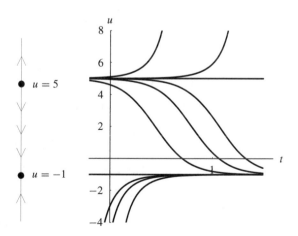

Figure 1.96
Phase line and graphs of solutions for
$du/dt = u^2 - 4u - 5$.

$u = -1$ is an equilibrium point for the new equation. That is, the constant function $u_1(t) = -1$ is a solution of the new equation. But any solution of the new equation corresponds to a solution of the original equation. Hence

$$y_1(t) = u_1(t) + 2t = -1 + 2t$$

is a solution of the original equation.

Similarly, because $u_2(t) = 5$ is a solution of the new equation,

$$y_2(t) = u_2(t) + 2t = 5 + 2t$$

is a solution of the original equation. Every solution $u(t)$ of the new equation corresponds to a solution of the original equation by the change of variables $y(t) = u(t) + 2t$. This means that the graphs of solutions of the new equation in the tu-plane and the solutions of the original equation in the ty-plane are closely related. The ty-plane can be obtained from the tu-plane by adding $2t$ to every solution. In effect, this is the same as taking the tu-plane and shearing it upward with slope 2. The equilibrium solutions in u correspond to solutions whose graphs are lines with slope 2 in the ty-plane. Also, because $u = -1$ is a sink, solutions tend toward the solution $y_1(t) = -1 + 2t$ as t increases. Similarly, solutions tend away from the solution $y_2(t) = 5 + 2t$ because $u = 5$ is a source. So we may graph solutions in the ty-plane just from the corresponding graphs in the tu-plane (see Figure 1.97).

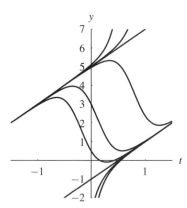

Figure 1.97
Graphs of solutions for the original equation

$$\frac{dy}{dt} = y^2 - 4ty + 4t^2 - 4y + 8t - 3.$$

Compare these graphs with the slope field in Figure 1.95.

Another example

For another example, consider the equation

$$\frac{dv}{dt} = v - \frac{v^2}{t} + \frac{v}{t}.$$

Although there might be several reasonable choices of changes of variables for this equation, we try to simplify the equation by getting rid of some of the denominators; that is, we try

$$u = \frac{v}{t} \quad \text{so} \quad tu = v.$$

The Product Rule yields

$$\frac{dv}{dt} = t\frac{du}{dt} + u,$$

and the new equation is

$$t\frac{du}{dt} + u = tu - \frac{(tu)^2}{t} + \frac{tu}{t}.$$

Simplifying, we find

$$t\frac{du}{dt} + u = tu - tu^2 + u$$

or

$$\frac{du}{dt} = (1 - u)u,$$

our old friend. We sketch the phase line and solutions for this equation (probably from memory) (see Figure 1.98). The equilibrium points $u = 0$ and $u = 1$ correspond to the constant solutions $u_0(t) = 0$ and $u_1(t) = 1$, respectively, and the corresponding solutions in t and v are $v_0(t) = tu_0(t) = t \cdot 0 = 0$ and $v_1(t) = tu_1(t) = t \cdot 1 = t$. Moreover, because $u = 0$ is a source, solutions in the tv-plane tend to move away from the solution $v_0(t)$ as t increases. Similarly, because $u = 1$ is a sink, solutions in the tv-plane tend toward the solution $v_1(t)$ as t increases (see Figure 1.99).

The graphs of solutions $v_0(t) = 0$ and $v_1(t) = t$ intersect at $t = 0$. This makes us very nervous because the Existence and Uniqueness Theorem says that solutions whose graphs intersect are actually the same. However, there is no contradiction here because the original equation is not defined when $t = 0$. The change of variables $u = v/t$ is also undefined when $t = 0$. When t is close to zero, a small value of v yields a big value of u, so the region near $t = 0$ on the tv-plane is "blown up" by the change of variables.

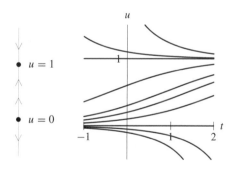

Figure 1.98
Phase line and graphs of solutions for the logistic equation $du/dt = (1 - u)u$.

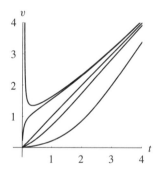

Figure 1.99
Graphs of solutions for the equation
$dv/dt = v - v^2/t + v/t$.

Mixing Problems Revisited

Consider the following situation. Suppose a 10-gallon vat contains 4 gallons of clean water. We start pouring sugar into the vat at time $t = 0$ at a rate of 0.25 pounds per minute. We also pour clean water into the vat at a rate of 2 gallons per minute. Finally, suppose that the water is kept well mixed so the concentration of sugar is uniform in

Figure 1.100
Schematic of the vat with sugar and
clean water being added and
sugar-water being removed.

the vat and that we remove 1 gallon per minute of sugar-water from the bottom of the
vat (see Figure 1.100). What will the concentration of sugar be at the moment when the
tank starts to overflow?

Let $S(t)$ be the amount of sugar in the tank at time t. The rate of change of S
is determined by the rate that sugar enters the vat (a positive term) and the rate that
sugar leaves the vat (a negative term). In this case the rate that sugar enters the vat is
a constant (0.25 pounds per minute), but the rate that sugar leaves the vat depends on
the concentration. At time t, the concentration is given as the amount of sugar in the
vat, $S(t)$, divided by the volume of liquid in the vat. Hence the differential equation
governing the change of the amount of sugar in the vat is

$$\frac{dS}{dt} = 0.25 - \frac{S}{4+t}.$$

The vat will overflow when $4 + t = 10$; that is, $t = 6$. We wish to determine
$S(6)/10$, given that $S(0) = 0$. The equation is linear, so we could use the technique
of Section 1.8: Find the integrating factor and do the integrals. We can also change
variables, which makes the problem separable.

In this case our choice of the new variable depends on physical considerations.
Since our goal is to determine the concentration at time $t = 6$, we let the new variable
be the concentration of sugar in the vat at time t. Let $C(t)$ denote the concentration at
time t in pounds of sugar per gallon, so

$$C = \frac{S}{4+t} \quad \text{or} \quad (4+t)C = S.$$

To find the new differential equation in terms of the concentration, we compute the
derivative of $(4+t)C = S$, obtaining

$$(4+t)\frac{dC}{dt} + C = \frac{dS}{dt}.$$

Using this we obtain

$$(4+t)\frac{dC}{dt} + C = \frac{dS}{dt}$$

$$= 0.25 - \frac{S}{4+t}$$

$$= 0.25 - C.$$

In terms of the concentration, the differential equation is

$$(4+t)\frac{dC}{dt} + C = 0.25 - C$$

or

$$\frac{dC}{dt} = \frac{0.25 - 2C}{4+t}.$$

This equation describes the same physical situation as the equation for dS/dt. It is really the same equation, expressed in terms of the concentration C instead of the amount of sugar S. The surprising thing is that this equation is separable. Separating variables, we find that

$$\frac{1}{0.25 - 2C} dC = \frac{1}{4+t} dt,$$

and integrating both sides yields

$$-\frac{1}{2}\ln(0.25 - 2C) = \ln(4+t) + c,$$

where c is the constant of integration. Since the water in the vat is free of sugar at $t = 0$, the initial condition is $C(0) = 0$, and therefore we can determine the value of c by

$$-\frac{1}{2}\ln(0.25) = \ln(4) + c$$

or

$$\frac{1}{2}\ln(4) = \ln(4) + c.$$

Thus, $c = -(1/2)\ln(4) = \ln(4^{-1/2}) = \ln(1/2)$. Now that we have determined c, we can solve for the concentration C. We have

$$-\frac{1}{2}\ln(0.25 - 2C) = \ln(4+t) + \ln(\tfrac{1}{2}) = \ln(\tfrac{1}{2}(4+t)) = \ln\left(2 + \frac{t}{2}\right).$$

Multiplying both sides by -2 yields

$$\ln(\tfrac{1}{4} - 2C) = -2\ln\left(2 + \frac{t}{2}\right) = \ln\left[\left(2 + \frac{t}{2}\right)^{-2}\right],$$

and exponentiating we obtain

$$\tfrac{1}{4} - 2C = \left(2 + \frac{t}{2}\right)^{-2}.$$

Solving for C gives

$$C = \frac{1}{8} - \frac{1}{2}\left(2 + \frac{t}{2}\right)^{-2}.$$

The concentration at the time of overflow is $C(6) = 0.105$ pounds of sugar per gallon.

Whether this equation was easier or harder to solve than the dS/dt equation depends on the person doing the arithmetic. Having two ways to find the solution is useful for any problem, since it is unlikely that an arithmetic error will give the same wrong answer with both methods. From a theoretical point of view, it is interesting that changing from amount of sugar to concentration makes such a large difference in the type of differential equation obtained and the technique of solution used.

A Linearization Problem

As we have noted, changing variables is not just a method for helping to find analytic solutions. It can also be very useful in isolating qualitative behavior of solutions. For example, consider the logistic population model

$$\frac{dP}{dt} = 0.06P\left(1 - \frac{P}{500}\right),$$

where $P(t)$ is the population at time t measured in hundreds or thousands of individuals. Note that the growth-rate parameter is 0.06 and the carrying capacity is 500.

The equilibrium points are a source at $P = 0$ and a sink at $P = 500$. All solutions with positive initial conditions will approach the sink as time increases (see Figure 1.101). We know that when P is small, the term $(1 - P/500)$ is close to one and solutions of this logistic equation are very close to solutions of the exponential growth model

$$\frac{dP}{dt} = 0.06P,$$

which have the form $P(t) = ke^{0.06t}$. So, for P small, the population grows at an exponential rate with exponent $0.06t$.

After a long time, we expect that the population will be near $P = 500$. Suppose that some unexpected event (not included in the model) pushes the population away from $P = 500$. Such an event could be a period of uncharacteristically harsh weather, unlawful poaching, or the unexpected immigration of a small number of individuals. Any of these events will give a population near, but not at, $P = 500$. We know that, if conditions return to those of the model, then the population will return to the carrying capacity of $P = 500$. A natural question is: How long will this recovery take? In other words, for $P(0)$ near 500, how will the solution behave?

The sort of answer we want is similar to the description of the behavior of small populations. We change variables, moving the point $P = 500$ to the origin. In the new variables, the behavior of solutions near the equilibrium point are easier to discover. Let

$$u = P - 500 \quad \text{or} \quad P = u + 500.$$

The new equation in terms of u is given by

$$\frac{du}{dt} = \frac{dP}{dt}$$

$$= 0.06P\left(1 - \frac{P}{500}\right)$$

$$= 0.06(u + 500)\left(1 - \frac{u + 500}{500}\right)$$

$$= 0.06(u + 500)\left(-\frac{u}{500}\right).$$

Collecting terms, this equation becomes

$$\frac{du}{dt} = -0.06u - 0.06\frac{u^2}{500}.$$

Figure 1.101

Phase line for the familiar equation

$$\frac{dP}{dt} = 0.06P\left(1 - \frac{P}{500}\right).$$

$\bullet$ $p = 0$

$\bullet$ $p = 500$

The equilibrium point $P = 500$ corresponds to the equilibrium point $u = 0$ in the new variable. For u very close to zero, the term containing u^2 will be the square of a very small number. Hence, for u near zero, the behavior of solutions of this equation will be very close to the behavior of solutions of the equation where we neglect the u^2 term, that is

$$\frac{du}{dt} = -0.06u.$$

Solutions of this equation are of the form $u(t) = ke^{-0.06t}$, so they decay toward $u = 0$ at an exponential rate with exponent $-0.06t$ (see Figure 1.102). Because the new dependent variable u is just a translation of the original variable P, solutions for the logistic equation that have initial condition near $P = 500$ approach $P = 500$ at an exponential rate, with exponent $-0.06t$ (see Figure 1.103).

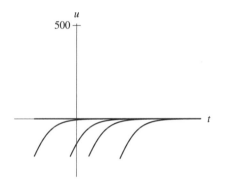

Figure 1.102
Solutions of the equation
$du/dt = -0.06u.$

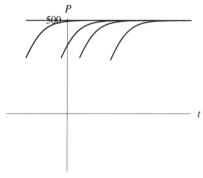

Figure 1.103
Solutions of the equation
$dP/dt = 0.06P(1 - P/500)$ with initial
condition near $P = 500$.

Linearization

In general, to find the behavior of a solution near an equilibrium point, we first change variables moving that equilibrium point to the origin. Next we neglect all the "nonlinear" terms in the equation. Since we are only concerned with values of the new variable near zero, any higher power of that variable will be so close to zero that it can be safely neglected, at least on first approximation. The equation that results is linear and hence (hopefully) easy to analyze.

This method involves an approximation and hence is only valid for values of the variable where that approximation is valid. Once the new variable becomes large, the nonlinear terms in the equation become significant and cannot be ignored. Exactly how small is small and how large is large depends on the particular equation considered (see the Exercises).

This technique is called **linearization**. The idea is to approximate a complicated equation with a simpler, linear equation. Hopefully, the linear equation will be simple enough that we understand the behavior of solutions. Linearization will be an important tool in our study of systems of differential equations and we will discuss it at length in Section 4.1.

Exercises for Section 1.9

In Exercises 1–4, change the dependent variable from y to u by the indicated change of variables. List the adjectives that describe the equation in the new variable (such as separable, linear, and so on).

1. $\dfrac{dy}{dt} = y - 4t + y^2 - 8yt + 16t^2 + 4$ let $u = y - 4t$

2. $\dfrac{dy}{dt} = \dfrac{y^2 + ty}{y^2 + 3t^2}$ let $u = \dfrac{y}{t}$

3. $\dfrac{dy}{dt} = t(y + ty^2) + \cos ty$ let $u = ty$

4. $\dfrac{dy}{dt} = e^y + \dfrac{t^2}{e^y}$ let $u = e^y$

In Exercises 5–7, find a change of variables that makes the equation autonomous. Sketch the phase plane for the autonomous equation you obtain, and use this to sketch solutions for the original equations.

5. $\dfrac{dy}{dt} = (y - t)^2 - (y - t) - 1$ **6.** $\dfrac{dy}{dt} = \dfrac{y^2}{t} + 2y - 4t + \dfrac{y}{t}$

7. $\dfrac{dy}{dt} = y \cos(ty) - \dfrac{y}{t}$

In Exercises 8–10, change variables as indicated, and then find the general solution of the resulting equation. Use this information to give the general solution of the original equation.

8. $\dfrac{dy}{dt} = \dfrac{ty}{2} + \dfrac{e^{t^2/2}}{2y}$ let $y = \sqrt{u}$

9. $\dfrac{dy}{dt} = \dfrac{y}{1+t} - \dfrac{y}{t} + t^2(1 + t)$ let $u = \dfrac{y}{1+t}$

10. $\dfrac{dy}{dt} = y^2 - 2yt + t^2 + y - t + 1$ let $u = y - t$

In Exercises 11–13, create your own change of variables problems. Starting with the simple equation $du/dt = (1 - u)u$, change variables as indicated below. Sketch the solutions of the equations you create using what you know about solutions of the original equation.

11. $y = u + 3t$ **12.** $y = \sqrt{u}$ **13.** $y = u^2$

14. Consider a 20-gallon vat that at time $t = 0$ contains 5 gallons of clean water. Suppose water is entering the vat from two pipes. From the first pipe, saltwater containing 2 pounds of salt per gallon is entering the vat at a rate of 3 gallons per minute.

From the second pipe, saltwater containing 0.5 pounds of salt per gallon is entering the vat at a rate of 4 gallons per minute. Suppose the water is kept well mixed and saltwater is removed from the vat at a rate of 2 gallons per minute.

(a) Derive a differential equation for the rate of change of the total pounds of salt $S(t)$ in the vat at time t.

(b) Convert this equation to a differential equation for the concentration $C(t)$ of salt in the vat at time t.

(c) Find the concentration of salt in the vat at the time when the vat first starts to overflow.

15. Consider a 20-gallon vat that at time $t = 0$ contains 10 gallons of clean water. Suppose water is entering the vat from two pipes. From the first pipe, saltwater containing 2 pounds of salt per gallon is entering the vat at a rate of 1 gallon per minute. From the second pipe, saltwater containing 0.5 gallons per minute is entering the vat at a rate of 2 gallons per minute. Suppose the water is kept well mixed and saltwater is removed from the vat at a rate of 3 gallons per minute.

(a) Derive a differential equation for the rate of change of the total pounds of salt $S(t)$ in the vat at time t.

(b) Convert this equation to a differential equation for the concentration $C(t)$ of salt in the vat at time t.

(c) What will be the concentration of salt in the vat after a very long time?

(d) Find the concentration of salt in the vat at the time $t = 5$.

16. Suppose a 20-gallon vat contains 10 gallons of clean water at time $t = 0$. Water containing 0.3 pounds per gallon of salt is dumped into the vat at a rate of 2 gallons per minute. Suppose clean water is being dumped into the vat at a rate of 1 gallon per minute. Finally, suppose the vat is well mixed and saltwater is being removed from the vat at a rate of 0.5 gallons per minute.

(a) Develop a differential equation for the rate of change of the concentration $C(t)$ of salt in the vat at time t. Be sure to justify each term in your equation carefully. [*Hint*: Consider how each pipe changes the concentration. In particular, does the removal of saltwater affect the concentration?]

(b) Take the equation for the rate of change of the concentration of salt in the vat and convert it into an equation for the amount of salt in the vat at time t. [*Hint*: This should provide a check of your equation for the rate of change of the concentration.]

In Exercises 17–20,

(a) Identify the equilibrium points.

(b) For each equilibrium point, give the change of variables that moves the equilibrium point to the origin, and give the linear approximation to the new system around the origin. Identify the equilibrium point as a source, a sink, or a node.

(c) For each equilibrium point, describe quantitatively the behavior of solutions that begin very close to the equilibrium point (that is, how quickly a solution that starts close to the equilibrium point approaches or recedes from the equilibrium point as t increases). In particular, approximate the amount of time it takes a point to halve or double its distance from the equilibrium point as t increases.

17. $\dfrac{dy}{dt} = 10y^3 - 1$

18. $\dfrac{dy}{dt} = (y + 1)(y - 3)$

19. $\dfrac{dy}{dt} = (y + 1)(3 - y)$

20. $\dfrac{dy}{dt} = y^3 - 3y^2 + y$

21. Consider the differential equation

$$\frac{dy}{dt} = -3 \sin y.$$

    (a) What is the Taylor polynomial for $\sin y$ up to order three? (Work it out or look it up in your calculus book.)

    (b) What is the linearization of this equation near the equilibrium point at $y = 0$?

    (c) Use the result of part (b) to approximate $y(2)$ for the solution with $y(0) = 0.04$.

22. Consider the population model (see Section 1.6, page 84)

$$\frac{dS}{dt} = 0.05S \left(\frac{S}{20} - 1 \right) \left(1 - \frac{S}{100} \right),$$

where S is measured in thousands of individuals. Suppose the population is near the carrying capacity of the system, and a tornado suddenly reduces the population by 2000 individuals.

    (a) Approximately how long will it take for the population to return to within 1000 individuals of the carrying capacity?

    (b) Approximately how long will it take for the population to return to within 100 individuals of the carrying capacity?

23. Consider the logistic population model with small periodic harvesting

$$\frac{dP}{dt} = 0.05P \left(1 - \frac{P}{100} \right) - 0.02(1 + \sin t).$$

    (a) Change variables using $u = P - 100$, moving the carrying capacity to zero.

    (b) Find a linear equation that is close to your equation from part (a) if u is close to zero (that is, P is close to 100).

    (c) Solve the equation you found in part (b).

    (d) What do you conclude about the behavior of the population with small periodic harvesting given by the equation above? (For example, how far do you predict the population will vary from the value $P = 100$?)

24. Consider the differential equation

$$\frac{dy}{dt} = -y + \frac{y^2}{4}.$$

    (a) What is the linear approximation of this equation near the equilibrium point $y = 0$?

    (b) If $y(0) = y_0$ is very close to zero, approximately how long will it take the solution to reach $y_0/2$?

(c) If we let $f(y) = -y + y^2/4$, then for what interval of y's near $y = 0$ is $-1.1y < f(y) < -0.9y$?

(d) For how large a value of $y(0) = y_0$ will your answer to part (b) apply to the nonlinear equation? Justify your answer.

25. Consider the differential equation $dy/dt = f(y)$, where $f(y)$ is a smooth function (it can be differentiated as many times as we like). Suppose $y = y_0$ is an equilibrium point for this equation.

(a) Let $u = y - y_0$ and write the differential equation in terms of the new dependent variable u.

(b) Show that the linear approximation of the differential equation for du/dt near the equilibrium point at $u = 0$ is given by

$$\frac{du}{dt} = f'(y_0)u.$$

26. So far we have only changed the dependent variable. It is also possible to change the independent variable to obtain a new equation. This is analogous to changing the way time is measured. Consider the differential equation

$$\frac{dy}{dt} = 3t^2y + 3t^5.$$

Suppose we define a new time variable by $s = t^3$, so $t = s^{1/3}$.

(a) Verify that

$$\frac{dt}{ds} = \frac{1}{3}s^{-2/3}.$$

(b) Verify that

$$\frac{dy}{ds} = \frac{dy}{dt}\frac{dt}{ds}.$$

(c) Verify that the differential equation

$$\frac{dy}{dt} = 3t^2y + 3t^5$$

in terms of the new time variable s is given by

$$\frac{dy}{ds} = y + s.$$

(d) Find the general solution of the differential equation with time variable s.

(e) What is the general solution of the original dy/dt equation?

Lab 1.1   Exponential and Logistic Population Models

In the text we modeled the U.S. population over the last 200 years using both an exponential growth model and a logistic growth model. For this lab project, we ask that you model the population growth of a particular state. Population data for several states are given below. (Your instructor will assign the state(s) you are to consider.)

We have also discussed three general approaches that can be employed to study a differential equation: numerical techniques yield graphs of approximate solutions, geometric/qualitative techniques provide predictions of the long-term behavior of the solution, and in special cases analytic techniques provide explicit formulas for the solution. In your report, you should employ as many of these techniques as is appropriate to help understand the models.

Your report should address the following items:

1. Using an exponential growth model, determine as accurate a prediction as possible for the population of your state in the Year 2000. How much does your prediction differ from the prediction that comes from linear extrapolation using the populations in 1980 and 1990? To what extent do solutions of your model agree with the historical data?

2. Produce a logistic growth model for the population of your state. What is the carrying capacity for your model? Using Euler's method, predict the population in the Years 2000 and 2050. Using analytic techniques, obtain a formula for the population function $P(t)$ that satisfies your model. To what extent do solutions of your model agree with the historical data?

3. Comment on how much confidence you have in your predictions of the future populations. Discuss which model, exponential or logistic growth, is better for your data and why (and if neither is very good, suggest alternatives).

Your report: The body of your report should address the three items above, one at a time, in the form of a short essay. For each model, you must choose specific values for certain parameters (the growth-rate parameter and the carrying capacity). Be sure to give a complete justification of why you made the choices that you did. You should include pictures and graphs of data and of solutions of your models *as appropriate*. (Remember that one carefully chosen picture can be worth a thousand words, but a thousand pictures aren't worth anything.)

Table 1.9 Population (in thousands) of selected states (Data from 1994 World Almanac).

Year	Massachusetts	New York	North Carolina	Alaska	Florida	California	Montana	Hawaii
1790	379	340	394					
1800	423	589	478	1				
1810	472	959	556	9				
1820	523	1373	638	127				
1830	610	1919	738	309	35			
1840	738	2429	753	591	54			
1850	995	3097	869	772	87	93		
1860	1231	3881	993	964	140	380		
1870	1457	4383	1071	996	188	560	20	
1880	1783	5083	1399	1262	269	865	39	
1890	2239	6003	1618	1513	391	1213	143	
1900	2805	7269	1893	1829	529	1485	243	154
1910	3366	9114	2206	2138	753	2378	376	192
1920	3852	10385	2559	2348	968	3427	549	256
1930	4250	12588	3170	2646	1468	5677	538	368
1940	4317	13479	3571	2832	1897	6907	559	423
1950	4691	14830	4061	3062	2771	10586	591	500
1960	5149	16782	4556	3267	4952	15717	675	633
1970	5689	18241	5084	3444	6791	19971	694	770
1980	5737	17558	5880	3894	9747	23668	787	965
1990	6016	17990	6628	4040	12938	29760	799	1108

Lab 1.2   Logistic Population Models with Harvesting

In this lab we consider logistic models of population growth that have been modified to include terms that account for "harvesting." In particular, you should imagine a fish population subject to various degrees and types of fishing. The differential equation models are given below. (Your instructor will indicate the values of the parameters k, N, a_1 and a_2 you should use. Several possible choices are listed in the table below.) In your report, you should include a discussion of the meaning of each variable and parameter and an explanation of why the equation is written the way it is.

We have discussed three general approaches that can be employed to study a differential equation: numerical techniques yield graphs of approximate solutions, geometric/qualitative techniques provide predictions of the long-term behavior of the solution and in special cases analytic techniques provide explicit formulas for the solution. In your report, you should employ as many of these techniques as is appropriate to help understand the models.

Your report should consider the following equations:

1. (Logistic growth with constant harvesting) The equation

$$\frac{dp}{dt} = kp\left(1 - \frac{p}{N}\right) - a$$

represents a logistic model of population growth with constant harvesting at a rate a. For $a = a_1$, what will happen to the fish population for various initial conditions? (Note: This equation is autonomous, so you can take advantage of the special techniques that are available for autonomous equations.)

2. (Logistic growth with periodic harvesting) The equation

$$\frac{dp}{dt} = kp\left(1 - \frac{p}{N}\right) - a(1 + \sin(bt))$$

is a nonautonomous equation that considers periodic harvesting. What do the parameters a and b represent? Let $b = 1$. If $a = a_1$, what will happen to the fish population for various initial conditions?

3. Consider the same equation as in part 2 above, but let $a = a_2$. What will happen to the fish population for various initial conditions with this value of a?

Your report: In your report you should address the three questions above, one at a time, in the form of a short essay. Begin questions 1 and 2 with a description of the meaning of each of the variables and parameters and an explanation of why the differential equation is the way it is. You should include pictures and graphs of data and of solutions of your models *as appropriate*. (Remember that one carefully chosen picture can be worth a thousand words, but a thousand pictures aren't worth anything.)

Table 1.10 Possible choices for the parameters.

Choice	k	N	a_1	a_2
1	$k = 0.25$	$N = 4$	$a_1 = 0.16$	$a_2 = 0.25$
2	$k = 0.5$	$N = 2$	$a_1 = 0.21$	$a_2 = 0.25$
3	$k = 0.2$	$N = 5$	$a_1 = 0.21$	$a_2 = 0.25$
4	$k = 0.2$	$N = 5$	$a_1 = 0.16$	$a_2 = 0.25$
5	$k = 0.25$	$N = 4$	$a_1 = 0.09$	$a_2 = 0.25$
6	$k = 0.2$	$N = 5$	$a_1 = 0.09$	$a_2 = 0.25$
7	$k = 0.5$	$N = 2$	$a_1 = 0.16$	$a_2 = 0.25$
8	$k = 0.2$	$N = 5$	$a_1 = 0.24$	$a_2 = 0.25$
9	$k = 0.25$	$N = 4$	$a_1 = 0.21$	$a_2 = 0.25$
10	$k = 0.5$	$N = 2$	$a_1 = 0.09$	$a_2 = 0.25$

Lab 1.3   Rate of Memorization Model

Human learning is, to say the least, an extremely complicated process. The biology and chemistry of learning is far from understood. While simple models of learning can not hope to encompass this complexity, they can illuminate limited aspects of the learning process. In this lab we study a simple model of the process of memorization of lists. (Lists of nonsense syllables or techniques of integration.)

The model is based on the assumption that the rate of learning is proportional to the amount left to be learned. We let $L(t)$ be the fraction of the list already committed to memory at time t. So $L = 0$ corresponds to knowing none of the list and $L = 1$ corresponds to knowing the entire list. The differential equation is

$$\frac{dL}{dt} = k(1 - L).$$

Different people take different amounts of time to memorize a list. According to the model this means that each person has their own personal value of k. The value of k for a given individual must be determined by experiment.

Carry out the following steps:

1. Four lists of three-digit numbers are given below, and additional lists can be generated by a random number generator on a computer. Collect the data necessary to determine your personal k value as follows:

   (i) Spend one minute studying one of the lists of numbers given below. (Measure the time carefully. A friend can help.)

   (ii) Quiz yourself on how many of the numbers you have memorized by writing down as many of the numbers as you remember in their correct order. (You may skip over numbers you don't remember and obtain "credit" for numbers you remember later in the list.) Put your quiz aside to be graded later.

   (iii) Spend another minute studying the same list.

   (iv) Quiz yourself again.

   Repeat the process ten times (or until you have learned the entire list). Grade your quizzes, (a correct answer is having a correct number in its correct position in the list). Compile your data in a graph with t, the amount of time spent studying on the horizontal axis, and L, the fraction of the list learned, on the vertical axis.

2. Use this data to approximate your personal k value and compare your data with the predictions of the model. You may use numeric or analytic methods, but be sure to carefully explain your work. Estimate how long it would take you to learn a list of fifty and one hundred three-digit numbers.

3. Repeat the process in part 1 on two of the other lists and compute your k value on these lists. Is your personal k value really constant or does it improve with practice? If k does improve with practice, how would you modify the model to include this?

Your Report: In your report you should give your data in parts 1 and 3 neatly and clearly. Your answer to the questions in parts 2 and 3 should be in the form of short essays. You should include graphs (hand or computer drawn) of your data and solutions of the model *as appropriate*. (Remember that one carefully chosen picture can be worth a thousand words, but a thousand pictures aren't worth anything.)

Table 1.11 Four lists of random three-digit numbers.

	List 1	List 2	List 3	List 4
1	457	167	733	240
2	938	603	297	897
3	363	980	184	935
4	246	326	784	105
5	219	189	277	679
6	538	846	274	011
7	790	040	516	020
8	895	891	051	013
9	073	519	925	144
10	951	306	102	209
11	777	424	826	419
12	300	559	937	191
13	048	911	182	551
14	918	439	951	282
15	524	140	643	587
16	203	155	434	609
17	719	847	921	391
18	518	245	820	364
19	130	752	017	733
20	874	552	389	735

Lab 1.4

Euler's Method and Finding Roots of Polynomials

We have seen in the text how, given an autonomous differential equation

$$\frac{dy}{dt} = f(y),$$

we can use qualitative methods to obtain a description of the solutions. In order to apply the qualitative techniques, we must be able to work with the function $f(y)$. For example, to sketch the phase line for the differential equation, we must be able to find, or at least approximate, the equilibrium points. These correspond to the roots of f, that is, the values of y where $f(y) = 0$.

For some functions f, finding the roots analytically can be difficult (or even impossible). In this case we turn to numerical techniques. We show below how Euler's method can be used to find approximations of the roots of f.

In particular, given a function $f(y)$ that is continuously differentiable, we know that solutions of the differential equation

$$\frac{dy}{dt} = f(y)$$

cannot cross an equilibrium point. We also know that every solution tends monotonically to an equilibrium point or to $\pm\infty$ as t increases and as t decreases. In your report, address the following questions:

1. Use Euler's method to find approximate values of all the real roots of the polynomial

$$f(y) = -y^3 - 4y + 1.$$

Your approximate roots should be correct to at least two decimal places. Your report should include the values of the roots that you discovered and a description of your method. For example, the initial points and step sizes you used in Euler's method.

2. For a function $f(y)$, consider the differential equation

$$\frac{dy}{dt} = -\frac{f(y)}{f'(y)}.$$

What are the equilibrium points of this equation? For what values of y is the differential equation undefined? Are the equilibrium points sinks, sources or nodes?

3. Using the function

$$f(y) = -y^3 - 4y + 1$$

from part 1, form the differential equation

$$\frac{dy}{dt} = -\frac{f(y)}{f'(y)}$$

for this function. Use Euler's method on this equation to find the roots of f. Is this more efficient than the method of part 1, that is, can you get the roots to the same accuracy with with less computation than in part 1? If you make the step size equal to one, does the equation of Euler's method look familiar?

Your report: You should include a description of the results of your numerical computations in parts 1 and 3 along with a description of the details of your method (for example, the initial points and the step sizes used in Euler's method). You may include tables and graphs of approximate solutions if they aid your exposition (but a flood of tables and/or graphs is worse than useless). In part 2, include all steps in the algebra and all details of your arguments. Answer the questions in parts 2 and 3 in the form of short essays.

Lab 1.5   Growth of a Population of Mold

In the text we modeled the U.S. population using both an exponential growth model and a logistic growth model. The assumptions we used to create the models are easy to state. For the exponential model we assumed only that the growth of the population is proportional to the size of the population. For the logistic model we added the assumption that the ratio of the population to the growth rate decreases as the population increases. In this lab we apply these same principles to model the colonization of a piece of bread by mold.

Place a piece of mold free bread in a plastic bag with a small amount of water and leave the bread in a warm place. Each day, record the area of the bread which is covered with mold. (One way to do this is to trace the grid from a piece of graph paper onto a clear piece of plastic. Hold the plastic over the bread and count the number of squares that are mostly covered by mold.)

Warning: It takes at least two weeks to accumulate a reasonable amount of data. Some types of bread seem to be resistant to mold growth and the bread just dries out. If the mold grows, then after about a week the bread will look pretty disgusting. Take precautions to make sure your assignment isn't thrown out.

In your report, address the following questions:

1. Model the growth of mold using an exponential growth model. How accurately does the model fit the data? Be sure to explain carefully how you obtained the value for the growth-rate parameter.

2. Model the growth of mold using a logistic growth model. How accurately does the model fit the data? Be sure to explain carefully how you obtained the value of the growth-rate parameter and carrying capacity.

3. Discuss the models for mold growth population. Were there any surprises? Does it matter that we are measuring the area covered by the mold rather than the total weight of the mold? To what extent would you believe predictions of future mold populations based on these models?

Your report: You should include in your report the details of the type of bread used, where it was kept and how and how often the mold was measured. Your analysis of the models may include qualitative, numerical and analytic arguments and graphs of data and solutions of your models as appropriate. (Remember that well chosen picture can be worth a thousand words, but a thousand pictures aren't worth anything.) **Do not hand in the piece of bread.**

FIRST-ORDER SYSTEMS

2

Few phenomena are completely described by a single number. For example, the size of a population of rabbits can be represented using one number, but to know its rate of change, we should consider other quantities such as the size of predator populations and the availability of food.

In this chapter we begin the study of systems of differential equations — equations that involve more than one dependent variable. As with first-order equations, the techniques for studying these systems fall into three general categories: analytic, qualitative, and numeric. Only very special (and simple) systems of differential equations can be attacked using analytic methods, so in this chapter we focus primarily on qualitative and numerical methods. The main class of systems (linear systems) that can be handled analytically are the subject of Chapter 3.

We continue to study models involving differential equations by discussing models that have more than one dependent variable. Included is a system known as the harmonic oscillator. This particular model has numerous applications in many branches of science such as mechanics, electronics, and physics.

2.1 THE PREDATOR-PREY MODEL

The Predator-Prey System Revisited

We begin our study of systems of differential equations by considering several variations of the predator-prey model discussed in Section 1.1. Recall that $R(t)$ denotes the population (in thousands, or millions, or whatever) of prey present at time t and that $F(t)$ denotes the number of predators. We assume that both $R(t)$ and $F(t)$ are nonnegative. One system of differential equations that might govern the changes in the population of these two species is

$$\frac{dR}{dt} = 2R - 1.2RF$$

$$\frac{dF}{dt} = -F + 0.9RF$$

The $2R$ term in dR/dt represents exponential growth of the prey in the absence of predators, and the $-1.2RF$ term corresponds to the negative effect on the prey of predator-prey interaction. The $-F$ term in dF/dt corresponds to the assumption that the predators die off when there are no prey to eat, and the $0.9RF$ term corresponds to the positive effect on the predators of predator-prey interaction.

The constant coefficients 2, -1.2, -1, and 0.9 depend on the species involved. Similar systems with different constants are considered in the exercises. (We choose these values of the parameters in this example solely for convenience.)*

It is the presence of the RF terms in these equations that makes this system difficult to solve. In general, it is impossible to derive explicit formulas for the solutions. Nevertheless, there are some initial conditions that do yield simple solutions. For instance, suppose that both $R = 0$ and $F = 0$. Then the right-hand sides of both equations vanish ($dR/dt = dF/dt = 0$). As a consequence, the pair of constant functions $R(t) = 0$ and $F(t) = 0$ form a solution to the system. By analogy to first-order equations, we call such a constant solution an **equilibrium solution** to the system. This equilibrium solution makes perfect sense: If both the predator and prey populations vanish, we certainly do not expect the populations to grow at any later time.

We can also look for other values of R and F that correspond to constant solutions. We rewrite the system as

$$\frac{dR}{dt} = (2 - 1.2F)R$$

$$\frac{dF}{dt} = (-1 + 0.9R)F$$

and note that both equations vanish if $R = 1/0.9 \approx 1.11$ and $F = 2/1.2 \approx 1.67$. Thus $R(t) \approx 1.11$, $F(t) \approx 1.67$ is another equilibrium solution. A solution of this type means that, when the prey population is 1.11 and the predators number 1.67, the system is in perfect balance. There are just enough prey to support a constant predator population of 1.67, and similarly, there are neither too many predators (which would cause the population of prey to fall) nor too few (in which case the number of prey would rise). For both species the birth rate equals the death rate exactly, and constant populations are maintained. The system is in *equilibrium*.

---

*For more details on the development, use, and limitations of this system as a model of predator-prey interactions in the wild, we refer the reader to the excellent discussions in J.P. Dempster, *Animal Population Ecology* (New York: Academic Press, 1975) and M. Braun, *Differential Equations and Their Applications* (New York: Springer-Verlag, 1993).

For certain initial conditions, we can use the techniques that we have already developed for first-order equations to study systems. For example, if $R = 0$, the first equation in this system vanishes. Therefore the constant function $R(t) = 0$ satisfies this differential equation no matter what initial condition we choose for F. In this case the second differential equation reduces to

$$\frac{dF}{dt} = -F,$$

which we recognize as simply the exponential decay model for the predator population — an easy differential equation to solve. We see that the population of predators tends exponentially to zero. All of this is perfectly reasonable, since, if there are no prey at some time, then there never will be any prey no matter how many predators there are. Without a food supply, the predators must die out.

In similar fashion, note that the equation for dF/dt vanishes if $F = 0$, and the equation for dR/dt reduces to

$$\frac{dR}{dt} = 2R,$$

another exponential growth model. As we saw in Section 1.1, any nonzero prey population grows without bound under these assumptions. Again these conclusions make sense because there are no predators to control the growth of the prey population. On the other hand, we could make the more realistic assumption here that the prey population obeys a logistic growth law. An example later in this section incorporates this additional assumption.

The Phase Plane for the Predator-Prey System

In general we would like to be able to solve the predator-prey system for all meaningful initial conditions. In other words, if we are given initial values R_0 and F_0 of the two populations, we want functions $R(t)$ and $F(t)$ that give the populations of two species for all possible times t. So far, using special techniques, we have been able to treat those initial conditions that correspond to equilibrium points, $(R_0, F_0) = (0, 0)$, $(R_0, F_0) = (1.11, 1.67)$, and those initial conditions for which either $R_0 = 0$ or $F_0 = 0$. To simplify the study of solutions satisfying other initial conditions, we make use of vector notation.

Suppose for a moment that we have the two functions $R(t)$ and $F(t)$ that satisfy both the system of differential equations and a given initial condition. Then we can pair these functions, $(R(t), F(t))$, and view this pair as a parameterized curve in the RF-plane, the plane whose axes are given by R, the prey population, and F, the predator population. As t evolves, we obtain (different) points of the plane, and the coordinates of these points represent both the predator and prey populations at the associated time t.

More precisely, a curve of the form

$$\mathbf{P}(t) = \begin{pmatrix} R(t) \\ F(t) \end{pmatrix}$$

is called a **solution curve** for the system if both $R(t)$ and $F(t)$ satisfy the corresponding system of differential equations. In particular, the tangent vector to this curve is given by the right-hand side of the system of equations

$$\frac{d\mathbf{P}}{dt} = \begin{pmatrix} \dfrac{dR}{dt} \\ \dfrac{dF}{dt} \end{pmatrix} = \begin{pmatrix} 2R - 1.2RF \\ -F + 0.9RF \end{pmatrix}.$$

We therefore interpret the right-hand side of our system of differential equations as a **vector field**. To each point in the RF-plane, the system of equations assigns the vector

$$\mathbf{V}(R, F) = (2R - 1.2RF, -F + 0.9RF).$$

For example, to the point $(R, F) = (2, 1)$, the system assigns the vector

$$\mathbf{V}(2, 1) = \Big(2(2) - 1.2(2)(1), -(1) + 0.9(2)(1)\Big) = (1.6, 0.8).$$

Similarly, the vector $\mathbf{V}(1, 1) = (0.8, -0.1)$ is assigned to the point $(R, F) = (1, 1)$. Note that the zero vector is assigned to $(R, F) = (1.11, 1.67)$ as well as to the other equilibrium point, $(R, F) = (0, 0)$. That is, $\mathbf{V}(1.11, 1.67) = (0, 0)$ and $\mathbf{V}(0, 0) = (0, 0)$. Note that we sometimes write vectors vertically and, at other times, horizontally. The meaning will always be clear from the context.

Using this notation, the system of differential equations can be rewritten in vector notation as

$$\frac{d\mathbf{P}}{dt} = \mathbf{V}(R, F) = \mathbf{V}(\mathbf{P}).$$

We visualize this vector field as a "field" of arrows, one based at each point in the RF-plane. A solution curve for the system is a curve that threads its way through this field, with the tangent vector to the curve given by the vector field at each point along the **solution curve**. In Figure 2.1 we illustrate a number of the vectors in the vector field associated with this system. We also sketch one solution curve. Note that the solution curve is always tangent to the vector field at any point on the curve.

It is often helpful to view a solution curve to a system of differential equations not merely as a set of points in the plane but, rather in more dynamic fashion, as a

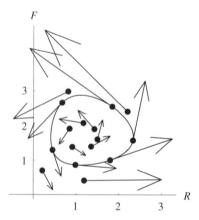

Figure 2.1
Selected vectors in the vector field for the predator-prey model, along with one solution curve in the RF-plane.

point following the curve whose speed at any given time is given by the magnitude of the tangent vector at that point and whose direction is indicated by the direction of the vector. We imagine the point moving quickly when the tangent vectors are long and moving more slowly when the vectors are short. This dynamic view of solution curves is easy to see using technology but is more difficult to capture on paper.

Equilibrium solutions are special. Since they are constant functions, the solution curves are represented by points. As in the case of first-order equations, we view such a solution curve as a point at rest at the equilibrium point. Its velocity is zero; the point never moves.

In Figure 2.2 we display several solution curves for the predator-prey system. We cannot derive explicit formulas for the solution curves marked B and C, but we are able to approximate them using numerical approximation schemes similar to Euler's method. More details about how these curves are plotted are described in Section 2.5. Note that we often place an arrowhead on a solution curve to indicate the evolution of the system as t increases.

In Figure 2.2 the point marked A is the equilibrium point at $R = 1.11$, $F = 1.67$. The curve marked B is a curve in the plane of the form $\mathbf{B}(t) = (R(t), F(t))$, where $R(0) = 1.0$, $F(0) = 0.5$. Thus $\mathbf{B}(t)$ is the solution curve to this system of differential equations that goes through the point $(R, F) = (1.0, 0.5)$ at time $t = 0$. It appears that this solution curve is a closed curve. That is, assuming we trust what the computer tells us, the solution curve seems to return to its initial value and then retrace itself as $t \to \infty$. We can therefore conclude that both the prey and the predator populations are cyclic.

Once we know the shape of the solution curve, we can surmise what the model predicts will happen to the predator and prey populations. At the initial value marked B in Figure 2.2, there are too few predators to maintain the balance between the populations. Thus the prey population increases. As the prey become abundant, the predator population also begins to increase. Eventually the number of predators becomes so large that the prey have a difficult time avoiding the predators, and so the prey population begins to fall. This in turn affects the predator population; some time later, the number of predators begins to decline. This means that more prey will survive, so eventually their population begins to rise. This in turn benefits the predators, so eventually their population begins to rise. In fact it is at just this moment that both populations have returned to their initial values, so the entire process begins again.

In Figure 2.2 the curve marked C is the solution curve through $R = 3$ and $F = 4$.

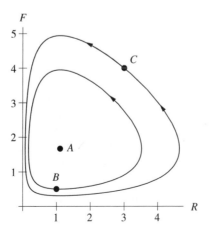

Figure 2.2

Several solution curves for the predator-prey system. The point A is an equilibrium point. The points B and C are initial values for the periodic solutions $B(t)$ and $C(t)$.

The model predicts that at first the predator population will increase and the prey population will decrease. Then there is a crash in the populations of both predator and prey. Thereafter the populations rise and fall cyclically, just as in the case of the previous solution curve.

All of this is derived from the solution curve. As a point travels around the solution curve B, both $R(t)$ and $F(t)$ increase initially. The rabbit population $R(t)$ reaches a maximum and begins to decrease, while $F(t)$ continues to increase. Later $F(t)$ also reaches a maximum and then begins to decline. Eventually both $R(t)$ and $F(t)$ return to their original values, and the solution curve retraces itself. Solution curves with this property are called **periodic** solutions.

This type of geometric representation of the solutions of a system of differential equations is called the **phase portrait** or **phase plane**. In the phase plane we indicate the location of all equilibrium points as well as a representative sample of other solution curves together with their behavior as $t \to \pm\infty$. The phase plane for the predator-prey system is shown in Figure 2.3. Note that the computer suggests that all solution curves (that do not lie on the R- or F-axes) for this system are periodic. Our computational techniques cannot tell us if the actual solutions are periodic or simply close to being periodic, but we will develop mathematical techniques that confirm our suspicion that all solution curves are actually periodic in this case.

Another important image associated with a system of differential equations is the **direction field**. This picture is closely related to the vector field. However, only the directions of the vectors are plotted, not the magnitudes. In most examples the vectors in the vector field tend to overlap, and thus the paths of the solution curves are hard to visualize. By shortening the vectors, we get a better sense of how the solution curves behave. Figure 2.4 depicts the direction field for the predator-prey equation.

A solution curve, plotted in the RF-plane, provides a good qualitative indication of what happens to R and F as t evolves. However, with this representation, it is difficult to see the dependence of R and F on the independent variable t. The computational techniques that provide sketches of the solution curve also provide a very different type of picture, the $R(t)$- **and** $F(t)$-**graphs**. These images are graphs of $R(t)$ or $F(t)$ plotted versus t for a specific initial condition $R(0) = R_0$, $F(0) = F_0$. In

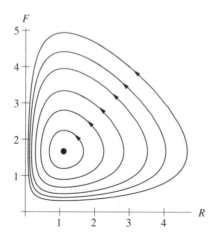

Figure 2.3
The phase plane for the predator-prey system.

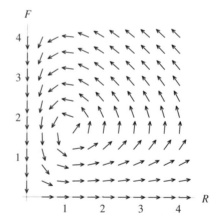

Figure 2.4
The direction field for the predator-prey equation.

Figures 2.5 and 2.6, we display the $R(t)$- and $F(t)$-graphs for the initial condition $(R(0), F(0)) = (1.0, 0.5)$. Recall that the corresponding solution curve is $\mathbf{B}(t)$, as shown in Figure 2.2. Note that both $R(t)$ and $F(t)$ appear to be periodic functions of t.

Note that in these graphs we recover the same information about $R(t)$ and $F(t)$ as we derived from the solution curve. We often find it useful to plot both graphs using the same set of axes, and Figure 2.7 illustrates how we combine the graphs in Figures 2.5 and 2.6.

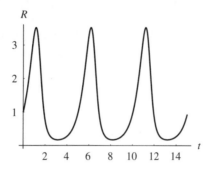

Figure 2.5
The $R(t)$-graph if $R(0) = 1.0$ and $F(0) = 0.5$.

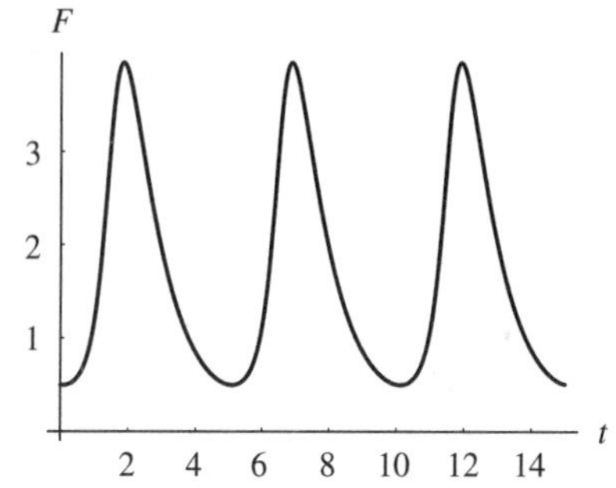

Figure 2.6
The $F(t)$-graph if $R(0) = 1.0$ and $F(0) = 0.5$.

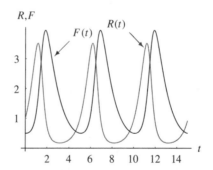

Figure 2.7
The $R(t)$- and $F(t)$-graphs given by the initial condition $R(0) = 1.0$ and $F(0) = 0.5$.

A Modified Predator-Prey Model

We now consider a modification of the predator-prey model in which we assume that, in the absence of predators, the prey population obeys a logistic rather than an exponential growth model. One such model for this situation is provided by the system of differential equations given by

$$\frac{dR}{dt} = 2R\left(1 - \frac{R}{2}\right) - 1.2RF$$
$$\frac{dF}{dt} = -F + 0.9RF.$$

In this system, when the predators are not present (that is, $F = 0$), the prey population obeys a logistic growth model with carrying capacity 2. The behavior of solution curves and therefore the predictions made by this model are quite different from those made for the model presented earlier in this section.

First, let's find the equilibrium solutions for this system. Recall that these solutions occur at points where the right-hand sides of both of the differential equations vanish. As before, $(R, F) = (0, 0)$ is one equilibrium solution. There are two other equilibria given by $(R, F) = (2, 0)$ and $(R, F) \approx (1.11, 0.74)$.

As before, if there are no prey present, the predator population declines exponentially. In other words, if $R = 0$, then $dR/dt = 0$, so $R(t) = 0$. Then the equation for dF/dt reduces to the familiar exponential decay model

$$\frac{dF}{dt} = -F.$$

In the absence of predators, the situation is somewhat different. If $F = 0$, we have $dF/dt = 0$ and the equation for R simplifies to the familiar logistic model

$$\frac{dR}{dt} = 2R\left(1 - \frac{R}{2}\right).$$

From this equation we see that the growth coefficient for low populations of prey is 2 and the carrying capacity is 2. Thus, if $F = 0$, we expect any nonzero initial population of prey eventually to approach 2.

When both R and F are nonzero, the populations change in a very different manner. In Figure 2.8 we plot three solution curves for $t \geq 0$. Note that, in all cases, the solutions tend to the equilibrium point $(R, F) = (1.11, 0.74)$, which is marked A. The solution curve through B originates at a point for which there is an overabundance of both predators and prey. From the shape of the solution curve with initial point B, we can make conclusions about what the model predicts will happen to the populations. Starting at the initial point B, the predator population initially rises. This is reasonable because the prey population is large, so it is easy for the predators to find food. After a sufficient number of prey have been killed, however, the predator population begins to decline and eventually approaches equilibrium. On the other hand, after an initial decline, the prey population eventually begins to recover, and this population also tends to stabilize at the equilibrium value.

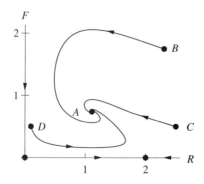

Figure 2.8
The equilibria and three solution curves
for the logistic predator-prey model.

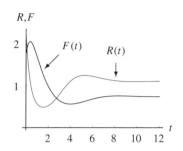

Figure 2.9
The $R(t)$- and $F(t)$-graphs for the
solution curve B in Figure 2.8.

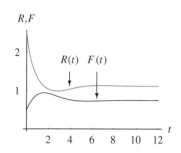

Figure 2.10
The $R(t)$- and $F(t)$-graphs for the
solution curve C in Figure 2.8.

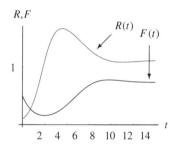

Figure 2.11
The $R(t)$- and $F(t)$-graphs for the
solution curve D in Figure 2.8.

Similar interpretations may be assigned to the other solution curves pictured. In Figures 2.9–2.11, we sketch the corresponding $R(t)$- and $F(t)$-graphs for each of these three solution curves. Note that the graphs of both $F(t)$ and $R(t)$ tend to the equilibrium values $R = 1.11$ and $F = 0.74$. We can predict this from the solution curves in the phase plane (see Figure 2.8).

The phase plane for this model, shown in Figure 2.12, indicates that all solutions through initial values that do not lie on one of the axes tend to the equilibrium solution $(R, F) = (1.11, 0.74)$ as time increases. Note how the solution curves are tangent to the direction field at each point.

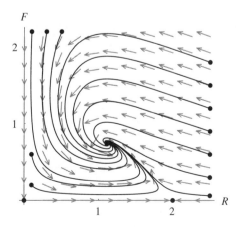

Figure 2.12
The phase plane and the direction field for
the logistic predator-prey model

$$\frac{dR}{dt} = 2R\left(1 - \frac{R}{2}\right) - 1.2RF$$

$$\frac{dF}{dt} = -F + 0.9RF.$$

Exercises for Section 2.1

Exercises 1–6 refer to the following systems of equations:

(i)
$$\frac{dx}{dt} = 10x \left(1 - \frac{x}{10}\right) - 20xy$$
$$\frac{dy}{dt} = -5y + \frac{xy}{20}$$

(ii)
$$\frac{dx}{dt} = 0.3x - \frac{xy}{100}$$
$$\frac{dy}{dt} = 15y \left(1 - \frac{y}{15}\right) + 25xy.$$

1. In one of these systems, the prey are very large animals and the predators are very small animals, such as elephants and mosquitoes. Thus it takes many predators to eat one prey, but each prey eaten is a tremendous benefit for the predator population. The other system has very large predators and very small prey. Determine which system is which and provide a justification for your answer.

2. Find all equilibrium points for the two systems. Explain the significance of these points in terms of the predator and prey populations.

3. Suppose that the predators are extinct at time $t_0 = 0$. For each system, verify that the predators remain extinct for all time.

4. For each system, describe the behavior of the prey population if the predators are extinct. (Sketch the phase line for the prey population assuming that the predators are extinct, and sketch the graphs of the prey population as a function of time for several solutions. Then interpret these graphs for the prey population.)

5. For each system, suppose that the prey are extinct at time $t_0 = 0$. Verify that the prey remain extinct for all time.

6. For each system, describe the behavior of the predator population if the prey are extinct. (Sketch the phase line for the predator population assuming that the prey are extinct, and sketch the graphs of the predator population as a function of time for several solutions. Then interpret these graphs for the predator population.)

7. Consider the system of predator-prey equations

$$\frac{dR}{dt} = 2 \left(1 - \frac{R}{3}\right) R - RF$$
$$\frac{dF}{dt} = -2F + 4RF.$$

The figure below shows a computer-generated plot of a solution curve for this system in the RF-plane. What can you say about the fate of the prey (R) and predator (F) populations based on this image?

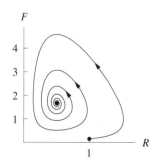

8. Consider the following predator-prey system and solution curves in the phase plane:

$$\frac{dR}{dt} = 2R\left(1 - \frac{R}{2.5}\right) - 1.5RF$$

$$\frac{dF}{dt} = -F + 0.8RF.$$

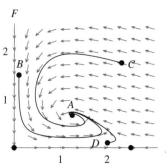

(a) Sketch the $R(t)$- and $F(t)$-graphs for the solutions with initial points A, B, C, and D.

(b) Interpret each of these solution curves in terms of the behavior of the populations over time.

Exercises 9–14 refer to the predator-prey and the modified predator-prey systems discussed in the text (repeated here for convenience):

(i)
$$\frac{dR}{dt} = 2R - 1.2RF$$
$$\frac{dF}{dt} = -F + 0.9RF$$

(ii)
$$\frac{dR}{dt} = 2R\left(1 - \frac{R}{2}\right) - 1.2RF$$
$$\frac{dF}{dt} = -F + 0.9RF.$$

9. How would you modify these systems to include the effect of hunting of the prey at a rate of α units of prey per unit of time?

10. How would you modify these systems to include the effect of hunting of the predators at a rate proportional to the number of predators?

11. Suppose the predators discover a second, unlimited source of food, but they still prefer to eat prey when they can catch them. How would you modify these systems to include this assumption?

12. Suppose the predators found a second food source that is limited in supply. How would you modify these systems to include this fact?

13. Suppose predators migrate to an area if there are five times as many prey as predators in that area (that is, if $5R > F$), and they move away if there are fewer than five times as many prey as predators. How would you modify these systems to take this into account?

14. Suppose prey move out of an area at a rate proportional to the number of predators in the area. How would you modify these systems to take this into account?

15. Consider the two systems of differential equations

(i)
$$\frac{dx}{dt} = 0.3x - 0.1xy$$
$$\frac{dy}{dt} = -0.1y + 2xy$$

(ii)
$$\frac{dx}{dt} = 0.3x - 3xy$$
$$\frac{dy}{dt} = -2y + 0.1xy.$$

One of these systems refers to a predator-prey system with very inefficient predators — predators who seldom catch their prey but who can live for a long time on a single prey (for example, boa constrictors). The other system refers to a very active and efficient predator that requires many prey to stay healthy (such as a small cat). The prey in each case is the same. Identify which system is which and justify your answer.

16. Consider the system of predator-prey equations

$$\frac{dR}{dt} = 2\left(1 - \frac{R}{3}\right)R - RF$$
$$\frac{dF}{dt} = -16F + 4RF.$$

The figure below shows a computer-generated plot of a solution curve for this system in the RF-plane. What can you say about the fate of the rabbit and fox populations based on this image?

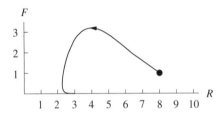

17. Pesticides that kill all insect species are not only bad for the environment, they can also be inefficient at controlling pest species. Suppose a pest insect species in a particular field has population $R(t)$ at time t, and suppose its primary predator is another insect species with population $F(t)$ at time t. Suppose the populations of these species are accurately modeled by the system

$$\frac{dR}{dt} = 2R - 1.2RF$$
$$\frac{dF}{dt} = -F + 0.9RF$$

studied in this section. Finally, suppose that at time $t = 0$ a pesticide is applied to the field that reduces both the pest and predator populations to very small but nonzero numbers.

(a) Using Figures 2.3 and 2.4, predict what will happen as t increases to the population of the pest species.

(b) Write a short essay, in nontechnical language, warning of the possibility of the "paradoxical" effect that pesticide application can have on pest populations.

18. Some predator species seldom capture healthy adult prey, eating only injured or weak prey. Because weak prey consume resources but are not as successful at reproduction, the harsh reality is that their removal from the population increases the population of prey. Discuss how you would modify a predator-prey system to model this sort of interaction.

Exercises 19–24 refer to a situation in which models similar to the predator-prey population models arise. Suppose A and B represent two substances that can combine to form a new substance C (chemists would write A + B → C). Suppose we have a container with a solution containing low concentrations of substances A and B, and A and B molecules react only when they happen to come close to each other. If $a(t)$ and $b(t)$ represent the amount of A and B in the solution, respectively, then the chance that a molecule of A is close to a molecule of B at time t is proportional to the product $a(t) \cdot b(t)$. Hence the rate of reaction of A and B to form C is proportional to ab. Suppose C precipitates out of the solution as soon as it is formed, and the solution is always kept well mixed.

19. Write a system of differential equations that models the evolution of $a(t)$ and $b(t)$. Be sure to identify and describe any parameters you introduce.

20. Describe an experiment you could perform to determine an approximate value for the parameter(s) in the system you developed in Exercise 19. Include the calculations you would perform using the data from your experiment to determine the parameter(s).

21. Suppose substances A and B are added to the solution at constant (perhaps unequal) rates. How would you modify your system to include this assumption?

22. Suppose A and B are being added to the solution at constant (perhaps unequal) rates, and, in addition to the A + B → C reaction, a reaction A + A → D also can occur when two molecules of A are close and substance D precipitates out of the solution. How would you modify your system of equations to include these assumptions?

23. Suppose A and B are being added to the solutions at constant (perhaps unequal) rates, and, in addition to the A + B → C reaction, a reaction B + B → A can also occur when two molecules of B are close. How would you modify your system of equations to include these assumptions?

24. Suppose A and B are being added to the solution at constant (perhaps unequal) rates, and, in addition to the A + B → C reaction, a reaction A + 2B → D can occur when two B and one A molecules are close. Suppose substance D precipitates out of the solution. How would you modify your system of equations to include these assumptions?

25. Suppose Tom and Ray are driving down a highway. Let $x(t)$ denote Tom's speed at time t, and $y(t)$ denote Ray's speed at time t. Suppose Tom wants to go at least as fast but not faster than Ray, but Tom also doesn't want to go faster than 60 miles per hour. Ray wants to go at least as fast as Tom, but because of previous traffic tickets, Ray is very nervous about going faster than 55 miles per hour.

Consider the following model systems for the speeds:

(i)
$$\frac{dx}{dt} = (y - x)(60 - x)$$
$$\frac{dy}{dt} = (x - y)(55 - y)$$

(ii)
$$\frac{dx}{dt} = (y - x) + (60 - x)$$
$$\frac{dy}{dt} = (x - y) + (55 - y).$$

(a) For each of the systems above, what are the equilibrium points? If (x, y) is an equilibrium point, what does that mean about the speeds of the cars?

(b) Which system is a better model for the speeds from the description above? Does either system completely capture the behavior described for Tom and Ray? Justify your answer.

2.2 SYSTEMS OF DIFFERENTIAL EQUATIONS

As we saw in the last section, a first-order system of differential equations involves two or more quantities (dependent variables) that are functions of the same independent variable. In the predator-prey example, the dependent variables are the populations $R(t)$ and $F(t)$, which are both functions of time t. A system of differential equations specifies the rates of change of the dependent variables in terms of the values of the independent variable and all of the dependent variables. In the predator-prey system, both dR/dt and dF/dt depend on both R and F.

Adjectives for Systems of Differential Equations

Just as for first-order equations, the type of techniques we can use to study a system of differential equations depends on the form of the "right-hand sides." The following terms are used to describe different types of first-order systems.

Dimension

The **dimension** of a first-order system of differential equations is the number of dependent variables. For example, the predator-prey system of Section 2.1,

$$\frac{dR}{dt} = 2R - 1.2RF$$

$$\frac{dF}{dt} = -F + 0.9RF$$

is a two-dimensional system because there are two dependent variables, R and F. The system

$$\frac{dx}{dt} = \alpha(x - y)$$

$$\frac{dy}{dt} = rx - y - xz$$

$$\frac{dz}{dt} = \beta z + xy$$

has three dependent variables x, y, and z, so it is three dimensional. The independent variable is t, and α, β, and r are parameters. This system is called the **Lorenz system** and was first studied by the meteorologist E. N. Lorenz in the 1960s. It is one of the simplest systems that exhibits what is called "chaotic" behavior. We discuss this system in detail in Section 2.7.

Linear and nonlinear systems

A system is called **linear** if each term on the right-hand side contains at most one x or y (or whatever the dependent variables are) raised to the first power. That is, the right-hand sides of the system are linear functions of the dependent variables. For example, the system

$$\frac{dx}{dt} = 3x + 2\sin(t^2 + 3)y + 4t^5$$

$$\frac{dy}{dt} = 2x - y + \cos t$$

is a linear system because each term contains at most a single factor of x or y. Note that we say that this system is linear even though the dependence on the independent variable t is not linear.

The preceding predator-prey and Lorenz systems are both nonlinear because they contain products of the dependent variables. The system

$$\frac{d\theta}{dt} = \omega$$

$$\frac{d\omega}{dt} = -\sin\theta$$

is also nonlinear because the term $-\sin\theta$ is a nonlinear function of θ. This system models the motion of a frictionless pendulum where θ is the angle of the pendulum arm with respect to the vertical, and ω is the "angular velocity." We will study this system carefully in Chapter 4.

Autonomous and nonautonomous systems

A system is called **autonomous** if the independent variable t does not appear in the expressions for the rate of change of the dependent variables. For example, the system

$$\frac{dy}{dt} = v$$

$$\frac{dv}{dt} = -6y - 5v$$

is autonomous, whereas the system

$$\frac{dy}{dt} = v$$

$$\frac{dv}{dt} = -6y - 5v + \cos 3t$$

is nonautonomous due to the $\cos 3t$ term.

As we saw in Chapter 1, the type of tools we can use depends strongly on whether an equation is autonomous or nonautonomous. For autonomous equations, we can quickly find and classify the equilibrium points and sketch the phase line. In the nonautonomous case, we must use the entire slope field.

For systems of differential equations, it turns out that there are many qualitative and analytical techniques that work in the autonomous case but not for nonautonomous equations. To develop intuition for the qualitative behavior of systems, we will restrict our attention to autonomous systems until Chapter 5; but be warned that not all systems are autonomous.

Vector Notation

When dealing with systems of differential equations, we can use vectors to simplify the notation. This notation also leads to useful geometric representations for the solutions. Consider an autonomous system with two dependent variables of the form

$$\frac{dx}{dt} = f(x, y)$$
$$\frac{dy}{dt} = g(x, y)$$

We introduce the vector

$$\mathbf{Y}(t) = (x(t), y(t))$$

and the function

$$\mathbf{F}(\mathbf{Y}) = \mathbf{F}(x, y) = (f(x, y), g(x, y)).$$

Sometimes we write vectors as column vectors; other times, we write them as row vectors. The meaning will always be clear from the context. However, we are careful to write vectors in boldface to help distinguish them from scalars. Note that $\mathbf{F}$ is a vector-valued function, or **vector field**, that accepts as input the vector (x, y) and returns as output the vector $(f(x, y), g(x, y))$. With this notation, the system of two equations may be written in the compact form

$$\frac{d\mathbf{Y}}{dt} = \mathbf{F}(\mathbf{Y}).$$

Here $d\mathbf{Y}/dt$ means the derivative of the vector-valued function $\mathbf{Y}(t) = (x(t), y(t))$. Recall that $d\mathbf{Y}/dt = (dx/dt, dy/dt)$. Thus the vector equation $d\mathbf{Y}/dt = \mathbf{F}(\mathbf{Y})$ is exactly the same as our original system of scalar differential equations.

One advantage of the vector notation is that it extends easily to systems with an arbitrary number of dependent variables. For example, the three-dimensional system

$$\frac{dx}{dt} = f(x, y, z)$$
$$\frac{dy}{dt} = g(x, y, z)$$
$$\frac{dz}{dt} = h(x, y, z)$$

can be written as

$$\frac{d\mathbf{Y}}{dt} = \mathbf{F}(\mathbf{Y})$$

where $\mathbf{Y}(t) = \big(x(t), y(t), z(t)\big)$ and $\mathbf{F}(\mathbf{Y}) = \big(f(x, y, z), g(x, y, z), h(x, y, z)\big)$.

Solutions of a System

To solve a two-dimensional autonomous system of differential equations

$$\frac{d\mathbf{Y}}{dt} = \mathbf{F}(\mathbf{Y}),$$

we must find functions $\mathbf{Y}(t) = (x(t), y(t))$ that satisfy this vector equation. Geometrically this equation says that the tangent vectors for the vector-valued function $\mathbf{Y}(t)$ always agree with the vectors specified by $\mathbf{F}(\mathbf{Y})$. An initial condition for the system at time t_0 is a point $\mathbf{Y}_0 = (x_0, y_0)$, and we say that $\mathbf{Y}(t)$ satisfies the initial condition if $\mathbf{Y}(t_0) = \mathbf{Y}_0$ — both $x(t_0) = x_0$ and $y(t_0) = y_0$. The following definition is a more formal statement of this concept.

Definition: Suppose $d\mathbf{Y}/dt = \mathbf{F}(\mathbf{Y})$ is an autonomous system of differential equations. Let t_0 be an initial time and let $\mathbf{Y}_0$ be an initial value. Then a **solution** of the system through $\mathbf{Y}_0$ at time t_0 is a vector-valued function $\mathbf{Y}(t)$ such that

$$\mathbf{Y}(t_0) = \mathbf{Y}_0 \quad \text{and} \quad \frac{d\mathbf{Y}}{dt} = \mathbf{F}(\mathbf{Y}(t)) \quad \text{for all } t.$$

Whenever we find such a function $\mathbf{Y}(t)$, we say that we have solved the **initial-value problem**

$$\frac{d\mathbf{Y}}{dt} = \mathbf{F}(\mathbf{Y}), \quad \mathbf{Y}(t_0) = \mathbf{Y}_0.$$

Finding formulas for solutions of an initial-value problem for a system is, in general, even more daunting a task than it is for equations with one dependent variable. In fact, the systems of equations for which we can find formulas for solutions are very few and far between. (We will spend Chapter 3 discussing the most important class of such systems.)

However, given an initial-value problem, checking whether a given vector-valued function is a solution is simply a matter of checking that the initial condition is satisfied and that the velocity vector of the curve equals the vector field for all t. That is, just as with one dependent variable, we substitute the candidate solution into both sides of all equations in the system and see whether they are equal for all t.

Checking solutions

Consider the system

$$\frac{dx}{dt} = -x + y$$
$$\frac{dy}{dt} = -3x - 5y.$$

We can rewrite this system in vector notation using the standard $\mathbf{Y}(t) = (x(t), y(t))$ and the vector function $\mathbf{F}(x, y) = (-x + y, -3x - 5y)$. Using this function $\mathbf{F}$, the system becomes

$$\frac{d\mathbf{Y}}{dt} = \mathbf{F}(\mathbf{Y}).$$

We claim that the vector-valued function

$$\mathbf{Y}(t) = (x(t), y(t)) = (e^{-4t} - 3e^{-2t}, -3e^{-4t} + 3e^{-2t})$$

is a solution to this system.

To verify this claim, we must compute both derivatives

$$\frac{dx}{dt} = \frac{d(e^{-4t} - 3e^{-2t})}{dt} = -4e^{-4t} + 6e^{-2t}$$
$$\frac{dy}{dt} = \frac{d(-3e^{-4t} + 3e^{-2t})}{dt} = 12e^{-4t} - 6e^{-2t}.$$

(We must check both $dx/dt = -x + y$ and $dy/dt = -3x - 5y$.) Using $x(t) = e^{-4t} - 3e^{-2t}$ and $y(t) = -3e^{-4t} + 3e^{-2t}$ from the formula for $\mathbf{Y}(t)$, the right-hand sides $-x + y$ and $-3x - 5y$ are

$$-x + y = -(e^{-4t} - 3e^{-2t}) + (-3e^{-4t} + 3e^{-2t}) = -4e^{-4t} + 6e^{-2t}$$

and

$$-3x - 5y = -3(e^{-4t} - 3e^{-2t}) - 5(-3e^{-4t} + 3e^{-2t}) = 12e^{-4t} - 6e^{-2t}.$$

Thus, for both equations in the system, the left-hand side equals the right-hand side for all t. Hence $\mathbf{Y}(t) = (x(t), y(t)) = (e^{-4t} - 3e^{-2t}, -3e^{-4t} + 3e^{-2t})$ is a solution.

Note that $\mathbf{Y}(0) = (-2, 0)$. Consequently, we have checked that $\mathbf{Y}(t)$ is a solution of the initial-value problem

$$\frac{d\mathbf{Y}}{dt} = \mathbf{F}(\mathbf{Y}), \quad \mathbf{Y}(0) = (-2, 0).$$

As a second example, consider the system

$$\frac{dx}{dt} = 2x - y$$
$$\frac{dy}{dt} = x - 2y,$$

and suppose we want to see if the function $\mathbf{Y}(t) = (e^{-t}, 3e^{-t})$ is a solution that satisfies the initial condition $\mathbf{Y}(0) = (1, 3)$.

It is easy to check that this function satisfies the initial condition; we simply evaluate it at $t = 0$. We have $\mathbf{Y}(0) = (e^{-0}, 3e^{-0}) = (1, 3)$. Next we check to see if the first equation of the system is satisfied. We have

$$\frac{dx}{dt} = \frac{d(e^{-t})}{dt} = -e^{-t},$$

and the right-hand side of the dx/dt equation gives

$$2x - y = 2e^{-t} - 3e^{-t} = -e^{-t}.$$

Thus the first equation holds for all t. Finally, we must check the second equation in the system. We have

$$\frac{dy}{dt} = \frac{d(3e^{-t})}{dt} = -3e^{-t},$$

and

$$x - 2y = e^{-t} - 2(3e^{-t}) = -5e^{-t}.$$

Since the second equation is not satisfied, the function $\mathbf{Y}(t) = (e^{-t}, 3e^{-t})$ is not a solution of the initial-value problem.

The moral of these two examples is a very important one, and one that is often overlooked. Given a formula for a function $\mathbf{Y}(t)$, we can always check to see if that function satisfies the system simply by direct computation. This type of computation is certainly not the most exciting part of the subject, but it is straightforward. We can immediately determine if a given vector-valued function is a solution.

Geometry of Autonomous Systems

Much of the language and many of the concepts from the predator-prey models in Section 2.1 are used for all systems. Consider the autonomous system

$$\frac{d\mathbf{Y}}{dt} = \mathbf{F}(\mathbf{Y}).$$

A solution is a vector-valued function $\mathbf{Y}(t) = (x(t), y(t))$, and if we view such a function as the parameterization of a curve in the plane, then $d\mathbf{Y}/dt$ can be interpreted as a tangent vector to the curve. In fact, if we think of $\mathbf{Y}(t)$ as describing the motion of a particle throughout the plane, then $d\mathbf{Y}/dt$ is the velocity vector of that particle.

With this interpretation, the left-hand side of the system $d\mathbf{Y}/dt = \mathbf{F}(\mathbf{Y})$ is the velocity vector of a solution curve, and the right-hand side is a function that specifies that velocity vector in terms of the particle's location in the plane. This observation is crucial to much of what we do with systems. A solution is a curve that winds through the plane according to the velocities specified by the function $\mathbf{F}(\mathbf{Y})$ (see Figures 2.13 and 2.14).

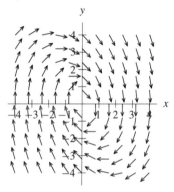

Figure 2.13
An example of a direction field.

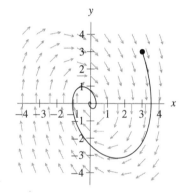

Figure 2.14
The same direction field and a solution with initial condition at $(3, 3)$.

Metaphor of the parking lot

To help visualize autonomous systems of differential equations from this point of view, imagine an infinite, perfectly flat parking lot. At each point in the parking lot, an arrow is painted on the pavement. These arrows come from the function $\mathbf{F}(\mathbf{Y})$. You are in a car at a point $\mathbf{Y}_0$, the initial point, and your instructions are to look out your window at the ground and drive so that your velocity vector always agrees with the arrow on the ground. (Imagine professional drivers in a closed parking lot.) Your direction is given by the direction of the arrow and your speed is given by its length. As you move, the arrow outside your window changes, so you must adjust the speed and direction of the car accordingly. The path you follow is the solution $\mathbf{Y}(t)$ to the initial-value problem

$$\frac{d\mathbf{Y}}{dt} = \mathbf{F}(\mathbf{Y}), \quad \mathbf{Y}(0) = \mathbf{Y}_0.$$

Vector Fields and Direction Fields

Armed with this geometric interpretation, we see that the solutions of the system are determined by the vectors given by the right-hand side $\mathbf{F}(\mathbf{Y})$ of the system (the vectors painted on the pavement). Thus we can view the system as a vector field in exactly the same way we did in Section 2.1. To each point $\mathbf{Y} = (x, y)$ in the plane, we associate the vector $\mathbf{F(Y)} = \mathbf{F}(x, y)$. Solutions are curves whose velocity vectors are given by the vector field at every point. Thus given a picture of the vector field $\mathbf{F(Y)}$, we can sketch solutions by drawing curves whose tangent vectors agree with the vector field.

As an example, consider the autonomous system

$$\frac{dx}{dt} = \frac{x}{2}$$
$$\frac{dy}{dt} = -\frac{y}{3}.$$

The associated vector field is given by the vector function $\mathbf{F}(x, y) = (x/2, -y/3)$. Selected vectors from this vector field are shown in Figure 2.15. The vector function $\mathbf{F}(x, y)$ assigns a vector based at the point (x, y). Note that $\mathbf{F}$ associates the vector $(1, -2)$ to the point $(x, y) = (2, 6)$, the vector $(3, 0)$ to the point $(x, y) = (6, 0)$ and so on.

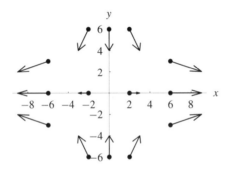

Figure 2.15
Vectors from the vector field $\mathbf{F}(x, y) = (x/2, -y/3)$.

If we try to draw a figure that includes several of the vectors in the vector field, we get something that looks like Figure 2.16. To avoid confusion of overlapping vectors, we scale each nonzero vector in the vector field so that all of the vectors in the picture have the same length. This scaled picture is called a **direction field**. Figure 2.17 illustrates the direction field associated with the vector field $\mathbf{F}(x, y) = (x/2, -y/3)$. Compare Figures 2.16 and 2.17.

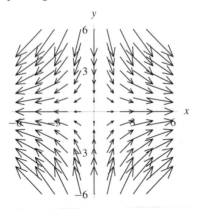

Figure 2.16
The vector field associated with $\mathbf{F}(x, y) = (x/2, -y/3)$.

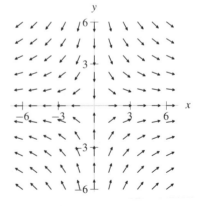

Figure 2.17
The direction field associated with $\mathbf{F}(x, y) = (x/2, -y/3)$.

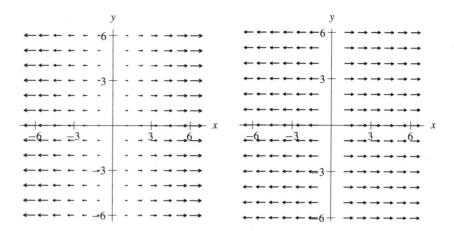

Figure 2.18
Scaled vector field for $\mathbf{F}(x, y) = (x, 0)$.

Figure 2.19
Direction field for the vector field
$\mathbf{F}(x, y) = (x, 0)$.

Simple, but important examples

The vector field

$$\mathbf{F}(x, y) = (x, 0)$$

always points in the horizontal direction if $x \neq 0$. It points to the right if $x > 0$, to the left if $x < 0$, and vanishes if $x = 0$ (see Figures 2.18 and 2.19).

The vector field

$$\mathbf{G}(x, y) = (x, y)$$

always points away from the origin, whereas the vector field

$$\mathbf{H}(x, y) = (-x, -y)$$

always points toward the origin (see Figures 2.20 and 2.21).

The direction field for $\mathbf{K}(x, y) = (2y, -x)$ is shown in Figure 2.22.

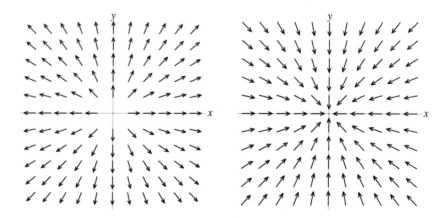

Figure 2.20
The direction field for $\mathbf{G}(x, y) = (x, y)$.

Figure 2.21
The direction field for
$\mathbf{H}(x, y) = (-x, -y)$.

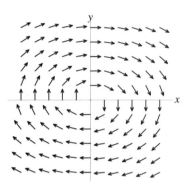

Figure 2.22
The direction field for

$$\mathbf{K}(x, y) = (2y, -x).$$

Unlike our other examples, this vector field "spirals" about the origin. When we study linear systems in Chapter 3, we will see that this phenomenon is typical of linear systems whose solutions oscillate over time.

Equilibrium Solutions

Viewing systems in terms of vector fields enables us to develop a qualitative analysis of systems that is very similar to the theory of phase lines from Section 1.6. In particular, we start our analysis by finding those solutions that are constant — the equilibrium solutions.

Definition: Suppose $d\mathbf{Y}/dt = \mathbf{F}(\mathbf{Y})$ is an autonomous system of differential equations. The vector $\mathbf{Y}_0$ is an **equilibrium point** for the system if $\mathbf{F}(\mathbf{Y}_0) = 0$. The constant function $\mathbf{Y}(t) = \mathbf{Y}_0$ is an **equilibrium solution**.

Equilibrium points are simply points at which the right-hand side of the differential equation vanishes. If $\mathbf{Y}_0$ is an equilibrium point, then the constant function $\mathbf{Y}(t) = \mathbf{Y}_0$ for all t is a solution of the system. To verify this claim, note that the constant function has $d\mathbf{Y}/dt = (0, 0)$ for all t, and $\mathbf{F}(\mathbf{Y}(t)) = \mathbf{F}(\mathbf{Y}_0) = (0, 0)$ at an equilibrium point. Hence equilibrium points in the vector field correspond to constant solutions.

Computation of equilibrium points

The linear system

$$\frac{dx}{dt} = 3x + y$$
$$\frac{dy}{dt} = x - y$$

has only one equilibrium point, the origin $(0, 0)$. We see this by simultaneously solving the two linear equations

$$\begin{cases} 3x + y = 0 \\ x - y = 0. \end{cases}$$

(Add the first equation to the second to see that $x = 0$, then use either equation to conclude that $y = 0$.) If we look at the vector field for this system, we see that the vectors are getting short near the origin (see Figure 2.23). Solutions passing near the origin will move slowly. All the vectors of the direction field are the same length by definition; however, we can still tell that there must be an equilibrium point at the origin because the direction of the vectors in the direction field change radically near $(0, 0)$ (see Figure 2.24).

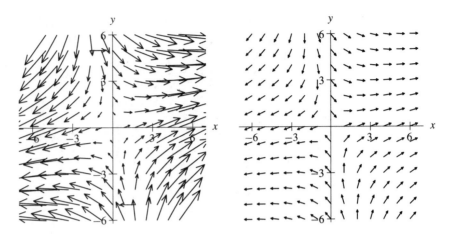

Figure 2.23
Vector field for the system

Figure 2.24
Direction field for the system

$$\frac{dx}{dt} = 3x + y$$
$$\frac{dy}{dt} = x - y.$$

$$\frac{dx}{dt} = 3x + y$$
$$\frac{dy}{dt} = x - y.$$

As a second example, consider the nonlinear system

$$\frac{dy}{dt} = v$$
$$\frac{dv}{dt} = -y + y^3 - v.$$

This is called the **Duffing system**. The vector field for this system is given by

$$\mathbf{F}(y, v) = (v, -y + y^3 - v).$$

We can find the equilibrium points by solving

$$\begin{cases} v = 0 \\ -y + y^3 - v = 0. \end{cases}$$

Since the first equation is $v = 0$, the second equation becomes $-y + y^3 = 0$, which has solutions ± 1 and 0. Hence the system has three equilibrium points, $(0, 0)$, $(1, 0)$, and $(-1, 0)$. Again we emphasize that, because the vector field is continuous, the vectors are very short near the equilibrium points. Solutions move very slowly near the equilibrium points. We can locate the equilibrium points in the direction field for this system because the direction of the vectors changes radically near these points (see Figure 2.25).

In the next section we will continue the discussion of the qualitative theory of autonomous systems, using the direction field to obtain sketches of solutions.

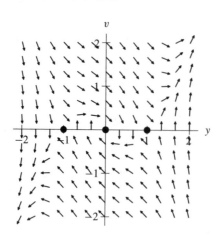

Figure 2.25
Direction field for the Duffing system

$$\frac{dy}{dt} = v$$

$$\frac{dv}{dt} = -y + y^3 - v.$$

There are equilibrium points at $(0, 0)$, $(1, 0)$ and $(-1, 0)$.

Exercises for Section 2.2

In Exercises 1–8, determine the dimension of the system, the independent variable, the dependent variables, and the parameters. Also determine whether the system is autonomous or nonautonomous and whether it is linear or nonlinear.

1.
$$\frac{dx}{dt} = x - 2xy$$
$$\frac{dy}{dt} = 3y - x$$

2.
$$\frac{dx}{dt} = t^3 x + y + t^2$$
$$\frac{dy}{dt} = \cos t$$

3.
$$\frac{dz}{dt} = z - \beta w^2$$
$$\frac{dw}{dt} = w - \epsilon \cos t$$

4.
$$\frac{dz}{dt} = 2qz + w$$
$$\frac{dw}{dt} = \sigma + w + z$$

5.
$$\frac{dx}{dt} = x^2 + y + \epsilon \cos t$$
$$\frac{dy}{dt} = y + 3x^2 + \epsilon \sin t$$

6.
$$\frac{dx}{dt} = \alpha x y^3$$
$$\frac{dy}{dt} = x^2 y$$

7.
$$\frac{dx}{ds} = x^2 + y + \epsilon \cos t$$
$$\frac{dy}{ds} = y + 3x^2 + \epsilon \sin t$$
$$\frac{dt}{ds} = 1$$

8.
$$\frac{dx}{dt} = \alpha x y^3$$
$$\frac{dy}{dt} = x^2 y$$
$$\frac{d\alpha}{dt} = 0$$

In Exercises 9–16, we consider the system

$$\frac{dx}{dt} = 2x + y$$
$$\frac{dy}{dt} = -y.$$

For the curves $\mathbf{Y}(t) = (x(t), y(t))$, check to see if $\mathbf{Y}(t)$ is a solution to the system.

9. $(x(t), y(t)) = (e^{2t} - e^{-t}, 3e^{-t})$ **10.** $(x(t), y(t)) = (e^{2t}, e^{-t})$

11. $(x(t), y(t)) = (e^{-t}, 3e^{-t})$ **12.** $(x(t), y(t)) = (3e^{2t}, 0)$

13. $(x(t), y(t)) = (0, 1)$ **14.** $(x(t), y(t)) = (4e^{2t} - e^{-t}, 3e^{-t})$

15. $(x(t), y(t)) = (2e^{2t}, 2)$ **16.** $(x(t), y(t)) = (2e^{2t} - 2e^{-t}, 6e^{-t})$

In Exercises 17–20, match the direction field with one of the following eight systems. Provide justification for your choices.

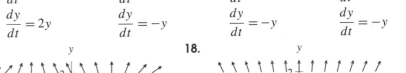

i.
$$\frac{dx}{dt} = -x$$
$$\frac{dy}{dt} = y - 1$$

ii.
$$\frac{dx}{dt} = x^2 - 1$$
$$\frac{dy}{dt} = y$$

iii.
$$\frac{dx}{dt} = x + 2y$$
$$\frac{dy}{dt} = -y$$

iv.
$$\frac{dx}{dt} = 2x$$
$$\frac{dy}{dt} = y$$

v.
$$\frac{dx}{dt} = x$$
$$\frac{dy}{dt} = 2y$$

vi.
$$\frac{dx}{dt} = x - 1$$
$$\frac{dy}{dt} = -y$$

vii.
$$\frac{dx}{dt} = x^2 - 1$$
$$\frac{dy}{dt} = -y$$

viii.
$$\frac{dx}{dt} = x - 2y$$
$$\frac{dy}{dt} = -y$$

17.

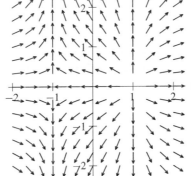

18.

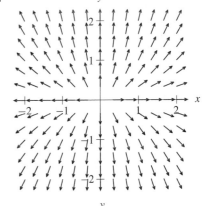

19.

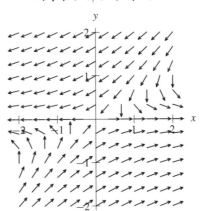

20.

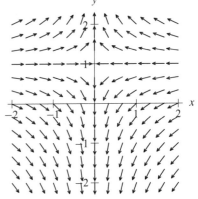

In Exercises 21–23, we consider the system

$$\frac{dx}{dt} = -2x + 3y$$
$$\frac{dy}{dt} = -3x.$$

For the curves $\mathbf{Y}(t) = (x(t), y(t))$, check to see if $\mathbf{Y}(t)$ is a solution to the system.

21. $(x(t), y(t)) = (e^{-t}\cos(2\sqrt{2}t) + 2\sqrt{2}e^{-t}\sin(2\sqrt{2}t), 3e^{-t}\cos(2\sqrt{2}t))$

22. $(x(t), y(t)) = (e^{-t}\sin(2\sqrt{2}t) - 2\sqrt{2}e^{-t}\cos(2\sqrt{2}t), 3e^{-t}\sin(2\sqrt{2}t))$

23. $(x(t), y(t)) = (e^{-t}\cos(2\sqrt{2}t), e^{-t}\sin(2\sqrt{2}t))$

In Exercises 24–27, we consider the system

$$\frac{dx}{dt} = 2y^2 - x + 1$$
$$\frac{dy}{dt} = \frac{x-1}{2y} - \frac{y}{2}.$$

For the curves $\mathbf{Y}(t) = (x(t), y(t))$, check to see if $\mathbf{Y}(t)$ is a solution to the system.

24. $(x(t), y(t)) = (1 + \sin t + \cos t, \sqrt{\cos t})$

25. $(x(t), y(t)) = (9, 2)$

26. $(x(t), y(t)) = (\cos t, \sin^2 t)$

27. $(x(t), y(t)) = (1 + \sqrt{2}\cos t, \sqrt{\cos t + \sin t}/\sqrt[4]{2})$

In Exercises 28–35,

   (a) find the equilibrium points of the system;

   (b) sketch the direction fields of the system.

[It is definitely worth the effort to use either a computer or graphing calculator for part (b). To obtain a respectable direction field, you must do a great deal of arithmetic.]

28. $$\frac{dx}{dt} = y$$
$$\frac{dy}{dt} = x - x^3 - y$$

29. $$\frac{dR}{dt} = 4R - 7F - 1$$
$$\frac{dF}{dt} = 3R + 6F - 12$$

30. $$\frac{dx}{dt} = (x^2 - 2x + 1)(y^2 - 1)$$
$$\frac{dy}{dt} = (x - 1)(y + 1)$$

31. $$\frac{dz}{dt} = \cos w$$
$$\frac{dw}{dt} = -z + w$$

32. $$\frac{dx}{dt} = y(x^2 + y^2 - 1)$$
$$\frac{dy}{dt} = -x(x^2 + y^2 - 1)$$

33. $$\frac{dx}{dt} = y$$
$$\frac{dy}{dt} = -\cos x - y$$

34. $\dfrac{dx}{dt} = \cos 2y$

$\dfrac{dy}{dt} = -x + y$

35. $\dfrac{dx}{dt} = 4x - 7y + 2$

$\dfrac{dy}{dt} = 3x + 6y - 1$

36. Discuss how the systems in Exercises 5 and 7 are related. For example, if you know a solution to one of these systems, can you determine a solution to the other?

37. Discuss how the systems in Exercises 6 and 8 are related. For example, if you know a solution to one of these systems, can you determine a solution to the other?

2.3 GRAPHICAL REPRESENTATION OF SOLUTIONS OF SYSTEMS

A solution of an initial-value problem

$$\frac{d\mathbf{Y}}{dt} = \mathbf{F}(\mathbf{Y}), \quad \mathbf{Y}(0) = \mathbf{Y}_0,$$

is a curve $\mathbf{Y}(t) = (x(t), y(t))$ that satisfies both the differential equation and the initial condition. When we go to graph such functions, we run into a difficult problem. The most complete graph of such a solution is three dimensional. We need an axis for the independent variable t and an axis for each of the dependent variables x and y.

For example, we can easily check that $\mathbf{Y}(t) = (x(t), y(t)) = (\cos t, \sin t)$ is a solution of the system

$$\frac{dx}{dt} = -y$$

$$\frac{dy}{dt} = x$$

because

$$\frac{dx}{dt} = \frac{d(\cos t)}{dt} = -\sin t = -y$$

$$\frac{dy}{dt} = \frac{d(\sin t)}{dt} = \cos t = x.$$

[See Figure 2.26 for the graph of this curve in txy-space along with its three projections — one in the tx-plane, one in the ty-plane, and one in the xy-plane (the phase plane).] Drawing such a curve requires considerable artistic skill, even though this solution is made up of very familiar functions. Interpreting the pictures requires an even greater skill in visualization. Hence instead of three-dimensional pictures, we use various projections onto the two-dimensional coordinate planes.

The projection of the graphs of solutions onto the xy-plane is called the *phase plane* diagram of the system. The projections of the solution onto the tx- or ty-planes are called the $x(t)$- and $y(t)$-graphs. The $x(t)$- and $y(t)$-graphs are also called the **wave-form** and the **time-series** graphs. Each type of graph is useful, and being able to move easily from one to another is an extremely powerful tool in the analysis of solutions of systems.

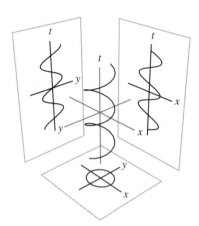

Figure 2.26
Graphs of the curve $(\cos t, \sin t)$ in txy-space along with its projections into the coordinate planes. Note that the $x(t)$-graph is the graph of the cosine function, the $y(t)$-graph is the graph of the sine function, and the solution curve in the phase plane is a circle.

Drawing Phase Planes

The **phase plane** or **phase portrait** for a two-dimensional, autonomous system of the form $d\mathbf{Y}/dt = \mathbf{F}(x, y)$ is a sketch of a number of solution curves in the xy-plane. The goal is to sketch enough of the solutions so that the general qualitative behavior of all solutions can be surmised by looking at the picture. A number of initial points are chosen, scattered over the xy-plane, and the solution curve through each of these initial points is sketched. Exactly which initial points and how many are needed to make a good phase plane diagram requires trial, error, and some artistic feeling. It is advisable to do this process numerically, that is, to use a computer or calculator to sketch the solution curves. A numerical technique (another version of Euler's method) for solving systems is discussed in Section 2.5.

As noted above, the phase plane is just one way of visualizing the solutions of systems of differential equations. Not all information about a particular solution can be seen by a glance at the phase plane. In particular, when we look at a picture of a solution curve in the phase plane, we do not see the time variable, so we don't know how fast the solution traverses the curve. The best way to get information about the time variable is to watch a computer sketch the solution curve in "real time." The next best thing is to give the solution in the phase plane along with the $x(t)$- and $y(t)$-graphs.

The phase portrait of a linear system

Consider the linear system

$$\frac{dy}{dt} = v$$

$$\frac{dv}{dt} = -2y - v.$$

The direction field indicates that solutions spiral toward $(0, 0)$, which is the only equilibrium point (see Figure 2.27). To draw the phase plane, we choose a number of initial points and sketch their solutions until they are very close to $(0, 0)$ (see Figure 2.28).

To obtain information about how fast the solutions are moving, we turn to the $y(t)$- and $v(t)$-graphs. These are given for the solution with initial condition $(1, 0)$ in Figure 2.29.

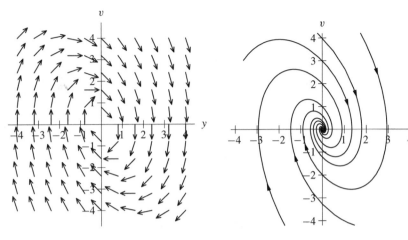

Figure 2.27
Direction field for the system

$$\frac{dy}{dt} = v$$

$$\frac{dv}{dt} = -2y - v.$$

Figure 2.28
Phase plane for the system

$$\frac{dy}{dt} = v$$

$$\frac{dv}{dt} = -2y - v.$$

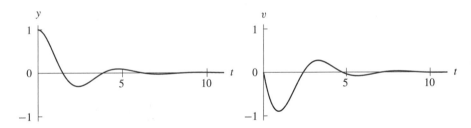

Figure 2.29
The $y(t)$- and $v(t)$-graphs of the solution with initial condition $(1, 0)$.

The phase portrait of a nonlinear system

As a second example, we consider the nonlinear system

$$\frac{dx}{dt} = 2x\left(1 - \frac{x}{2}\right) - xy$$

$$\frac{dy}{dt} = 3y\left(1 - \frac{y}{3}\right) - 2xy.$$

The direction field in the first quadrant (where most of the interesting solutions are) is shown in Figure 2.30. We can see from the direction field that there are four equilibrium points in the first quadrant (computing their exact location is left to the reader). To get a good phase plane picture, we must choose enough solutions so that we can see all the different types of solution curves, but not so many that the picture gets messy (see Figure 2.31).

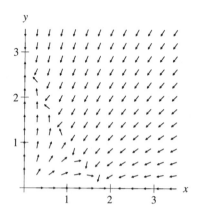

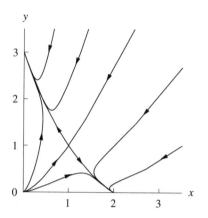

Figure 2.30
Direction field for the system
$$\frac{dx}{dt} = 2x\left(1 - \frac{x}{2}\right) - xy$$
$$\frac{dy}{dt} = 3y\left(1 - \frac{y}{3}\right) - 2xy.$$

Figure 2.31
Phase plane for the system
$$\frac{dx}{dt} = 2x\left(1 - \frac{x}{2}\right) - xy$$
$$\frac{dy}{dt} = 3y\left(1 - \frac{y}{3}\right) - 2xy.$$

In Figure 2.32, we see the $x(t)$- and $y(t)$-graphs for the indicated solution curves. Again it is important to notice how these two different types of representations of the solutions give different information about the behavior of the solution. For example, the $x(t)$- and $y(t)$-graphs cross in the lower-left figure. This means that the corresponding solution curve crosses the line $y = x$ at this moment.

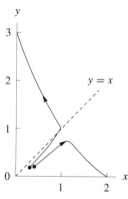

Figure 2.32
Two solutions in the phase plane for

$$\frac{dx}{dt} = 2x\left(1 - \frac{x}{2}\right) - xy$$
$$\frac{dy}{dt} = 3y\left(1 - \frac{y}{3}\right) - 2xy$$

(left); and the $x(t)$- and $y(t)$-graphs of these solutions (below).

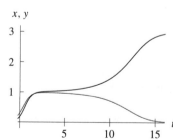

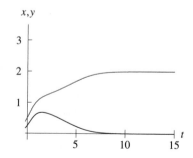

Systems with Computable Solutions

We have said many times how hard it is to find solutions analytically for systems of differential equations in general. However, a few systems are simple enough that we can compute formulas for solutions. In Chapter 3 we will see how to compute formulas for solutions of autonomous linear systems. Occasionally we can use the techniques of Chapter 1 as well, as shown in the following examples.

A decoupled system

Consider the system

$$\frac{dx}{dt} = -2x$$
$$\frac{dy}{dt} = -y$$

Since the equation for dx/dt involves only x and the equation for dy/dt involves only y, we can solve the two equations separately. When this happens we say the system is "decoupled." The general solution of the equation for x is $x(t) = k_1 e^{-2t}$, where k_1 is any constant. The general solution of the equation for y is $y(t) = k_2 e^{-t}$, where k_2 is any constant. We can put these together to find the **general solution** of the system

$$(x(t), y(t)) = (k_1 e^{-2t}, k_2 e^{-t}).$$

This general solution has two undetermined constants, k_1 and k_2. These constants can be adjusted so that any given initial condition is satisfied.

For example, given the initial condition $\mathbf{Y}(0) = (1, 1)$, we can choose $k_1 = k_2 = 1$ to obtain the solution

$$\mathbf{Y}(t) = \begin{pmatrix} e^{-2t} \\ e^{-t} \end{pmatrix}.$$

In Figure 2.33 we plot this curve along with the direction field associated to the vector field $\mathbf{F}(x, y) = (-2x, -y)$. From the formula for $\mathbf{Y}(t)$, we note that $\mathbf{Y}(t)$ gives a parameterization of the upper half of the curve $x = y^2$ in the plane because

$$(y(t))^2 = (e^{-t})^2 = e^{-2t} = x(t).$$

We obtain only the upper half of this parabola because $y(t) > 0$ for all t.

The plot of the solution curve in the phase plane hides the behavior of our solution with respect to the independent variable t. The solution actually tends toward the origin. Since we have the formulas for $x(t)$ and $y(t)$, it is not difficult to sketch the $x(t)$- and $y(t)$-graphs (see Figure 2.34).

Computing solutions for a special nonlinear system

As a second example, consider the nonlinear system

$$\frac{dx}{dt} = xy$$
$$\frac{dy}{dt} = -y + 1.$$

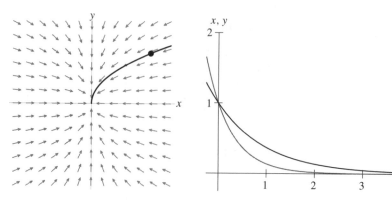

Figure 2.33
The curve $\mathbf{Y}(t) = (e^{-2t}, e^{-t})$.

Figure 2.34
The $x(t)$- and $y(t)$-graphs for the solution
$(x(t), y(t)) = (e^{-2t}, e^{-t})$.

For this system, we note that the equation for dy/dt depends only on y (and not on x). In fact the dy/dt equation is separable and linear, so we have a choice of techniques for finding the general solution. The general solution for y is

$$y(t) = 1 + ce^{-t},$$

where c is any constant.

Substituting $y = 1 + ce^{-t}$ into the equation for dx/dt, we have

$$\frac{dx}{dt} = (1 + ce^{-t})x.$$

This equation is also separable and linear, and we find by integration that the general solution for x is

$$x(t) = ke^{t - ce^{-t}},$$

where k is any constant. The general solution for the system is therefore

$$(x(t), y(t)) = (ke^{t - ce^{-t}}, 1 + ce^{-t}),$$

where c and k are constants which we can adjust to satisfy any given initial condition. For example, if $(x(0), y(0)) = (1, 2)$, then we must have the simultaneous equations

$$\begin{cases} 1 = x_0 = ke^{0 - ce^{-0}} = ke^{-c} \\ 2 = y_0 = 1 + ce^{-0} = 1 + c. \end{cases}$$

Hence $c = 1$ and $k = e^1 = e$. Thus the solution satisfying this initial condition is

$$(x(t), y(t)) = (e \cdot e^{t - e^{-t}}, 1 + e^{-t}) = (e^{1 + t - e^{-t}}, 1 + e^{-t}).$$

The moral here is that even fairly simple-looking nonlinear systems can have solutions that are very complicated functions. It was only by luck (or rather careful choice of example) that we could find the general solution at all. In general, equations will not decouple. On the other hand, we can obtain a qualitative picture of the behavior of solutions from the direction field and phase plane (see Figure 2.35).

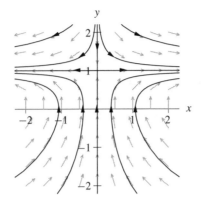

Figure 2.35
Direction field and sketches of solutions for the system

$$\frac{dx}{dt} = xy$$

$$\frac{dy}{dt} = -y + 1.$$

Existence and Uniqueness

As with first-order differential equations, given a system of differential equations and an initial condition we are faced with the question: Is there a solution of the system with the given initial condition, and is that solution unique? The answer to this fundamental question is given by the Existence and Uniqueness Theorem for systems. We state this theorem for arbitrary systems (that is, including nonautonomous systems), since it applies to all systems of ordinary differential equations.

Existence and Uniqueness Theorem

Let

$$\frac{d\mathbf{Y}}{dt} = \mathbf{F}(t, \mathbf{Y})$$

be a system of differential equations. Suppose that t_0 is an initial time and $\mathbf{Y}_0$ is an initial value. Suppose also that the component functions of $\mathbf{F}$ are continuously differentiable. Then there is an $\epsilon > 0$ and a function $\mathbf{Y}(t)$ defined for $t_0 - \epsilon < t < t_0 + \epsilon$, such that $\mathbf{Y}(t)$ satisfies the initial-value problem $d\mathbf{Y}/dt = \mathbf{F}(t, \mathbf{Y})$ and $\mathbf{Y}(t_0) = \mathbf{Y}_0$. Moreover, for t in this interval, this solution is unique.

We emphasize that, just as in the case of differential equations, we must first verify the hypotheses of this theorem to apply it properly. In other words, we must be sure that the components of the vector field on the right-hand side of the system have continuous partial derivatives. If the hypotheses do not hold, then the theorem gives no information. In this case solutions may not exist, or they may not be unique.

The existence part of the theorem is mostly just reassuring. If we are studying a certain system, then it is nice to know that what we are studying exists. The uniqueness part is useful in a much more practical way. Roughly speaking, the Uniqueness Theorem says that two different solutions cannot start at the same place at the same time.

For autonomous systems

$$\frac{d\mathbf{Y}}{dt} = \mathbf{F}(\mathbf{Y}),$$

we obtain even more information from the Uniqueness Theorem. Since the vector field does not change with time, if we start a solution at a point $\mathbf{Y}_0$ in the phase plane at time t_1 and another solution at the same point at a later time t_2, then the solution curves will look exactly the same. Informally, since the rules of the game don't change with time, it doesn't matter what time we start playing. More formally, if $\mathbf{Y}_1(t)$ is the solution with $\mathbf{Y}_1(t_1) = \mathbf{Y}_0$ and $\mathbf{Y}_2(t)$ is the solution with $\mathbf{Y}_2(t_2) = \mathbf{Y}_0$, then $\mathbf{Y}_2(t) = \mathbf{Y}_1(t - (t_2 - t_1))$ for all t, or in other words, solution $\mathbf{Y}_2(t)$ is always $t_2 - t_1$ time units behind solution $\mathbf{Y}_1(t)$.

An application of this observation is that, for an autonomous system, distinct solution curves in the phase plane never cross. If two solution curves ever touch, then they must differ only by a time delay. We have used this rule without comment while drawing the figures of phase planes throughout this section. The advantage of having a precisely stated theorem is that we can incorporate the idea into theoretical arguments on the behavior of solutions.

Exercises for Section 2.3

In Exercises 1–8,

(a) find the equilibrium points of the system;

(b) sketch the direction field of the system;

(c) using the direction field, sketch the phase plane of the system.

[It is definitely worth the effort to use either a computer or graphing calculator for parts (b) and (c). To obtain a respectable picture of a direction field and/or phase plane, you must do a great deal of arithmetic. Also see Exercises 28–35 in Section 2.2.]

1.
$$\frac{dx}{dt} = y$$
$$\frac{dy}{dt} = x - x^3 - y$$

2.
$$\frac{dR}{dt} = 4R - 7F - 1$$
$$\frac{dF}{dt} = 3R + 6F - 12$$

3.
$$\frac{dz}{dt} = \cos w$$
$$\frac{dw}{dt} = -z + w$$

4.
$$\frac{dx}{dt} = (x^2 - 2x + 1)(y^2 - 1)$$
$$\frac{dy}{dt} = (x - 1)(y + 1)$$

5.
$$\frac{dx}{dt} = y(x^2 + y^2 - 1)$$
$$\frac{dy}{dt} = -x(x^2 + y^2 - 1)$$

6.
$$\frac{dx}{dt} = y$$
$$\frac{dy}{dt} = -\cos x - y$$

7.
$$\frac{dx}{dt} = \cos 2y$$
$$\frac{dy}{dt} = -x + y$$

8.
$$\frac{dx}{dt} = 4x - 7y + 2$$
$$\frac{dy}{dt} = 3x + 6y - 1$$

In Exercises 9–12,

(a) find the general solution for the system;

(b) find the particular solution with the given initial condition;

(c) sketch the $x(t)$- and $y(t)$-graphs and the solution curve in the xy-phase plane;

(d) for the autonomous systems, sketch the direction field for each system and use it to check your sketch of the solution in the phase plane. [Sketching direction fields and solutions is a good job for a computer, but a bad job for a person.]

9.
$$\frac{dx}{dt} = xy$$
$$\frac{dy}{dt} = 2y$$
$$\mathbf{Y}_0 = (1, 1)$$

10.
$$\frac{dx}{dt} = 2x + 3y$$
$$\frac{dy}{dt} = y + 2$$
$$\mathbf{Y}_0 = (0, 0)$$

11.
$$\frac{dx}{dt} = xy^2$$
$$\frac{dy}{dt} = y + 2t$$
$$\mathbf{Y}_0 = (0, 1)$$

12.
$$\frac{dx}{dt} = \frac{1}{x}$$
$$\frac{dy}{dt} = x^2 + 2y$$
$$\mathbf{Y}_0 = (1, 2)$$

In Exercises 13–16,

(a) solve the given initial-value problem;

(b) sketch the $x(t)$- and $y(t)$-graphs of the solution, and sketch the solution curve in the xy-phase plane;

(c) sketch the direction field for the system, and use it to check your sketch of the solution curve.

13.
$$\frac{dx}{dt} = x^2 - xy^2$$
$$\frac{dy}{dt} = 3y - xy$$
$$\mathbf{Y}_0 = (0, 1)$$
[*Hint*: What is dx/dt when $x = 0$?]

14.
$$\frac{dx}{dt} = (x - 1)^2(2y + x)$$
$$\frac{dy}{dt} = xy^2$$
$$\mathbf{Y}_0 = (1, 2)$$

15.
$$\frac{dx}{dt} = xy - x^2 - 3$$
$$\frac{dy}{dt} = y^2 - 4y + 4$$
$$\mathbf{Y}_0 = (0, 2)$$

16.
$$\frac{dx}{dt} = -4x + y$$
$$\frac{dy}{dt} = 2x - 5y$$
$$\mathbf{Y}_0 = (2, 2)$$
[*Hint*: Note that when $x = y$, $dx/dt = dy/dt$.]

In Exercises 17–21, we consider the model of an arms race between countries Ytterbium and Zirconium. The model is

$$\frac{dy}{dt} = \left(z - \frac{y}{2}\right) - y^2$$
$$\frac{dz}{dt} = \left(y - \frac{z}{2}\right) - z^2,$$

where $y(t)$ and $z(t)$ represent the amount of spending on arms (in billions of dollars per year) by countries Ytterbium and Zirconium respectively. The term in parentheses in each equation represents the desire of each country to spend twice as much on arms as the other country. The squared term represents the economic pressure against spending too much on arms. In each Exercise, modifications of the assumptions that lead to the system above are given. For each of these modifications,

(a) make the corresponding modification to the system above;

(b) sketch the direction field for the new system;

(c) find the equilibrium points (either approximately, by inspection of the direction field, or exactly, if the algebra is not too difficult);

(d) sketch the phase plane for the new system;

(e) describe in a brief essay what the new system predicts will happen if initially Ytterbium has no arms and Zirconium has only a small number of arms. Compare this to what the system predicts will happen if initially Zirconium has no arms and Ytterbium has only a small number of arms.

[Take advantage of whatever technology you have available for drawing direction fields and phase planes.]

17. Suppose both countries decide that they need to have only as large a stockpile of arms as the other country. (Other assumptions remain the same; that is, both countries feel economic pressure against having too large an arms stockpile).

18. Suppose Ytterbium decides it needs to have an arms stockpile only equal to that of Zirconium, but Zirconium continues to want an arms stockpile twice the size of Ytterbium's.

19. Suppose the economic pressure against stockpiling arms is unchanged in Zirconium but doubles in Ytterbium.

20. Suppose the desire of each country to have twice the arms stockpile of the other is unchanged, but gold is discovered in Zirconium and it becomes so rich that there is no economic pressure against stockpiling arms.

21. Suppose both countries become so rich that there is no economic pressure in either country against stockpiling arms.

In Exercises 22–26,

(a) sketch the direction field;

(b) find the equilibrium points (either approximately, by inspection of the direction field, or exactly, if the algebra is not too difficult);

(c) sketch the phase plane;

(d) describe in a brief essay the possible behaviors of solutions with initial conditions $a > 0$ and $b > 0$.

These systems model the chemical reaction in a solution of substance A and substance B into substance C which precipitates out of the solutions $(A + B \rightarrow C)$. The amounts of A and B in the solutions at time t are $a(t)$ and $b(t)$, respectively. The rate at which molecules of A and B react is proportional to the chance that they encounter each other in a solution, which is proportional to ab. Models for these reactions were developed in Exercises 19–24 in Section 2.1.

22.

$$\frac{da}{dt} = -\frac{ab}{2}$$
$$\frac{db}{dt} = -\frac{ab}{2}$$

In this model no chemicals are added to the solution. The proportionality constant for the reaction rate of A and B is $1/2$.

23.

$$\frac{da}{dt} = 2 - \frac{ab}{2}$$
$$\frac{db}{dt} = \frac{3}{2} - \frac{ab}{2}$$

This system models the same chemical reaction with A added to the solution at a rate of 2 units per unit of time and B added to the solution at a rate of $3/2$ units per unit of time.

24.

$$\frac{da}{dt} = 2 - \frac{ab}{2} - \frac{a^2}{3}$$
$$\frac{db}{dt} = \frac{3}{2} - \frac{ab}{2}$$

This system models the chemical reaction described above with A and B added to the solution at rates of 2 and $3/2$ units per unit time, and with the additional reaction A + A $\rightarrow$ C occurring at a rate proportional to the chance that two A molecules are close with a proportionality constant of $1/3$.

25.

$$\frac{da}{dt} = 2 - \frac{ab}{2} + \frac{b^2}{3}$$
$$\frac{db}{dt} = \frac{3}{2} - \frac{ab}{2} - \frac{b^2}{3}$$

This system models the chemical reaction described above with A and B added to the solution at rates of 2 and $3/2$ units per unit of time, and with the additional reaction B + B $\rightarrow$ A occurring at a rate proportional to the chance that two B molecules are close with proportionality constant of $1/3$.

26.

$$\frac{da}{dt} = 2 - \frac{ab}{2} - \frac{2ab^2}{3}$$
$$\frac{db}{dt} = \frac{3}{2} - \frac{ab}{2} - \frac{ab^2}{3}$$

This system models the chemical reaction described above with A and B added to the solution at rates of 2 and $3/2$ units per unit of time, and with the additional reaction A + 2 B $\rightarrow$ C occurring at a rate proportional to the chance that one A and two B molecules are close with proportionality constant $1/3$.

27. Consider the system

$$\frac{dx}{dt} = -\frac{7x}{3} + \frac{2y}{3}$$
$$\frac{dy}{dt} = -\frac{2x}{3} - \frac{2y}{3}.$$

Show that, if $\mathbf{Y}(t) = (x(t), y(t))$ is a solution with $1/2 \leq y(0)/x(0) \leq 2$, then $1/2 \leq y(t)/x(t) \leq 2$ for all t. [*Hint*: What can you say about the curves $\mathbf{Y}_1(t) = (2e^{-2t}, e^{-2t})$ and $\mathbf{Y}_2(t) = (e^{-t}, 2e^{-t})$?]

28. For the predator-prey system

$$\frac{dR}{dt} = 2R\left(1 - \frac{R}{2}\right) - 1.2RF$$

$$\frac{dF}{dt} = -F + 0.9RF$$

discussed in Section 2.1,

  (a) show that, if $R(0) \geq 0$ and $F(0) \geq 0$ for a solution $(R(t), F(t))$, then $R(t) \geq 0$ and $F(t) \geq 0$ for all t;

  (b) show that, if $F = 5 - R$ and both $R \geq 0$ and $F \geq 0$, then the vector field at (R, F) points to the left (that is, the R component is negative).

  (c) use the Uniqueness Theorem to show that, if $R(0) \geq 0$, $F(0) \geq 0$, and $F(0) \leq 5 - R(0)$ for a solution $(R(t), F(t))$, then $R(t) \geq 0$, $F(t) \geq 0$, and $F(t) \leq 5 - R(t)$ for all t.

In Exercises 29–31, consider the linear system

$$\frac{dx}{dt} = -x + 3y$$

$$\frac{dy}{dt} = -3x - y.$$

29. Verify that $\mathbf{Y}_1(t) = (e^{-t}\sin(3t), e^{-t}\cos(3t))$ is a solution of this system.

30. Verify that $\mathbf{Y}_2(t) = (e^{-(t-1)}\sin(3(t-1)), e^{-(t-1)}\cos(3(t-1)))$ is a solution.

31. Sketch $\mathbf{Y}_1(t)$ and $\mathbf{Y}_2(t)$ in the xy-phase plane. Why don't $\mathbf{Y}_1(t)$ and $\mathbf{Y}_2(t)$ contradict the Uniqueness Theorem?

32. Consider the system

$$\frac{dx}{dt} = x^2 + y$$

$$\frac{dy}{dt} = x^2 y^2.$$

Show that, for the solution $(x(t), y(t))$ with initial condition $x(0) = y(0) = 1$, there is a time t_* such that $x(t) \to \infty$ as $t \to t_*$. In other words, the solution blows up in finite time. [*Hint*: Note that $dy/dt \geq 0$ for all x and y.]

2.4 SECOND-ORDER EQUATIONS AND THE HARMONIC OSCILLATOR

The progression from first-order equations to first-order systems is quite natural. We just add one (or more) new dependent variables. However, it is also natural to proceed from first-order equations to second-order equations. These types of differential equations involve only one dependent variable, but unlike systems, they involve the second as well as the first derivative of the dependent variable. Second-order equations arise naturally and frequently in applications because the second derivative of a displacement with respect to time is its acceleration. Newton's law says that force is the product of mass and acceleration, so examples of second-order differential equations are extremely common in physics and engineering.

In this section we show how a second-order equation can be converted into a first-order system. After the conversion, we can study second-order equations qualitatively using the language of direction fields and phase planes. In the next section we will see that the conversion of second-order equations into first-order systems is also necessary to study them numerically.

The Harmonic Oscillator

We begin by deriving a second-order differential equation that models the motion of a mass attached to a spring. The mass-spring apparatus is called the **harmonic oscillator** and is one of the most important models in science. Beyond its applications in mechanics, this equation also arises in physics (particles in potential wells), electronics (RLC circuits), and many other disciplines.

We begin with a mass that is attached to a spring and that slides on a table (see Figure 2.36). We wish to understand the motion of the mass when the spring is stretched (or compressed) and then released. For simplicity, we assume that the only forces acting on the mass are the force of the spring and the drag or damping as the mass moves back and forth. However, we assume that the table itself is frictionless.

There are three key quantities in this model — a quantity that measures the displacement from a "rest position," a "restoring force" that accelerates the mass toward the rest position, and a resisting or "damping force" that slows its motion. We wish to determine the position of the mass as a function of time as the mass moves on a horizontal line, so we let $y(t)$ denote the position of the mass at time t. It is convenient to let $y = 0$ represent the rest position of the mass (see Figure 2.37). At the rest position, the spring is neither stretched nor compressed, and it exerts no force on the mass. We specify that $y(t) < 0$ if the spring is compressed and $y(t) > 0$ if the spring is stretched, using whatever units are convenient (see Figures 2.37–2.39).

The main idea from physics needed to derive the differential equation describing this device is Newton's law of motion,

$$\text{Force } F = \text{mass} \times \text{acceleration}.$$

The acceleration is, of course, d^2y/dt^2. We let m denote the mass of the spring, which is a parameter for this differential equation. Our second-order differential equation is therefore

$$F = m\frac{d^2y}{dt^2}.$$

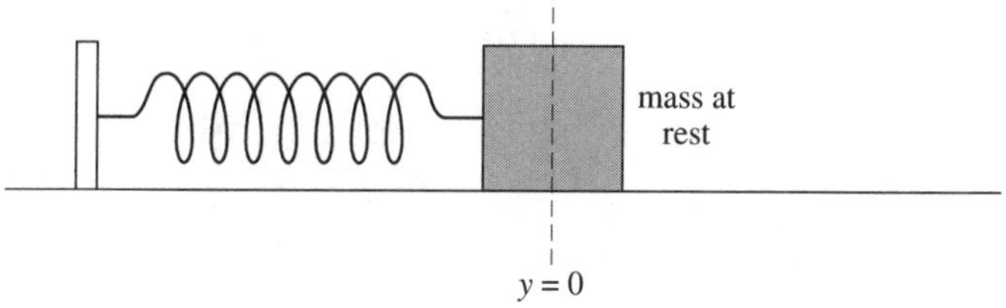

Figure 2.36
The mass-spring device.

mass at rest

$y = 0$

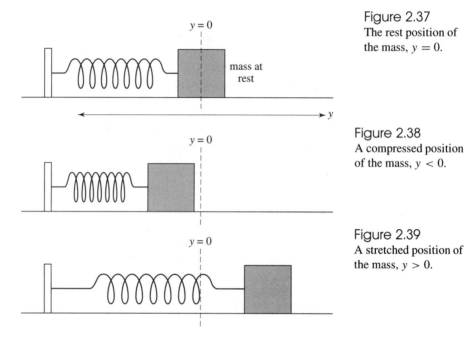

Figure 2.37
The rest position of the mass, $y = 0$.

Figure 2.38
A compressed position of the mass, $y < 0$.

Figure 2.39
A stretched position of the mass, $y > 0$.

To complete the model, we must find expressions for the forces. We use **Hooke's law of springs** as our model for the restoring force F_s of the spring:

The restoring force exerted by a spring is linearly proportional to the spring's displacement from its rest position and is directed toward the rest position.

Therefore we have

$$F_s = -k_s\, y,$$

where $k_s > 0$ is a constant of proportionality called the **spring constant** — a parameter we can adjust by changing springs. We model damping by assuming that the damping force F_d is proportional to the velocity and that it slows the motion. The idea behind this assumption is that the faster the mass travels, the more resistance it meets. Thus we have

$$F_d = -k_d \frac{dy}{dt},$$

where $k_d > 0$ is the **coefficient of damping**. The parameter k_d can be adjusted by adjusting the viscosity of the medium through which the mass is moving (for example, putting the whole mechanism in the bathtub).

Hence the differential equation that models the harmonic oscillator is

$$m\frac{d^2y}{dt^2} = F = F_d + F_s = -k_d \frac{dy}{dt} - k_s\, y$$

or

$$m\frac{d^2y}{dt^2} + k_d \frac{dy}{dt} + k_s\, y = 0.$$

The next step is to see how to relate this second-order differential equation to a first-order system of differential equations.

Second-Order Equations and First-Order Systems

If we divide by m and simplify the notation in the harmonic oscillator equation above by letting $q = k_s/m$ and $p = k_d/m$, we obtain the second-order equation

$$\frac{d^2y}{dt^2} + p\frac{dy}{dt} + qy = 0,$$

where p and q are constants. This equation is called a **second-order**, **constant-coefficient**, **homogeneous**, **linear** differential equation. The term "second-order" refers to the fact that the equation contains second derivatives (and no higher derivatives). The equation has "constant coefficients" because p and q are constants; they do not depend on the independent variable t. "Homogeneous" means that after the y, dy/dt, and d^2y/dt^2 terms are moved to the left-hand side of the equation, the right-hand side is zero. Nonhomogeneous equations have right-hand sides that include constant terms or terms that are functions of t. (We will study some important nonhomogeneous equations in Chapter 5). Finally, the term "linear" means that each summand contains only y, dy/dt, or d^2y/dt^2 to the first power.

To convert this equation into a first-order system, we rewrite it in the form

$$\frac{d^2y}{dt^2} = -p\frac{dy}{dt} - qy.$$

Thinking of the second derivative of y as its "acceleration," we define a new dependent variable $v(t)$ by

$$v = \frac{dy}{dt},$$

and we think of v as the velocity of y. Then

$$\frac{dv}{dt} = \frac{d^2y}{dt^2} = -p\frac{dy}{dt} - qy = -pv - qy$$

or

$$\frac{dv}{dt} = -qy - pv.$$

We now have equations for dy/dt and dv/dt, which we can group together to form the system

$$\frac{dy}{dt} = v$$

$$\frac{dv}{dt} = -qy - pv.$$

We must accept that although this process replaces the second derivative with first derivatives (which are easier to handle), it also increases the number of dependent variables from one to two. So the equations become simpler in one aspect but more complicated in another. (This is called *conservation of misery*.) The switch to a first-order system is necessary for qualitative and numerical study of a second-order equation.

As an example, consider the second-order differential equation

$$\frac{d^2y}{dt^2} + \frac{dy}{dt} + 3y = 0.$$

We rewrite this as

$$\frac{d^2y}{dt^2} = -3y - \frac{dy}{dt}.$$

As above, we introduce the new variable $v = dy/dt$. But then

$$\frac{dv}{dt} = \frac{d^2y}{dt^2} = -3y - \frac{dy}{dt},$$

which we can write as

$$\frac{dv}{dt} = -3y - v.$$

Hence the second-order differential equation

$$\frac{d^2y}{dt^2} + \frac{dy}{dt} + 3y = 0$$

can be rewritten as the system

$$\frac{dy}{dt} = v$$

$$\frac{dv}{dt} = -3y - v.$$

Solutions of Second-Order Equations and Systems

Since the second-order equation

$$\frac{d^2y}{dt^2} + p\frac{dy}{dt} + qy = 0$$

and the system

$$\frac{dy}{dt} = v$$

$$\frac{dv}{dt} = -qy - pv$$

are closely related, finding solutions for one should be equivalent to finding solutions for the other.

Suppose $(y_1(t), v_1(t))$ is a solution of the system. Then we know that

$$\frac{dy_1}{dt} = v_1$$

$$\frac{dv_1}{dt} = -qy_1 - pv_1$$

for all t. We can relate this to the second-order equation by noting that

$$\frac{d^2y_1}{dt^2} = \frac{dv_1}{dt} = -qy_1 - pv_1 = -qy_1 - p\frac{dy_1}{dt},$$

so

$$\frac{d^2 y_1}{dt} + p\frac{dy_1}{dt} + qy_1 = 0$$

for all t. Thus $y_1(t)$, the y-coordinate of the solution of the system, satisfies the second-order differential equation. A solution of the system automatically gives us a solution of the corresponding second-order equation.

Likewise, suppose $y_2(t)$ satisfies the second-order equation; that is

$$\frac{d^2 y_2}{dt} + p\frac{dy_2}{dt} + qy_2 = 0$$

for all t. If we let $v_2 = dy_2/dt$, then $(y_2(t), v_2(t))$ is a solution of the system. This follows since

$$\frac{dy_2}{dt} = v_2$$

by definition of v_2, and

$$\frac{dv_2}{dt} = \frac{d^2 y_2}{dt} = -qy_2 - p\frac{dy_2}{dt} = -qy_2 - pv_2$$

since $y_2(t)$ is a solution of the second-order equation. Hence finding solutions of the second-order equation is the same as finding solutions of the system.

Initial Conditions

As an application of this observation, recall that to specify a solution of the system

$$\frac{dy}{dt} = v$$

$$\frac{dv}{dt} = -qy - pv,$$

we must specify a point (y_0, v_0) in the yv-plane and a time t_0. The "initial condition" for a solution $(y(t), v(t))$ is given by $(y(t_0), v(t_0)) = (y_0, v_0)$. The solution of the corresponding second-order equation

$$\frac{d^2 y}{dt^2} + p\frac{dy}{dt} + qy = 0$$

satisfies $y(t_0) = y_0$ and $y'(t_0) = v(t_0) = v_0$. Hence, to specify a solution of the second-order equation, we must know the initial position and the initial velocity.

This agrees with common sense. For example, to specify the motion of a mass-spring system modeled by the harmonic oscillator equation, we must know both the initial position of the spring and the velocity.

A Harmonic Oscillator With Small Damping

Consider the harmonic oscillator with mass $m = 1$, spring constant $k_s = 2$, and damping constant $k_d = 1$. The second-order equation model is

$$\frac{d^2 y}{dt^2} + \frac{dy}{dt} + 2y = 0.$$

This corresponds to the system

$$\frac{dy}{dt} = v$$
$$\frac{dv}{dt} = -2y - v.$$

The direction field for the system is shown in Figure 2.40. Figure 2.41 shows sketches of solutions for a number of different initial conditions using the direction field. Note that all solution curves tend toward $(0, 0)$, which implies that all solutions $(y(t), v(t))$ have both $y(t)$ and $v(t)$ limiting to zero as $t \to \infty$.

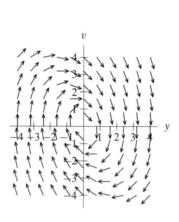

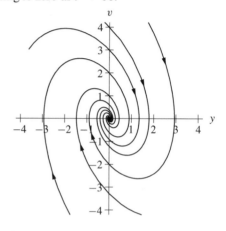

Figure 2.40
Direction field for the system

$$\frac{dy}{dt} = v$$
$$\frac{dv}{dt} = -2y - v.$$

Figure 2.41
Phase plane for the harmonic oscillator system

$$\frac{dy}{dt} = v$$
$$\frac{dv}{dt} = -2y - v.$$

From the phase plane picture, we can also sketch the $y(t)$-graphs of solutions. These solutions oscillate about $y = 0$ with an amplitude that tends to zero.

It is important to note that although the phase plane contains sketches of the solution curves, it says nothing about the speed at which solutions traverse these curves. So although the phase plane above indicates that all solutions tend to $(0, 0)$ as t increases, this picture gives no information about how fast the solutions tend to $(0, 0)$. We can see this information in the $y(t)$- and $v(t)$-graphs of the solutions. (These graphs can be created either numerically using the Euler's method of the next section or analytically using the methods of Chapter 3 — see Figure 2.42.)

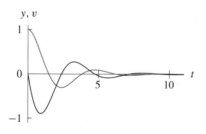

Figure 2.42
The $y(t)$- and $v(t)$-graphs of the solution of the system

$$\frac{dy}{dt} = v$$
$$\frac{dv}{dt} = -2y - v$$

with initial conditions $y(0) = 1$ and $v(0) = 0$.

Reality Check

Before leaving this example, we should do a "reality check." This system of differential equations is supposed to model an actual mechanical system consisting of a mass attached to a spring. We should check to make sure that what we have claimed for the motion of the system agrees with common sense.

Our model predicts that the mass will oscillate with decreasing amplitude. This is reasonable because the spring always pulls the mass back toward the rest position. Because the damping constant is relatively small, it is reasonable to think that the mass will "overshoot" the rest position repeatedly, oscillating around $y = 0$. As long as the mass is moving, damping will push against the motion. However it moves, its energy is decreasing.

It is important to realize that this is a "hindsight" check. It would be impossible to make this prediction from just the values of m, k_s, and k_d without looking at the direction field or the phase plane (or doing the sort of computations we do in Chapter 3). We are only checking that the predictions of our model are reasonable for the physical system.

An Undamped Harmonic Oscillator

As a second example, consider the harmonic oscillator with mass $m = 1$, spring constant $k_s = 3$, and no damping ($k_d = 0$). The second-order equation is

$$\frac{d^2y}{dt^2} = -3y,$$

and the corresponding system is

$$\frac{dy}{dt} = v$$
$$\frac{dv}{dt} = -3y.$$

The direction field for this system is shown in Figure 2.43.

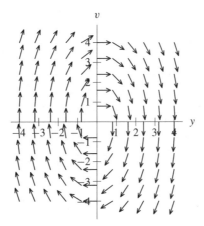

Figure 2.43
Direction field for the system

$$\frac{dy}{dt} = v$$
$$\frac{dv}{dt} = -3y.$$

It is clear from the direction field that solutions circulate around the origin (which is the only equilibrium point). However, it is very difficult to tell if the solutions are periodic solutions, returning to their initial conditions, or if they spiral very slowly toward the origin. Since every numerical method has some error, a numerical study cannot answer this question.

In this case we can use the "lucky guess" method to find the general solution of the second-order equation (and hence the general solution of the system). We already know functions whose second derivatives are negatives of themselves, namely sine and cosine. Moreover, since

$$\frac{d^2(\sin \omega t)}{dt^2} = -\omega^2 \sin \omega t,$$

we guess that $y_1(t) = \sin \sqrt{3}\,t$ is a solution of the second-order equation. This is easily checked:

$$\frac{d^2 y_1}{dt^2} = \frac{d^2(\sin \sqrt{3}\,t)}{dt^2} = -3 \sin \sqrt{3}\,t = -y_1.$$

In exactly the same manner, we can show that $y_2(t) = \cos \sqrt{3}\,t$ is also a solution of the second-order equation.

Because differentiation is a linear operation (that is, the derivative of the sum is the sum of the derivatives and the derivative of a constant times a function is the constant times the derivative of the function), we make an even more dramatic guess. We claim that

$$y(t) = k_1 y_1(t) + k_2 y_2(t) = k_1 \sin \sqrt{3}\,t + k_2 \cos \sqrt{3}\,t$$

is a solution of the second-order equation for any choice of constants k_1 and k_2. (In Chapter 3, we will see why we made this guess here.) We can check that this is a solution as usual by substituting back into the second-order equation.

Hence for the corresponding system

$$\frac{dy}{dt} = v$$

$$\frac{dv}{dt} = -3y,$$

we have that $(y(t), v(t))$ is a solution, where

$$y(t) = k_1 \sin \sqrt{3}\,t + k_2 \cos \sqrt{3}\,t$$

$$v(t) = \frac{dy}{dt} = k_1 \sqrt{3} \cos \sqrt{3}\,t - k_2 \sqrt{3} \sin \sqrt{3}\,t$$

for any constants k_1 and k_2. In the exercises we verify that this is the general solution of the system. Hence every solution returns to its original position every $t = 2\pi/\sqrt{3}$ time units, so every solution is periodic. If we want to look at the specific solution with initial conditions $y(0) = 1$ and $v(0) = 0$, which corresponds to starting the mass at $y = 1$ with zero velocity at time $t = 0$, then we must choose $k_1 = 0$ and $k_2 = 1$. That is, $y(t) = \cos \sqrt{3}\,t$ and $v(t) = -\sqrt{3} \sin \sqrt{3}\,t$. The solution curve in the yv-phase plane is an ellipse because

$$y(t)^2 + \frac{v(t)^2}{3} = \cos^2 \sqrt{3}\,t + \frac{(\sqrt{3} \sin \sqrt{3}\,t)^2}{3} = 1$$

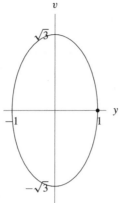

Figure 2.44
Solution curve $(y(t), v(t)) = (\cos \sqrt{3}\, t, -\sqrt{3} \sin \sqrt{3}\, t)$ for the system

$$\frac{dy}{dt} = v$$

$$\frac{dv}{dt} = -3y.$$

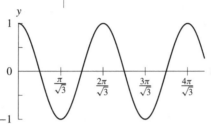

Figure 2.45
Plot of the y-coordinate of a solution with $y(0) = 1$ and $v(0) = 0$ of the system

$$\frac{dy}{dt} = v$$

$$\frac{dv}{dt} = -3y.$$

for all t (see Figure 2.44). If we sketch the $y(t)$-graph for this solution, then we see that $y(t)$ oscillates about $y = 0$ with constant amplitude (see Figure 2.45) and period $2\pi/\sqrt{3}$.

Our model predicts that, if the spring is stretched to length 1 from the rest position and then released with zero initial velocity, it will oscillate with constant amplitude forever. This sounds dangerously like perpetual motion, but if we remember that we originally said that there was no damping, then this result is reasonable. If there is no damping, then nothing takes energy from the system, so it moves forever. Any real system would have some friction. If the friction is very, very small, we might choose to model it with equations that do not include any damping. So the general qualitative form of the solution is reasonable: The period of the oscillation is $2\pi/\sqrt{3}$ — something we couldn't have predicted from common sense.

Common Sense versus Computation

We have made a point of comparing our conclusions about the qualitative behavior of the harmonic
oscillator system with what common sense says about masses and springs. This sort of check of our conclusions against common sense is very important. An arithmetic error can easily give results that are far from the actual solution. If we make a prediction that is wildly different from what common sense tells us the behavior of the system should be, then we have either made a remarkable discovery or we have made a mistake that could be very embarrassing.

Exercises for Section 2.4

In Exercises 1–4, convert the higher-order differential equation into a first-order system. Identify the independent variable, the dimension of the system, the dependent variables, the parameters, whether the system is autonomous or nonautonomous and whether it is linear or nonlinear.

1. $\dfrac{d^2y}{dt^2} + 4\dfrac{dy}{dt} + 3y = 0$

2. $\dfrac{d^2y}{dt^2} + \beta\dfrac{dy}{dt} - 3y^2 = \cos 2t$

[*Hint*: Follow the same procedure setting $dy/dt = v$ and then get the equation for $dv/dt = d^2y/dt^2$.]

3. $\dfrac{d^3y}{dt^3} - \dfrac{d^2y}{dt^2} + 2t\dfrac{dy}{dt} - 3y + 6 = 0$

[*Hint*: This is a third-order equation (third derivatives), so you will need two new variables. Set $dy/dt = v$, $d^2y/dt^2 = dv/dt = w$ so $d^3y/dt = dw/dt$, and we can use the equation to determine dw/dt.]

4. $\dfrac{d^4y}{dt^4} + 4y = 0$

[*Hint*: For this fourth-order equation, you will need three new dependent variables.]

In Exercises 5–7, consider a mass suspended from the ceiling of a room by a spring so that gravity pulls the mass straight down, as shown below.

Let g be the gravitational constant, m the mass, k_s the spring constant, and k_d the coefficient of damping.

5. Write a second-order differential equation for the position of the mass at time t. [*Hint*: One possible approach is to start with the harmonic oscillator system and add the force of gravity. Your answer depends on your choice of the origin.]

6. Convert your second-order equation from Exercise 5 into a first-order system using $v = dy/dt$ for the velocity.

7. Compare your system in Exercise 6 to the harmonic oscillator system. If it is the same as a harmonic oscillator system, explain why; if it is not the same, can you make it the same by adjusting the choice of the origin $y = 0$? [*Hint*: There are two natural choices for the position $y = 0$. These correspond to the rest position of the spring when it is horizontal (not effected by gravity) and when it is vertical (lengthened by gravity).]

In Exercises 8–11, we give parameters for a harmonic oscillator. In each case,

    (a) write the second-order differential equation and the corresponding first-order system;

    (b) sketch the direction field for the system;

    (c) sketch the solution curve for the given initial condition in the phase plane;

    (d) sketch the $y(t)$-graph for the solution with the given initial condition; and

    (e) describe in a paragraph the motion of the mass for the given initial condition.

8. Let $m = 1$, $k_s = 3$, $k_d = 1$, and initial condition $y(0) = 0$, $y'(0) = -2$.

9. Let $m = 2$, $k_s = 6$, $k_d = 2$, and initial condition $y(0) = 2$, $y'(0) = 2$.

10. Let $m = 1$, $k_s = 0.2$, $k_d = 4$, and initial condition $y(0) = -4$, $y'(0) = 4$.

11. Let $m = 1$, $k_s = 9$, $k_d = 0.05$, and initial condition $y(0) = 5$, $y'(0) = 0$.

In Exercises 12–14, consider a mass sliding on a frictionless table between two walls 1 unit apart and connected to both walls with springs, as shown below.

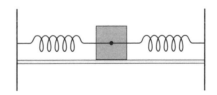

Let k_1 and k_2 be the spring constants of the left and right spring, respectively, let m be the mass, and let k_d be the damping coefficient of the medium the spring is sliding through. Suppose L_1 and L_2 are the rest lengths of the left and right springs, respectively.

12. Write a second-order differential equation for the position of the mass at time t. [*Hint*: The first step is to pick an origin, that is, a point to set the position equal to 0. Halfway between the walls is a natural choice, but remember at this point the springs may be exerting forces, depending on their rest lengths.]

13. Convert the second-order equation of Exercise 12 into a first-order system.

14. Compare the first-order system for this two-spring system with the harmonic oscillator system. How do they compare?

15. A mattress company advertises that its mattresses have "nonlinear" springs that give easily when you press lightly (for comfort), but compress much less easily when pushed harder (for support). Which of the following equations is a better model for this spring (where α and β are positive and $y < 0$ corresponds to compressed springs)? What do the parameters α and β tell you about the nonlinear spring? Justify your answer.

    (a) $\dfrac{d^2 y}{dt^2} = -\alpha y - \beta y^3$
                (b) $\dfrac{d^2 y}{dt^2} = -\alpha y + \beta y^3$

In Exercises 16–20, we consider the behavior of a cable of a suspension bridge (see also Section 5.4). Each cable acts as a spring when it is stretched, exerting a force pulling the roadbed of the bridge back toward the rest position. However, if the roadbed of the bridge is above its rest position, the cable is slack and provides no downward force. Only gravity pulls the roadbed back down. To model this, we let $y(t)$ denote the position of a section of the bridge above ($y > 0$) or below ($y < 0$) the rest position. Assuming a unit mass, Newton's laws give

$$\frac{d^2y}{dt^2} = f(y),$$

where $f(y)$ is the force, so

$$f(y) = -g - ky \text{ if } y < 0$$

and

$$f(y) = -g \text{ if } y > 0,$$

where g is the gravitational constant and k is the spring constant for the stretched cable. This yields the system

$$\frac{dy}{dt} = v$$
$$\frac{dv}{dt} = f(y).$$

16. Sketch the direction field for this system. (For the sake of the picture, take $k = 10$ and $g = 1$ (that is, choose units so the acceleration of gravity is 1.) Do the precise values of k and g matter to the qualitative behavior of solutions in the phase plane?)

17. Sketch the phase plane for this system (with $k = 10$ and $g = 1$).

18. Do the hypotheses of the Existence and Uniqueness Theorem hold for this system? Justify your answer.

19. Do solutions exist for every initial condition? Are solutions unique? [*Hint*: Consider two cases $y \leq 0$ and $y > 0$, and then consider what happens when a solution moves from one case to the other.]

20. Compare the phase plane of this system with the phase plane for the harmonic oscillator

$$\frac{dy}{dt} = v$$
$$\frac{dv}{dt} = -ky.$$

What are the similarities and differences between the solutions for these systems?

2.5 EULER'S METHOD FOR AUTONOMOUS SYSTEMS

Many of our examples in this chapter include some type of plot of solutions, either as curves in the phase plane or as $x(t)$- or $y(t)$-graphs. And in most cases, these plots are provided without any indication of how we obtain them. Occasionally the solutions are line segments or circles or ellipses, and we are able to verify this analytically. But more often, the solutions do not lie on familiar curves. For example, consider the predator-prey type system

$$\frac{dx}{dt} = 2x - 1.2xy$$

$$\frac{dy}{dt} = -y + 1.2xy$$

and the solution that satisfies the initial condition $(x(0), y(0)) = (1.75, 1.0)$. Figure 2.46 shows this solution in the phase plane, the xy-plane, and Figure 2.47 contains the corresponding $x(t)$- and $y(t)$-graphs. Figure 2.46 suggests that this solution is a closed curve, but the curve is certainly neither circular nor elliptical. Similarly, the $x(t)$- and $y(t)$-graphs appear to be periodic, although they do not seem to be graphs of any of the standard periodic functions (sine, cosine, etc.) So how did we compute these graphs?

The answer to this question is essentially the same as the answer to the analogous question for first-order equations. We use a dependable numerical technique and the aid of a computer. In this section we define Euler's method for first-order systems. Other numerical methods are discussed in Chapter 7. We confine our discussion to numerical approximation of the solutions of autonomous equations, although it is not difficult to generalize the method to nonautonomous equations.

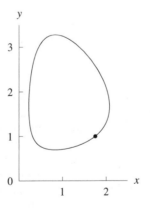

Figure 2.46
A solution curve corresponding to the initial condition
$(x_0, y_0) = (1.75, 1.0)$.

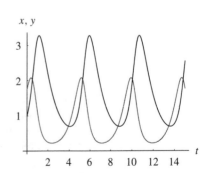

Figure 2.47
The corresponding $x(t)$- and $y(t)$-graphs for the solution curve in Figure 2.46.

Derivation of Euler's method

Consider the first-order, autonomous system

$$\frac{dx}{dt} = f(x, y)$$
$$\frac{dy}{dt} = g(x, y),$$

along with the initial condition $(x(t_0), y(t_0)) = (x_0, y_0)$. We have seen that we can use vector notation to rewrite this system as

$$\frac{d\mathbf{Y}}{dt} = \mathbf{F}(\mathbf{Y}),$$

where $\mathbf{Y} = (x, y)$, $d\mathbf{Y}/dt = (dx/dt, dy/dt)$, and $\mathbf{F}(\mathbf{Y}) = (f(x, y), g(x, y))$. The vector-valued function $\mathbf{F}$ yields a vector field, and a solution is a curve whose tangent vector at any point on the curve agrees with the vector field (see Figure 2.48). In other words, the "velocity" vector for the curve is equal to the vector $\mathbf{F}(x(t), y(t))$.

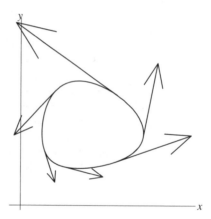

Figure 2.48
A solution curve is a curve that is
everywhere tangent to the vector field.

As we saw in Section 1.4, Euler's method for a first-order equation is based on the idea of approximating the graph of a solution by line segments whose slopes are obtained from the differential equation. Euler's approximation scheme for systems is the same basic idea interpreted in a vector framework.

Given an initial condition (x_0, y_0), how can we use the vector field $\mathbf{F}(x, y)$ to approximate the solution curve? Just as for equations, we first pick a step size Δt. The vector $\mathbf{F}(x_0, y_0)$ is the velocity vector of the solution through (x_0, y_0), so we begin our approximate solution by using $\Delta t \, \mathbf{F}(x_0, y_0)$ to form the first "step." In other words, we step from (x_0, y_0) to (x_1, y_1), where the point (x_1, y_1) is given by

$$(x_1, y_1) = (x_0, y_0) + \Delta t \, \mathbf{F}(x_0, y_0)$$

(see Figure 2.49). This corresponds to traveling for time Δt with velocity $\mathbf{F}(x_0, y_0)$.

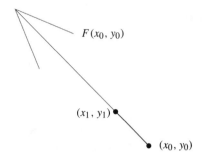

Figure 2.49
The vector at (x_0, y_0) and the point (x_1, y_1) obtained from one step of Euler's method.

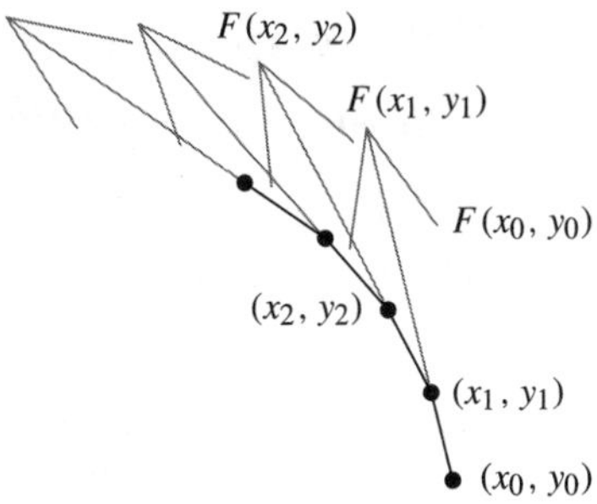

Figure 2.50
The approximate solution curve obtained from five Euler steps.

Having calculated a point (x_1, y_1) on the approximate solution curve, we calculate the new velocity vector $\mathbf{F}(x_1, y_1)$. The second step in the approximation is

$$(x_2, y_2) = (x_1, y_1) + \Delta t\, \mathbf{F}(x_1, y_1).$$

We repeat this scheme and obtain an approximate solution curve (see Figure 2.50).

In practice we choose a step size Δt that is small enough to provide an accurate solution over the given time interval. (See Chapter 7 for a technical discussion of how small is small enough for Δt and how small is too small).

Euler's method for Systems

Euler's method for systems can be written without the vector notation as follows. Given the system

$$\frac{dx}{dt} = f(x, y)$$
$$\frac{dy}{dt} = g(x, y),$$

the initial condition (x_0, y_0), and the step size Δt, we calculate the Euler approximation by repeating the calculations:

$$m_k = f(x_k, y_k)$$
$$n_k = g(x_k, y_k),$$

$$x_{k+1} = x_k + m_k\, \Delta t$$
$$y_{k+1} = y_k + n_k\, \Delta t.$$

Euler's Method Applied to the Van der Pol Equation

For example, consider the second-order differential equation

$$\frac{d^2x}{dt^2} - (1 - x^2)\frac{dx}{dt} + x = 0.$$

This equation is called the **Van der Pol equation**. To study it numerically, we must convert it into a first-order system by letting $y = dx/dt$. The resulting system is

$$\frac{dx}{dt} = y$$

$$\frac{dy}{dt} = -x + (1 - x^2)y.$$

Suppose we want to find the approximate solution for the initial condition $(x(0), y(0)) = (1, 1)$. We do a few calculations by hand to see how Euler's method works, and then turn to the computer for the repetitive part. The method is best illustrated by doing a calculation with a relatively large step size, although in practice we would never use such a large value for Δt.

Let $\Delta t = 0.25$. Given the initial condition $(x_0, y_0) = (1, 1)$, we compute the vector field

$$\mathbf{F}(x, y) = (y, -x + (1 - x^2)y)$$

at $(1, 1)$. We obtain the vector $\mathbf{F}(1, 1) = (1, -1)$. Thus our first step starts at $(1, 1)$ and ends at

$$(x_1, y_1) = (x_0, y_0) + \Delta t \, \mathbf{F}(x_0, y_0)$$

$$= (1, 1) + 0.25 \, (1, -1)$$

$$= (1.25, 0.75).$$

In other words, since $\Delta t = 0.25$, we obtain (x_1, y_1) from (x_0, y_0) by stepping one-quarter of the way along the displacement vector $(1, -1)$ (see Figure 2.51).

The next step is obtained by computing the vector field at (x_1, y_1). We have $\mathbf{F}(1.25, 0.75) = (0.75, -1.67)$ (to 2 decimal places). Consequently, our next step starts at $(1.25, 0.75)$ and ends at

$$(x_2, y_2) = (1.25, 0.75) + 0.25 \, (0.75, -1.67)$$

$$= (1.44, 0.33)$$

(see Figure 2.51).

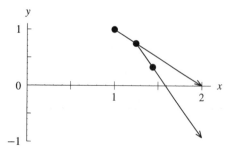

Figure 2.51
Two steps of Euler's method applied to the Van der Pol equation with initial condition $(x_0, y_0) = (1, 1)$ and step size $\Delta t = 0.25$.

Table 2.1 Ten steps of Euler's method

i	x_i	y_i	m_i	n_i
0	1	1	1	-1
1	1.25	0.75	0.75	-1.671875
2	1.4375	0.332031	0.332031	-1.791580
3	1.520507	-0.115864	-0.115864	-1.368501
4	1.491542	-0.457989	-0.457989	-0.930644
5	1.377045	-0.690650	-0.690650	-0.758048
6	1.204382	-0.880162	-0.880162	-0.807837
7	0.984342	-1.082121	-1.082121	-1.017965
8	0.713811	-1.336613	-1.336613	-1.369384
9	0.379658	-1.678959	-1.678959	-1.816611
10	-0.04008	-2.133112		

Table 2.1 illustrates the computations necessary to calculate ten steps of Euler's method, starting at the initial condition $(x_0, y_0) = (1, 1)$ with $\Delta t = 0.25$. The resulting approximate solution curve is shown in Figure 2.52.

As we mentioned above, $\Delta t = 0.25$ is much larger than the typical step size, so let's repeat our calculations with $\Delta t = 0.1$. Since we are going to use a computer to do these calculations, we might as well do more steps, too. Figure 2.53 shows the result of this calculation. In this figure we show both the points obtained in the calculation as well as a graph of an approximate solution curve obtained by joining successive points by line segments. Note that the curve is hardly a "standard" shape and that it is almost a closed curve.

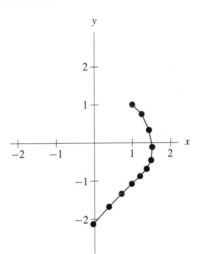

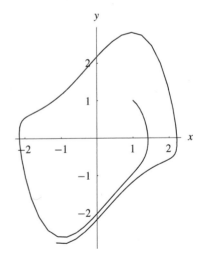

Figure 2.52
Ten steps of Euler's method applied to the Van der Pol equation with initial condition $(x_0, y_0) = (1, 1)$ and step size $\Delta t = 0.25$.

Figure 2.53
One hundred steps of Euler's method applied to the Van der Pol equation with initial condition $(x_0, y_0) = (1, 1)$ and step size $\Delta t = 0.1$.

Table 2.2 Ten steps of Euler's method with $t_0 = 0$

i	t_i	x_i	y_i	m_i	n_i
0	0	1	1	1	−1
1	0.25	1.25	0.75	0.75	−1.671875
2	0.50	1.4375	0.332031	0.332031	−1.791580
3	0.75	1.520507	−0.115864	−0.115864	−1.368501
4	1.00	1.491542	−0.457989	−0.457989	−0.930644
5	1.25	1.377045	−0.690650	−0.690650	−0.758048
6	1.50	1.204382	−0.880162	−0.880162	−0.807837
7	1.75	0.984342	−1.082121	−1.082121	−1.017965
8	2.00	0.713811	−1.336613	−1.336613	−1.369384
9	2.25	0.379658	−1.678959	−1.678959	−1.816611
10	2.50	−0.04008	−2.133112		

To show the $x(t)$- and $y(t)$-graphs for this approximate solution, we must include information about the independent variable t in our Euler's method table. If we assume that the initial condition $(x_0, y_0) = (1, 1)$ corresponds to the initial time $t_0 = 0$, we can augment that table by adding the corresponding times (see Table 2.2). Thus we are able to produce $x(t)$- and $y(t)$-graphs of approximate solutions (see Figures 2.54 and 2.55). Figures 2.56 and 2.57 illustrate how the "almost" closed solution curve in the phase plane (the xy-plane) corresponds to the functions $x(t)$ and $y(t)$, which are essentially periodic.

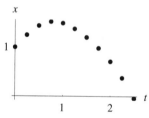

Figure 2.54
The $x(t)$-graph corresponding to the approximate solution curve obtained in Table 2.2.

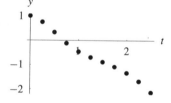

Figure 2.55
The $y(t)$-graph corresponding to the approximate solution curve obtained in Table 2.2.

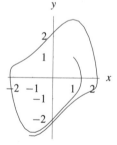

Figure 2.56
The approximate solution curve in the xy-plane.

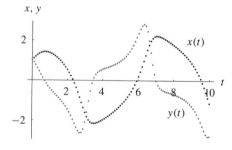

Figure 2.57
The corresponding $x(t)$- and $y(t)$-graphs.

A Swaying Skyscraper

Modern skyscrapers are built to be flexible. In strong gusts of wind or in earthquakes, these buildings tend to sway back and forth to absorb the shocks. Oscillations with an amplitude of several feet and periods on the order of 5 to 10 seconds are common. *

As an application of Euler's method, let's see how we can analyze two simple differential equations models of swaying buildings.

The model

To describe the swaying motion of the skyscraper, let $y(t)$ be a measure of how far the building is bent — the displacement (in meters) of the top of the building with $y = 0$ corresponding to the perfectly vertical position. When y is not zero, the building is bent and the structure applies a strong restoring force back toward the vertical (see Figure 2.58). This is reminiscent of the harmonic oscillator described in Section 2.4. Therefore, as a very crude first approximation of the motion of a swaying building, we can use the damped harmonic oscillator equation

$$\frac{d^2y}{dt^2} + p\frac{dy}{dt} + qy = 0.$$

Here the constants q and p are chosen to reflect the characteristics of the particular building being studied. For the sake of definiteness, we fix the constants $p = 0.2$ and $q = 0.25$ and consider the second-order equation

$$\frac{d^2y}{dt^2} + 0.2\frac{dy}{dt} + 0.25y = 0$$

and the corresponding system

$$\frac{dy}{dt} = v$$
$$\frac{dv}{dt} = -0.25y - 0.2v.$$

These numbers are chosen for the purpose of demonstrating the behavior of solutions (and do not refer to any building, currently standing or not).

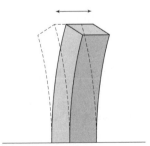

Figure 2.58
Schematic of a skyscraper swaying.

---

*For more information, see Matthys Levy and Mario Salvadori, *Why Buildings Fall Down*, New York: W.W. Norton and Co., 1992, pp. 109–120.

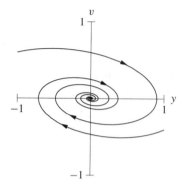

Figure 2.59
Phase plane for the harmonic oscillator

$$\frac{dy}{dt} = v$$

$$\frac{dv}{dt} = -0.25y - 0.2v.$$

For the harmonic oscillator system, we could use analytical techniques to obtain an exact solution (see Chapter 3). But in practice, even when other techniques are available, we often start by getting an idea of the behavior of solutions using numerical methods. Using Euler's method, we can sketch the phase plane for the system (see Figure 2.59). All solutions spiral toward the origin in the yv-plane. This means that once displaced from the vertical, the building will sway back and forth, and the amplitude of the oscillation decreases with each oscillation. If we sketch the $y(t)$-graph for several different initial conditions (see Figure 2.60), we see that these oscillations always have the same frequency no matter what the amplitude or initial condition. This frequency is called the **natural frequency** of the harmonic oscillator and is a fundamental property of solutions of these equations (see Section 3.4 and Chapter 5).

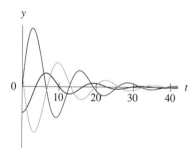

Figure 2.60
Graphs of $y(t)$ for three different solutions of the system

$$\frac{dy}{dt} = v$$

$$\frac{dv}{dt} = -0.25y - 0.2v.$$

The P-Delta Effect

The model given above is extremely crude. We do not claim that the forces present in a swaying building are identical to those of a spring. The harmonic oscillator is only a first approximation of a complicated physical system. To extend the usefulness of this model, we must consider other factors that govern the motion of a swaying building.

One aspect of the model of the swaying building that we have not yet included is the effect of gravity. When the building undergoes small oscillations, gravity does not play a very important role. However, if the oscillations become large, then gravity can have a significant effect. When $y(t)$ is at its maximum value, a portion of the building is not directly above any other part of the building (see Figure 2.61). Therefore, gravity pulls downward on this portion of the building and this force tends to bend the building farther. This is called the "P-Delta" effect ("Delta" is the overhang distance and "P" is

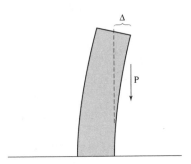

Figure 2.61
The force of gravity on a bent building.

the force of gravity).*

To include this effect in our model in a way that is quantitatively accurate requires knowledge of the density of the building and the flexibility of the construction materials. Without going into these specific details, we can construct a much simplified model that is a caricature of the P-Delta effect.

The P-Delta effect is very small when y is small, much smaller than the restoring force. As y increases, the P-Delta effect becomes quite large. As a first model, we may assume that the force provided by the P-Delta effect is proportional to y^3. Adding this force corresponds to adding a term to the expression for the acceleration of y, that is, adding a term proportional to y^3 to the right-hand side of the second-order differential equation. For the system, this corresponds to adding a term proportional to y^3 to the expression for dv/dt, since this term gives the acceleration (rate of change or velocity).

To study the qualitative behavior of solutions, we assume that the proportionality constant on the y^3 term is 1 (again, with no particular building in mind). Hence our new model is

$$\frac{d^2y}{dt^2} + 0.2\frac{dy}{dt} + 0.25y = y^3,$$

or

$$\frac{dy}{dt} = v$$
$$\frac{dv}{dt} = -0.25y + y^3 - 0.2v.$$

To study this system, we could use qualitative methods (see Sections 2.6 and 3.4). However, numerical techniques give us a good picture of the behavior of solutions. Again using Euler's method, we can compute solutions on the phase plane and $y(t)$- and $v(t)$-graphs (see Figures 2.62 and 2.63). The behavior of solutions is somewhat unnerving. If the initial condition is sufficiently close to zero, then the solution spirals toward the origin as in the case of our original harmonic oscillator model. When the initial condition is sufficiently far from the origin, however, the behavior is quite different; the solution in the phase plane moves away from the origin.

The interpretation of the behavior of these solutions in terms of the behavior of the building yields dramatic results. For small oscillations the building sways with decreasing amplitude and eventually returns to its rest position. However, if the initial

---

*See Levy and Salvadori, *Why Buildings Fall Down*, New York: W.W. Norton and Co., 1992, p. 109, for a lively description of the costs incurred when this effect was not completely considered during the construction of the John Hancock Tower in Boston.

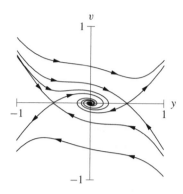

Figure 2.62
Phase plane for the system

$$\frac{dy}{dt} = v$$

$$\frac{dv}{dt} = -0.25y + y^3 - 0.2v.$$

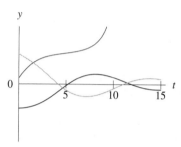

Figure 2.63
Graphs of $y(t)$ for three different solutions of the system

$$\frac{dy}{dt} = v$$

$$\frac{dv}{dt} = -0.25y + y^3 - 0.2v.$$

displacement exceeds a threshold distance, then the amplitude of the solution quickly moves away from zero. The building sways more and more violently, and the result is a disaster. In the exercises and labs we consider similar behavior models in more detail.

Reality Check

Again we must emphasize that this model is only a caricature of the actual dynamics of a swaying building. However, the model does teach an important lesson. Solutions with initial conditions in one region of the phase space may behave very differently from solutions in another region. The transition between different types of solutions can occur abruptly as the initial conditions are varied. Just because a physical system is "stable" with respect to small initial displacements does not imply that it will be stable with respect to all initial conditions. If this simple model can behave in such a radical way, then we should not be surprised to find such bizarre behavior in an actual building.

Exercises for Section 2.5

In Exercises 1–4, a system, an initial condition, a step size, and an integer n are given. Also the direction field for the system is provided.

(a) Calculate the approximate solution given by Euler's method for the given system with the given initial condition and step size for n steps.

(b) Plot your approximate solution on the direction field. Make sure that your approximate solution is consistent with the direction field.

(c) Using a computer or a graphing calculator, obtain a more detailed sketch of the phase portrait for the system.

(d) Make sure that your approximate solution curve is consistent with the phase portrait that you have obtained.

1.

$$\frac{dx}{dt} = y$$

$$\frac{dy}{dt} = -2x - 3y$$

$(x_0, y_0) = (1, 1)$

$\Delta t = 0.25$

$n = 5$

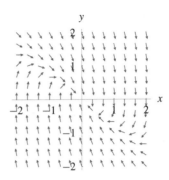

2.

$$\frac{dx}{dt} = y$$

$$\frac{dy}{dt} = -\sin x$$

$(x_0, y_0) = (1, 1)$

$\Delta t = 0.25$

$n = 6$

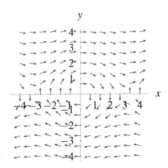

3.

$$\frac{dx}{dt} = y + y^2$$

$$\frac{dy}{dt} = -x + \frac{y}{5} - xy + \frac{6y^2}{5}$$

$(x_0, y_0) = (1, 1)$

$\Delta t = 0.25$

$n = 5$

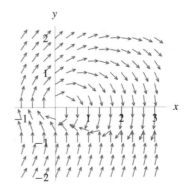

4.

$$\frac{dx}{dt} = y + y^2 \qquad\qquad (x_0, y_0) = (-0.5, 0)$$

$$\frac{dy}{dt} = -\frac{x}{2} + \frac{y}{5} - xy + \frac{6y^2}{5} \qquad\qquad \Delta t = 0.25$$

$$n = 7$$

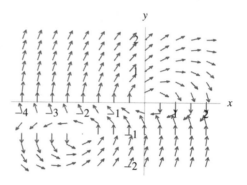

5. For the system

$$\frac{dx}{dt} = -y$$

$$\frac{dy}{dt} = x,$$

the curve $\mathbf{Y}(t) = (\cos t, \sin t)$ is a solution. This solution is periodic. Its initial position is $\mathbf{Y}(0) = (1, 0)$, and it returns to this position when $t = 2\pi$. So $\mathbf{Y}(2\pi) = (1, 0)$ and $\mathbf{Y}(t + 2\pi) = \mathbf{Y}(t)$ for all t.

(a) Check that $\mathbf{Y}(t) = (\cos t, \sin t)$ is a solution.

(b) Use Euler's method with step size 0.5 to approximate this solution, and check how closely the approximate solution is to the real solution when $t = 4$, $t = 6$, and $t = 10$.

(c) Use Euler's method with step size 0.1 to approximate this solution, and check how closely the approximate solution is to the real solution when $t = 4$, $t = 6$, and $t = 10$.

(d) The points on the solution curve $\mathbf{Y}(t)$ are all 1 unit distance from the origin. Is this true of the approximate solutions? Are they too far from the origin or too close to it? What will happen for other step sizes (that is, will approximate solutions formed with other step sizes be too far or too close to the origin)?

[It is well worthwhile to find a computer or calculator to help with the arithmetic of Euler's method.]

6. For the system

$$\frac{dx}{dt} = 2x$$

$$\frac{dy}{dt} = y$$

we claim that the curve $\mathbf{Y}(t) = (e^{2t}, 3e^t)$ is a solution. Its initial position is $\mathbf{Y}(0) = (1, 3)$.

(a) Check that $\mathbf{Y}(t) = (e^{2t}, 3e^t)$ is a solution.

(b) Use Euler's method with step size $\Delta t = 0.5$ to approximate this solution, and check how closely the approximate solution is to the real solution when $t = 2$, $t = 4$, and $t = 6$.

(c) Use Euler's method with step size $\Delta t = 0.1$ to approximate this solution, and check how closely the approximate solution is to the real solution when $t = 2$, $t = 4$, and $t = 6$.

(d) Discuss how and why the Euler approximations differ from the solution.

[It is well worthwhile to find a computer or calculator to help with the arithmetic of Euler's method.]

In Exercises 7– 10, we consider the two systems given by

(i) $\dfrac{dy}{dt} = v$ (ii) $\dfrac{dy}{dt} = v$

$\dfrac{dv}{dt} = -y - 0.2v$ $\dfrac{dv}{dt} = -y - 0.2v + y^3.$

In each exercise, we give a table containing numerical approximations of two solutions generated using Euler's method with step size $\Delta t = 0.1$ for one of the systems above. (In order to save space, only every fifth step of Euler's method is given in the table.) In each case, state which of the two systems the approximate solutions come from and justify your answer. [*Hint*: The two approximate solutions in each problem come from the same system. You may want to compare the two solutions to determine which system they come from.]

7.

Table 2.3 Two approximate solutions for either system (i) or (ii) via Euler's method with $\Delta t = 0.1$.

k	t_k	(y_k, v_k)	(y_k, v_k)	k	t_k	(y_k, v_k)	(y_k, v_k)
0	0	(0.5,0)	(1,0)	35	3.5	(-0.40,0.02)	(1.0,0.0)
5	0.5	(0.46,-0.18)	(1.0,0.0)	40	4.0	(-0.36,0.18)	(1.0,0.0)
10	1.0	(0.34,-0.33)	(1.0,0.0)	45	4.5	(-0.24,0.30)	(1.0,0.0)
15	1.5	(0.16,-0.42)	(1.0,0.0)	50	5.0	(-0.08,0.35)	(1.0,0.0)
20	2.0	(-0.05,-0.41)	(1.0,0.0)	55	5.5	(0.09,0.32)	(1.0,0.0)
25	2.5	(-0.24,-0.30)	(1.0,0.0)	60	6.0	(0.24,0.22)	(1.0,0.0)
30	3.0	(-0.36,-0.15)	(1.0,0.0)	65	6.5	(0.32,0.08)	(1.0,0.0)

8.

Table 2.4 Two approximate solutions for either system (i) or (ii) via Euler's method with $\Delta t = 0.1$.

k	t_k	(y_k, v_k)	(y_k, v_k)
0	0	(0,-0.5)	(0,-2)
5	.5	(-0.24,-0.40)	(-0.94,-1.7)
10	1.0	(-0.41,-0.23)	(-1.8,-2.0)
15	1.5	(-0.48,-0.04)	(-3.4,-7.1)
20	2.0	(-0.46,0.14)	(-16.0,-154.4)

9.

Table 2.5 Two approximate solutions for either system (i) or (ii) via Euler's method with $\Delta t = 0.1$.

k	t_k	(y_k, v_k)	(y_k, v_k)	k	t_k	(y_k, v_k)	(y_k, v_k)
0	0	(0.25,0)	(0.9,0)	35	3.5	(-0.20,0.07)	(-0.73,0.27)
5	0.5	(0.23,-0.12)	(0.81,-0.42)	40	4.0	(-0.14,0.16)	(-0.53,0.56)
10	1.0	(0.15,-0.20)	(0.53,-0.72)	45	4.5	(-0.06,0.19)	(-0.22,0.71)
15	1.5	(0.04,-0.23)	(0.14,-0.84)	50	5.0	(0.04,0.19)	(0.14,0.67)
20	2.0	(-0.07,-0.21)	(-0.27,-0.74)	55	5.5	(0.12,0.13)	(0.44,0.48)
25	2.5	(-0.16,-0.13)	(-0.59,-0.47)	60	6.0	(0.17,0.05)	(0.62,0.18)
30	3.0	(-0.21,-0.03)	(-0.76,-0.10)	65	6.5	(0.18,-0.04)	(0.65,-0.15)

10.

Table 2.6 Two approximate solutions for either system (i) or (ii) via Euler's method with $\Delta t = 0.1$.

k	t_k	(y_k, v_k)	(y_k, v_k)	k	t_k	(y_k, v_k)	(y_k, v_k)
0	0	(0.25,0)	(0.9,0)	35	3.5	(-0.21,0.06)	(-0.31,-0.46)
5	0.5	(0.23,-0.11)	(0.88,-0.8)	40	4.0	(-0.16,0.14)	(-0.50,-0.26)
10	1.0	(0.15,-0.19)	(0.82,-0.18)	45	4.5	(-0.08,0.19)	(-0.59,-0.05)
15	1.5	(0.05,-0.23)	(0.71,-0.31)	50	5.0	(0.02,0.19)	(-0.58,0.13)
20	2.0	(-0.06,-0.21)	(0.52,-0.46)	55	5.5	(0.10,0.14)	(-0.47,0.31)
25	2.5	(-0.15,-0.14)	(0.26,-0.58)	60	6.0	(0.16,0.07)	(-0.29,0.43)
30	3.0	(-0.20,-0.04)	(-0.03,-0.59)	65	6.5	(0.18,-0.02)	(-0.06,0.49)

11. Suppose you are studying the behavior of a swaying building. Let $y(t)$ represent the displacement from the vertical and $v(t)$ be the velocity of motion. Consider the two systems below which model this situation (these are the same as the systems considered in the section except the coefficient of damping has been set to zero). The first assumes that the "P-delta" effect is negligible (that is, can be safely left out of the model), while the second includes a term for the P-delta effect. What observations would you make to determine which model is more accurate? [*Hint:* You want to observe the building under situations when there is no danger that it will fall down. Thus, no matter which model is more accurate, only solutions with $y(t)$ fairly small will be available.]

(i) $\dfrac{dy}{dt} = v$

$\dfrac{dv}{dt} = -y$

(ii) $\dfrac{dy}{dt} = v$

$\dfrac{dv}{dt} = -y + y^3$

2.6 QUALITATIVE ANALYSIS

As we have emphasized many times already, finding formulas for solutions of systems of differential equations is difficult at best and frequently impossible. Even when formulas for solutions are available, it is difficult to extract an understanding of the behavior of the solution from a complicated formula. In this section we continue to develop qualitative techniques and use them in combination with numerics to understand some examples.

Competing Species

Let x and y denote the populations of two species that compete for space or resources. An increase in either of the species causes a decrease in the growth rate of the other species. An example of such a system of differential equations is

$$\frac{dx}{dt} = 2x \left(1 - \frac{x}{2}\right) - xy$$

$$\frac{dy}{dt} = 3y \left(1 - \frac{y}{3}\right) - 2xy.$$

The coefficients are chosen for purposes of illustration rather than for modeling any particular species. The effect of altering the parameters is considered in the exercises.

For a given value of x, if y increases then the $-xy$ term causes dx/dt to decrease. Similarly, for a given value of y, if x increases then the $-2xy$ term causes dy/dt to decrease. An increase in the population of either species causes a decrease in the rate of growth of the other species. Thus this system models the populations of competing species.

We begin our analysis of this system by noting that if $y = 0$, we have $dy/dt = 0$; that is, if the y's are extinct, they stay extinct. In this case

$$\frac{dx}{dt} = 2x \left(1 - \frac{x}{2}\right),$$

which is a logistic population model with a carrying capacity of 2. The phase line of this equation agrees with the x-axis of the phase plane. In particular, $(0, 0)$ and $(2, 0)$ are equilibrium points of the system. Similarly, if $x = 0$, we have $dx/dt = 0$, so the phase line of

$$\frac{dy}{dt} = 3y \left(1 - \frac{y}{3}\right)$$

agrees with the y-axis of the phase plane of the system, and $(0,3)$ is another equilibrium point.

By the Uniqueness Theorem, solutions with initial conditions in the first quadrant (x and $y > 0$) must remain in the first quadrant for all time. That is, the axes are made up of solutions of the system, and the Uniqueness Theorem guarantees that solutions can't cross. Because this model refers to populations and negative populations do not make much sense (they are not "physically relevant"), we can limit our attention to the first quadrant only.

We find equilibrium points by solving for x and y in the system of equations

$$\begin{cases} 2x \left(1 - x/2\right) - xy = 0 \\ 3y \left(1 - y/3\right) - 2xy = 0, \end{cases}$$

which can be rewritten in the form

$$\begin{cases} x(2 - x - y) = 0 \\ y(3 - y - 2x) = 0. \end{cases}$$

The first equation is satisfied if $x = 0$ or if $2 - x - y = 0$. The second equation is satisfied if $y = 0$ or if $3 - y - 2x = 0$. With $x = 0$, the equation $y = 0$ yields the equilibrium point at the origin, and the equation $3 - y - 2x = 0$ yields the equilibrium point at $(0, 3)$. With $2 - x - y = 0$, the equation $y = 0$ yields the equilibrium point at $(2, 0)$, and the equation $3 - y - 2x = 0$ yields the equilibrium point at $(1, 1)$. (Subtract $2 - x - y = 0$ from $3 - y - 2x = 0$ to obtain $x = 1$, and then solve for y using either equation.) Hence the equilibrium points are $(0, 0)$, $(0, 3)$, $(2, 0)$, and $(1, 1)$. The existence of the equilibrium point $(1, 1)$ indicates that it is possible for these two species to coexist in equilibrium.

Nullclines

The direction field for the competing species system is sketched in Figure 2.64. We can use this picture to provide a rough sketch of solution curves, but with a bit more qualitative analysis, we can give a much more complete picture of the behavior of solutions. One tool for this analysis is the **nullcline**.

Definition: For the system

$$\frac{dx}{dt} = f(x, y)$$

$$\frac{dy}{dt} = g(x, y),$$

the *x-nullcline* is the set of points (x, y) where $f(x, y)$ is zero — that is, the level curve where $f(x, y)$ is zero. The *y-nullcline* is the set of points where $g(x, y)$ is zero.

Along the x-nullcline, the x-component of the vector field is zero, and consequently, the vector field is vertical. It points either straight up or straight down. Similarly, on the

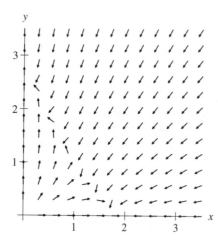

Figure 2.64
Direction field for the competing species

$$\frac{dx}{dt} = 2x \left(1 - \frac{x}{2}\right) - xy$$

$$\frac{dy}{dt} = 3y \left(1 - \frac{y}{3}\right) - 2xy.$$

y-nullcline, the y-component of the vector field is zero, so the vector field is horizontal. It points either left or right. Because both $f(x, y)$ and $g(x, y)$ must be zero at an equilibrium point, the intersections of the nullclines are the equilibrium points.

To show how nullclines can be used to help in the qualitative analysis of systems, we consider our competing species example

$$\frac{dx}{dt} = f(x, y) = 2x\left(1 - \frac{x}{2}\right) - xy$$

$$\frac{dy}{dt} = g(x, y) = 3y\left(1 - \frac{y}{3}\right) - 2xy.$$

Recall that we are only interested in what happens in the first quadrant ($x, y \geq 0$) since this is a population model. The x-nullcline is the set of points (x, y) that satisfy

$$2x\left(1 - \frac{x}{2}\right) - xy = x(2 - x - y) = 0.$$

This set consists of two lines, $x = 0$ and $y = -x + 2$. On these lines the x-component of the vector field is zero. Thus, the vector field is vertical along these lines. In the remainder of the phase plane the x-component of the vector field is either positive or negative. If $dx/dt = f(x, y) > 0$, then solutions move toward the right; if $dx/dt = f(x, y) < 0$, then solutions move toward the left. In Figure 2.65 we mark the x-nullcline with vertical lines as a reminder that the vector field is vertical on the nullcline. We can label the regions off the x-nullcline as either "right" or "left" depending on whether $dx/dt = f(x, y)$ is positive or negative.

Similarly, the y-nullcline is the set of points where $dy/dt = g(x, y) = 0$ or the set of points satisfying

$$3y\left(1 - \frac{y}{3}\right) - 2xy = y(3 - y - 2x) = 0.$$

This set also consists of two lines, $y = 0$ and $y = -2x + 3$. On these lines the y-component of the vector field is zero, so the vector field is horizontal. On the rest of the phase plane either $dy/dt > 0$ and solutions move upward, or $dy/dt < 0$ and solutions move downward. In Figure 2.66 we mark the y-nullcline with horizontal segments as a reminder that the vector field is horizontal on the nullcline. We label the regions off the y-nullcline as either "up" or "down" depending on whether $dy/dt = g(x, y)$ is positive or negative.

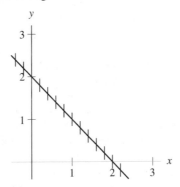

Figure 2.65
The x-nullcline for the system
$dx/dt = 2x(1 - x/2) - xy,\quad dy/dt = 3y(1-y/3)-2xy.$

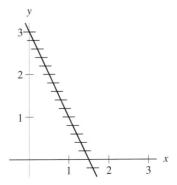

Figure 2.66
The y-nullcline for the system
$dx/dt = 2x(1 - x/2) - xy,\quad dy/dt = 3y(1-y/3)-2xy.$

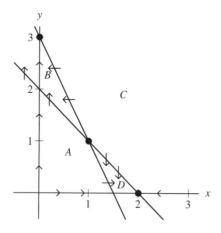

Figure 2.67
The x- and y-nullclines for

$$\frac{dx}{dt} = 2x\left(1 - \frac{x}{2}\right) - xy$$

$$\frac{dy}{dt} = 3y\left(1 - \frac{y}{3}\right) - 2xy.$$

The nullclines divide the first quadrant into four regions marked A, B, C and D.

We combine the x- and y-nullclines in Figure 2.67. The equilibrium points occur at the intersections of the x- and y-nullclines. The nullclines divide the first quadrant into four regions labeled A, B, C and D. We can use this picture to give a detailed analysis of the behavior of solutions of this system.

First consider the triangular region B (see Figure 2.67). The segment $[0, 3]$ on the y-axis is a solution curve and the vector field on the other two sides of B point into B. Hence, a solution that begins in region B at time zero will remain in B for all positive time. There is no way the solution can leave. Region B lies in the "left" ($dx/dt < 0$) and the "up" ($dy/dt > 0$) regions in the phase plane, so we can label it "left-up." That is, as t increases, solutions in region B must move toward the upper-left corner of the region — toward the equilibrium point $(0, 3)$. Similarly, solutions cannot leave region D, and they move "right-down" as t increases. Hence all solutions in region D tend to the equilibrium point $(2, 0)$ as t increases. In Figure 2.68, we display two solutions of the competing species model together with their $x(t)$- and $y(t)$-graphs.

Solutions in region A move "right-up." There are three possibilities for what happens to these solutions. They may leave region A and enter B, they may leave A and enter D, or they may approach the equilibrium point $(1, 1)$. Since some solutions in A enter B and some enter D, there must be a least one solution curve in A that tends toward the equilibrium point $(1, 1)$. We will see in Chapter 4 that there is in fact exactly one curve of solutions in A that tend to $(1, 1)$. These solutions separate the solutions that end up in B from those that end up in D, so they are called **separatrix** solutions.

Similarly, solutions in region C move left-down, tending toward regions B, D, or the equilibrium point $(1, 1)$. Again there are separatrix solutions that tend to the equilibrium point $(1, 1)$. These solutions divide the solutions that reach B and subsequently approach $(0, 3)$ from those that enter D and approach $(2, 0)$.

This analysis gives us a fairly complete qualitative picture of the behavior of solutions of the competing species model. We know that many solutions tend to an equilibrium population with one species extinct and the other at its carrying capacity (see Figure 2.68). A few exceptional solutions tend to the equilibrium point at $(1, 1)$ where both species coexist. We do not know exactly how many solutions reach this point, but we know that there must be some that do. To refine this picture and determine precisely the solutions that enter regions B or D, we will have to resort to further numerical and analytic techniques. We will revisit this analysis in Chapter 4.

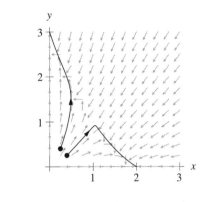

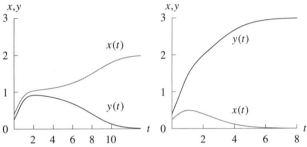

Figure 2.68
Two solutions in the phase plane for the competing species system above with their corresponding $x(t)$- and $y(t)$-graphs.

Be careful

In this example each of the nullclines is a straight line. This is a fairly special situation. In the next examples we will see that nullclines need not be straight lines. Also note that there were two very different kinds of nullclines in this example. For the nullclines along the x- and y-axes, the vector field was tangent to these lines. Consequently, solutions that start on these lines stay on them forever. We can use techniques from Chapter 1 to analyze these solutions completely, since these nullclines really are phase lines for a one-dimensional equation.

The vector field is not tangent to the other nullclines. The vector field points at an angle to these lines, and so these nullclines only give us information about the direction of solutions. They are not solution curves, so we cannot use earlier techniques here.

Nullclines That Are Not Lines

As a slightly more complicated example of how nullclines are used, consider

$$\frac{dx}{dt} = 2x \left(1 - \frac{x}{2}\right) - xy$$
$$\frac{dy}{dt} = y \left(\frac{9}{4} - y^2\right) - x^2 y.$$

We can still interpret this as a competing species model because the growth rate of each species decreases when the population of the other increases. In this model, the additional complications in the dy/dt equation are for the sake of illustration of our techniques. As is usual with population models, we consider only solutions in the first quadrant.

The x-nullcline is given by

$$x(2 - x - y) = 0,$$

which consists of the two lines $x = 0$ and $y = -x + 2$. The y-nullcline is now the set of points (x, y) that satisfy

$$y\left(-x^2 - y^2 + \tfrac{9}{4}\right) = 0.$$

These points lie either on the line $y = 0$ or the circle $x^2 + y^2 = 9/4 = (3/2)^2$. The intersection points of the x- and y-nullclines give the equilibrium points $(0, 0)$, $(0, 3/2)$, $(1 + \sqrt{2}/4, 1 - \sqrt{2}/4) \approx (1.35, 0.65)$, $(1 - \sqrt{2}/4, 1 + \sqrt{2}/4) \approx (0.65, 1.35)$, and $(2, 0)$ (see Figure 2.69). (The point $(0, -3/2)$ is also an equilibrium point, but since it lies outside the first quadrant, it is not important to our analysis.)

As in the previous example, the x- and y-axes consist of solution curves. If $y = 0$, then $dy/dt = 0$ and we have

$$\frac{dx}{dt} = 2x\left(1 - \frac{x}{2}\right),$$

which is the same logistic model as in the previous example. If $x = 0$, we have $dx/dt = 0$ and

$$\frac{dy}{dt} = y\left(-y^2 + \frac{9}{4}\right).$$

The phase line for $dy/dt = y(-y^2 + 9/4)$ has equilibrium points at $y = 0$, $y = 3/2$ (and $y = -3/2$ but we are not considering $y < 0$). On the phase line, $y = 0$ is a source and $y = 3/2$ is a sink (see Figure 2.70).

The x- and y-nullclines divide the first quadrant into 5 regions labeled A, B, C, D, and E in Figure 2.69. A solution that enters in regions B, C, or D remains in that region as t increases since the vector field on the boundaries of these regions never points out of these regions. By labeling the regions with the direction of the vector field (such as "right-up" in region A), we see that solutions in B and C tend toward the equilibrium point $(0.65, 1.35)$, whereas those in region D tend toward $(2, 0)$. In Figure 2.70 we

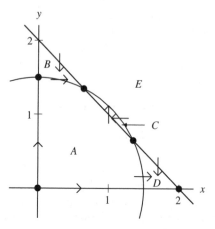

Figure 2.69
Nullclines for the system

$$\frac{dx}{dt} = 2x\left(1 - \frac{x}{2}\right) - xy$$
$$\frac{dy}{dt} = y\left(\frac{9}{4} - y^2\right) - x^2 y.$$

sketch the phase plane and $x(t)$- and $y(t)$-graphs for two of these solutions. Solutions in regions A and E either enter one of the regions B, C, or D, or else they tend to one of the equilibrium points (0.65, 1.35) or (1.35, 0.65). Again we must have separatrix solutions that divide those solutions that tend to (0.65, 1.35) from those that tend to (2, 0) (see Section 4.1 for techniques that allow us to analyze these separatrix solutions).

Qualitative analysis using the nullclines enables us to conclude that all solutions in this model tend toward equilibrium points as t increases, as in the previous model. The choice of which equilibrium point the solution tends toward depends on the location of the initial condition. Unlike the previous example, there is a large set of solutions that tend to the equilibrium point (0.65, 1.35), where the populations of both species are positive (mutual coexistence). The only solutions that approach the equilibrium (0, 3/2) are those on the y-axis (where $x = 0$). On the other hand, there is a large set of initial conditions that give solutions approaching the equilibrium point (2, 0), where the y-species is extinct and the x-species is at its carrying capacity.

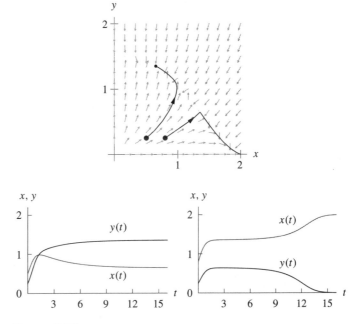

Figure 2.70
Phase plane for modified competing species system with $x(t)$- and $y(t)$-graphs for indicated solutions.

Using All Our Tools

Consider the system

$$\frac{dx}{dt} = x + y - x^3$$
$$\frac{dy}{dt} = -0.5x.$$

Systems of this form arise in the study of nerve cells. Roughly speaking, the variable $x(t)$ represents voltage across the boundary of a nerve cell at time t and $y(t)$ represents the permeability of the cell wall at time t. A rapid change in x corresponds to the nerve cell "firing".

Information from the nullclines

The x-nullcline for this system is the curve $y = -x + x^3$. Above this curve, the x-component of the vector field is positive and below it the x-component is negative. Hence, above the x-nullcline, solutions move to the right, and below the x-nullcline, solutions move to the left. The y-nullcline is the line $x = 0$, the y-axis. To the right of the y-nullcline, we have $dy/dt < 0$, and to the left of the y-nullcline, we have $dy/dt > 0$. Hence, in the right half of the phase plane, solutions move down, and in the left half plane, solutions move up (see Figure 2.71). The x- and y-nullclines divide the the phase plane into four regions. Using the above qualitative analysis, we can conclude that all solutions must circulate clockwise around the origin — the only equilibrium point of the system (see Figure 2.72).

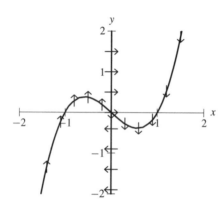

Figure 2.71
The x- and y-nullclines for the system

$$\frac{dx}{dt} = x + y - x^3$$
$$\frac{dy}{dt} = -0.5x.$$

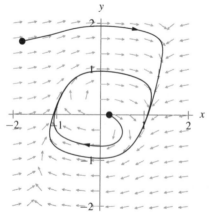

Figure 2.72
Solutions of the system

$$\frac{dx}{dt} = x + y - x^3$$
$$\frac{dy}{dt} = -0.5x.$$

Note that solutions with initial conditions close to the origin spiral outward while those with initial conditions far from the origin spiral inward.

Information from numerical approximations of solutions

To get a more detailed idea of the behavior of solutions of this system, we use numerical methods to find some approximate solutions. If we take an initial condition near the origin, then the resulting solution spirals outward. Likewise, if we take an initial condition far from the origin, then the resulting solution spirals inward.

From the Uniqueness Theorem, we know that solutions never cross. Hence, the solutions that start near the origin must eventually stop spiraling outward, otherwise they would cross the inward spiraling solutions. Between the outward and inward spiraling solutions there must be at least one solution that neither spirals outward nor inward. This solution is periodic. Numerical evidence indicates that there is only one periodic solution and that all other solutions (except the equilibrium solution at the origin) spiral toward this periodic solution. The $x(t)$- and $y(t)$-graphs of two solutions are shown in Figure 2.73. From this picture we see that both of these solutions converge toward the same type of periodic behavior.

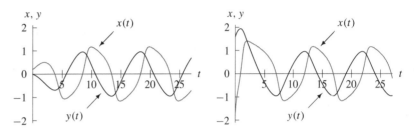

Figure 2.73
The $x(t)$- and $y(t)$-graphs for two solutions of the system
$dx/dt = x + y - x^3$, $dy/dt = -0.5x$.

Mathematical Maturity

In this section we have analyzed three first-order systems of differential equations, employing qualitative and numerical techniques. As we have seen, the analysis of systems of differential equations is much more difficult than the analysis of differential equations with only one dependent variable. We need many different types of techniques to handle systems, and we must be willing to use whichever technique is appropriate. The ability to choose the appropriate technique for a particular problem is part of what is called *mathematical maturity*.

Exercises for Section 2.6

In Exercises 1–3, sketch the x- and y-nullclines of the system specified. Then find all equilibrium points. Using the direction of the vector field between the nullclines, describe the possible fate of the solution curves corresponding to the initial conditions a, b, and c.

1. $\dfrac{dx}{dt} = 2 - x - y$

$\dfrac{dy}{dt} = y - x^2$

(a) $x_0 = 2$, $y_0 = 1$

(b) $x_0 = 0$, $y_0 = -1$

(c) $x_0 = 0$, $y_0 = 0$

2. $\dfrac{dx}{dt} = 2 - x - y$

$\dfrac{dy}{dt} = y - |x|$

(a) $x_0 = -1$, $y_0 = 1$

(b) $x_0 = 2$, $y_0 = 1$

(c) $x_0 = 2$, $y_0 = 2$

3. $\dfrac{dx}{dt} = x(x - 1)$

$\dfrac{dy}{dt} = x^2 - y$

(a) $x_0 = -1$, $y_0 = 0$

(b) $x_0 = 0.832$, $y_0 = 0$

(c) $x_0 = 1.832$, $y_0 = 0$

In Exercises 4–6 we consider the **Volterra-Lotka system** of differential equations is the system

$$\frac{dx}{dt} = x(-Ax - By + C)$$

$$\frac{dy}{dt} = y(-Dx - Ey + F),$$

where $x, y \geq 0$ and where the parameters A through F are all nonnegative. For each of the systems,

(a) find all equilibria;

(b) discuss the fate of solutions on the x- and y-axes;

(c) draw the nullclines; and

(d) use the nullclines to decide on the fate of solutions in regions between the null-clines wherever possible.

4. $\frac{dx}{dt} = x(-x - 3y + 150)$

$\frac{dy}{dt} = y(-2x - y + 100)$

5. $\frac{dx}{dt} = x(10 - x - y)$

$\frac{dy}{dt} = y(30 - 2x - y)$

6. $\frac{dx}{dt} = x(100 - x - 2y)$

$\frac{dy}{dt} = y(150 - x - 6y)$

7. For the system

$$\frac{dx}{dt} = 1 - x - y$$

$$\frac{dy}{dt} = 2(1 - x - y),$$

show that every non-equilibrium solution lies on a straight line. Discuss the fate of solution curves on these lines.

8. Consider again the Volterra-Lotka system modeling competitive species

$$\frac{dx}{dt} = x(-Ax - By + C)$$

$$\frac{dy}{dt} = y(-Dx - Ey + F),$$

where $x, y \geq 0$ and where the parameters A through F are all positive.

(a) Can there ever be more than one equilibrium point for which both $x > 0$ and $y > 0$ (that is, where the species "coexist in equilibrium")? If so, give an example of such values of A through F. If not, why not?

(b) What condition on parameters A through F guarantees that there is at least one equilibrium point with $x > 0$ and $y > 0$?

9. The Volterra-Lotka system of differential equations for competitive species is

$$\frac{dx}{dt} = x(-Ax - By + C)$$
$$\frac{dy}{dt} = y(-Dx - Ey + F),$$

where $x, y \geq 0$ and the parameters A through F are all nonnegative.

Some species live in a "cooperative" manner, each species helping the other to survive and prosper (for example, flowers and honey bees).

(a) How would you alter the Volterra-Lotka system described above to give a general form for a set of systems modeling cooperative species?

(b) What do the nullclines look like for your cooperative system? Are there equilibrium points? Are there conditions on the parameters that guarantee that there are equilibrium points with both x and y positive?

In Exercises 10–13, draw the nullclines and find all equilibrium points for the given competitive system. Use this information to determine the fate of as many solution curves as possible.

10. $\dfrac{dx}{dt} = x(-x - y + 100)$

$\dfrac{dy}{dt} = y(-x^2 - y^2 + 2500)$

11. $\dfrac{dx}{dt} = x(-x - y + 40)$

$\dfrac{dy}{dt} = y(-x^2 - y^2 + 2500)$

12. $\dfrac{dx}{dt} = x(-4x - y + 160)$

$\dfrac{dy}{dt} = y(-x^2 - y^2 + 2500)$

13. $\dfrac{dx}{dt} = x(-8x - 6y + 480)$

$\dfrac{dy}{dt} = y(-x^2 - y^2 + 2500)$

Exercises 14–18 refer to the models of chemical reactions created in Exercises 19–24 of Section 2.1 and studied in Exercises 22–26 of Section 2.3. Here $a(t)$ is the amount of substance A in a solution and $b(t)$ is the amount of substance B in a solution at time t. We will need to consider only $a, b \geq 0$.

For each system,

(a) sketch the nullclines and draw in the vector field along the nullcline;

(b) label the regions in the ab-phase plane created by the nullclines and determine which general direction the vector field points in each region (that is, increasing/decreasing a, increasing/decreasing b); and

(c) identify the regions that solutions cannot leave and determine the fate of solutions in these regions as time increase.

14. $\dfrac{da}{dt} = -\dfrac{ab}{2}$

$\dfrac{db}{dt} = -\dfrac{ab}{2}$

15. $\dfrac{da}{dt} = 2 - \dfrac{ab}{2}$

$\dfrac{db}{dt} = \dfrac{3}{2} - \dfrac{ab}{2}$

16. $\dfrac{da}{dt} = 2 - \dfrac{ab}{2} - \dfrac{a^2}{3}$

$\dfrac{db}{dt} = \dfrac{3}{2} - \dfrac{ab}{2}$

17. $\dfrac{da}{dt} = 2 - \dfrac{ab}{2} + \dfrac{b^2}{3}$

$\dfrac{db}{dt} = \dfrac{3}{2} - \dfrac{ab}{2} - \dfrac{b^2}{3}$

18. $\dfrac{da}{dt} = 2 - \dfrac{ab}{2} - \dfrac{2ab^2}{3}$

$\dfrac{db}{dt} = \dfrac{3}{2} - \dfrac{ab}{2} - \dfrac{ab^2}{3}$

Exercises 19–21 refer to the system

$$\frac{dx}{dt} = y$$
$$\frac{dy}{dt} = x - x^2.$$

19. Sketch the nullclines and find the direction of the vector field along the nullclines.

20. Show that there is at least one solution in each of the second and fourth quadrants that tends to the origin as $t \to \infty$. Similarly, show that there is at least one solution in each of the first and third quadrants that tends to the origin as $t \to -\infty$.

21. Describe the behavior of solutions near the equilibrium point $(1, 0)$.

2.7 THE LORENZ EQUATIONS

We have already seen that the behavior of solutions of autonomous systems of differential equations can be much more interesting and complicated than solutions of single autonomous equations. For autonomous equations with one dependent variable, the solutions live on a phase line, and their behavior is completely governed by the position and nature of the equilibrium points. Solutions of systems with two dependent variables live in two-dimensional phase planes. A plane has much more "room" than a line, so solutions in a phase plane can do many more interesting things. This includes forming loops (periodic solutions) and approaching and retreating from equilibrium points.

However, there are still severe restrictions on the types of phase plane pictures that are possible for systems. The Uniqueness Theorem says that solution curves in the phase plane do not cross. So, for example, if there is a periodic solution that forms a loop in the phase plane, then solutions with initial conditions inside the loop must stay inside for all time (see Figure 2.74). Also, two or three solutions can fit together to divide the phase plane, and from the Uniqueness Theorem, solutions have to stay in the same region as their initial point (see Figure 2.75).

If we raise the number of dependent variables to three, the situation becomes much more complicated. A solution of an autonomous system with three dependent variables

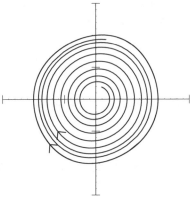

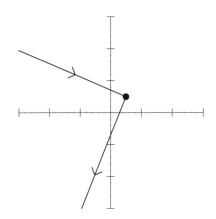

Figure 2.74
Solutions with initial conditions inside a periodic solution must stay inside for all time.

Figure 2.75
Two solutions and an equilibrium point cut the phase plane into regions. Solutions with initial conditions in one region must stay in that region for all time.

is a curve in a three-dimensional "phase space." The Uniqueness Theorem still applies so solution curves do not cross, but in three dimensions this restriction is not nearly so confining as it is in two dimensions. In Figure 2.76 we see examples of curves in three dimensions that do not cross. These curves can knot and link in very complicated ways.

The first to realize the possible complications of three-dimensional systems was Henri Poincaré. In the 1890s, while working on the Newtonian three-body problem, Poincaré realized that systems with three dependent variables can have behavior so complicated that he did not even want to attempt to draw them. With a computer, we can easily draw numerical approximations of complicated solution curves. The problem now is to make sense of the pictures. This is an active area of current research in dynamical systems, and the complete story for systems with three dependent variables is still far from being written.

In this section we study a particular three-dimensional system known as the Lorenz equations. This system was first studied by Ed Lorenz in 1963 in an effort to model the

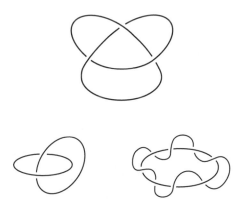

Figure 2.76
A knot and two links in three-space.

weather. It is important because the vector field is formed by very simple equations, yet solutions are very complicated curves.

The Lorenz System

The behavior of a physical system like the weather on earth is (obviously) extremely complicated. To predict the weather, many mathematical models have been developed. The readings from weather stations and satellites are used as initial conditions, and numerical approximations of solutions are used to obtain predictions.

The success of long-range weather forecasts (that is, more than five days in the future) is very limited. This might be because the model equations are inaccurate in their representation of some aspect of the evolution of the weather. It is also possible that the model is accurate but that some property of the equations makes prediction difficult. Consequently it is important to study these models theoretically as well as numerically.

Since the weather is so complicated, it is necessary to start the theoretical study by looking at simplifications. Through a process of simplification, meteorologist Ed Lorenz came up with the following system,

$$\frac{dx}{dt} = \sigma(y - x)$$

$$\frac{dy}{dt} = \rho x - y - xz$$

$$\frac{dz}{dt} = -\beta z + xy,$$

where x, y, and z are dependent variables and σ, ρ, and β are parameters. This system is so much simpler than the equations used for modeling the weather that they have nothing to tell us about tomorrow's temperature. However, by studying this system, Lorenz helped start a scientific revolution by making scientists and engineers aware of the field of mathematics now called "Chaos Theory."

The Vector Field

Lorenz chose to study the system with the parameter values $\sigma = 10$, $\beta = 8/3$, and $\rho = 28$,

$$\frac{dx}{dt} = 10(y - x)$$

$$\frac{dy}{dt} = 28x - y - xz$$

$$\frac{dz}{dt} = -\frac{8z}{3} + xy.$$

The right-hand sides of these equations define a vector field in three-dimensional space

$$\mathbf{F}(x, y, z) = (10(y - x), 28x - y - xz, -\tfrac{8}{3}z + xy),$$

which assigns a three-component vector to each point (x, y, z). Just as in two dimensions, the equilibrium points are the points (x, y, z) for which the vector field is zero, that is, $\mathbf{F}(x, y, z) = (0, 0, 0)$. By direct computation (see Exercise 1) we find that the equilibrium points are $(0, 0, 0)$, $(6\sqrt{2}, 6\sqrt{2}, 27)$, and $(-6\sqrt{2}, -6\sqrt{2}, 27)$.

An initial point for a solution must include values for the three coordinates x, y, and z, so it is a point in space. Given an initial point, we can sketch the corresponding solution by drawing the vector field and sketching a curve whose velocity vector equals the vector field at each point. For example, suppose we choose the initial point $(x(0), y(0), z(0)) = (1, 0, 0)$. Then the vector field at this point is $\mathbf{F}(1, 0, 0) = (-10, 28, 0)$. Hence at time $t = 0$, the x-coordinate of the solution is decreasing, the y-coordinate is increasing, and the z-coordinate has zero velocity. As we move along the solution, we compute the vector field and adjust the velocity vector accordingly. In Figure 2.77 we sketch a small portion of the solution with initial condition $(1, 0, 0)$ using this method. As we can see from this picture, sketching solutions using the vector field is an artistically challenging project.

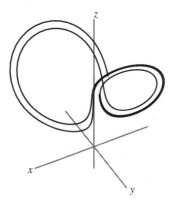

Figure 2.77
Beginning of the solution curve of the Lorenz system with initial point $(1, 0, 0)$.

With the exception of the equilibrium points and solutions with initial conditions on the z-axis (see Exercises 1 and 4), there is little hope of finding a formula for solutions. It is natural to turn to numerical methods. Euler's method for three-dimensional systems works exactly the same as in two dimensions. The approximate solution is constructed by following the vector field for short time steps. Lorenz began his study of this system by finding numerical approximations of solutions, so we follow in his footsteps.

Numerical Approximation of Solutions

We begin by looking at a numerical approximation of the solution with initial condition $(0, 1, 0)$ (that is, $x(0) = 0$, $y(0) = 1$, $z(0) = 0$ — see Figure 2.78). The $x(t)$-, $y(t)$-, and $z(t)$-graphs of this solution are shown in Figure 2.80. Clearly something interesting is going on. The solution does not seem to have any particular pattern. For example,

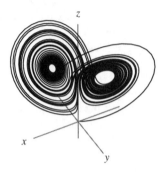

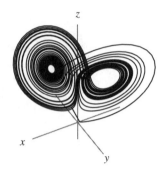

Figure 2.78
Solution of Lorenz system with initial point $(0, 1, 0)$.

Figure 2.79
Solution of Lorenz system with initial point $(0, 1.001, 0)$.

the x-coordinate jumps from positive to negative values in an "unpredictable" way. Although this is somewhat unnerving, we see something even more interesting if we look at the solution with initial condition $(0, 1.001, 0)$ (see Figure 2.79).

This solution starts out very close to the previous one, with x, y, and z oscillating unpredictably. However, if we compare the two solutions, we see that after a short time they become quite far apart (see Figure 2.81). So a very small change in initial condition leads, after a short time, to a big change in the solution.

To get a better view of what is going on, we can look at a solution in the xyz-phase

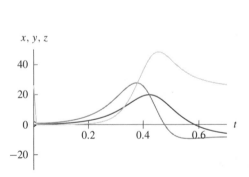

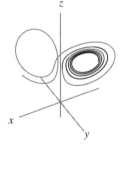

Figure 2.80
$x(t)$-, $y(t)$-, and $z(t)$-graphs of the Lorenz system with initial condition $(0, 1, 0)$. The x-coordinate is black, y is medium gray and z is light gray.

Figure 2.81
Comparison of solutions of Lorenz system with initial points $(0, 1, 0)$ (black line) and $(0, 1.001, 0)$ (gray line) where $44 \leq t \leq 47$.

space where a solution $(x(t), y(t), z(t))$ is a curve in space. We show this curve from two different angles to better display its three-dimensional character (see Figure 2.82 and 2.83).

It turns out that this strange behavior occurs for almost every solution curve. The graphs of $x(t)$, $y(t)$, and $z(t)$ oscillate in an "unpredictable" and unique way, but the graph of the solution curve in three dimensions fills in a figure that is the same for every solution.

The solution curves we have drawn seem to loop around the equilibrium points above $z = 0$ in increasing spirals. Once the radius gets too large, the solution passes close to $(0, 0, 0)$ and then gets "reinjected" toward one of the two equilibria. (It is very instructive to watch a movie of a solution moving in real time through the phase space.) In the next chapter we develop the tools to study the behavior near the equilibrium points.

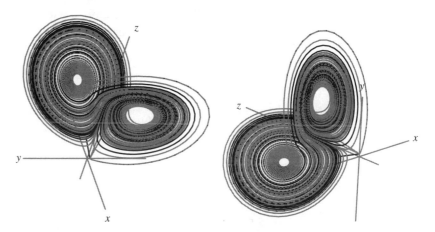

Figure 2.82
Two views of the solution curve with initial point $(0, 1, 0)$ in the three-dimensional phase space.

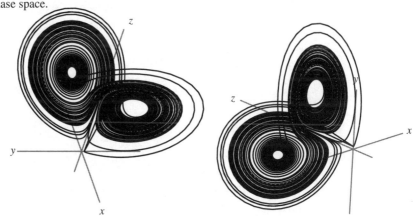

Figure 2.83
Two views of the solution curve with initial point $(0, 1.001, 0)$ in the three-dimensional phase space.

Chaos

The qualitative analysis of this system is a difficult undertaking that must wait until Chapters 4 and 6. However, there is a moral to what we have seen so far that is having an important effect on many different branches of science. The Lorenz system has two important properties. The first is that a small change in initial conditions leads fairly quickly to large differences in the corresponding solutions. If a system as simple as the Lorenz equations can have this property, it is entirely reasonable to think that much more complicated systems (like the weather) might have it as well. This means that any small error in the initial conditions will lead quickly to a big error in prediction of the solution. This might well be why physical systems like the weather are so hard to predict.

The second property of the Lorenz system is that, although the details of individual solutions are quite different, the pictures of the solution curves in the three-dimensional phase space look remarkably alike. Many solutions seem to be "filling in" the same region in three dimensions. So the solutions of the Lorenz system still have structure that we can study. We don't have to give up studying the solutions of the Lorenz equations, we just have to ask the right questions.

Exercises for Section 2.7

1. **(a)** Verify that $(0, 0, 0)$, $(6\sqrt{2}, 6\sqrt{2}, 27)$, and $(-6\sqrt{2}, -6\sqrt{2}, 27)$ are equilibrium points of the Lorenz system

$$\frac{dx}{dt} = 10(y - x)$$
$$\frac{dy}{dt} = 28x - y - xz$$
$$\frac{dz}{dt} = -\frac{8}{3}z + xy.$$

   (b) Verify that these are the only equilibrium points of the Lorenz system.

2. Suppose that in the Lorenz system we fix $\sigma = 10$ and $\beta = 8/3$, as in the text, but we leave ρ a parameter. For the resulting system,

$$\frac{dx}{dt} = 10(y - x)$$
$$\frac{dy}{dt} = \rho x - y - xz$$
$$\frac{dz}{dt} = -\frac{8}{3}z + xy,$$

   (a) show that if $\rho > 1$, there are three equilibrium points, and compute their locations; and

   (b) show that if $\rho \leq 1$, there is just one equilibrium point, and compute its location.

3. A system with three dependent variables has nullclines that are surfaces. For the Lorenz system with $\sigma = 10$, $\rho = 28$, and $\beta = 8/3$,

   (a) find the surface in xyz-space where $dx/dt = 0$ and sketch it;

(b) find the surface in xyz-space where $dy/dt = 0$ and sketch it; and

(c) find the surface in xyz-space where $dz/dt = 0$ and sketch it.

[It may be worthwhile to seek out a mathematical software package with good graphing capabilities.]

4. For the Lorenz system with $\sigma = 10$, $\rho = 28$, and $\beta = 8/3$,

    (a) verify that if $(x(t), y(t), z(t))$ is a solution with $x(0) = y(0) = 0$, then $x(t) = y(t) = 0$ for all t;

    (b) find the solution with initial condition $(0, 0, 1)$; and

    (c) find the solution with initial condition $(0, 0, z_0)$, where z_0 is any constant, and sketch these solutions in the three-dimensional phase space.

5. Close to the origin, where x, y, and z are very small, the quadratic terms $-xz$ and $+xy$ will be very, very small. So near $x = y = z = 0$, we can approximate the Lorenz system with the linear system

$$\frac{dx}{dt} = 10(y - x)$$
$$\frac{dy}{dt} = 28x - y$$
$$\frac{dz}{dt} = -\frac{8}{3}z.$$

(This is called *linearization* at the origin and will be studied in detail in Chapter 4.) Notice that the dx/dt and dy/dt equations do not contain any z's and the equation for dz/dt does not contain x or y; that is, the system "decouples" into a two-dimensional system and a one-dimensional equation.

    (a) Sketch the nullclines for the planar system

$$\frac{dx}{dt} = 10(y - x)$$
$$\frac{dy}{dt} = 28x - y.$$

    (b) Sketch the direction field and the phase plane for the planar system in part (a).

    (c) Sketch the phase line for the planar system

$$\frac{dz}{dt} = -\frac{8}{3}z.$$

    (d) Sketch solutions in the three-dimensional phase space for the linear system above.

Remark: This picture gives the behavior of the Lorenz system *near* the origin $(0, 0, 0)$.

6. With the help of whatever computing equipment you have available, find numerical approximations of solutions of the Lorenz system with $\sigma = 10$, $\rho = 28$, and $\beta = 8/3$ for the given initial conditions. Comment on how long it takes two solutions with initial conditions close together to separate.

(a) $(1, 0, 0)$	**(b)** $(0, 1.1, 0)$	**(c)** $(0, 1.01, 0)$
(d) $(1.001, 0, 0)$	**(e)** $(0.001, 1, 0)$	**(f)** $(0, 1, -0.001)$

Lab 2.1

Cooperative and Competitive Species Population Models

In this chapter we have focused on first-order autonomous systems of differential equations such as the predator-prey systems described in Section 2.1. In particular, we have seen how such systems can be studied using vector fields and phase plane analysis and how solution curves in the phase plane relate to the $x(t)$- and $y(t)$-graphs of the solutions. In this lab project, you will use these concepts and related numerical computations to study the behavior of the solutions to two different systems.

We have discussed predator-prey systems at length. These are systems in which one species benefits while the other species is harmed by the interaction of the two species. In this lab, you will study two other types of systems — competitive and cooperative systems. A competitive system is one in which both species are harmed by interaction, for example, cars and pedestrians. A cooperative system is one in which both species benefit from interaction, for example, bees and flowers. Your overall goal is to understand what happens in both systems for all possible nonnegative initial conditions. Several pairs of cooperative and competitive systems are given below. (Your instructor will tell you which pair(s) of systems you should study.) The analytic techniques that are appropriate to analyze these systems have not been discussed so far, so you will employ mostly geometric/qualitative and numeric techniques to establish your conclusions. Since these are population models you need only consider x and y in the first quadrant ($x \geq 0$ and $y \geq 0$).

Your report should include:

1. A brief discussion of all terms in each system. For example, what does the coefficient to the x term in equation for dx/dt represent? Which system is cooperative and which is competitive?

2. For each system, determine all relevant equilibrium points and analyze the behavior of solutions whose initial conditions satisfy either $x_0 = 0$ or $y_0 = 0$. Determine the nullclines for the vector field and the direction of the vector field on the nullclines and in the regions between the nullclines.

3. For each system, describe all possible population evolution scenarios using the phase portrait as well as $x(t)$- and $y(t)$-graphs. Give special attention to the interpretation of the computer output in terms of the long-term behavior of the populations.

Your report: The text of your report should address the three items above, one at a time, in form of a short essay. You should include a description of all "hand" computations that you did. You may include a limited number of pictures and graphs. (You should spend some time organizing the qualitative and numerical information since a few, well-organized figures are much more useful than a long catalog.)

Systems:

Pair (1):

A. $\dfrac{dx}{dt} = -5x + 2xy$
   $\dfrac{dy}{dt} = -4y + 3xy$

B. $\dfrac{dx}{dt} = 6x - x^2 - 4xy$
   $\dfrac{dy}{dt} = 5y - 2xy - 2y^2$

Pair (2):

$$A. \frac{dx}{dt} = -3x + 2xy \qquad B. \frac{dx}{dt} = 5x - x^2 - 3xy$$

$$\frac{dy}{dt} = -5y + 3xy \qquad \frac{dy}{dt} = 8y - 3xy - 3y^2$$

Pair (3):

$$A. \frac{dx}{dt} = -4x + 3xy \qquad B. \frac{dx}{dt} = 5x - 2x^2 - 4xy$$

$$\frac{dy}{dt} = -3y + 2xy \qquad \frac{dy}{dt} = 7y - 4xy - 3y^2$$

Pair (4):

$$A. \frac{dx}{dt} = -5x + 3xy \qquad B. \frac{dx}{dt} = 9x - 2x^2 - 4xy$$

$$\frac{dy}{dt} = -3y + 2xy \qquad \frac{dy}{dt} = 8y - 5xy - 3y^2$$

Lab 2.2   The Harmonic Oscillator with Modified Damping

Autonomous second-order differential equations are studied numerically by reducing them to first-order systems with two dependent variables. In this lab, you will use the computer to analyze three somewhat related second-order equations. In particular, you will analyze phase planes and $y(t)$- and $v(t)$-graphs to describe the long-term behavior of the solutions.

In Section 2.4, we discuss the most classic of all second-order equations, the harmonic oscillator. The harmonic oscillator is

$$m\frac{d^2y}{dt^2} + k_d\frac{dy}{dt} + k_s y = 0.$$

It is an example of a second-order, homogeneous, linear equation with constant coefficients. In the text, we explain how this equation is used to model the motion of a spring. The force due to the spring is assumed to obey Hooke's law (the force is proportional to the amount the spring is compressed or stretched). The force due to damping is assumed to be proportional to the velocity. In your report, you should describe the motion of the spring assuming certain values of m, k_d, and k_s. (A table of values of the parameters is given below. Your instructor will tell you what values of m, k_d and k_s to consider.)

Your report should discuss the following:

1. (Harmonic oscillator with no damping) The first equation that you should study is the harmonic oscillator with no damping. That is, $k_d = 0$ and with $k_s \neq 0$. Examine solutions using both their graphs and the phase plane. Are the solutions periodic? If so, what does the period seem to be? Describe the behavior of three different solutions that have especially different initial conditions and be specific about the physical interpretation of the different initial conditions. (Remark: Analytic methods to answer these questions are discussed in Chapter 3. For now, work numerically.)

2. (Harmonic oscillator with damping) Repeat part 1 using the equation

$$m\frac{d^2y}{dt^2} + k_d\left(\frac{dy}{dt}\right)^2 + k_s y = 0$$

in place of the usual harmonic oscillator equation. (The force of damping is proportional to the square of the velocity.)

3. (Nonlinear second-order equation) Finally consider a somewhat related second-order equation where the damping coefficient k_d is replaced by the factor $(y^2 - \alpha)$, that is,

$$m\frac{d^2y}{dt^2} + (y^2 - \alpha)\frac{dy}{dt} + k_s y = 0.$$

Is it reasonable to interpret this factor as some type of damping? Provide a complete description of the long-term behavior of the solutions. Are the solutions periodic? If so, what does the period seem to be? Explain why this equation is not a good model for something like a mass-spring system.

Your report: Address the questions in each item above in the form of a short essay. Be particularly sure to describe the behavior of the solution and the corresponding behavior of the mass-spring system. You may use the phase planes and graphs of $y(t)$ to illustrate the points you make in your essay. (However, please remember that, although one good illustration may be worth 1000 words, 1000 illustrations are usually worth nothing.)

Table 2.7 Possible choices for the parameters.

Choice	m	k_s	k_d	α
1	$m = 2$	$k_s = 5$	$k_d = 2$	$\alpha = 3$
2	$m = 3$	$k_s = 5$	$k_d = 3$	$\alpha = 3$
3	$m = 5$	$k_s = 5$	$k_d = 4$	$\alpha = 3$
4	$m = 2$	$k_s = 6$	$k_d = 3$	$\alpha = 5$
5	$m = 3$	$k_s = 6$	$k_d = 3$	$\alpha = 5$
6	$m = 5$	$k_s = 6$	$k_d = 3$	$\alpha = 5$
7	$m = 5$	$k_s = 4$	$k_d = 4$	$\alpha = 2$
8	$m = 5$	$k_s = 5$	$k_d = 4$	$\alpha = 2$
9	$m = 5$	$k_s = 6$	$k_d = 4$	$\alpha = 2$
10	$m = 5$	$k_s = 4$	$k_d = 4$	$\alpha = 2$

Lab 2.3   Swaying Building Models

In the text we studied simple models of the swaying motion of a tall building. The first model we used is the harmonic oscillator

$$\frac{d^2y}{dt^2} + p\frac{dy}{dt} + qy = 0$$

where y is the displacement of the top of the building from vertical, q is determined by the restoring force that pushes the building back toward vertical when it is bent and p is determined by the damping force which slows the motion of the building. The second model included a term for the effect of gravity (the P-Delta effect)

$$\frac{d^2 y}{dt^2} + p\frac{dy}{dt} + qy = ky^3,$$

where k is a parameter measuring the strength of this effect. Listed below are three sets of data from three (hypothetical) buildings.

In your report, address the following items:

1. For each of the three sets of data, which of the two equations above forms the better model?

2. Having decided which equation is better for each set of data, determine approximate values of the parameters for each of the sets of data.

3. Comment on how well the models with your choice of parameters agree with the data. Is there any danger that these (hypothetical) buildings will fall down in a heavy wind storm?

Your report: Describe your work on each of the items above in the form of a short essay. Be sure to give justification for each statement that you make. In part 2, make every effort to be systematic, that is, think about the effect of adjusting each of the parameters on the behavior of solutions (avoid randomly choosing parameters to find a good fit). Include your methods in your essay. You may include a limited number of graphs of data and solution curves to illustrate your work (but pictures should not replace words, they should enhance them).

Table 2.8 Displacement as a function of time for hypothetical building number 1.

time	$y_1(t)$	$y_2(t)$
0	0.21	.074
2	0.11	0.70
4	-0.10	0.51
6	-0.20	0.05
8	-0.09	-0.43
10	0.10	-0.57
12	0.18	-0.38
14	0.07	0.08
16	-0.11	0.44
18	-0.17	0.46
20	-0.06	0.14
22	0.11	-0.29

Table 2.9 Displacement as a function of time for hypothetical building number 2.

time	$y_1(t)$	$y_2(t)$
0	0.14	0.50
2	0.38	0.40
4	0.21	-0.12
6	-0.18	-0.49
8	-0.37	-.033
10	-0.15	0.18
12	0.21	0.48
14	0.34	0.25
16	0.10	-0.23
18	-0.24	-0.45
20	-0.31	-0.18
22	-0.05	0.27

Table 2.10 Displacement as a function of time for hypothetical building number 3.

time	$y_1(t)$	$y_2(t)$
0	0.10	0.50
2	0.21	0.48
4	0.11	0.05
6	-0.12	-0.46
8	-0.23	-0.56
10	-0.10	-0.23
12	0.14	0.33
14	0.24	0.60
16	0.08	0.42
18	-0.17	-0.13

LINEAR SYSTEMS

3

In Chapter 2 we focused on qualitative and numerical techniques for studying systems of differential equations. We did this because we can rarely find explicit formulas for solutions of a system with two or more dependent variables. The only exceptions to this are linear systems.

In this chapter we show how to use the algebraic and geometric form of the vector field to give the general solution of autonomous linear systems. Along the way, we find that understanding the qualitative behavior of a linear system is much easier than finding its general solution. The description of the qualitative behavior of linear systems leads to a classification scheme for these systems which is very useful in applications. We also continue our study of models that yield linear systems, particularly the harmonic oscillator system.

3.1 PROPERTIES OF LINEAR SYSTEMS AND THE LINEARITY PRINCIPLE

In this chapter we investigate the behavior of the simplest types of systems of differential equations — autonomous linear systems. These systems are important both in their own right and as a tool in the study of nonlinear systems. We are able to classify the linear systems by their qualitative behavior and even give formulas for solutions.

Throughout this chapter we use two models repeatedly to illustrate the techniques we develop. One is the harmonic oscillator, the most important of all second-order linear equations. We derived this model in Section 2.4. Now, using the techniques of this chapter, we can give a complete description of its solutions for all possible values of the parameters. The other model is an artificial one that we present to solely illustrate all possibilities that can arise for planar linear systems. Study our analysis, but don't invest any money based on it.

The Harmonic Oscillator

The harmonic oscillator is a model for (among other things) the motion of a mass attached to a spring. The spring provides a restoring force that obeys Hooke's Law, and the only other force considered is that due to damping. Let $y(t)$ be the position of the mass at time t with $y = 0$ corresponding to the rest position of the spring. Newton's law of motion,

$$\text{force} = \text{mass} \times \text{acceleration},$$

when applied to a mass-spring system, yields the second-order differential equation

$$-k_s y - k_d \frac{dy}{dt} = m \frac{d^2 y}{dt^2},$$

where m is the mass, k_s is the spring constant, and k_d is coefficient of damping (see Section 2.4). This second-order, linear equation is more commonly written as

$$m \frac{d^2 y}{dt^2} + k_d \frac{dy}{dt} + k_s y = 0.$$

As we did in Section 2.4, we can convert this equation into a linear system by letting $v = dy/dt$ be the velocity at time t. We obtain

$$\frac{dy}{dt} = v$$
$$\frac{dv}{dt} = -\frac{k_s}{m} y - \frac{k_d}{m} v.$$

This is a linear system because the derivatives dy/dt and dv/dt depend linearly on y and v. As we will see, the behavior of solutions depends on the values of the parameters m, k_s, and k_d.

Two CD stores

Harmonic oscillators do not display all possible behaviors for solutions to autonomous linear systems, so we present the following microeconomics model to demonstrate the full spectrum of behaviors. We are not going to base a marketing plan on this model; it is just a way to interpret the behavior of solutions to linear systems in everyday language.

After retiring from writing differential equations textbooks full of silly examples, Paul and Bob both decide to open small stores selling compact discs. "Paul's Rock and Roll CDs" and "Bob's Opera Only Discs" are located on the same block, and Paul and Bob soon become concerned about the effect the stores have on each other. On the one hand, having two CD stores on the same block might help both stores by attracting customers to the neighborhood. On the other hand, the stores may compete with each other for limited customers. Paul and Bob argue about this until they are so sick of arguing that they hire the famous mathematician Glen to settle the matter. Glen follows his own advice about keeping models as simple as possible and suggests the following system.

Let

$$x(t) = \text{profit/day of Paul's store at time } t$$
$$[\text{if } x(t) > 0, \text{Paul's store is making money but, if } x(t) < 0,$$
$$\text{then Paul's store is losing money}];$$
$$y(t) = \text{profit/day of Bob's store at time } t.$$

Since there is no hard information yet about how the profits of each store affect the change in profits of the other, Glen formulates the simplest possible model that allows each store to affect the other — a linear model. The system is

$$\frac{dx}{dt} = ax + by$$
$$\frac{dy}{dt} = cx + dy,$$

where a, b, c, and d are parameters. This is the most general form of an autonomous, linear system. The rate of change of Paul's profits depends linearly on both Paul's profits and Bob's profits (and nothing else). The same assumptions apply to Bob's profits. In Chapter 4, we'll see that using a linear model is usually justified as long as both stores are operating near the break-evenpoint.

We can not yet use this model to predict future profits because we do not know the values of the parameters a, b, c, and d. However, we can develop a basic understanding of the significance of the signs and magnitudes of the parameters. Consider, for example, the parameter a. It measures the effect of Paul's profits on the rate of change dx/dt of that profit. Suppose, for instance, that a is positive. If Paul is making money, then $x > 0$ and so $ax > 0$. The ax term contributes positively to dx/dt, and so Paul makes more money in that case. In other words, Paul hopes that $a > 0$. On the other hand, being profitable $(x > 0)$ could conceivably have a negative effect on Paul's profits (for example, the store might be crowded, and customers might go elsewhere). In this case profits would decrease, and under this assumption the parameter a should be negative in our model.

The parameter b measures the effect of Bob's profits on the rate of change of Paul's profits. If $b > 0$ and Bob makes money $(y > 0)$, then Paul's profits also benefit because the by term contributes positively to dx/dt. On the other hand, if $b < 0$ then, when Bob makes money $(y > 0)$, Paul's profits suffer. Conceivably we could interpret b as a measure of Bob's stealing customers from Paul if $b < 0$.

Similarly, both Paul's profits and Bob's profits affect the rate of change of Bob's profits, and the parameters c and d have similar interpretations relative to dy/dt. This model assumes that only the profit of the two stores influences the change in those profits. This is clearly a vast oversimplification. However, it does give us a simple situation for which we can interpret the solutions of various linear systems.

In Figure 3.1 we plot the phase portrait for the linear system

$$\frac{dx}{dt} = ax + by$$

$$\frac{dy}{dt} = cx + dy,$$

assuming that $a = d = 0$, $b = 1$, and $c = -1$. We see that we get periodic solution curves. In Figure 3.2 we consider the case where $a = -1$, $b = 4$, $c = -3$, and $d = -1$. In this case the solution curves are spirals that approach the origin.

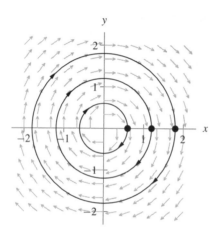

Figure 3.1

The direction field and three solution curves for the linear system

$$\frac{dx}{dt} = y$$

$$\frac{dy}{dt} = -x.$$

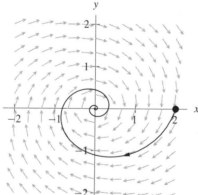

Figure 3.2

The direction field and a solution curve for the linear system

$$\frac{dx}{dt} = -x + 4y$$

$$\frac{dy}{dt} = -3x - y.$$

In terms of the model, Figure 3.1 implies that both Paul's and Bob's profits periodically oscillate between making money and losing money. The solution curve in Figure 3.2 suggests that the profits oscillate while tending toward the point $(0, 0)$, which is the break-even point for both stores. The corresponding $x(t)$- and $y(t)$-graphs illustrate this behavior (see Figures 3.3 and 3.4). There are a number of other possible phase plane pictures for a linear system, depending on the values of the parameters a, b, c, and d. In this chapter we develop techniques to handle all possibilities.

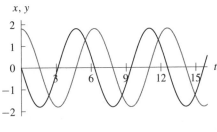

Figure 3.3
The $x(t)$- and $y(t)$-graphs corresponding to the solution curve in Figure 3.1, with initial condition $(x_0, y_0) = (1.8, 0)$.

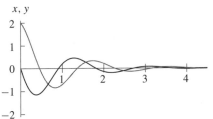

Figure 3.4
The $x(t)$- and $y(t)$-graphs corresponding to the solution curve in Figure 3.2 with initial condition $(x_0, y_0) = (2, 0)$.

Linear Systems and Matrix Notation

The system

$$\frac{dx}{dt} = ax + by$$
$$\frac{dy}{dt} = cx + dy$$

is called a **linear system with constant coefficients**. It is also *autonomous*. The coefficients are a, b, c, and d. This system is a *planar* or *two-dimensional* because it has two dependent variables. Usually we call these systems *planar linear systems*.

We use matrix and vector notation to write this system much more efficiently. Let **A** be the "2-by-2" square matrix (or 2×2 matrix)

$$\mathbf{A} = \begin{pmatrix} a & b \\ c & d \end{pmatrix},$$

and let

$$\mathbf{Y} = \begin{pmatrix} x \\ y \end{pmatrix}$$

denote the column vector of dependent variables. Then the **product** of a 2×2 matrix **A** and a column vector **Y** is the column vector **AY** given by

$$\mathbf{AY} = \begin{pmatrix} a & b \\ c & d \end{pmatrix} \begin{pmatrix} x \\ y \end{pmatrix} = \begin{pmatrix} ax + by \\ cx + dy \end{pmatrix}.$$

For example,

$$\begin{pmatrix} 5 & 2 \\ -1 & 3 \end{pmatrix} \begin{pmatrix} 3 \\ 4 \end{pmatrix} = \begin{pmatrix} 5 \cdot 3 + 2 \cdot 4 \\ -1 \cdot 3 + 3 \cdot 4 \end{pmatrix} = \begin{pmatrix} 23 \\ 9 \end{pmatrix}$$

and

$$\begin{pmatrix} (2-a) & \pi \\ e & y \end{pmatrix} \begin{pmatrix} y \\ 2v \end{pmatrix} = \begin{pmatrix} (2-a)y + 2\pi v \\ ey + 2yv \end{pmatrix}.$$

As in Chapter 2, if x and y are dependent variables, then we write

$$\mathbf{Y}(t) = \begin{pmatrix} x(t) \\ y(t) \end{pmatrix} \quad \text{and} \quad \frac{d\mathbf{Y}}{dt} = \begin{pmatrix} \dfrac{dx}{dt} \\ \dfrac{dy}{dt} \end{pmatrix}.$$

Using this matrix notation we can write the two-dimensional linear system

$$\frac{dx}{dt} = ax + by$$

$$\frac{dy}{dt} = cx + dy$$

as

$$\begin{pmatrix} \dfrac{dx}{dt} \\ \dfrac{dy}{dt} \end{pmatrix} = \begin{pmatrix} ax + by \\ cx + dy \end{pmatrix} = \begin{pmatrix} a & b \\ c & d \end{pmatrix} \begin{pmatrix} x \\ y \end{pmatrix},$$

or more compactly as

$$\frac{d\mathbf{Y}}{dt} = \mathbf{AY},$$

where

$$\mathbf{A} = \begin{pmatrix} a & b \\ c & d \end{pmatrix} \quad \text{and} \quad \mathbf{Y} = \begin{pmatrix} x \\ y \end{pmatrix}.$$

The matrix $\mathbf{A}$ of coefficients of the system is called the **coefficient matrix**. This matrix notation also gives us some handy algebraic tools that we happily exploit throughout the chapter.

Vector notation can be extended to include systems with any number n of dependent variables $y_1, y_2, \ldots, y_n$. The linear system with n dependent variables is

$$\frac{dy_1}{dt} = a_{11}y_1 + a_{12}y_2 + \cdots + a_{1n}y_n$$

$$\frac{dy_2}{dt} = a_{21}y_1 + a_{22}y_2 + \cdots + a_{2n}y_n$$

$$\vdots \qquad\qquad \vdots$$

$$\frac{dy_n}{dt} = a_{n1}y_1 + a_{n2}y_2 + \cdots + a_{nn}y_n.$$

In this case the coefficients of this system are $a_{11}, a_{12}, \ldots, a_{nn}$. Let

$$\mathbf{Y} = \begin{pmatrix} y_1 \\ y_2 \\ \vdots \\ y_n \end{pmatrix}, \quad \text{so} \quad \frac{d\mathbf{Y}}{dt} = \begin{pmatrix} \dfrac{dy_1}{dt} \\ \dfrac{dy_2}{dt} \\ \vdots \\ \dfrac{dy_n}{dt} \end{pmatrix}.$$

The coefficient matrix is the $n \times n$ matrix

$$\mathbf{A} = \begin{pmatrix} a_{11} & a_{12} & \cdots & a_{1n} \\ a_{21} & a_{22} & \cdots & a_{2n} \\ \vdots & \vdots & \ddots & \vdots \\ a_{n1} & a_{n2} & \cdots & a_{nn} \end{pmatrix},$$

and we have

$$\frac{d\mathbf{Y}}{dt} = \mathbf{A}\mathbf{Y} = \begin{pmatrix} a_{11} & a_{12} & \cdots & a_{1n} \\ a_{21} & a_{22} & \cdots & a_{2n} \\ \vdots & \vdots & \ddots & \vdots \\ a_{n1} & a_{n2} & \cdots & a_{nn} \end{pmatrix} \begin{pmatrix} y_1 \\ y_2 \\ \vdots \\ y_n \end{pmatrix}$$

$$= \begin{pmatrix} a_{11}y_1 + a_{12}y_2 + \ldots + a_{1n}y_n \\ a_{21}y_1 + a_{22}y_2 + \ldots + a_{2n}y_n \\ \vdots \\ a_{n1}y_1 + a_{n2}y_2 + \ldots + a_{nn}y_n \end{pmatrix}.$$

The number of dependent variables is called the **dimension** of the system, so this system is n-**dimensional**. For example, the three-dimensional system

$$\frac{dx}{dt} = \sqrt{2}\,x + y$$

$$\frac{dy}{dt} = z$$

$$\frac{dz}{dt} = -x - y + 2z$$

can be written as

$$\frac{d\mathbf{Y}}{dt} = \mathbf{A}\mathbf{Y},$$

where

$$\mathbf{Y} = \begin{pmatrix} x \\ y \\ z \end{pmatrix} \quad \text{and} \quad \mathbf{A} = \begin{pmatrix} \sqrt{2} & 1 & 0 \\ 0 & 0 & 1 \\ -1 & -1 & 2 \end{pmatrix}.$$

In this text we deal with primarily planar, or 2-dimensional, systems. However, readers familiar with linear algebra will recognize that many of the concepts we discuss carry over to higher dimensional systems with little or no modification.

Linear systems are like other systems of differential equations, only simpler. All of the methods of Chapter 2 apply, and we use these methods to understand the associated vector fields, direction fields, and graphs of solutions. In addition, because linear systems are so simple algebraically, it is reasonable to hope that we can "read off" the behavior of solutions just from the coefficients. That is, if we are given the planar linear system

$$\frac{d\mathbf{Y}}{dt} = \mathbf{A}\mathbf{Y}, \quad \text{where} \quad \mathbf{A} = \begin{pmatrix} a & b \\ c & d \end{pmatrix}$$

is the coefficient matrix, then we would like to understand the system completely if we know the four numbers a, b, c, and d. In fact we might even hope to come up with an explicit formula for the general solution. This is the goal of this chapter. From the four numbers a, b, c, and d, we are able to give a geometric description of the behavior of the solutions in the xy-plane, describe the $x(t)$- and $y(t)$-graphs of solutions, and even give a formula for the general solution. Hence we are able to produce explicit formulas that solve any initial-value problem.

Equilibrium Points of Linear Systems and the Determinant

We start by looking for the simplest solutions, the equilibrium points. Recall that a point $\mathbf{Y}_0 = (x_0, y_0)$ is an equilibrium point of a system if and only if the vector field at $\mathbf{Y}_0$ vanishes. Since the vector field of a system at the point $\mathbf{Y}_0$ is given by the right-hand side of the differential equation evaluated at that point, we know that the vector field at $\mathbf{Y}_0$ for a linear system is given by the matrix product $\mathbf{AY}_0$. Consequently, the equilibrium points are the points $\mathbf{Y}_0$ such that

$$\mathbf{AY}_0 = \begin{pmatrix} 0 \\ 0 \end{pmatrix}.$$

That is,

$$\begin{pmatrix} a & b \\ c & d \end{pmatrix} \begin{pmatrix} x_0 \\ y_0 \end{pmatrix} = \begin{pmatrix} 0 \\ 0 \end{pmatrix},$$

or

$$\begin{pmatrix} ax_0 + by_0 \\ cx_0 + dy_0 \end{pmatrix} = \begin{pmatrix} 0 \\ 0 \end{pmatrix}.$$

Written in scalar form, this vector equation is a pair of simultaneous, linear equations

$$\begin{cases} ax_0 + by_0 = 0 \\ cx_0 + dy_0 = 0. \end{cases}$$

Clearly $(x_0, y_0) = (0, 0)$ is a solution to these equations. As a result, $\mathbf{Y}_0 = (0, 0)$ is an equilibrium point, and the constant function $\mathbf{Y}(t) = (0, 0)$ for all t is a solution to the linear system. (Because this was easy to establish, some people call this the **trivial solution** to the system.) Note that this computation does not depend on the values of the coefficients a, b, c, and d. In other words, the origin is always an equilibrium point if the system is linear.

Any other equilibrium points are also solutions of

$$\begin{cases} ax_0 + by_0 = 0 \\ cx_0 + dy_0 = 0. \end{cases}$$

To find them, assume for the moment that $a \neq 0$. Then using the first equation, we get

$$x_0 = -\frac{b}{a} y_0.$$

The second equation then yields

$$c\left(-\frac{b}{a}\right) y_0 + d y_0 = 0,$$

which can be rewritten as

$$(ad - bc) y_0 = 0.$$

Hence either $y_0 = 0$ or $ad - bc = 0$. If $y_0 = 0$, then $x_0 = 0$, and once again we've got the trivial solution. We see that a linear system has nontrivial equilibrium points only if $ad - bc = 0$. This quantity, $ad - bc$, is a particularly important number associated with the 2×2 matrix $\mathbf{A}$.

Definition: Given a 2×2 matrix

$$\mathbf{A} = \begin{pmatrix} a & b \\ c & d \end{pmatrix},$$

the **determinant** of $\mathbf{A}$ is the quantity $ad - bc$. It is denoted

$$\det \mathbf{A} = ad - bc.$$

With this definition, we are able to summarize the results of the above computation for equilibrium points of linear systems.

Theorem   *If $\mathbf{A}$ is a matrix with $\det \mathbf{A} \neq 0$, then the only equilibrium point for the linear system $d\mathbf{Y}/dt = \mathbf{AY}$ is the origin.*

The argument above proves this theorem, provided the upper left-hand corner entry, a, of $\mathbf{A}$ is nonzero. The other cases are similar and are left as exercises.

As an example, if

$$\mathbf{A} = \begin{pmatrix} 2 & 1 \\ -4 & 0.3 \end{pmatrix},$$

then $\det \mathbf{A} = (2)(0.3) - (1)(-4) = 4.6$. Since $\det \mathbf{A} \neq 0$, the only equilibrium point for linear system $d\mathbf{Y}/dt = \mathbf{AY}$ is the origin, $(0, 0)$.

An Important Property of the Determinant

The determinant is a quantity that pops up repeatedly throughout this chapter. For us its significance usually is whether or not it is zero. If we pick four numbers a, b, c, and d at random, then it is unlikely that the number $ad - bc$ is exactly zero. Thus matrices whose determinant is zero are often called **singular** or **degenerate**.

From above we know that if a linear system $d\mathbf{Y}/dt = \mathbf{AY}$ is **nondegenerate** ($\det \mathbf{A} \neq 0$), it has exactly one equilibrium point and that point is located at the origin. In other words, an initial value at $(0, 0)$ corresponds to a solution curve that sits at $(0, 0)$ for all time. Any other initial condition yields a solution that changes with time.

In the next sectionwe need to use the determinant again, so it is important to understand exactly what we verified above. For linear systems, equilibria correspond to points $\mathbf{Y}_0$ for which

$$\mathbf{AY}_0 = \begin{pmatrix} 0 \\ 0 \end{pmatrix},$$

which is just the simultaneous system of linear equations

$$\begin{cases} ax_0 + by_0 = 0 \\ cx_0 + dy_0 = 0. \end{cases}$$

What we really verified is that this system of equations has nontrivial solutions if and only if $\det \mathbf{A} = 0$.

The Linearity Principle

The solutions of linear systems have special properties that solutions of arbitrary systems do not have. These properties are so useful that we take advantage of them repeatedly. In fact they are exactly the reason that we are so successful in our analysis of linear systems.

However, a note of caution is in order: It is important to make sure that the system under consideration actually *is* a linear system before you use any of these special properties. This is the equivalent of making sure that the car is in reverse before trying to back out of the garage. If the car is in drive instead of reverse when you start backing out of the garage, there can be dire consequences.

The most important property of linear systems is the **Linearity Principle**.

Linearity Principle

Suppose $d\mathbf{Y}/dt = \mathbf{AY}$ is a linear system of differential equations.

1. *If $\mathbf{Y}(t)$ is a solution of this system and k is any constant, then $k\mathbf{Y}(t)$ is also a solution.*

2. *If $\mathbf{Y}_1(t)$ and $\mathbf{Y}_2(t)$ are solutions of this system, then $\mathbf{Y}_1(t) + \mathbf{Y}_2(t)$ is also a solution.*

Using the Linearity Principle (also called the **Principle of Superposition**, we can manufacture infinitely many new solutions from any given solution or pair of solutions. Taken together, the two parts of the Linearity Principle imply that, if $\mathbf{Y}_1(t)$ and $\mathbf{Y}_2(t)$ are solutions of the system and if k_1 and k_2 are any constants, then $k_1 \mathbf{Y}_1(t) + k_2 \mathbf{Y}_2(t)$ is also a solution. A solution of the form $k_1 \mathbf{Y}_1(t) + k_2 \mathbf{Y}_2(t)$ is called a **linear combination** of the solutions $\mathbf{Y}_1(t)$ and $\mathbf{Y}_2(t)$. Given two solutions we can, by taking linear combinations, find infinitely many more solutions.

For example, consider the linear system

$$\frac{d\mathbf{Y}}{dt} = \begin{pmatrix} 2 & 3 \\ 0 & -4 \end{pmatrix} \mathbf{Y}.$$

We can check that

$$\mathbf{Y}_1(t) = \begin{pmatrix} e^{2t} \\ 0 \end{pmatrix} \quad \text{and} \quad \mathbf{Y}_2(t) = \begin{pmatrix} -e^{-4t} \\ 2e^{-4t} \end{pmatrix}$$

are solutions to this system by directly calculating both sides, $d\mathbf{Y}/dt$ and $\mathbf{AY}$, of the differential equation. For $\mathbf{Y}_1(t)$ we have

$$\frac{d\mathbf{Y}_1}{dt} = \begin{pmatrix} 2e^{2t} \\ 0 \end{pmatrix} \quad \text{and} \quad \mathbf{AY}_1 = \begin{pmatrix} 2 & 3 \\ 0 & -4 \end{pmatrix} \begin{pmatrix} e^{2t} \\ 0 \end{pmatrix} = \begin{pmatrix} 2e^{2t} \\ 0 \end{pmatrix},$$

so $\mathbf{Y}_1(t)$ is a solution. The verification that $\mathbf{Y}_2(t)$ is a solution is similar. (You should try it yourself.)

The solution curves for $\mathbf{Y}_1(t)$ and $\mathbf{Y}_2(t)$ are shown in Figure 3.5. Note that they form straight line segments that approach the origin in the xy-plane. In the next section we exploit the geometry of these solutions to derive them directly from the matrix.

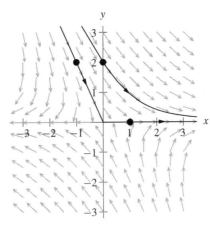

Figure 3.5
The function $\mathbf{Y}_1(t) + \mathbf{Y}_2(t)$ is a solution because it is the sum of the two solutions $\mathbf{Y}_1(t)$ and $\mathbf{Y}_2(t)$.

The Linearity Principle tell us that any function of the form $k_1\mathbf{Y}_1(t) + k_2\mathbf{Y}_2(t)$ is also a solution to this system for any constants k_1 and k_2. To illustrate this fact, we check directly that $\mathbf{Y}_3(t) = -2\mathbf{Y}_1(t) + 5\mathbf{Y}_2(t)$ is a solution. Note that

$$\mathbf{Y}_3(t) = -2\mathbf{Y}_1(t) + 5\mathbf{Y}_2(t) = \begin{pmatrix} -2e^{2t} - 5e^{-4t} \\ 10e^{-4t} \end{pmatrix},$$

and therefore

$$\frac{d\mathbf{Y}_3}{dt} = \begin{pmatrix} -4e^{2t} + 20e^{-4t} \\ -40e^{-4t} \end{pmatrix}.$$

Also we compute

$$\mathbf{AY}_3 = \begin{pmatrix} 2 & 3 \\ 0 & -4 \end{pmatrix} \begin{pmatrix} -2e^{2t} - 5e^{-4t} \\ 10e^{-4t} \end{pmatrix}$$

$$= \begin{pmatrix} 2(-2e^{2t} - 5e^{-4t}) + 3(10e^{-4t}) \\ -4(10e^{-4t}) \end{pmatrix}$$

$$= \begin{pmatrix} -4e^{2t} + 20e^{-4t} \\ -40e^{-4t} \end{pmatrix}.$$

Since both computations yield the same function, this linear combination of the two solutions $\mathbf{Y}_1(t)$ and $\mathbf{Y}_2(t)$ is also a solution (see Figure 3.5). (Of course, in the future we will not bother to double-check the consequences of the Linearity Principle.) Note that the solution curves of $\mathbf{Y}_1(t)$ and $\mathbf{Y}_2(t)$ are straight line segments, but the solution curve of their sum $\mathbf{Y}_1(t) + \mathbf{Y}_2(t)$ is not straight.

Verification of the Linearity Principle

To show that the Linearity Principle holds in general, we first verify the following two algebraic properties of matrix multiplication:

1. If $\mathbf{A}$ is a matrix and $\mathbf{Y}$ is a vector, then

$$\mathbf{A}(k\mathbf{Y}) = k\mathbf{A}\mathbf{Y}$$

for any constant k.

2. If $\mathbf{A}$ is a matrix and $\mathbf{Y}_1$ and $\mathbf{Y}_2$ are vectors, then

$$\mathbf{A}(\mathbf{Y}_1 + \mathbf{Y}_2) = \mathbf{A}\mathbf{Y}_1 + \mathbf{A}\mathbf{Y}_2.$$

We can verify these two facts for 2×2 matrices and 2-dimensional vectors by direct computation. For example, to verify Property 2, let

$$\mathbf{A} = \begin{pmatrix} a & b \\ c & d \end{pmatrix}$$

be an arbitrary 2×2 matrix and let

$$\mathbf{Y}_1 = \begin{pmatrix} x_1 \\ y_1 \end{pmatrix} \quad \text{and} \quad \mathbf{Y}_2 = \begin{pmatrix} x_2 \\ y_2 \end{pmatrix}$$

be arbitrary vectors. Then

$$
\begin{aligned}
\mathbf{A}(\mathbf{Y}_1 + \mathbf{Y}_2) &= \begin{pmatrix} a & b \\ c & d \end{pmatrix} \begin{pmatrix} x_1 + x_2 \\ y_1 + y_2 \end{pmatrix} \\
&= \begin{pmatrix} a(x_1 + x_2) + b(y_1 + y_2) \\ c(x_1 + x_2) + d(y_1 + y_2) \end{pmatrix} \\
&= \begin{pmatrix} ax_1 + ax_2 + by_1 + by_2 \\ cx_1 + cx_2 + dy_1 + dy_2 \end{pmatrix}
\end{aligned}
$$

and

$$
\begin{aligned}
\mathbf{A}\mathbf{Y}_1 + \mathbf{A}\mathbf{Y}_2 &= \begin{pmatrix} a & b \\ c & d \end{pmatrix} \begin{pmatrix} x_1 \\ y_1 \end{pmatrix} + \begin{pmatrix} a & b \\ c & d \end{pmatrix} \begin{pmatrix} x_2 \\ y_2 \end{pmatrix} \\
&= \begin{pmatrix} ax_1 + by_1 \\ cx_1 + dy_1 \end{pmatrix} + \begin{pmatrix} ax_2 + by_2 \\ cx_2 + dy_2 \end{pmatrix} \\
&= \begin{pmatrix} ax_1 + ax_2 + by_1 + by_2 \\ cx_1 + cx_2 + dy_1 + dy_2 \end{pmatrix}.
\end{aligned}
$$

Thus Property 2 holds. Given these algebraic properties of matrix multiplication, the Linearity Principle is easy to verify using the standard rules of differentiation. Suppose $\mathbf{Y}_1(t)$ and $\mathbf{Y}_2(t)$ are solutions to $d\mathbf{Y}/dt = \mathbf{A}\mathbf{Y}$; that is,

$$\frac{d\mathbf{Y}_1}{dt} = \mathbf{A}\mathbf{Y}_1 \quad \text{and} \quad \frac{d\mathbf{Y}_2}{dt} = \mathbf{A}\mathbf{Y}_2 \quad \text{for all } t.$$

For any constant k we have

$$\frac{d(k\mathbf{Y}_1)}{dt} = k\frac{d\mathbf{Y}_1}{dt} = k\mathbf{A}\mathbf{Y}_1 = \mathbf{A}(k\mathbf{Y}_1),$$

so $k\mathbf{Y}_1(t)$ is a solution to the system. Also

$$\frac{d(\mathbf{Y}_1 + \mathbf{Y}_2)}{dt} = \frac{d\mathbf{Y}_1}{dt} + \frac{d\mathbf{Y}_2}{dt} = \mathbf{A}\mathbf{Y}_1 + \mathbf{A}\mathbf{Y}_2 = \mathbf{A}(\mathbf{Y}_1 + \mathbf{Y}_2) \quad \text{for all } t.$$

As a result, $\mathbf{Y}_1(t) + \mathbf{Y}_2(t)$ is also a solution and the Linearity Principle is verified. We can see the advantage of the matrix and vector notation. To write out the above equations showing all the components would be a tedious exercise — in fact, it is a tedious exercise at the end of this section (see Exercise 30).

Linearly Independent Solutions and the General Solution of a Linear System

From the Linearity Principle we know that given two solutions $\mathbf{Y}_1(t)$ and $\mathbf{Y}_2(t)$, we can make many more solutions of the form $k_1\mathbf{Y}_1(t) + k_2\mathbf{Y}_2(t)$ for any constants k_1 and k_2. This type of expression is called a **two-parameter family of solutions** since we have two constants, k_1 and k_2, that we can adjust to obtain various solutions. It is reasonable to ask whether these are all of the solutions or, put another way, if each solution is one of this form.

To see how the Linearity Principle is used to solve initial-value problems, we return to the differential equation

$$\frac{d\mathbf{Y}}{dt} = \begin{pmatrix} 2 & 3 \\ 0 & -4 \end{pmatrix}\mathbf{Y}$$

that was discussed earlier in this section. Suppose we want to find the solution $\mathbf{Y}(t)$ of this system with initial value $\mathbf{Y}(0) = (2, -3)$. We already know that

$$\mathbf{Y}_1(t) = \begin{pmatrix} e^{2t} \\ 0 \end{pmatrix} \quad \text{and} \quad \mathbf{Y}_2(t) = \begin{pmatrix} -e^{-4t} \\ 2e^{-4t} \end{pmatrix}$$

are solutions. By direct evaluation we know that

$$\mathbf{Y}_1(0) = \begin{pmatrix} 1 \\ 0 \end{pmatrix} \quad \text{and} \quad \mathbf{Y}_2(0) = \begin{pmatrix} -1 \\ 2 \end{pmatrix},$$

so neither $\mathbf{Y}_1(t)$ nor $\mathbf{Y}_2(t)$ is the solution to the initial-value problem

$$\frac{d\mathbf{Y}}{dt} = \begin{pmatrix} 2 & 3 \\ 0 & -4 \end{pmatrix}\mathbf{Y}, \quad \mathbf{Y}(0) = \begin{pmatrix} 2 \\ -3 \end{pmatrix}.$$

But the Linearity Principle says that we can form any linear combination of $\mathbf{Y}_1(t)$ and $\mathbf{Y}_2(t)$ and still have a solution. Hence we seek k_1 and k_2 so that $k_1 \mathbf{Y}_1(0) + k_2 \mathbf{Y}_2(0) = (2, -3)$ or, written as column vectors,

$$k_1 \begin{pmatrix} 1 \\ 0 \end{pmatrix} + k_2 \begin{pmatrix} -1 \\ 2 \end{pmatrix} = \begin{pmatrix} 2 \\ -3 \end{pmatrix}.$$

This vector equation is equivalent to the simultaneous scalar equations

$$\begin{cases} k_1 - k_2 = 2 \\ 2k_2 = -3. \end{cases}$$

The second equation yields $k_2 = -3/2$, and consequently, the first equation yields $k_1 = 1/2$. Given the equation

$$\tfrac{1}{2}\mathbf{Y}_1(0) - \tfrac{3}{2}\mathbf{Y}_2(0) = \begin{pmatrix} 2 \\ -3 \end{pmatrix},$$

we consider the function

$$\mathbf{Y}(t) = \tfrac{1}{2}\mathbf{Y}_1(t) - \tfrac{3}{2}\mathbf{Y}_2(t)$$

$$= \tfrac{1}{2} \begin{pmatrix} e^{2t} \\ 0 \end{pmatrix} - \tfrac{3}{2} \begin{pmatrix} -e^{-4t} \\ 2e^{-4t} \end{pmatrix}$$

$$= \begin{pmatrix} \tfrac{1}{2}e^{2t} + \tfrac{3}{2}e^{-4t} \\ -3e^{-4t} \end{pmatrix}.$$

This function has the correct initial condition, and by the Linearity Principle we know that it must be a solution to the system. The Uniqueness Theorem tells us that this is the only function that solves the initial-value problem.

In this example we found the solution to the initial-value problem using the two solutions of the system that were already available plus a little arithmetic (and no calculus). By taking the appropriate linear combination of the two known solutions, we were able to find a solution with the desired initial conditions.

Maybe we were just lucky. Will we always be able to find the appropriate k_1 and k_2 no matter what initial condition we have? To check, suppose we consider the same initial-value problem with an arbitrary initial condition,

$$\frac{d\mathbf{Y}}{dt} = \begin{pmatrix} 2 & 3 \\ 0 & -4 \end{pmatrix} \mathbf{Y}, \quad \mathbf{Y}(0) = \begin{pmatrix} x_0 \\ y_0 \end{pmatrix},$$

and consider the two solutions $\mathbf{Y}_1(t)$ and $\mathbf{Y}_2(t)$ with which we started. To solve the initial-value problem, we need to find k_1 and k_2 so that

$$k_1 \mathbf{Y}_1(0) + k_2 \mathbf{Y}_2(0) = \mathbf{Y}(0) = \begin{pmatrix} x_0 \\ y_0 \end{pmatrix}.$$

In other words, given arbitrary x_0 and y_0, can we always find k_1 and k_2 such that

$$k_1 \begin{pmatrix} 1 \\ 0 \end{pmatrix} + k_2 \begin{pmatrix} -1 \\ 2 \end{pmatrix} = \begin{pmatrix} x_0 \\ y_0 \end{pmatrix}?$$

This vector equation is equivalent to the simultaneous system of equations

$$\begin{cases} k_1 - k_2 = x_0 \\ \qquad 2k_2 = y_0. \end{cases}$$

Since the second equation is so simple, it is clear that we can always find k_1 and k_2 given x_0 and y_0. We use the second equation to find k_2 first, and then we find k_1 using this value of k_2 in the first equation.

Since we are able to solve every possible initial-value problem for the system

$$\frac{d\mathbf{Y}}{dt} = \begin{pmatrix} 2 & 3 \\ 0 & -4 \end{pmatrix} \mathbf{Y}$$

using a linear combination of

$$\mathbf{Y}_1(t) = \begin{pmatrix} e^{2t} \\ 0 \end{pmatrix} \quad \text{and} \quad \mathbf{Y}_2(t) = \begin{pmatrix} -e^{-4t} \\ 2e^{-4t} \end{pmatrix},$$

we have found the general solution to this system. It is the two-parameter family

$$\mathbf{Y}(t) = k_1 \mathbf{Y}_1(t) + k_2 \mathbf{Y}_2(t) = k_1 \begin{pmatrix} e^{2t} \\ 0 \end{pmatrix} + k_2 \begin{pmatrix} -e^{-4t} \\ 2e^{-4t} \end{pmatrix}.$$

Using vector addition, this two-parameter family can be written as one vector function

$$\mathbf{Y}(t) = \begin{pmatrix} k_1 e^{2t} - k_2 e^{-4t} \\ 2k_2 e^{-4t} \end{pmatrix}.$$

Note that, in this example, we used the Linearity Principle to produce infinitely many solutions starting with two given solutions $\mathbf{Y}_1(t)$ and $\mathbf{Y}_2(t)$. Then because we were able to express an arbitrary initial condition as a linear combination of the initial conditions $\mathbf{Y}_1(0)$ and $\mathbf{Y}_2(0)$, we could use $\mathbf{Y}_1(t)$ and $\mathbf{Y}_2(t)$ to form the general solution.

Expressing arbitrary vectors as linear combinations of given vectors is a fundamental topic in linear algebra. In the two-dimensional case the key property that ensures that an arbitrary vector can be written as some linear combination of the given vectors (x_1, y_1) and (x_2, y_2) is that they do *not* lie on the same line through the origin. [Note that in the previous example the initial conditions $(1, 0)$ and $(-1, 2)$ do not lie on the same line through the origin — see Figure 3.6.] We say that two vectors (x_1, y_1) and (x_2, y_2) are **linearly independent** if they do not lie on the same line through the origin or, equivalently, if neither one is a multiple of the other.

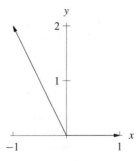

Figure 3.6
The vectors $(1, 0)$ and $(-1, 2)$ are linearly independent.

Theorem *Suppose (x_1, y_1) and (x_2, y_2) are two linearly independent vectors in the plane. Then given any vector (x_0, y_0), there exist k_1 and k_2 so that*

$$k_1 \begin{pmatrix} x_1 \\ y_1 \end{pmatrix} + k_2 \begin{pmatrix} x_2 \\ y_2 \end{pmatrix} = \begin{pmatrix} x_0 \\ y_0 \end{pmatrix}.$$

Two linearly independent vectors can be combined via addition and scalar multiplication to form any other vector in the plane.

Note that the equation

$$k_1 \begin{pmatrix} x_1 \\ y_1 \end{pmatrix} + k_2 \begin{pmatrix} x_2 \\ y_2 \end{pmatrix} = \begin{pmatrix} x_0 \\ y_0 \end{pmatrix}$$

is really a simultaneous system of two linear equations

$$\begin{cases} x_1 k_1 + x_2 k_2 = x_0 \\ y_1 k_1 + y_2 k_2 = y_0 \end{cases}$$

in the two unknowns k_1 and k_2. We are given the x's and the y's, and we must solve for the k's. We can show that there are solutions by writing down formulas for k_1 and k_2 in terms of the x's and y's. As long as the denominators in these formulas are nonzero, the solutions exist. Solving systems of equations of this form involves the same sort of algebra as finding equilibrium points for linear systems. Hence, it is not surprising that the determinant plays a role here too (see Exercises 31 and 32.)

The General Solution

The previous example and theorem illustrate the way in which we approach every linear system. Therefore it is worth summarizing the discussion.

Theorem *Suppose $\mathbf{Y}_1(t)$ and $\mathbf{Y}_2(t)$ are solutions of the linear system $d\mathbf{Y}/dt = \mathbf{AY}$. If $\mathbf{Y}_1(0)$ and $\mathbf{Y}_2(0)$ are linearly independent, then for any initial condition $\mathbf{Y}(0) = (x_0, y_0)$ we can find constants k_1 and k_2 so that $k_1 \mathbf{Y}_1(t) + k_2 \mathbf{Y}_2(t)$ is the solution to the initial-value problem*

$$\frac{d\mathbf{Y}}{dt} = \mathbf{AY}, \quad \mathbf{Y}(0) = \begin{pmatrix} x_0 \\ y_0 \end{pmatrix}.$$

In this situation we say that the two-parameter family $k_1 \mathbf{Y}_1(t) + k_2 \mathbf{Y}_2(t)$, where k_1 and k_2 are arbitrary constants, is the **general solution** of the system.

By the Existence and Uniqueness Theorem for systems, we know that each initial-value problem for a linear system has exactly one solution (a linear vector field certainly satisfies the hypotheses of the Existence and Uniqueness Theorem). Given any two solutions $\mathbf{Y}_1(t)$ and $\mathbf{Y}_2(t)$ of a linear system with linearly independent initial conditions $\mathbf{Y}_1(0)$ and $\mathbf{Y}_2(0)$, we can form the general solution to the system by forming the two-parameter family $k_1 \mathbf{Y}_1(t) + k_2 \mathbf{Y}_2(t)$. By adjusting the constants k_1 and k_2, we can obtain the solution that satisfies any given initial condition.

This is really excellent progress. We now know that to find all the solutions to a linear system, we need to find only two particular solutions with linearly independent initial positions. Two solutions $\mathbf{Y}_1(t)$ and $\mathbf{Y}_2(t)$ of a linear system for which $\mathbf{Y}_1(0)$ and $\mathbf{Y}_2(0)$ are linearly independent are called **linearly independent solutions** of the linear system. [We will see in the exercises that if $\mathbf{Y}_1(t)$ and $\mathbf{Y}_2(t)$ are linearly independent solutions of a linear system, then the vectors $\mathbf{Y}_1(t)$ and $\mathbf{Y}_2(t)$ are linearly independent for all t; see Exercise 35.] The next step is to find a way to come up with the two solutions $\mathbf{Y}_1(t)$ and $\mathbf{Y}_2(t)$. We start to tackle this issue in the next section.

An Undamped Harmonic Oscillator

In Section 2.4, we studied the undamped harmonic oscillator given by the second-order differential equation

$$\frac{d^2y}{dt^2} = -3y.$$

We obtained the general solution of this equation by considering the two functions $y_1(t) = \sin(\sqrt{3}\,t)$ and $y_2(t) = \cos(\sqrt{3}\,t)$. Once we realize that these functions are relevant, we can see that they are solutions to the second-order equation simply by computing their second derivatives and comparing the result to the original functions. At that point we gave a somewhat "inspired" justification that the general solution to this equation is

$$y(t) = k_1 \sin(\sqrt{3}\,t) + k_2 \cos(\sqrt{3}\,t).$$

Now that we have the Linearity Principle at our disposal, we can explain the source of our inspiration.

The second-order equation can be reduced to the first-order system

$$\frac{dy}{dt} = v$$

$$\frac{dv}{dt} = -3y,$$

which is a linear system. Using vector notation, we write

$$\mathbf{Y}(t) = \begin{pmatrix} y(t) \\ v(t) \end{pmatrix},$$

and the system can be represented as

$$\frac{d\mathbf{Y}}{dt} = \begin{pmatrix} 0 & 1 \\ -3 & 0 \end{pmatrix} \mathbf{Y}.$$

Recall that the second component of the vector function $\mathbf{Y}(t)$ is $v = dy/dt$. Therefore we form the two vector functions

$$\mathbf{Y}_1(t) = \begin{pmatrix} \sin(\sqrt{3}\,t) \\ \sqrt{3}\cos(\sqrt{3}\,t) \end{pmatrix} \quad \text{and} \quad \mathbf{Y}_2(t) = \begin{pmatrix} \cos(\sqrt{3}\,t) \\ -\sqrt{3}\sin(\sqrt{3}\,t) \end{pmatrix}.$$

Note that $\mathbf{Y}_1(t)$ is a solution to the system because

$$\frac{d\mathbf{Y}_1}{dt} = \begin{pmatrix} \sqrt{3}\cos(\sqrt{3}\,t) \\ -3\sin(\sqrt{3}\,t) \end{pmatrix}$$

and

$$\begin{pmatrix} 0 & 1 \\ -3 & 0 \end{pmatrix} \mathbf{Y}_1(t) = \begin{pmatrix} 0 & 1 \\ -3 & 0 \end{pmatrix} \begin{pmatrix} \sin(\sqrt{3}\,t) \\ \sqrt{3}\,\cos(\sqrt{3}\,t) \end{pmatrix}$$

$$= \begin{pmatrix} \sqrt{3}\,\cos(\sqrt{3}\,t) \\ -3\,\sin(\sqrt{3}\,t) \end{pmatrix}.$$

Similarly the solution $y_2(t)$ to the second-order equation yields a vector function

$$\mathbf{Y}_2(t) = \begin{pmatrix} \cos(\sqrt{3}\,t) \\ -\sqrt{3}\,\sin(\sqrt{3}\,t) \end{pmatrix},$$

which is also a solution to the first-order system (you should check this).

We have a first-order, linear system with two dependent variables. Therefore we need two linearly independent solutions to obtain the general solution. We consider the initial conditions

$$\mathbf{Y}_1(0) = \begin{pmatrix} 0 \\ \sqrt{3} \end{pmatrix} \quad \text{and} \quad \mathbf{Y}_2(0) = \begin{pmatrix} 1 \\ 0 \end{pmatrix}$$

of the two known solutions $\mathbf{Y}_1(t)$ and $\mathbf{Y}_2(t)$. The initial condition $\mathbf{Y}_1(0)$ sits on the v-axis and the initial condition $\mathbf{Y}_2(t)$ lies on the y-axis. Consequently, these initial conditions are linearly independent, and the general solution to the first-order system is

$$\mathbf{Y}(t) = k_1 \begin{pmatrix} \sin(\sqrt{3}\,t) \\ \sqrt{3}\,\cos(\sqrt{3}\,t) \end{pmatrix} + k_2 \begin{pmatrix} \cos(\sqrt{3}\,t) \\ -\sqrt{3}\,\sin(\sqrt{3}\,t) \end{pmatrix}.$$

However, remembering that

$$\mathbf{Y}(t) = \begin{pmatrix} y(t) \\ v(t) \end{pmatrix},$$

we obtain the general solution to the original second-order equation using the first component of $\mathbf{Y}(t)$. The result is

$$y(t) = k_1 \sin(\sqrt{3}\,t) + k_2 \cos(\sqrt{3}\,t),$$

which is what we asserted in Section 2.4.

Exercises for Section 3.1

Recall the model for Paul and Bob's CD stores,

$$\frac{dx}{dt} = ax + by$$

$$\frac{dy}{dt} = cx + dy,$$

where $x(t)$ is Paul's profit per day, $y(t)$ is Bob's profit per day, and a, b, c, and d are parameters governing how the profit per day of each store affects the other. In Exercises 1–4, different choices of the parameters a, b, c, and d are specified. For each exercise write a brief paragraph describing the interaction between the stores given the specified parameter values. [For example, suppose $a = 1$, $c = -1$, and $b = d = 0$. If Paul's store is making a profit ($x > 0$), then Paul's profit tends to increase more quickly (because $ax > 0$). However, if Paul makes a profit, then this hurts Bob's store (because $cx < 0$). Since $b = d = 0$, Bob's current profits have no impact on his or Paul's future profits.]

1. $a = 1$, $b = -1$, $c = 1$, and $d = -1$ **2.** $a = 2$, $b = -1$, $c = 0$, and $d = 0$

3. $a = 1$, $b = 0$, $c = 2$, and $d = 1$ **4.** $a = -1$, $b = 2$, $c = 2$, and $d = -1$

In Exercises 5–7, rewrite the specified linear system in matrix form.

5.
$$\frac{dx}{dt} = 2x + y$$
$$\frac{dy}{dt} = x + y$$

6.
$$\frac{dx}{dt} = 3y$$
$$\frac{dy}{dt} = 3\pi y - 0.3x$$

7.
$$\frac{dp}{dt} = 3p - 2q - 7r$$
$$\frac{dq}{dt} = -2p + 6r$$
$$\frac{dr}{dt} = 7.3q + 2r$$

In Exercises 8–9, rewrite the specified linear system in component form.

8.
$$\begin{pmatrix} \dfrac{dx}{dt} \\ \dfrac{dy}{dt} \end{pmatrix} = \begin{pmatrix} -3 & 2\pi \\ 4 & -1 \end{pmatrix} \begin{pmatrix} x \\ y \end{pmatrix}$$

9.
$$\begin{pmatrix} \dfrac{dx}{dt} \\ \dfrac{dy}{dt} \end{pmatrix} = \begin{pmatrix} 0 & \beta \\ \gamma & -1 \end{pmatrix} \begin{pmatrix} x \\ y \end{pmatrix}$$

For the linear systems given in Exercises 10–13, sketch (using whatever technology you have available) the direction fields, several solutions, and the $x(t)$- and $y(t)$-graphs for the solution with initial condition $(x, y) = (1, 1)$.

10.
$$\frac{dx}{dt} = 2x + y$$
$$\frac{dy}{dt} = x + y$$

11.
$$\frac{dx}{dt} = -2x + y$$
$$\frac{dy}{dt} = -x - 2y$$

12.
$$\begin{pmatrix} \dfrac{dx}{dt} \\ \dfrac{dy}{dt} \end{pmatrix} = \begin{pmatrix} -3 & 2\pi \\ 4 & -1 \end{pmatrix} \begin{pmatrix} x \\ y \end{pmatrix}$$

13.
$$\begin{pmatrix} \dfrac{dx}{dt} \\ \dfrac{dy}{dt} \end{pmatrix} = \begin{pmatrix} 5 & 1 \\ -1 & 6 \end{pmatrix} \begin{pmatrix} x \\ y \end{pmatrix}$$

14. Let

$$\mathbf{A} = \begin{pmatrix} a & b \\ c & d \end{pmatrix}$$

be a nonsingular matrix ($\det \mathbf{A} \neq 0$). Show that the only equilibrium point for the system $d\mathbf{Y}/dt = \mathbf{A}\mathbf{Y}$ is the origin if:

(a) $b \neq 0$ **(b)** $c \neq 0$ **(c)** $d \neq 0$

Note that, along with the verification given in the section, these results show that if $\det \mathbf{A} \neq 0$, then the only equilibrium point for the system $d\mathbf{Y}/dt = \mathbf{A}\mathbf{Y}$ is the origin.

15. Let

$$
\mathbf{A} = \begin{pmatrix} a & b \\ c & d \end{pmatrix}
$$

be a nonzero matrix. That is, suppose that at least one of its entries is nonzero. Show that, if $\det \mathbf{A} \neq 0$, then the system $d\mathbf{Y}/dt = \mathbf{A}\mathbf{Y}$ has an entire line of equilibria. [*Hint*: First consider the case where $a \neq 0$. The x-component of the vector field at a point (x, y) is $ax + by$. If the x-component is zero then $x = -by/a$. Show that any point of the form $(-by/a, y)$ is an equilibrium point. What about the other cases?]

16. The general form of a linear, homogeneous, second-order equation with constant coefficients is

$$
\frac{d^2 y}{dt^2} + p\frac{dy}{dt} + qy = 0.
$$

(a) Write the first-order system for this equation, and write this system in matrix form.

(b) Show that if $q \neq 0$, then the origin is the only equilibrium point of the system.

(c) Show that if $q \neq 0$, then the only solution of the second-order equation with y constant is $y(t) = 0$ for all t.

17. Consider the linear system corresponding to the second-order equation

$$
\frac{d^2 y}{dt^2} + p\frac{dy}{dt} + qy = 0.
$$

(a) If $q = 0$ and $p \neq 0$, find all the equilibrium points,

(b) If $q = p = 0$, find all the equilibrium points.

18. Convert the second-order equation

$$
\frac{d^2 y}{dt^2} = 0
$$

into a first-order system using $v = dy/dt$ as usual.

(a) Find the general solution for the dv/dt equation.

(b) Substitute this solution into the dy/dt equation, and find the general solution of the system.

(c) Sketch the phase plane of the system.

19. Three of the following six linear systems come from second-order equations. Determine which three and give the corresponding second-order equations.

(i) $\dfrac{d\mathbf{Y}}{dt} = \mathbf{A}\mathbf{Y}$ where $\mathbf{A} = \begin{pmatrix} 1 & 1 \\ 1 & 1 \end{pmatrix}$

(ii) $\dfrac{d\mathbf{Y}}{dt} = \mathbf{B}\mathbf{Y}$ where $\mathbf{B} = \begin{pmatrix} 0 & 1 \\ -2 & -3 \end{pmatrix}$

(iii) $\dfrac{d\mathbf{Y}}{dt} = \mathbf{C}\mathbf{Y}$ where $\mathbf{C} = \begin{pmatrix} 2 & 1 \\ 1 & 0 \end{pmatrix}$

(iv) $\dfrac{d\mathbf{Y}}{dt} = \mathbf{DY}$ where $\mathbf{D} = \begin{pmatrix} 2 & 0 \\ 0 & 3 \end{pmatrix}$

(v) $\dfrac{d\mathbf{Y}}{dt} = \mathbf{FY}$ where $\mathbf{F} = \begin{pmatrix} 0 & 1 \\ 1 & 0 \end{pmatrix}$

(vi) $\dfrac{d\mathbf{Y}}{dt} = \mathbf{GY}$ where $\mathbf{G} = \begin{pmatrix} 2 & 3 \\ 1 & 8 \end{pmatrix}$

In Exercises 20–23 we consider the following model of the market for single-family housing in a community. Let $S(t)$ equal the number of sellers at time t and $B(t)$ equal the number of buyers at time t. We assume that there are natural equilibrium levels of buyers and sellers (made up of people who retire, change job locations, or wish to move for family reasons). The equilibrium level of sellers is S_0, and the equilibrium level of buyers is B_0.

However, market forces can entice people to buy or sell under various conditions. For example, if the price of a house is very high, then house owners are tempted to sell their homes. If prices are very low, extra buyers enter the market looking for bargains. We let $b(t) = B(t) - B_0$ denote the deviation of the number of buyers from equilibrium at time t. So if $b(t) > 0$, then there are more buyers than usual, and we say it is a "seller's market." Presumably, the competition of the extra buyers for the same number of houses for sale will force the prices up (the law of supply and demand).

Similarly, we let $s(t) = S(t) - S_0$ denote the deviation of the number of sellers from the equilibrium level. If $s(t) > 0$, then there are more sellers on the market than usual; and if the number of buyers is low, there are too many houses on the market and prices decrease, which in turn affects decisions to buy or sell.

We can give a simple model of this situation as follows:

$$\frac{d\mathbf{Y}}{dt} = \mathbf{AY} = \begin{pmatrix} \alpha & \beta \\ \gamma & \delta \end{pmatrix}\begin{pmatrix} b \\ s \end{pmatrix}, \quad \text{where} \quad \mathbf{Y} = \begin{pmatrix} b \\ s \end{pmatrix}.$$

The exact values of the parameters α, β, γ, and δ depend on the economy of a particular community. Nevertheless if we assume that everybody wants to get a bargain when they are buying a house and to get top dollar when they are selling a house, then we can hope to predict whether the parameters are positive or negative even though we can't predict their exact values. In Exercises 20–23, use the information given above to obtain information about the parameters α, β, γ, and δ. Be sure to justify your answers.

20. If there are more than the usual number of buyers competing for houses, we would expect the price of houses to go up. This would make it less likely that new potential buyers will start to look for houses. What does this say about the parameter α?

21. If there are fewer than expected buyers competing for the houses available for sale, then we would expect the price of houses to decrease. This would make it less likely that any potential sellers will place their houses on the market. What does this imply about the parameter γ?

22. By considering the effect on the price of having $s(t) > 0$ and the subsequent effect on buyers and sellers, determine the sign of the parameter β.

23. Determine the most reasonable sign for the parameter δ.

24. Consider the linear system

$$\frac{d\mathbf{Y}}{dt} = \begin{pmatrix} 2 & 0 \\ 1 & 1 \end{pmatrix} \mathbf{Y}.$$

(a) Show that the two functions

$$\mathbf{Y}_1(t) = \begin{pmatrix} 0 \\ e^t \end{pmatrix} \quad \text{and} \quad \mathbf{Y}_2(t) = \begin{pmatrix} e^{2t} \\ e^{2t} \end{pmatrix}$$

are solutions to the differential equation.

(b) Solve the initial-value problem

$$\frac{d\mathbf{Y}}{dt} = \begin{pmatrix} 2 & 0 \\ 1 & 1 \end{pmatrix} \mathbf{Y}, \quad \mathbf{Y}(0) = \begin{pmatrix} -2 \\ -1 \end{pmatrix}.$$

25. Consider the linear system

$$\frac{d\mathbf{Y}}{dt} = \begin{pmatrix} 1 & -1 \\ 1 & 3 \end{pmatrix} \mathbf{Y}.$$

(a) Show that the function

$$\mathbf{Y}(t) = \begin{pmatrix} te^{2t} \\ -(t+1)e^{2t} \end{pmatrix}$$

is a solution to the differential equation.

(b) Solve the initial-value problem

$$\frac{d\mathbf{Y}}{dt} = \begin{pmatrix} 1 & -1 \\ 1 & 3 \end{pmatrix} \mathbf{Y}, \quad \mathbf{Y}(0) = \begin{pmatrix} 0 \\ 2 \end{pmatrix}.$$

In Exercises 26–29, a coefficient matrix for the linear system

$$\frac{d\mathbf{Y}}{dt} = \mathbf{AY} \quad \text{where} \quad \mathbf{Y}(t) = \begin{pmatrix} x(t) \\ y(t) \end{pmatrix}$$

is specified. Also, an initial condition and two curves are given. For each system:

(a) Check that the two curves are solutions of the system; if they are not solutions, then stop.

(b) Check that the two solutions are linearly independent; if they are not linearly independent, then stop.

(c) Find the solution to the linear system with the given initial value.

(d) Sketch the solution with the given initial condition on the xy-phase plane.

(e) Sketch the $x(t)$- and the $y(t)$-graphs for the solution with the given initial condition.

26.
$$A = \begin{pmatrix} -2 & -1 \\ 2 & -5 \end{pmatrix}$$

Curves: $Y_1(t) = (e^{-3t}, e^{-3t})$
$\qquad Y_2(t) = (e^{-4t}, 2e^{-4t})$
Initial condition: $Y(0) = (2, 3)$

27.
$$A = \begin{pmatrix} -2 & -1 \\ 2 & -5 \end{pmatrix}$$

Curves: $Y_1(t) = (e^{-3t} - 2e^{-4t}, e^{-3t} - 4e^{-4t})$
$\qquad Y_2(t) = (2e^{-3t} + e^{-4t}, 2e^{-3t} + 2e^{-4t})$
Initial condition: $Y(0) = (2, 3)$

28.
$$A = \begin{pmatrix} 2 & -1 \\ 1 & 3 \end{pmatrix}$$

Curves: $Y_1(t) = e^{5t/2}(-\cos(\sqrt{3}t/2) - \sqrt{3}\sin(\sqrt{3}t/2), 2\cos(\sqrt{3}t/2))$
$\qquad Y_2(t) = e^{5t/2}(-\sin(\sqrt{3}t/2) + \sqrt{3}\cos(\sqrt{3}t/2), 2\sin(\sqrt{3}t/2))$
Initial condition: $Y(0) = (2, 3)$

29.
$$A = \begin{pmatrix} 2 & 3 \\ 1 & 0 \end{pmatrix}$$

Curves: $Y_1(t) = (-e^{-t} + 12e^{3t}, e^{-t} + 4e^{3t})$
$\qquad Y_2(t) = (-e^{-t}, 2e^{-t})$
Initial condition: $Y(0) = (2, 3)$

30. Using scalar notation, write out and verify the Linearity Principle. (Aren't matrices nice?)

31. Show that the vectors (x_1, y_1) and (x_2, y_2) are linearly dependent — that is, not linearly independent — if any of the following conditions are satisfied:

(a) If $(x_1, y_1) = (0, 0)$.

(b) If $(x_1, y_1) = \lambda(x_2, y_2)$ for some constant λ.

(c) If $x_1 y_2 - x_2 y_1 = 0$ *Hint*: Assume x_1 is not zero; then $y_2 = x_2 y_1/x_1$. But, $x_2 = x_2 x_1/x_1$, and we can use part b. The other cases are similar. Note that the quantity $x_1 y_2 - x_2 y_1$ is the determinant of the matrix

$$\begin{pmatrix} x_1 & y_1 \\ x_2 & y_2 \end{pmatrix}.$$

32. Given the vectors (x_1, y_1) and (x_2, y_2), show that they are linearly independent if the quantity $x_1 y_2 - x_2 y_1$ is nonzero. [*Hint*: : Suppose $x_2 \neq 0$. If (x_1, y_1) and (x_2, y_2) are on the same line through $(0, 0)$, then $(x_1, y_1) = \lambda(x_2, y_2)$ for some λ. But then $\lambda = x_1/x_2$ and $\lambda = y_1/y_2$. What does this say about x_1/x_2 and y_1/y_2? What if $x_2 = 0$?]

33. Given that $Y_1(t) = (-e^{-t}, e^{-t})$ is a solution to a linear system $dY/dt = AY$, for which of the following initial vectors can you give the explicit solution?
(a) $Y(0) = (-2, 2)$ (b) $Y(0) = (3, 4)$ (c) $Y(0) = (0, 0)$ (d) $Y(0) = (3, -3)$

34. The Linearity Principle is a fundamental property of linear systems. However, don't assume that it is true for systems that are not linear, no matter how simple. For example, consider the system

$$\frac{dx}{dt} = 1$$
$$\frac{dy}{dt} = x.$$

(a) Show that $\mathbf{Y}(t) = (t, t^2/2)$ is a solution to this system.

(b) Show that $2\mathbf{Y}(t)$ is *not* a solution.

35. Given solutions $\mathbf{Y}_1(t) = (x_1(t), y_1(t))$ and $\mathbf{Y}_2(t) = (x_2(t), y_2(t))$ to the system

$$\frac{d\mathbf{Y}}{dt} = \mathbf{A}\mathbf{Y} \quad \text{where} \quad \mathbf{A} = \begin{pmatrix} a & b \\ c & d \end{pmatrix},$$

we define the **Wronskian** of $\mathbf{Y}_1(t)$ and $\mathbf{Y}_2(t)$ to be the (scalar) function

$$W(t) = x_1(t)y_2(t) - x_2(t)y_1(t).$$

(a) Compute dW/dt.

(b) Use the fact that $\mathbf{Y}_1(t)$ and $\mathbf{Y}_2(t)$ are solutions of the linear system to show that

$$\frac{dW}{dt} = (a + d)W(t).$$

(c) Find the general solution of the differential equation $dW/dt = (a + d)W(t)$.

(d) Suppose that $\mathbf{Y}_1(t)$ and $\mathbf{Y}_2(t)$ are solutions to the system $d\mathbf{Y}/dt = \mathbf{A}\mathbf{Y}$. Verify that if $\mathbf{Y}_1(0)$ and $\mathbf{Y}_2(0)$ are linearly independent, then $\mathbf{Y}_1(t)$ and $\mathbf{Y}_2(t)$ are also linearly independent for every t.

3.2 STRAIGHT-LINE SOLUTIONS

In Section 3.1, we discussed solutions of linear systems without giving any explanation of how we came up with them (the rabbit-out-of-the-hat method). To find these solutions, we used the time-honored method known as "guess and test." That is, we made a guess, then substituted the guess back into the equation and checked to see whether it satisfied the system. However, the guess-and-test method is somewhat unsatisfying because it does not give us any understanding of where the formulas came from in the first place. In this section we use the geometry of the vector field to find special solutions of linear systems.

Geometry of Straight-Line Solutions

We begin by reconsidering an example from the previous section. The direction field for the linear system

$$\frac{d\mathbf{Y}}{dt} = \mathbf{A}\mathbf{Y} \quad \text{where} \quad \mathbf{A} = \begin{pmatrix} 2 & 3 \\ 0 & -4 \end{pmatrix}$$

is shown in Figure 3.7. Looking at the direction field, we see that there are two special lines through the origin. The first is the x-axis on which the vectors in the direction field all point directly away from the origin. The other special line runs from the second quadrant to the fourth quadrant. Along this line the vectors of the direction field all point directly toward the origin.

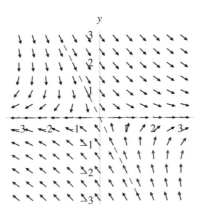

Figure 3.7

The direction field for the linear system

$$\frac{d\mathbf{Y}}{dt} = \begin{pmatrix} 2 & 3 \\ 0 & -4 \end{pmatrix} \mathbf{Y}.$$

There are two special lines through the origin. On the x-axis, the vectors in the direction field all point directly away from the origin. On the distinguished line that runs from the second quadrant to the fourth quadrant, all vectors of the direction field point directly toward the origin.

Because solutions of the system follow curves that are always tangent to the direction field, a solution that has its initial condition on the positive x-axis moves to the right, directly away from the origin. A solution with an initial condition on the negative x-axis moves to the left, directly away from the origin. Similarly, a solution with an initial condition in the second quadrant on the other special line moves directly toward the origin and a solution with an initial condition in the fourth quadrant on this line moves directly toward the origin. Thus, careful examination of the direction field suggests that there are solutions to this system that lie on straight lines through the origin in the phase plane.

In the previous section we used the rabbit-out-of-the-hat method to show that the system $d\mathbf{Y}/dt = \mathbf{A}\mathbf{Y}$ has two linearly independent solutions,

$$\mathbf{Y}_1(t) = \begin{pmatrix} e^{2t} \\ 0 \end{pmatrix} \quad \text{and} \quad \mathbf{Y}_2(t) = \begin{pmatrix} -e^{-4t} \\ 2e^{-4t} \end{pmatrix}.$$

These are formulas for solutions that lie along these special straight lines. To see this, note that $\mathbf{Y}_1(t)$ always lies on the positive x-axis since its y-coordinate is 0. Moreover, $\mathbf{Y}_1(t) \to \infty$ along the x-axis as $t \to \infty$ and $\mathbf{Y}_2(t)$ tends to the origin as $t \to -\infty$. So this is a solution that tends directly away from the origin on the x-axis.

For $\mathbf{Y}_2(t)$, it is convenient to rewrite this solution in the form

$$\mathbf{Y}_2(t) = e^{-4t} \begin{pmatrix} -1 \\ 2 \end{pmatrix}.$$

This representation tells us that, as t varies, $\mathbf{Y}_2(t)$ is always a (positive) scalar multiple of the vector $(-1, 2)$. Since positive scalar multiples of a fixed vector always lie on the same ray from the origin, we see that $\mathbf{Y}_2(t)$ parameterizes this ray. As $t \to \infty$, $e^{-4t} \to 0$, so this solution tends to the origin.

Consequently, we see that the formulas for $\mathbf{Y}_1(t)$ and $\mathbf{Y}_2(t)$ confirm what we guessed by looking at the direction field. There are solutions of this system that lie on the two distinguished straight lines in the phase plane.

Straight-line solutions are the simplest solutions (next to equilibrium points) for systems of differential equations. As these solutions move in the xy-plane along straight lines, it is important to remember that the speed at which they move depends on their position on the line. In this example, solutions go to $(0, 0)$ or ∞ at an exponential rate, and this can be seen in the $x(t)$- and $y(t)$-graphs for the solutions (see Figures 3.8 and 3.9).

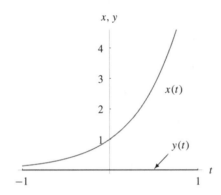

Figure 3.8
The $x(t)$- and $y(t)$-graphs of the straight-line solution

$$\mathbf{Y}_1(t) = \begin{pmatrix} e^{2t} \\ 0 \end{pmatrix}.$$

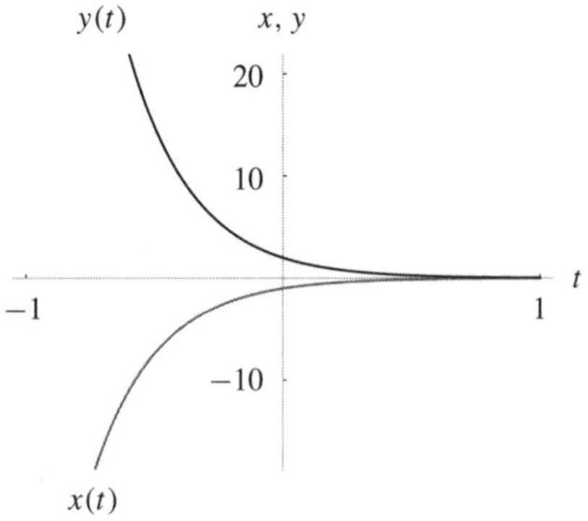

Figure 3.9
The $x(t)$- and $y(t)$-graphs of the straight-line solution

$$\mathbf{Y}_2(t) = \begin{pmatrix} -e^{-4t} \\ 2e^{-4t} \end{pmatrix}.$$

From the geometry to the algebra of straight-line solutions

Now we turn our attention to finding formulas for straight-line solutions, assuming that the system has straight-line solutions (sadly, not all linear systems do). The basic geometric observation is that, at each point on a straight-line solution through the origin, the vector field must point either directly toward or directly away from $(0, 0)$ (see Figure 3.7). That is, if $\mathbf{Y}_0 = (x_0, y_0)$ is on a straight-line solution, then the vector field at (x_0, y_0) must point in either the same direction or in exactly the opposite direction as the vector from $(0,0)$ to (x_0, y_0). This is the basic geometric observation.

We must turn this observation into an equation that we can solve to find points on straight-line solutions. For a linear system of the form $d\mathbf{Y}/dt = \mathbf{AY}$, the vector field at $\mathbf{Y}_0 = (x_0, y_0)$ is the product $\mathbf{AY}_0$ of $\mathbf{A}$ and $\mathbf{Y}_0$, which in this example is

$$\begin{pmatrix} 2 & 3 \\ 0 & -4 \end{pmatrix} \begin{pmatrix} x_0 \\ y_0 \end{pmatrix}.$$

Hence we seek points $\mathbf{Y}_0 = (x_0, y_0)$ such that $\mathbf{AY}_0$ points in the same or in the opposite direction as the vector from $(0, 0)$ to (x_0, y_0) or, equivalently, for which there is some number λ such that

$$\mathbf{A} \begin{pmatrix} x_0 \\ y_0 \end{pmatrix} = \lambda \begin{pmatrix} x_0 \\ y_0 \end{pmatrix}.$$

This is the key equation for finding straight-line solutions of the linear system $d\mathbf{Y}/dt = \mathbf{AY}$. If $\lambda > 0$, then the vector field points in the same direction as (x_0, y_0) and so away from $(0, 0)$. If $\lambda < 0$, the vector field points in the opposite direction or toward $(0, 0)$.

In our example, we seek points $\mathbf{Y}_0 = (x_0, y_0)$ such that

$$\mathbf{A} \begin{pmatrix} x_0 \\ y_0 \end{pmatrix} = \begin{pmatrix} 2 & 3 \\ 0 & -4 \end{pmatrix} \begin{pmatrix} x_0 \\ y_0 \end{pmatrix} = \lambda \begin{pmatrix} x_0 \\ y_0 \end{pmatrix}$$

or

$$\begin{pmatrix} 2x_0 + 3y_0 \\ -4y_0 \end{pmatrix} = \lambda \begin{pmatrix} x_0 \\ y_0 \end{pmatrix}.$$

We can rewrite this equation in the form

$$\begin{pmatrix} 2x_0 + 3y_0 \\ -4y_0 \end{pmatrix} - \lambda \begin{pmatrix} x_0 \\ y_0 \end{pmatrix} = \begin{pmatrix} 0 \\ 0 \end{pmatrix},$$

which is equivalent to the system of simultaneous equations

$$\begin{cases} (2 - \lambda)x_0 + 3y_0 = 0 \\ (-4 - \lambda)y_0 = 0. \end{cases}$$

There is one obvious solution to this system of equations, namely the trivial solution $(x_0, y_0) = (0, 0)$. But we already know that the origin is an equilibrium solution of this system, so this solution definitely does not give us a straight-line solution. What we need is a nonzero solution of this system of equations (that is, one where at least one of x_0 or y_0 is nonzero).

To find this nonzero solution, notice that our equations really have three unknowns, x_0, y_0, and λ. To eliminate this problem, we temporarily fix λ. Then our simultaneous system of equations becomes a system of linear equations and we know when such a system has nontrivial solutions. Recall from Section 3.1, page 232, that a system of linear equations of the form

$$\begin{cases} (2 - \lambda)x_0 + 3y_0 = 0 \\ (-4 - \lambda)y_0 = 0 \end{cases}$$

or

$$\begin{pmatrix} 2 - \lambda & 3 \\ 0 & -4 - \lambda \end{pmatrix} \begin{pmatrix} x_0 \\ y_0 \end{pmatrix} = \begin{pmatrix} 0 \\ 0 \end{pmatrix}$$

has nontrivial solutions if and only if

$$\det \begin{pmatrix} 2 - \lambda & 3 \\ 0 & -4 - \lambda \end{pmatrix} = 0.$$

Computing this determinant, we find that this system has nontrivial solutions if and only if

$$(2 - \lambda)(-4 - \lambda) = 0.$$

Therefore, we need only consider the special cases where $\lambda = 2$ and $\lambda = -4$. All other λ-values do not yield straight-line solutions.

Incidentally, recall that our two straight-line solutions involve exponentials of the form e^{2t} and e^{-4t}. In a moment we will see that the appearance of the λ-values in the exponential terms is no accident.

If $\lambda = -4$, then the simultaneous system of equations becomes simply

$$\begin{cases} 6x_0 + 3y_0 = 0 \\ 0 = 0. \end{cases}$$

The second equation always holds, so we need only choose x_0 and y_0 satisfying

$$6x_0 + 3y_0 = 0$$

or

$$y_0 = -2x_0.$$

There is an entire line of vectors (x_0, y_0) that satisfy these equations. One possible choice is $(x_0, y_0) = (-1, 2)$. This is exactly the initial condition we found above that gave us the straight-line solution

$$\mathbf{Y}_2(t) = e^{-4t} \begin{pmatrix} -1 \\ 2 \end{pmatrix}.$$

For the $\lambda = 2$ case, the equations become

$$\begin{cases} 3y_0 = 0 \\ -2y_0 = 0. \end{cases}$$

If $y_0 = 0$, then both equations are satisfied. Thus any point of the form $(x_0, 0)$ with $x_0 \neq 0$ gives a nontrivial solution with $\lambda = 2$. That is, at any point along the x-axis, the vector field points directly away from $(0, 0)$ since $\lambda > 0$. One vector on this line is $(1, 0)$. This is the vector that gave us the straight-line solution

$$\mathbf{Y}_1(t) = e^{2t} \begin{pmatrix} 1 \\ 0 \end{pmatrix}.$$

Eigenvalues and Eigenvectors

We return to this example in a moment, but first we generalize these computations so that we can apply them to any linear system. Consider the general planar linear system

$$\frac{d\mathbf{Y}}{dt} = \mathbf{AY}.$$

To find straight-line solutions through the origin, we must find points $\mathbf{Y}_0 = (x_0, y_0) \neq (0, 0)$ such that the vector field at $\mathbf{Y}_0$ points in the same direction as $\mathbf{Y}_0 = (x_0, y_0)$. So we seek points $\mathbf{Y}_0 = (x_0, y_0)$ that are nonzero and that satisfy

$$\mathbf{AY}_0 = \lambda\mathbf{Y}_0$$

for some scalar λ. This inspires the following definition.

Definition: Given a matrix $\mathbf{A}$, a number λ is called an **eigenvalue** of $\mathbf{A}$ if there is a nonzero vector $\mathbf{Y}_0 = (x_0, y_0)$ for which

$$\mathbf{AY}_0 = \mathbf{A}\begin{pmatrix} x_0 \\ y_0 \end{pmatrix} = \lambda\begin{pmatrix} x_0 \\ y_0 \end{pmatrix} = \lambda\mathbf{Y}_0.$$

The vector $\mathbf{Y}_0$ is called an **eigenvector** corresponding to the eigenvalue λ.

The word *eigen* is German for "own" or "self." An eigenvector is a vector where the vector field points in the same or opposite direction as the vector itself.

For example, consider the matrix

$$\mathbf{A} = \begin{pmatrix} 4 & 3 \\ -1 & 0 \end{pmatrix}.$$

The vector $(6, -2)$ is an eigenvector with the eigenvalue 3 because

$$\mathbf{A}\begin{pmatrix} 6 \\ -2 \end{pmatrix} = \begin{pmatrix} 4 & 3 \\ -1 & 0 \end{pmatrix}\begin{pmatrix} 6 \\ -2 \end{pmatrix} = \begin{pmatrix} 18 \\ -6 \end{pmatrix} = 3\begin{pmatrix} 6 \\ -2 \end{pmatrix}$$

Also, the vector $(-1, 1)$ is an eigenvector with eigenvalue 1 because

$$\mathbf{A}\begin{pmatrix} -1 \\ 1 \end{pmatrix} = \begin{pmatrix} 4 & 3 \\ -1 & 0 \end{pmatrix}\begin{pmatrix} -1 \\ 1 \end{pmatrix} = \begin{pmatrix} -1 \\ 1 \end{pmatrix} = 1\begin{pmatrix} -1 \\ 1 \end{pmatrix}.$$

It is important to remember that being an eigenvector is a special property. For a typical matrix, most vectors are not eigenvectors. For example, $(2, 3)$ is not an eigenvector for $\mathbf{A}$ because

$$\mathbf{A}\begin{pmatrix} 2 \\ 3 \end{pmatrix} = \begin{pmatrix} 4 & 3 \\ -1 & 0 \end{pmatrix}\begin{pmatrix} 2 \\ 3 \end{pmatrix} = \begin{pmatrix} 17 \\ -2 \end{pmatrix},$$

and $(17, -2)$ is not a multiple of $(2, 3)$.

Lines of eigenvectors

Given a matrix $\mathbf{A}$, if $\mathbf{Y}_0$ is an eigenvector for eigenvalue λ, then any scalar multiple $k\mathbf{Y}_0$ is also an eigenvector for λ. To verify this, we compute

$$\mathbf{A}(k\mathbf{Y}_0) = k\mathbf{AY}_0 = k(\lambda\mathbf{Y}_0) = \lambda(k\mathbf{Y}_0)$$

where the first equality is a property of matrix multiplication and the second equality uses the fact that $\mathbf{Y}_0$ is an eigenvector. Hence, given an eigenvector $\mathbf{Y}_0$ for the eigenvalue λ, the entire line of vectors through $\mathbf{Y}_0$ and the origin are also eigenvectors for λ.

Computation of Eigenvalues

To find straight-line solutions of linear systems, we must find the eigenvalues and eigenvectors of the corresponding coefficient matrix. That is, we need to find the points $\mathbf{Y}_0 = (x_0, y_0)$ such that

$$\mathbf{AY}_0 = \mathbf{A} \begin{pmatrix} x_0 \\ y_0 \end{pmatrix} = \lambda \begin{pmatrix} x_0 \\ y_0 \end{pmatrix} = \lambda \mathbf{Y}_0.$$

If

$$\mathbf{A} = \begin{pmatrix} a & b \\ c & d \end{pmatrix},$$

then we have

$$\begin{pmatrix} a & b \\ c & d \end{pmatrix} \begin{pmatrix} x_0 \\ y_0 \end{pmatrix} = \lambda \begin{pmatrix} x_0 \\ y_0 \end{pmatrix},$$

which is written in components as

$$\begin{cases} ax_0 + by_0 = \lambda x_0 \\ cx_0 + dy_0 = \lambda y_0. \end{cases}$$

Thus we want nonzero solutions (x_0, y_0) to

$$\begin{cases} (a - \lambda)x_0 + by_0 = 0 \\ cx_0 + (d - \lambda)y_0 = 0. \end{cases}$$

As we have seen, this system has nontrivial solutions if and only if

$$\det \begin{pmatrix} a - \lambda & b \\ c & d - \lambda \end{pmatrix} = 0.$$

We will encounter this matrix each time we compute eigenvalues and eigenvectors, so we introduce some shorthand for it. The **identity matrix** is the 2×2 matrix

$$\mathbf{I} = \begin{pmatrix} 1 & 0 \\ 0 & 1 \end{pmatrix}.$$

This matrix is called the identity matrix because $\mathbf{IV} = \mathbf{V}$ for any vector $\mathbf{V}$. Also, $\lambda \mathbf{I}$ represents the matrix

$$\lambda \mathbf{I} = \begin{pmatrix} \lambda & 0 \\ 0 & \lambda \end{pmatrix}.$$

If we compute the difference between the matrices $\mathbf{A}$ and $\lambda \mathbf{I}$ by subtracting corresponding entries, we find

$$\mathbf{A} - \lambda \mathbf{I} = \begin{pmatrix} a - \lambda & b \\ c & d - \lambda \end{pmatrix}.$$

Thus, our determinant condition for a nontrivial solution of the equation $\mathbf{AY}_0 = \lambda \mathbf{Y}_0$ may be written in the compact form

$$\det(\mathbf{A} - \lambda \mathbf{I}) = 0.$$

It is important to remember that the matrix $\mathbf{A} - \lambda \mathbf{I}$ is shorthand for the matrix $\mathbf{A}$ with λ's subtracted from the upper left and lower right entries.

The Characteristic Polynomial

To find the eigenvalues of our system, we must find the roots of the equation

$$\det(\mathbf{A} - \lambda\mathbf{I}) = 0.$$

If we write out this equation explicitly, we find

$$\det(\mathbf{A} - \lambda\mathbf{I}) = \det\begin{pmatrix} a - \lambda & b \\ c & d - \lambda \end{pmatrix} = (a - \lambda)(d - \lambda) - bc = 0$$

or

$$\lambda^2 - (a + d)\lambda + (ad - bc) = 0.$$

The expression $\det(\mathbf{A} - \lambda\mathbf{I})$ is a quadratic polynomial in λ. This polynomial is called the **characteristic polynomial** of the system. Its roots are the eigenvalues of the system.

A quadratic polynomial always has two roots, but these roots need not be real numbers, nor distinct. The characteristic polynomial can have non-real roots (called complex eigenvalues). We will study the behavior of systems with complex eigenvalues in Section 3.4.

As a familiar example, consider the matrix

$$\mathbf{A} = \begin{pmatrix} 2 & 3 \\ 0 & -4 \end{pmatrix}$$

that we discussed earlier. This matrix has the characteristic polynomial

$$\det(\mathbf{A} - \lambda\mathbf{I}) = (2 - \lambda)(-4 - \lambda) - (3)(0) = \lambda^2 + 2\lambda - 8,$$

which has roots $\lambda_1 = 2$ and $\lambda_2 = -4$. As we saw earlier, these numbers are the eigenvalues of this matrix.

Computation of Eigenvectors

The next step in the process of finding straight-line solutions of a system of differential equations is to find the eigenvectors associated to the eigenvalues. Suppose we are given a matrix

$$\mathbf{A} = \begin{pmatrix} a & b \\ c & d \end{pmatrix}$$

and we know that λ is an eigenvalue. To find a corresponding eigenvector, we must solve the equation

$$\mathbf{AY}_0 = \lambda\mathbf{Y}_0$$

for the vector $\mathbf{Y}_0$. If we write

$$\mathbf{Y}_0 = \begin{pmatrix} x_0 \\ y_0 \end{pmatrix},$$

then $\mathbf{AY}_0 = \lambda\mathbf{Y}_0$ becomes a simultaneous system of linear equations in two unknowns, x_0 and y_0. In fact, the equations are

$$\begin{cases} ax_0 + by_0 = \lambda x_0 \\ cx_0 + dy_0 = \lambda y_0. \end{cases}$$

Since λ is an eigenvalue, we know that there is at least an entire line through the origin of eigenvectors (x_0, y_0) that satisfy this system of equations. Each of the equations defines a line in the plane, so one of the two equations is always redundant. That is, either one of the equations is a multiple of the other or one of the equations is always satisfied.

For example, suppose we are given the matrix

$$\mathbf{B} = \begin{pmatrix} 2 & 2 \\ 1 & 3 \end{pmatrix}.$$

We find the eigenvalues of $\mathbf{B}$ by finding the roots of the characteristic polynomial

$$\det(\mathbf{B} - \lambda \mathbf{I}) = (2 - \lambda)(3 - \lambda) - 2 = 0,$$

which yields the quadratic equation

$$\lambda^2 - 5\lambda + 4 = 0.$$

The roots are $\lambda_1 = 4$ and $\lambda_2 = 1$, so 1 and 4 are the eigenvalues of $\mathbf{B}$.

To find an eigenvector $\mathbf{V}_1$ for $\lambda_1 = 4$, we must solve

$$\mathbf{B} \begin{pmatrix} x_1 \\ y_1 \end{pmatrix} = \begin{pmatrix} 2 & 2 \\ 1 & 3 \end{pmatrix} \begin{pmatrix} x_1 \\ y_1 \end{pmatrix} = 4 \begin{pmatrix} x_1 \\ y_1 \end{pmatrix}.$$

Rewritten in terms of components, this equation is

$$\begin{cases} 2x_1 + 2y_1 = 4x_1 \\ x_1 + 3y_1 = 4y_1, \end{cases}$$

or, equivalently,

$$\begin{cases} -2x_1 + 2y_1 = 0 \\ x_1 - y_1 = 0. \end{cases}$$

Note that these equations are redundant (multiply both sides of the second by -2 to get the first). So any vector (x_1, y_1) that satisfies

$$x_1 - y_1 = 0$$

is an eigenvector. This is a straight line in the plane. Any nonzero vector on this line is an eigenvector of $\mathbf{B}$ corresponding to the eigenvalue $\lambda_1 = 4$. For example, the vectors $(1, 1)$ and $(-\pi, -\pi)$ are two of the infinitely many eigenvectors for $\mathbf{B}$ corresponding to $\lambda_1 = 4$.

For $\lambda_2 = 1$ we must solve

$$\mathbf{B} \begin{pmatrix} x_2 \\ y_2 \end{pmatrix} = \begin{pmatrix} 2 & 2 \\ 1 & 3 \end{pmatrix} \begin{pmatrix} x_2 \\ y_2 \end{pmatrix} = 1 \begin{pmatrix} x_2 \\ y_2 \end{pmatrix}.$$

In terms of coordinates, this vector equation is the same as the system

$$\begin{cases} 2x_2 + 2y_2 = x_2 \\ x_2 + 3y_2 = y_2 \end{cases}$$

or, equivalently,

$$\begin{cases} x_2 + 2y_2 = 0 \\ x_2 + 2y_2 = 0. \end{cases}$$

Again these equations are redundant. So the eigenvectors corresponding to eigenvalue $\lambda_2 = 1$ are the nonzero vectors (x_2, y_2) that lie on the line $y_2 = -x_2/2$.

Straight-Line Solutions

After all of the algebra of the last few pages, it is time to return to the study of differential equations. To summarize what we have accomplished so far, suppose we are given a linear system of differential equations,

$$\frac{d\mathbf{Y}}{dt} = \mathbf{AY}.$$

To find straight-line solutions, we first find the eigenvalues of $\mathbf{A}$ and then their associated eigenvectors. Once we have this information, we have determined the straight-line solutions.

To do this, suppose that λ is an eigenvalue with associated eigenvector $\mathbf{Y}_0 = (x_0, y_0)$. Then consider the curve in the plane given by

$$\mathbf{Y}(t) = e^{\lambda t}\mathbf{Y}_0 = \begin{pmatrix} e^{\lambda t}x_0 \\ e^{\lambda t}y_0 \end{pmatrix}.$$

For each t, $\mathbf{Y}(t)$ is a scalar multiple of our eigenvector (x_0, y_0), so the curve $\mathbf{Y}(t)$ lies on the ray from the origin through this point. Moreover, $\mathbf{Y}(t)$ is a solution of the differential equation. All we have to do is check this by substituting $\mathbf{Y}(t)$ in the differential equation. We compute

$$\frac{d\mathbf{Y}}{dt} = \frac{d}{dt}\begin{pmatrix} e^{\lambda t}x_0 \\ e^{\lambda t}y_0 \end{pmatrix} = \begin{pmatrix} \lambda e^{\lambda t}x_0 \\ \lambda e^{\lambda t}y_0 \end{pmatrix} = \lambda\mathbf{Y}(t).$$

On the other hand, we have

$$\mathbf{AY}(t) = \mathbf{A}e^{\lambda t}\mathbf{Y}_0 = e^{\lambda t}\mathbf{AY}_0 = e^{\lambda t}\lambda\mathbf{Y}_0 = \lambda\mathbf{Y}(t)$$

since $\mathbf{Y}_0$ is an eigenvector of $\mathbf{A}$. Comparing the results of these two computations, we see that

$$\frac{d\mathbf{Y}}{dt} = \mathbf{AY}(t),$$

so $\mathbf{Y}(t)$ is indeed a solution.

This is an important observation: It tells us how to write down formulas for straight-line solutions using just the eigenvalues and eigenvectors of the matrix $\mathbf{A}$.

Sometimes we can even do better. Suppose we find two real, distinct eigenvalues λ_1 and λ_2 for the system and associated eigenvectors $\mathbf{V}_1$ and $\mathbf{V}_2$. These vectors are necessarily independent vectors. That is, any scalar multiple of $\mathbf{V}_1$ is an associated eigenvector corresponding to λ_1. Consequently, $\mathbf{V}_2$ does not lie on the line through the origin determined by $\mathbf{V}_1$, and $\mathbf{V}_1$ and $\mathbf{V}_2$ are independent. But this means that the solutions

$$\mathbf{Y}_1(t) = e^{\lambda_1 t}\mathbf{V}_1 \quad \text{and} \quad \mathbf{Y}_2(t) = e^{\lambda_2 t}\mathbf{V}_2$$

are linearly independent. Therefore, using the Linearity Principle, the general solution of the system is $k_1\mathbf{Y}_1(t) + k_2\mathbf{Y}_2(t)$.

Theorem

Suppose the matrix $\mathbf{A}$ has a real eigenvalue λ with associated eigenvector $\mathbf{Y}_0$. Then the linear system of differential equations

$$\frac{d\mathbf{Y}}{dt} = \mathbf{AY}$$

has the straight-line solution

$$\mathbf{Y}(t) = e^{\lambda t}\mathbf{Y}_0.$$

Moreover, if λ_1 and λ_2 are distinct, real eigenvalues with eigenvectors $\mathbf{V}_1$ and $\mathbf{V}_2$ respectively, then the solutions

$$\mathbf{Y}_1(t) = e^{\lambda t}\mathbf{V}_1 \quad \text{and} \quad \mathbf{Y}_2(t) = e^{\lambda t}\mathbf{V}_2$$

are linearly independent and

$$\mathbf{Y}(t) = k_1\mathbf{Y}_1(t) + k_2\mathbf{Y}_2(t)$$

is the general solution of the system.

This is a powerful theorem. It lets us find certain solutions of linear systems of differential equations using only algebra. All that is required is finding an eigenvalue and an associated eigenvector. There are no tedious or impossible integrations to perform. Of course, the caveat here is that the eigenvalue must be real; we tackle the more difficult case of complex eigenvalues in Section 3.4.

The theorem also explicitly provides the general solution of certain linear systems, namely those that have two distinct, real eigenvalues. We will also have to deal with the possibility that the eigenvalues of $\mathbf{A}$ are real but not distinct. We take up this issue in Section 3.5.

Putting Everything Together

Now let's combine the geometry of the direction field with the algebra of this section to produce the general solution of a linear system of differential equations. Consider the linear system

$$\frac{d\mathbf{Y}}{dt} = \mathbf{BY} = \begin{pmatrix} 2 & 2 \\ 1 & 3 \end{pmatrix}\mathbf{Y}.$$

The direction field for this system is depicted in Figure 3.10. There appear to be two straight-line solutions, one cutting diagonally through the first and third quadrants, and the second through the second and fourth quadrants. On each straight-line solution, the vector field points away from the origin.

To find these solutions explicitly, we must find the eigenvalues and eigenvectors of $\mathbf{B}$. We did this earlier in the section. We found the eigenvalues to be $\lambda_1 = 4$ and $\lambda_2 = 1$. Any eigenvector associated to λ_1 lies on the line $y_1 = x_1$. Similarly, we saw that an eigenvector associated to λ_2 is $(-2, 1)$. Thus we have two independent straight-line solutions, and the general solution is

$$\mathbf{Y}(t) = k_1 e^{4t}\begin{pmatrix} 1 \\ 1 \end{pmatrix} + k_2 e^{t}\begin{pmatrix} -2 \\ 1 \end{pmatrix}.$$

Each of these solutions moves away from the origin as time increases.

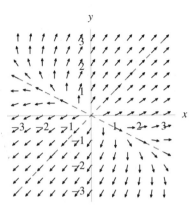

Figure 3.10
The direction field for the system

$$\frac{d\mathbf{Y}}{dt} = \mathbf{B}\mathbf{Y} = \begin{pmatrix} 2 & 2 \\ 1 & 3 \end{pmatrix} \mathbf{Y}.$$

Note the two distinguished lines of eigenvectors. The one in the first quadrant corresponds to the solution $\mathbf{Y}_1(t) = e^{4t}(1, 1)$ and the one in the second quadrant corresponds to the solution $\mathbf{Y}_2(t) = e^t(-2, 1)$.

A Harmonic Oscillator

Consider the harmonic oscillator with spring constant $k_s = 1$ and damping constant $k_d = 3$. The second-order equation is

$$\frac{d^2y}{dt^2} + 3\frac{dy}{dt} + y = 0,$$

and the corresponding system is

$$\frac{d\mathbf{Y}}{dt} = \mathbf{C}\mathbf{Y}, \quad \text{where} \quad \mathbf{C} = \begin{pmatrix} 0 & 1 \\ -1 & -3 \end{pmatrix} \quad \text{and} \quad \mathbf{Y} = \begin{pmatrix} y \\ v \end{pmatrix}.$$

The phase plane is shown in Figure 3.11. Again we see that there appear to be straight-line solutions for this system.

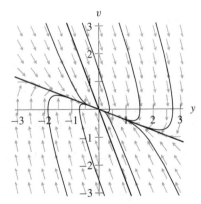

Figure 3.11
Phase plane for

$$\frac{d\mathbf{Y}}{dt} = \begin{pmatrix} 0 & 1 \\ -1 & -3 \end{pmatrix} \mathbf{Y}.$$

This linear system is obtained from the harmonic oscillator

$$\frac{d^2y}{dt^2} + 3\frac{dy}{dt} + y = 0,$$

where $\mathbf{Y} = (y, v)$ and $v = dy/dt$.

The characteristic polynomial for the system is

$$(-\lambda)(-3 - \lambda) + 1 = \lambda^2 + 3\lambda + 1,$$

and using the quadratic formula, we find that the eigenvalues are

$$\lambda = \frac{-3 \pm \sqrt{9-4}}{2}$$

or $\lambda = (-3 \pm \sqrt{5})/2$. Note that both of these eigenvalues are negative.

We compute the eigenvector for $\lambda_1 = (-3 + \sqrt{5})/2$ by solving

$$\begin{cases} y_0 = \left(\dfrac{-3 + \sqrt{5}}{2} \right) x_0 \\[3mm] -x_0 - 3y_0 = \left(\dfrac{-3 + \sqrt{5}}{2} \right) y_0 \end{cases}$$

As usual, these two equations are redundant (multiply the first by $(3 + \sqrt{5})/2$ to obtain the second). Setting $x_0 = 1$, we find that the vector $(1, (-3 + \sqrt{5})/2)$ is on this line. Similarly, we compute that an eigenvector for $\lambda_2 = (-3 - \sqrt{5})/2$ is $(1, (-3 - \sqrt{5})/2)$.

We can now write down the general solution for this system. It is

$$\mathbf{Y}(t) = k_1 e^{(-3+\sqrt{5})t/2} \begin{pmatrix} 1 \\[2mm] \dfrac{-3+\sqrt{5}}{2} \end{pmatrix} + k_2 e^{(-3-\sqrt{5})t/2} \begin{pmatrix} 1 \\[2mm] \dfrac{-3-\sqrt{5}}{2} \end{pmatrix}.$$

This formula looks formidable. However, with enough calculation, we can use it to find the exact position of the oscillator at any time. Moreover, we can also determine qualitative features of the model from these formulas. Each term in the expression for $\mathbf{Y}(t)$ contains an exponential of the form $e^{(-3 \pm \sqrt{5})t/2}$. Consequently, they tend to 0 as t increases. Of course, we see this immediately from the phase plane (see Figure 3.11), but it is comforting to see everything fit together so nicely. Since $\mathbf{Y}(t) = (y(t), v(t))$, we obtain the general solution to the second-order equation from the first component of $\mathbf{Y}(t)$. We have

$$y(t) = k_1 e^{(-3+\sqrt{5})t/2} + k_2 e^{(-3-\sqrt{5})t/2}.$$

The Method of the Lucky Guess

We have used intelligent guessing techniques in several different situations to find formulas for solutions of differential equations. It turns out that there is a "lucky guess" method that can be used to find at least some solutions of second-order, linear equations like the equation for the harmonic oscillator. We briefly describe this method here for two reasons. First, many readers will see this method in other subjects where second-order equations are used. Second, we feel it is important to contrast it with the more geometric eigenvalue approach.

We begin the description of the method by considering the second-order equation

$$\frac{d^2 y}{dt^2} + 3\frac{dy}{dt} + y = 0.$$

If we ignore everything we have done so far in this section, we have no idea of how to solve this equation analytically. Consequently, we use a classic mathematical technique. We guess.

Since the most common function we have seen so far in solutions of differential equations is the exponential, it is reasonable to start by guessing an exponential function. However, this immediately leads to another problem. We have no idea what the exponent should be. So we try an exponential of the form

$$y(t) = e^{st},$$

where s is some number we don't know. Hopefully, we can compute s as part of the process.

If this guess is a solution, then the differential equation must be satisfied if we substitute e^{st} for y. That is, we should have

$$\frac{d^2(e^{st})}{dt^2} + 3\frac{d(e^{st})}{dt} + e^{st} = 0,$$

which simplifies to

$$s^2 e^{st} + 3se^{st} + e^{st} = 0.$$

Now, e^{st} is never zero, so we can divide both sides by it to obtain

$$s^2 + 3s + 1 = 0,$$

a quadratic equation. Note that this quadratic is the same as the characteristic polynomial that we obtained from the linear system (with λ replaced by s). Thus, the exponents that we obtain from this guessing technique are the same as the eigenvalues of the linear system.

If s is a root of this quadratic, then the function $y(t) = e^{st}$ is a solution of the differential equation. The roots in this case are $s = (-3+\sqrt{5})/2$ and $s = (-3-\sqrt{5})/2$, and it is easy to check that the two functions

$$y_1(t) = e^{(-3+\sqrt{5})t/2} \quad \text{and} \quad y_2(t) = e^{(-3-\sqrt{5})t/2}$$

are solutions.

Now, for the general constant coefficient, linear, homogeneous second-order equation

$$\frac{d^2 y}{dt^2} + p\frac{dy}{dt} + qy = 0,$$

we can use the same method. We guess $y(t) = e^{st}$ and substitute it into the differential equation to obtain

$$\frac{d^2(e^{st})}{dt^2} + p\frac{d(e^{st})}{dt} + qy = 0$$

or

$$s^2 e^{st} + pse^{st} + qe^{st} = 0.$$

Dividing both sides by e^{st}, we obtain

$$s^2 + ps + q = 0.$$

This quadratic is called the **characteristic equation** for the second-order differential equation. If s is a root of the characteristic equation, then $y(t) = e^{st}$ is a solution. If there are two distinct, real roots s_1 and s_2 (such as in the example above), then we get two solutions $y_1(t) = e^{s_1 t}$ and $y_2(t) = e^{s_2 t}$, and we can obtain the general solution

$$y(t) = k_1 e^{s_1 t} + k_2 e^{s_2 t}$$

using the Linearity Principle.

The algebra involved in this approach is definitely easier than the algebra involved in the computation of eigenvalues. However, if we turn a second-order equation into a first-order system, we have access to numerical techniques that are not available to us with the guessing techniques.

Exercises for Section 3.2

In Exercises 1–8,

(a) sketch the direction field (using whatever technology is available to you, or do it "by hand"); and

(b) from your sketch of the direction field, decide whether the corresponding linear system has straight-line solutions and, if so, draw these solutions and describe their behavior. For example, do they tend toward the origin or away from the origin as t increases?

1. $\dfrac{d\mathbf{Y}}{dt} = \begin{pmatrix} 3 & 2 \\ 0 & -2 \end{pmatrix} \mathbf{Y}$

2. $\dfrac{dx}{dt} = 2x + y$

$\dfrac{dy}{dt} = x + y$

3. $\begin{pmatrix} \dfrac{dx}{dt} \\ \dfrac{dy}{dt} \end{pmatrix} = \begin{pmatrix} 4 & 2 \\ 1 & 3 \end{pmatrix} \begin{pmatrix} x \\ y \end{pmatrix}$

4. $\dfrac{d\mathbf{Y}}{dt} = \begin{pmatrix} -4 & -2 \\ -1 & -3 \end{pmatrix} \mathbf{Y}$

5. $\dfrac{dx}{dt} = x + 4y$

$\dfrac{dy}{dt} = -3x - y$

6. $\begin{pmatrix} \dfrac{dx}{dt} \\ \dfrac{dy}{dt} \end{pmatrix} = \begin{pmatrix} 2 & 1 \\ -1 & 4 \end{pmatrix} \begin{pmatrix} x \\ y \end{pmatrix}$

7. $\dfrac{d\mathbf{Y}}{dt} = \begin{pmatrix} -1/2 & 0 \\ 1 & -1/2 \end{pmatrix} \mathbf{Y}$

8. $\dfrac{dx}{dt} = 0.3x - y$

$\dfrac{dy}{dt} = 2x + 3.5y$

In Exercises 9–16,

(a) compute the eigenvalues and eigenvectors; and

(b) if there are two distinct, real eigenvalues, compute the general solution.

9. $\dfrac{d\mathbf{Y}}{dt} = \begin{pmatrix} 3 & 2 \\ 0 & -2 \end{pmatrix} \mathbf{Y}$

10. $\dfrac{dx}{dt} = 2x + y$

$\dfrac{dy}{dt} = x + y$

11. $\begin{pmatrix} \dfrac{dx}{dt} \\ \dfrac{dy}{dt} \end{pmatrix} = \begin{pmatrix} 4 & 2 \\ 1 & 3 \end{pmatrix} \begin{pmatrix} x \\ y \end{pmatrix}$

12. $\dfrac{d\mathbf{Y}}{dt} = \begin{pmatrix} -4 & -2 \\ -1 & -3 \end{pmatrix} \mathbf{Y}$

13. $\dfrac{dx}{dt} = x + 4y$

$\dfrac{dy}{dt} = -3x - y$

14. $\begin{pmatrix} \dfrac{dx}{dt} \\ \dfrac{dy}{dt} \end{pmatrix} = \begin{pmatrix} 2 & 1 \\ -1 & 4 \end{pmatrix} \begin{pmatrix} x \\ y \end{pmatrix}$

15. $\dfrac{d\mathbf{Y}}{dt} = \begin{pmatrix} -1/2 & 0 \\ 1 & -1/2 \end{pmatrix} \mathbf{Y}$

16. $\dfrac{dx}{dt} = 0.3x - y$

$\dfrac{dy}{dt} = 2x + 3.5y$

In Exercises 17–22,

(a) compute the eigenvalues and eigenvectors; and

(b) if there are two distinct, real eigenvalues, compute the general solution.

17. $\dfrac{d\mathbf{Y}}{dt} = \mathbf{AY}$ where $\mathbf{A} = \begin{pmatrix} -3 & 0 \\ -1 & 2 \end{pmatrix}$

18. $\dfrac{d\mathbf{Y}}{dt} = \mathbf{BY}$ where $\mathbf{B} = \begin{pmatrix} 3 & 0 \\ 1 & -2 \end{pmatrix}$

19. $\dfrac{d\mathbf{Y}}{dt} = \mathbf{CY}$ where $\mathbf{C} = \begin{pmatrix} -4 & 1 \\ 2 & -3 \end{pmatrix}$

20. $\dfrac{d\mathbf{Y}}{dt} = \mathbf{GY}$ where $\mathbf{G} = \begin{pmatrix} 4 & 3 \\ 2 & 1 \end{pmatrix}$

21. $\dfrac{d\mathbf{Y}}{dt} = \mathbf{DY}$ where $\mathbf{D} = \begin{pmatrix} -2 & 2 \\ 1 & -1 \end{pmatrix}$

22. $\dfrac{d\mathbf{Y}}{dt} = \mathbf{HY}$ where $\mathbf{H} = \begin{pmatrix} 4 & -3 \\ -2 & 1 \end{pmatrix}$

In Exercises 23–28, we specify initial conditions that go with the linear systems given in Exercises 17–22. (Note that the order is jumbled.)

- Find the particular solutions with the given initial conditions.
- Sketch the particular solution in the xy-plane and sketch the $x(t)$- and $y(t)$-graphs for these solutions.
- Compare these sketches (and the amount of work involved in making them) with the corresponding sketches from this section.

23. The system $d\mathbf{Y}/dt = \mathbf{BY}$ and the initial conditions:

(a) $\mathbf{Y}_0 = (1, 0)$ (b) $\mathbf{Y}_0 = (0, 1)$ (c) $\mathbf{Y}_0 = (2, 2)$

24. The system $d\mathbf{Y}/dt = \mathbf{AY}$ and the initial conditions:

(a) $\mathbf{Y}_0 = (1, 0)$ (b) $\mathbf{Y}_0 = (0, 1)$ (c) $\mathbf{Y}_0 = (-2, 1)$

25. The system $d\mathbf{Y}/dt = \mathbf{DY}$ and the initial conditions:

(a) $\mathbf{Y}_0 = (1, 0)$ (b) $\mathbf{Y}_0 = (2, 2)$ (c) $\mathbf{Y}_0 = (-2, 1)$

26. The system $d\mathbf{Y}/dt = \mathbf{CY}$ and the initial conditions:

(a) $\mathbf{Y}_0 = (1, 0)$ (b) $\mathbf{Y}_0 = (2, 1)$ (c) $\mathbf{Y}_0 = (-1, -2)$

27. The system $d\mathbf{Y}/dt = \mathbf{HY}$ and the initial conditions:

(a) $\mathbf{Y}_0 = (1, 0)$ (b) $\mathbf{Y}_0 = (1, 1)$ (c) $\mathbf{Y}_0 = (-2, 1)$

28. The system $d\mathbf{Y}/dt = \mathbf{GY}$ and the initial conditions:

(a) $\mathbf{Y}_0 = (1, 0)$ (b) $\mathbf{Y}_0 = (-1, 0)$ (c) $\mathbf{Y}_0 = (1, -1)$

29. Show that a is the only eigenvalue and that every vector is an eigenvector for the matrix

$$\mathbf{A} = \begin{pmatrix} a & 0 \\ 0 & a \end{pmatrix}.$$

30. A matrix of the form

$$\mathbf{A} = \begin{pmatrix} a & b \\ 0 & d \end{pmatrix}$$

is called **upper triangular**. Suppose that $b \neq 0$ and $a \neq d$. Find the eigenvalues and eigenvectors of $\mathbf{A}$.

31. A matrix of the form

$$\mathbf{B} = \begin{pmatrix} a & b \\ b & d \end{pmatrix}$$

is called **symmetric**. Show that $\mathbf{B}$ has real eigenvalues and that, if $b \neq 0$, then $\mathbf{B}$ has two distinct eigenvalues.

32. Consider the second-order equation

$$\frac{d^2y}{dt^2} + p\frac{dy}{dt} + qy = 0,$$

where p and q are positive.

(a) Write the corresponding system of equations.

(b) Find the eigenvalues.

(c) Under what conditions will the eigenvalues be two distinct real numbers?

(d) Verify that when the eigenvalues are real numbers, they are negative.

In Exercises 33–36, the parameter values and an initial condition for a harmonic oscillator are given.

(a) Write the second-order equation and the corresponding system.

(b) Find the general solution.

(c) Find the particular solution for the given initial condition.

33. $m = 1$, $k_s = 2$, $k_d = 3$, with initial condition $y(0) = 1$, $v(0) = 0$

34. $m = 2$, $k_s = 2$, $k_d = 5$, with initial condition $y(0) = 1$, $v(0) = 0$

35. $m = 1$, $k_s = 1$, $k_d = 4$, with initial condition $y(0) = -2$, $v(0) = 0$

36. $m = 1$, $k_s = 1$, $k_d = 6$, with initial condition $y(0) = -2$, $v(0) = -4$

37. Verify that the harmonic oscillator with $m = 1$, $k_s = 4$ and $k_d = 1$ does not have real eigenvalues. Does this tell you anything about the phase plane. [*Hint:* Sketch the direction field for this system.]

38. For the harmonic oscillator with $m = 1$, $k_s = 4$ and $k_d = 5$,

(a) compute the eigenvalues and eigenvectors;

(b) using Euler's method, compute solutions with initial conditions on each of the lines of eigenvectors [*Hint:* Use the eigenvector as the initial condition];

(c) draw the $y(t)$-graphs of these solutions; and

(d) describe, in a brief essay, the behavior of a mass-spring system with these parameters for the two solutions you have computed.

3.3 PHASE PLANES FOR LINEAR SYSTEMS WITH REAL EIGENVALUES

In the preceding section we saw that straight-line solutions play a dominant role in finding the general solution of certain linear systems of differential equations. To solve such a system, we first use algebra to compute the eigenvalues and eigenvectors of the coefficient matrix. When we find a real eigenvalue and an associated eigenvector, we can write down the corresponding straight-line solution. Moreover, in the special case where we find two real, distinct eigenvalues, we can write down an explicit formula for the general solution of the system.

The sign of the eigenvalue plays an important role in determining the behavior of the corresponding straight-line solutions. If the eigenvalue is negative, the solution tends to the origin as $t \to \infty$. If the eigenvalue is positive, the solution tends away from the origin as $t \to \infty$. In this section, we use the behavior of these straight-line solutions to determine the behavior of all solutions.

Saddles

One common type of linear system features both a positive and negative eigenvalue. For example, consider the linear system

$$\frac{d\mathbf{Y}}{dt} = \mathbf{A}\mathbf{Y} \quad \text{where} \quad \mathbf{A} = \begin{pmatrix} -3 & 0 \\ 0 & 2 \end{pmatrix}.$$

This is a particularly simple linear system, since it corresponds to the equations

$$\frac{dx}{dt} = -3x$$
$$\frac{dy}{dt} = 2y.$$

Note that dx/dt depends only on x and dy/dt depends only on y, so we can solve these two equations independently using methods from Chapter 1. However, in order to understand the geometry more fully, we use the methods of the previous two sections.

As usual, we first compute the eigenvalues of $\mathbf{A}$ by finding the roots of the characteristic polynomial

$$\det(\mathbf{A} - \lambda\mathbf{I}) = \det \begin{pmatrix} -3-\lambda & 0 \\ 0 & 2-\lambda \end{pmatrix} = (-3-\lambda)(2-\lambda) = 0.$$

The roots of this quadratic equation are $\lambda_1 = -3$ and $\lambda_2 = 2$, the eigenvalues of $\mathbf{A}$. Next we compute the eigenvectors. For $\lambda_1 = -3$, the equations that give the eigenvectors are

$$\begin{cases} (-3+3)x_1 = 0 \\ (2+3)y_1 = 0. \end{cases}$$

So any nonzero vector lying along the line $y = 0$ (the x-axis) in the plane serves as an eigenvector for $\lambda_1 = -3$. We choose $\mathbf{V}_1 = (1, 0)$. Therefore the solution

$$\mathbf{Y}_1(t) = e^{-3t}\mathbf{V}_1$$

is a straight-line solution lying on the x-axis. It tends to the origin as t increases.

In similar fashion we can check that any eigenvector corresponding to $\lambda_2 = 2$ lies along the y-axis (the line $x = 0$). We choose $\mathbf{V}_2 = (0, 1)$ and find a second solution

$$\mathbf{Y}_2(t) = e^{2t}\mathbf{V}_2.$$

The general solution is therefore

$$\mathbf{Y}(t) = k_1 e^{-3t}\mathbf{V}_1 + k_2 e^{2t}\mathbf{V}_2 = \begin{pmatrix} k_1 e^{-3t} \\ k_2 e^{2t} \end{pmatrix}.$$

In Figure 3.12 we display the phase plane for this system. The straight-line solutions lie on the axes, but all other solutions behave differently. In the figure we see that the other solutions seem to tend to infinity asymptotic to the y-axis and to come from infinity asymptotic to the x-axis. The fact that any other solution is a linear combination of our straight-line solutions explains this. If we have a solution $\mathbf{Y}(t)$ that is not one of our straight-line solutions, then both k_1 and k_2 must be nonzero. When t is large and positive, the term e^{-3t} is very small. Therefore the vector $e^{-3t}\mathbf{V}_1$ in the general solution is negligible, and we have

$$\mathbf{Y}(t) \approx k_2 e^{2t}\mathbf{V}_2 = \begin{pmatrix} 0 \\ k_2 e^{2t} \end{pmatrix}.$$

That is, for large positive values of t, our solution behaves like the straight-line solution on the y-axis.

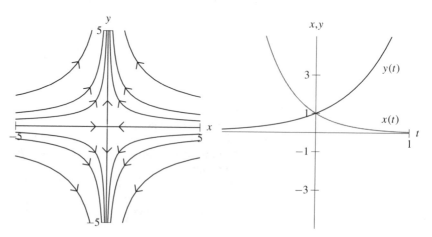

Figure 3.12
Phase plane for the system

$$\frac{d\mathbf{Y}}{dt} = \mathbf{AY} = \begin{pmatrix} -3 & 0 \\ 0 & 2 \end{pmatrix} \mathbf{Y}.$$

Figure 3.13
The $x(t)$- and $y(t)$-graphs for the solution with initial position $(1, 1)$.

The opposite is true when we consider large negative values of t. In this case, the term e^{2t} is very small, so we have

$$\mathbf{Y}(t) \approx k_1 e^{-3t} \mathbf{V}_1 = \begin{pmatrix} k_1 e^{-3t} \\ 0 \end{pmatrix},$$

which is the straight-line solution along the x-axis.

For example, the particular solution of this system that satisfies $\mathbf{Y}(0) = (1, 1)$ is given by

$$\mathbf{Y}(t) = \begin{pmatrix} e^{-3t} \\ e^{2t} \end{pmatrix}.$$

The x-coordinate of this solution tends to 0 as $t \to \infty$ and to infinity as $t \to -\infty$. The y-coordinate behaves in the opposite manner. This is reflected in the $x(t)$- and $y(t)$-graphs (see Figure 3.13).

Despite the fact that this example really consists of two one-dimensional differential equations, its phase plane is an entirely new type of picture. Along the axes we see the familiar phase lines for one-dimensional equations (a sink along the x-axis, a source on the y-axis). All other solutions tend to infinity as $t \to \pm\infty$. These solutions come from infinity in the direction of the eigenvector corresponding to the one-dimensional sink, and they tend back to infinity in the direction of the one-dimensional source.

Any linear system for which we have one positive and one negative eigenvalue has similar behavior. An equilibrium point of this form is called a **saddle**. This name is supposed to remind you of a horse saddle. The path followed by a drop of water on a horse's saddle resembles the path of a solution of this type of linear system. It approaches the seat in one direction and then veers off to the ground in another.

As a word of caution, note that this system is very special. The lines corresponding to the eigenvectors are also the nullclines for the system. This is not often the case. Along the eigenvectors, the vector field points toward the origin. Only when the eigenvectors lie on the axes does the vector field point in a horizontal or vertical direction.

Phase planes for other saddles

The previous example is special in that the eigenvectors lie on the x- and y-axes. In general, the eigenvectors for a saddle can lie on any two distinct lines through the origin. This makes the phase planes and the $x(t)$- and $y(t)$-graphs appear somewhat different in the general case.

For example, consider the system

$$\frac{d\mathbf{Y}}{dt} = \mathbf{AY} \quad \text{where} \quad \mathbf{A} = \begin{pmatrix} 8 & -11 \\ 6 & -9 \end{pmatrix}.$$

We first compute the eigenvalues of $\mathbf{A}$ by finding the roots of the characteristic polynomial

$$\det(\mathbf{A} - \lambda I) = \det \begin{pmatrix} 8 - \lambda & -11 \\ 6 & -9 - \lambda \end{pmatrix} = (8 - \lambda)(-9 - \lambda) + 66 = \lambda^2 + \lambda - 6 = 0.$$

The roots of this quadratic equation are $\lambda_1 = -3$ and $\lambda_2 = 2$, the eigenvalues of $\mathbf{A}$. These are exactly the same eigenvalues as in the previous example, so the origin is a saddle.

Next we compute the eigenvectors. For $\lambda_1 = -3$, the equations that give the eigenvectors are

$$\begin{cases} 8x_0 - 11y_0 = -3x_0 \\ 6x_0 - 9y_0 = -3y_0. \end{cases}$$

So any nonzero vector that lies along the line $y = x$ in the plane serves as an eigenvector for $\lambda_1 = -3$. We choose $\mathbf{V}_1 = (1, 1)$. Therefore the solution

$$\mathbf{Y}_1(t) = e^{-3t}\mathbf{V}_1$$

is a straight-line solution lying on the line $y = x$. It tends to the origin as t increases.

Similar computations yield eigenvectors corresponding to $\lambda_2 = 2$ lying along the line $6x - 11y = 0$, for example $\mathbf{V}_2 = (11, 6)$. This in turn gives a straight-line solution of the form

$$\mathbf{Y}_2(t) = e^{2t}\mathbf{V}_2$$

that tends away from the origin as t increases. Thus the general solution is

$$\mathbf{Y}(t) = k_1 e^{-3t}\mathbf{V}_1 + k_2 e^{2t}\mathbf{V}_2.$$

As above, we expect that, when k_1 and k_2 are nonzero, these solutions will come from infinity in the direction of $\mathbf{V}_1$ and will tend back to infinity in the direction of $\mathbf{V}_2$. In the phase plane we see the two straight-line solutions together with several other solutions (see Figure 3.14). The important point is that, once we have the eigenvalues and eigenvectors, we can immediately visualize the entire phase plane.

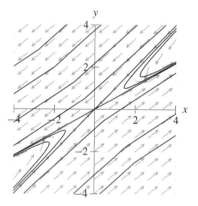

Figure 3.14
The direction field and phase plane for the system

$$\frac{d\mathbf{Y}}{dt} = \mathbf{AY} = \begin{pmatrix} 8 & -11 \\ 6 & -9 \end{pmatrix}\mathbf{Y}.$$

Sinks

Now consider the system of differential equations

$$\frac{d\mathbf{Y}}{dt} = \mathbf{BY} \quad \text{where} \quad \mathbf{B} = \begin{pmatrix} -2 & -2 \\ -1 & -3 \end{pmatrix}.$$

The matrix $\mathbf{B}$ has eigenvalues $\lambda_1 = -4$ and $\lambda_2 = -1$. For $\lambda_1 = -4$, one eigenvector is $\mathbf{V}_1 = (1, 1)$ and for $\lambda_2 = -1$, one eigenvector is $\mathbf{V}_2 = (-2, 1)$. (Checking this is a good review of eigenvalues and eigenvectors. You should be able to check that these vectors are eigenvectors without recomputing from scratch.)

Thus this vector field has two straight-line solutions that tend to the origin, and in fact the general solution is

$$\mathbf{Y}(t) = k_1 e^{-4t} \mathbf{V}_1 + k_2 e^{-t} \mathbf{V}_2$$

$$= k_1 e^{-4t} \begin{pmatrix} 1 \\ 1 \end{pmatrix} + k_2 e^{-t} \begin{pmatrix} -2 \\ 1 \end{pmatrix}$$

$$= \begin{pmatrix} k_1 e^{-4t} - 2k_2 e^{-t} \\ k_1 e^{-4t} + k_2 e^{-t} \end{pmatrix}.$$

Once we know that the eigenvalues for this system are -4 and -1, we know that every term in the general solution will have a factor of e^{-4t} or e^{-t}. Hence, every solution must tend to the origin as $t \to \infty$. The point is that the long-term behavior of solutions can be determined from the eigenvalues alone (without the eigenvectors or the formula for the general solution).

In Figure 3.15 we sketch the phase plane for this system. In this picture we clearly see the straight-line solutions. As predicted, all other solutions tend to the origin. This always happens when we have a system with two negative eigenvalues. In analogy with one-dimensional differential equations, we call such an equilibrium point a **sink**.

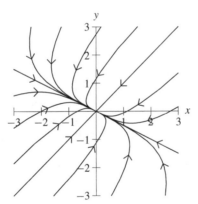

Figure 3.15
Phase plane for the system

$$\frac{d\mathbf{Y}}{dt} = \mathbf{BY} = \begin{pmatrix} -2 & -2 \\ -1 & -3 \end{pmatrix} \mathbf{Y}.$$

Direction of approach to the sink

In Figure 3.15 observe that all other solutions (except those on the straight-line solution corresponding to λ_1) tend toward the origin in a direction tangent to the line of eigenvectors corresponding to $\lambda_2 = -1$. That is, these solutions seem to tend to the origin tangentially to the line with slope $-1/2$.

To understand why this happens, we compute the slope of the tangent line to any solution curve and then ask what happens to this slope as $t \to \infty$. Each solution curve is given by

$$\begin{pmatrix} x(t) \\ y(t) \end{pmatrix} = \begin{pmatrix} k_1 e^{-4t} - 2k_2 e^{-t} \\ k_1 e^{-4t} + k_2 e^{-t} \end{pmatrix}$$

for some choice of constants k_1 and k_2. The tangent vector to this curve is given by $(dx/dt, dy/dt)$. The slope of the tangent line to the curve is given by the ratio of the rate of change of y to the rate of change of x, that is

$$\text{Slope} = \frac{dy/dt}{dx/dt}.$$

In our case, we have

$$\frac{dy/dt}{dx/dt} = \frac{-4k_1 e^{-4t} - k_2 e^{-t}}{-4k_1 e^{-4t} + 2k_2 e^{-t}}.$$

If we take the limit of this expression as $t \to \infty$, we end up with the indeterminate $\frac{0}{0}$. It is tempting to use L'Hôpital's Rule to handle this limit, but this approach is destined to fail since the derivatives all involve exponential terms as well. The way to handle this limit is to multiply both numerator and denominator by e^t. Then the new expression is

$$\frac{dy/dt}{dx/dt} = \frac{-4k_1 e^{-3t} - k_2}{-4k_1 e^{-3t} + 2k_2}.$$

When $t \to \infty$, both exponential terms in this quotient tend to 0 and we see that the limit is $-k_2/(2k_2) = -1/2$. That is, these solutions tend to the origin with slopes tending to $-1/2$ or, equivalently, tangent to the line of eigenvectors corresponding to λ_2. Technically, this limit is $-1/2$ only if $k_2 \neq 0$ in this expression, since we divide by k_2. But when $k_2 = 0$, our solution lies along the line of eigenvectors corresponding the eigenvalue -4. In this case our formula for the slope reads

$$\frac{dy/dt}{dx/dt} = \frac{-4k_1 e^{-4t} - 0 \cdot e^{-t}}{-4k_1 e^{-4t} + 0 \cdot e^{-t}} = 1,$$

which, of course, is exactly what the slope is supposed to be along this straight-line solution.

The discussion of the direction of approach to the equilibrium point may seem technical, but there really is a good qualitative reason why most solutions tend to $(0, 0)$ tangent to the eigenvector corresponding to the eigenvalue -1. Recall that the vector field on the line of eigenvectors corresponding to the eigenvalue λ is simply the scalar product of λ and the position vector. Because $-4 < -1$, the vector field on the line of eigenvectors for the eigenvalue -4 at a given distance from the origin is much longer than those on the line of eigenvectors for the eigenvalue -1. So solutions on the line of eigenvectors for -4 tend to zero much more quickly than those for the eigenvalue -1. In particular, the solution $e^{-4t} \mathbf{V}_1$ tends to $(0, 0)$ more quickly than $e^{-t} \mathbf{V}_2$.

In our general solution

$$\mathbf{Y}(t) = k_1 e^{-4t} \mathbf{V}_1 + k_2 e^{-t} \mathbf{V}_2,$$

if both k_1 and k_2 are nonzero, then the first term tends to the origin more quickly than the second. So when t is sufficiently large, the second term dominates and we see that most solutions tend to zero along the direction of the eigenvectors for the eigenvalue closer to zero. The only exceptions are the solutions on the line of eigenvectors for the eigenvalue which is more negative. So, as before, provided $k_2 \neq 0$, we can write $\mathbf{Y}(t) \approx k_2 e^{-t} \mathbf{V}_2$ as $t \to \infty$.

The case of an arbitrary sink with two eigenvalues $\lambda_1 < \lambda_2 < 0$ is entirely analogous. All solutions tend to the origin, and all (except those on the line of eigenvectors corresponding to λ_1) tend to $(0, 0)$ tangent to the λ_2-eigenvectors. We interpret what these solutions mean more precisely when we return to a harmonic oscillator example at the end of this section. Before that, however, we deal with the final possibility for the eigenvalues, namely when they are both positive.

Sources

Consider the system

$$\frac{d\mathbf{Y}}{dt} = \mathbf{BY} \quad \text{where} \quad \mathbf{B} = \begin{pmatrix} 2 & 2 \\ 1 & 3 \end{pmatrix}.$$

In the previous section we computed that the eigenvalues of this matrix are $\lambda_1 = 4$ and $\lambda_2 = 1$. Also, $\mathbf{V}_1 = (1, 1)$ is an eigenvector for the eigenvalue $\lambda_1 = 4$ and $\mathbf{V}_2 = (-2, 1)$ is an eigenvector for the eigenvalue 1, as is easily checked. Then we have two straight-line solutions $e^{4t}\mathbf{V}_1$ and $e^t\mathbf{V}_2$, and the general solution is

$$\mathbf{Y}(t) = k_1 e^{4t}\mathbf{V}_1 + k_2 e^t\mathbf{V}_2.$$

Since both eigenvalues of this system are positive, all nonzero solutions move away from the origin as $t \to \infty$.

The phase plane for this system is shown in Figure 3.16. As in the previous example, we see two straight-line solutions, and all other solutions leave the origin in a direction tangent to the line of eigenvectors corresponding to the eigenvalue $\lambda_2 = 1$. The reason for this is essentially the same as the reason given for sinks above. In fact, the astute reader will note that this vector field is just the negative of our previous example, so we have merely changed the direction of the arrows and not the geometry of the solution curves. For this system, instead of considering the behavior as $t \to \infty$, we consider the behavior as $t \to -\infty$. Now the eigenvalue 4 plays the role of the stronger eigenvalue. Solutions involving terms with e^{4t} tend to the origin much more quickly than those involving e^t as $t \to -\infty$.

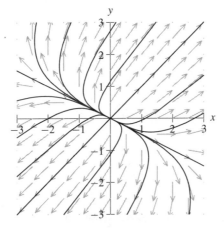

Figure 3.16
Phase plane for the system

$$\frac{d\mathbf{Y}}{dt} = \mathbf{CY} = \begin{pmatrix} 2 & 2 \\ 1 & 3 \end{pmatrix} \mathbf{Y}.$$

In general, once we know that both eigenvalues of a linear system are positive, we can conclude that all solutions tend away from the origin as t increases. We call the equilibrium point for a linear system with two positive eigenvalues a **source**. All solutions tend away from the equilibrium point as $t \to \infty$, and all except those on the line of eigenvectors corresponding to λ_1 leave the origin in a direction tangent to the line of eigenvectors corresponding to λ_2.

Stable and Unstable
Equilibrium Points

Before turning to specific examples, we summarize the behavior of a linear system with two real, distinct eigenvalues λ_1 and λ_2.

1. If $\lambda_1 < 0 < \lambda_2$, then the origin is a saddle. There are two straight-line solutions, one tending toward $(0, 0)$ as t increases, the other tending away from $(0, 0)$. All other solutions come from and go to infinity.

2. If $\lambda_1 < \lambda_2 < 0$, then the origin is a sink. All solutions tend to $(0, 0)$ as $t \to \infty$, and most tend to $(0, 0)$ in the direction of the λ_2-eigenvector.

3. If $\lambda_1 > \lambda_2 > 0$, then the origin is a source. All solutions tend away from $(0, 0)$ as $t \to \infty$, and most tend away in the direction of the λ_2-eigenvector.

A sink is said to be **stable** because nearby initial points give solutions that tend back toward the equilibrium point as time increases. So if the initial condition is "bumped" a little bit away from the sink, the resulting solution does not stray far away from the initial point. Saddle and source equilibrium points are called **unstable** because there are initial conditions arbitrarily close to the equilibrium point whose solutions move away. Hence a small bump to an initial condition can have dramatic consequences. For a source, every initial condition near the equilibrium point corresponds to a solution that moves away. For a saddle, every initial condition except those on the straight-line solution tending to the equilibrium point (so almost every initial condition) corresponds to a solution that moves away.

If we run time backwards, then a source looks like a sink with solutions tending toward it. Similarly, in backwards time a sink looks like a source with solutions moving away from it. This is analogous to the situation for phase lines.

The saddle is a new type of equilibrium point that can not occur in one-dimensional systems. Saddles need two dimensions in order to have one direction that is stable (corresponding to the negative eigenvalue) and another that is unstable (corresponding to the positive eigenvalue).

The Harmonic Oscillator

Recall the harmonic oscillator equation

$$\frac{d^2y}{dt^2} + 3\frac{dy}{dt} + y = 0$$

with corresponding system

$$\frac{d\mathbf{Y}}{dt} = \mathbf{CY} \quad \text{where} \quad \mathbf{C} = \begin{pmatrix} 0 & 1 \\ -1 & -3 \end{pmatrix} \quad \text{and} \quad \mathbf{Y} = \begin{pmatrix} y \\ v \end{pmatrix}$$

as discussed in Section 3.2. The eigenvalues of $\mathbf{C}$ are $(-3 \pm \sqrt{5})/2 \approx -1.5 \pm 1.12$, both of which are negative. Consequently, the origin is a sink. All solutions tend to $(0, 0)$ as $t \to \infty$. We can now complete the discussion of how these solutions tend to $(0, 0)$ and what this means for the mass-spring system.

We denote an eigenvector corresponding to the eigenvalue $(-3 + \sqrt{5})/2$ by $\mathbf{V}_1$ and one associated to the eigenvalue $(-3 - \sqrt{5})/2$ by $\mathbf{V}_2$. The straight-line solutions therefore assume the form

$$\mathbf{Y}_1(t) = e^{(-3+\sqrt{5})t/2}\mathbf{V}_1 \quad \text{and} \quad \mathbf{Y}_2(t) = e^{(-3-\sqrt{5})t/2}\mathbf{V}_2.$$

For each straight-line solution, the first coordinate (the position of the mass) tends directly to 0 as $t \to \infty$. We interpret this as meaning that the spring, once stretched, simply returns to its rest position, traveling ever more slowly as time increases. The spring does not oscillate around the rest position since the damping force is too large.

The solutions that do not lie on these straight lines also tend to the origin, but they do so in a slightly different way. As we have seen, they tend to $(0, 0)$ tangent to $\mathbf{V}_2$, an eigenvector corresponding to the eigenvalue $(-3 - \sqrt{5})/2$ (see Figure 3.17).

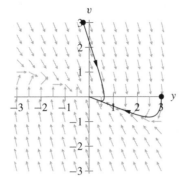

Figure 3.17

The direction field and selected solution curves for

$$\frac{d\mathbf{Y}}{dt} = \mathbf{CY} = \begin{pmatrix} 0 & 1 \\ -1 & -3 \end{pmatrix} \mathbf{Y}.$$

As a particular example, consider the solution to the system with initial position $(-0.25, 3)$. According to our model, this corresponds to the situation where the spring is compressed and then released with positive velocity toward the rest position. Looking at the solution curve through this point, we see that this curve crosses the y-axis and then turns and tends to the origin along the direction of $\mathbf{V}_2$. The graph of the position of the spring $y(t)$ for this initial condition is displayed in Figure 3.18. This graph shows that $y(t)$ initially increases and passes through $y = 0$. Then $y(t)$ reaches a maximum and slowly decreases to 0. This corresponds to the mass starting with the spring compressed and with initial velocity toward the right. The mass overshoots the $y = 0$ position once, reaches a maximum extension, then settles gently back to the rest position without overshooting $y = 0$ again.

On the other hand, solutions need not overshoot the rest position. For example, consider the solution curve through the initial position $(3, 0)$. This corresponds to the spring initially stretched and then released without any velocity. Now the solution curve simply tends to the origin without ever crossing the line $y = 0$. The graph of $y(t)$ decreases to 0 and $v(t)$ is always negative (see Figure 3.19). A harmonic oscillator system that has two negative real eigenvalues is called an **overdamped** system. The system has so much damping that solutions do not oscillate. They can overshoot the rest position at most once.

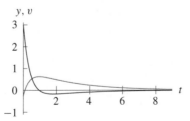

Figure 3.18
The graph of the $y(t)$ for the solution of the harmonic oscillator system with initial condition $(-0.25, 3)$

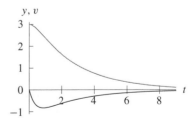

Figure 3.19
The $y(t)$- and $v(t)$-graphs for the solution of the harmonic oscillator system with initial condition $(3, 0)$.

Paul's and Bob's CD Stores

We conclude this section with one more example. Recall the model of Paul's and Bob's CD stores from the text and exercises of Section 3.1.

Suppose market research establishes that if a store becomes popular, then it becomes too crowded and profits tend to decrease. Also all stores near a popular store suffer from the effect of overcrowding and their profits also decrease. This means that if Paul's profits become positive, pofits of his store and of Bob's store tend to decrease, so parameters a and c should be negative. The same is true for Bob's store. As an example, we let $a = -2$, $b = -3$, $c = -3$, and $d = -2$, so the linear system is

$$\frac{d\mathbf{Y}}{dt} = \begin{pmatrix} -2 & -3 \\ -3 & -2 \end{pmatrix} \mathbf{Y}.$$

All the coefficients are negative. We might be tempted to say that this system should predict that profit for either store is impossible because, whenever one store starts to make money, it makes the rate of change of the profits of both stores smaller. However, we can't always trust guesses. We use the tools that we have developed to study this system carefully.

To give an accurate sketch of the phase plane, we first compute the eigenvalues and eigenvectors. The characteristic polynomial is

$$(-2 - \lambda)(-2 - \lambda) - 9 = \lambda^2 + 4\lambda - 5 = (\lambda - 1)(\lambda + 5).$$

The eigenvalues are the roots of the characteristic polynomial, $\lambda_1 = -5$ and $\lambda_2 = 1$. Because one eigenvalue is positive and one is negative, the origin is a saddle (see Figure 3.20). We find an eigenvector for the eigenvalue $\lambda_1 = -5$ by solving

$$\begin{cases} -2x_1 - 3y_1 = -5x_1 \\ -3x_1 - 2y_1 = -5y_1, \end{cases}$$

and these equations have a line of solutions given by $x_1 = y_1$. So $(1, 1)$ is an eigenvector for the eigenvalue $\lambda_1 = -5$.

For the other eigenvalue $\lambda_2 = 1$, we must solve

$$\begin{cases} -2x_2 - 3y_2 = x_2 \\ -3x_2 - 2y_2 = y_2 \end{cases}$$

These equations have a line of solutions given by $x_2 = -y_2$. So $(-1, 1)$ is an eigenvector for the eigenvalue $\lambda_2 = 1$. We could now use this information to write down the general solution, but it is more useful to use it to sketch the phase plane. We know that the diagonal $x_1 = y_1$ through the origin contains straight-line solutions, and that these solutions tend toward the origin because the eigenvalue $\lambda_1 = -5$ is negative. The other diagonal line through the origin, $x_2 = -y_2$, contains straight-line solutions that move away from $(0, 0)$ as t increases. Every other solution is a linear combination of these two. So the only solutions that tend to $(0, 0)$ are those on the line $x = y$. As $t \to \infty$, all other solutions eventually move away from the origin in either the second or fourth quadrants (see Figure 3.20).

Analysis of the model

This model leads to some startling predictions for Paul's and Bob's profits. Suppose that at $t = 0$ both Paul and Bob are making a profit [$x(0) > 0$ and $y(0) > 0$]. If it

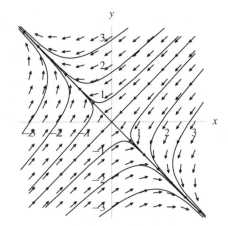

Figure 3.20

Phase plane for the system

$$\frac{d\mathbf{Y}}{dt} = \begin{pmatrix} -2 & -3 \\ -3 & -2 \end{pmatrix} \mathbf{Y}.$$

happens that Paul and Bob are making *exactly* the same amount, then $x(0) = y(0)$ and the initial point is on the $x = y$ line. The solution with this initial condition tends to the origin as t increases; that is, Paul and Bob both make less and less profit as time increases, both of them tending toward the break-even point $x = y = 0$.

Next consider the case $x(0) > y(0)$, even by just a tiny amount. Now the initial point is just below the $x = y$ diagonal. The corresponding solution at first tends toward $(0, 0)$, but it eventually turns and follows the straight-line solution along the line $x = -y$ into the fourth quadrant. In this case $x(t) \to \infty$ but $y(t) \to -\infty$. In other words, Paul eventually makes a fortune, but Bob loses his shirt. The vector field is very small near $(0, 0)$, so the solution moves slowly when it is near the origin. But eventually it turns the corner and Paul gets rich and Bob loses out (see Figure 3.20).

On the other hand, suppose $y(0)$ is slightly larger than $x(0)$. Then the initial point is just above the line $x = y$. In this case the solution first tends toward $(0, 0)$, but it eventually "turns the corner" and tends toward infinity along the line $x = -y$ in the second quadrant. In this case $x(t) \to -\infty$ (Paul goes broke) and $y(t) \to \infty$ (Bob gets rich; see Figure 3.20).

In this example a tiny change in the initial condition causes a large change in the long-term behavior of the system. We emphasize that the difference in behavior of solutions takes a long time to appear because solutions move very slowly near the equilibrium point. This sensitive dependence on the choice of initial condition is caused by the straight-line solution through the origin. The solutions with $x(0) = y(0) + 0.01$ and $x(0) = y(0) + 0.02$ are both on the same side of the diagonal, so they both behave the same way in the long run. It is only when a small change pushes the initial condition to the other side of the straight-line solution along the diagonal that the big jump in the long-term behavior occurs (see Figure 3.20). For this reason, a straight-line solution of a saddle corresponding to the negative eigenvalue is sometimes called a **separatrix** because it separates two different types of long-term behavior.

Common Sense versus Computation

The predictions of this model are not at all what we might have expected. The coefficient matrix

$$\begin{pmatrix} -2 & -3 \\ -3 & -2 \end{pmatrix}$$

consists of only negative numbers, so any increase in profits of either store has a negative effect on the rate of change of the profits. A common sense guess for the behavior of the solution is that neither store will ever show a profit. The behavior of the model is quite different.

The lesson to be learned is that although it is always wise to compare the predictions of a model with common sense, common sense does not replace computation. Models are most valuable when they predict something unexpected.

Exercises for Section 3.3

For Exercises 1–4:

(a) Find the eigenvalues of the coefficient matrix (if they are not real, then stop).

(b) Find the eigenvectors of the coefficient matrix (if there are not two linearly independent eigenvectors, then stop).

(c) Draw the straight-line solutions on the phase plane.

(d) Sketch the full phase plane.

1. $\dfrac{d\mathbf{Y}}{dt} = \mathbf{BY}$, where $\mathbf{B} = \begin{pmatrix} 3 & 0 \\ 1 & -2 \end{pmatrix}$ **2.** $\dfrac{d\mathbf{Y}}{dt} = \mathbf{CY}$, where $\mathbf{C} = \begin{pmatrix} -4 & 1 \\ 2 & -3 \end{pmatrix}$

3. $\dfrac{d\mathbf{Y}}{dt} = \mathbf{HY}$, where $\mathbf{H} = \begin{pmatrix} 4 & -3 \\ -2 & 1 \end{pmatrix}$ **4.** $\dfrac{d\mathbf{Y}}{dt} = \mathbf{GY}$, where $\mathbf{G} = \begin{pmatrix} 4 & 3 \\ 2 & 1 \end{pmatrix}$

In Exercises 5–8, we refer to the systems in Exercises 1–4. For each initial condition specified, sketch the solution in the xy-phase plane, and sketch the $x(t)$- and the $y(t)$-graphs.

5. The system $d\mathbf{Y}/dt = \mathbf{BY}$ and the initial conditions:

    (a) $\mathbf{Y}_0 = (1, 0)$ (b) $\mathbf{Y}_0 = (0, 1)$ (c) $\mathbf{Y}_0 = (2, 2)$

6. The system $d\mathbf{Y}/dt = \mathbf{CY}$ and the initial conditions:

    (a) $\mathbf{Y}_0 = (1, 0)$ (b) $\mathbf{Y}_0 = (2, 1)$ (c) $\mathbf{Y}_0 = (-1, -2)$

7. The system $d\mathbf{Y}/dt = \mathbf{HY}$ and the initial conditions:

    (a) $\mathbf{Y}_0 = (1, 0)$ (b) $\mathbf{Y}_0 = (1, 1)$ (c) $\mathbf{Y}_0 = (-2, 1)$

8. The system $d\mathbf{Y}/dt = \mathbf{GY}$ and the initial conditions:

    (a) $\mathbf{Y}_0 = (1, 0)$ (b) $\mathbf{Y}_0 = (-1, 0)$ (c) $\mathbf{Y}_0 = (1, -1)$

In Exercises 9–12, we consider a small pond inhabited by a species of fish. When left alone, the population of these fish settles into an equilibrium population. Suppose a few fish of another species are introduced to the pond. We would like to know if the new species survives and if the population of the native species changes much from its equilibrium population.

In order to determine the answers to these questions, we create a very simple model of the fish populations. Let $f(t)$ be the population of the native fish, and let f_0 denote the equilibrium population. We are interested in the change of the native fish population from its equilibrium level, so we let $x(t) = f(t) - f_0$, that is, $x(t)$ is the difference of the population of the native fish species from its equilibrium level. Let $y(t)$ denote the population of the introduced species. We note that, because $y(t)$ is an "absolute" population, it does not make sense to have $y(t) < 0$. So, if $y(t)$ is ever equal to zero, we say that the introduced species has gone extinct. On the other hand, $x(t)$ can assume both positive and negative values because this variable measures the difference of the native fish population from its equilibrium level.

We are concerned with the behavior of these population when both variables x and y are small, so the effects of terms involving x^2, y^2, xy or higher powers are very, very small. Consequently, we ignore them in this model. Also, we know that, if $x = y = 0$ then the native fish population is in equilibrium and there are none of the introduced species, so the population does not change, that is, $(x, y) = (0, 0)$ is an equilibrium point. Hence it is reasonable to use a linear system as a model.

For each model:

(a) Discuss briefly what sort of interaction between the species corresponds to the model, that is, do the introduced fish work to increase or decrease the native fish population, etc...

(b) Decide if the model agrees with the information above about the system. That is, will the population of the native species return to equilibrium if the introduced species is not present.

(c) Sketch the phase plane and describe the solutions of the linear system (using technology and information about eigenvalues and eigenvectors).

(d) State what predictions the model makes about what happens when a small number of the new species is introduced into the lake.

9. $\dfrac{d\mathbf{Y}}{dt} = \begin{pmatrix} -0.2 & -0.1 \\ 0.0 & -0.1 \end{pmatrix} \mathbf{Y}$

10. $\dfrac{d\mathbf{Y}}{dt} = \begin{pmatrix} -0.1 & 0.2 \\ 0.0 & 1.0 \end{pmatrix} \mathbf{Y}$

11. $\dfrac{d\mathbf{Y}}{dt} = \begin{pmatrix} -0.2 & 0.1 \\ 0.0 & -0.1 \end{pmatrix} \mathbf{Y}$

12. $\dfrac{d\mathbf{Y}}{dt} = \begin{pmatrix} 0.1 & 0.0 \\ -0.2 & 0.2 \end{pmatrix} \mathbf{Y}$

In Exercises 13–14, we (re)consider the model of Paul's and Bob's CD stores from Section 3.1. Suppose Paul and Bob are both operating at the break-even point $(x, y) = (0, 0)$. For the models given below, state what happens if one of the stores starts to earn or lose just a little bit. That is, will the profits return to 0 for both stores? If not, does it matter which store starts to earn money first?

13. $\begin{pmatrix} \dfrac{dx}{dt} \\ \dfrac{dy}{dt} \end{pmatrix} = \begin{pmatrix} 2 & 1 \\ 0 & -1 \end{pmatrix} \begin{pmatrix} x \\ y \end{pmatrix}$

14. $\begin{pmatrix} \dfrac{dx}{dt} \\ \dfrac{dy}{dt} \end{pmatrix} = \begin{pmatrix} -2 & -1 \\ -1 & -1 \end{pmatrix} \begin{pmatrix} x \\ y \end{pmatrix}$

In Exercises 15–18 we specify parameter values for the mass m, the spring constant k_s, and the coefficient of damping k_d for the harmonic oscillator. For each choice of parameters,

(a) write the second-order differential equation model and the corresponding system;

(b) find the eigenvalues and eigenvectors;

(c) sketch the phase plane; and

(d) sketch the graph of the y-coordinate for the solution with the given initial condition.

15. $m = 1, k_s = 1, k_d = 3$, for initial condition $y(0) = 2, v(0) = 0$

16. $m = 2, k_s = 1, k_d = 3$ for initial condition $y(0) = 0, v(0) = -2$

17. $m = 1, k_s = 2, k_d = 10$, with initial condition $y(0) = 2, v(0) = 2$

18. $m = 2, k_s = 4, k_d = 20$, with initial condition $y(0) = 2, v(0) = -2$

19. For the harmonic oscillator with $m = 1$, $k_s = 2$, $k_d = 3$, and initial condition $y(0) = -2$, $v(0) = -1$, compute the (approximate) times(s) t at which $y(t) = 0$. (The equation for the particular solution is so complicated that you may be able only to approximate the time(s) numerically.)

20. For the harmonic oscillator with $m = 1$, $k_s = 1$, $k_d = 4$, and initial condition $y(0) = 2$, find the initial velocity for which $y(t) > 0$ for all t and $y(t)$ reaches 0.1 the quickest. [*Hint*: It helps to look at the phase plane first.]

21. The slope field for the system

$$\frac{dx}{dt} = -2x + \frac{1}{2}y$$
$$\frac{dy}{dt} = -y$$

is shown below.

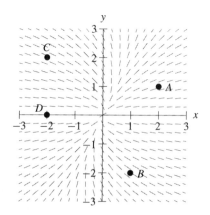

(a) Determine the type of the equilibrium point at the origin.

(b) Calculate all straight-line solutions.

(c) Plot the $x(t)$- and $y(t)$-graphs ($t \geq 0$) for the initial conditions $A = (2, 1)$, $B = (1, -2)$, $C = (-2, 2)$, and $D = (-2, 0)$.

22. The slope field for the system

$$\frac{dx}{dt} = 2x + 6y$$

$$\frac{dy}{dt} = 2x - 2y$$

is shown below.

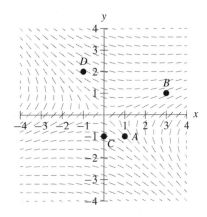

(a) Determine the type of the equilibrium point at the origin.

(b) Calculate all straight-line solutions.

(c) Plot the $x(t)$- and $y(t)$-graphs ($t \geq 0$) for the initial conditions $A = (1, -1)$, $B = (3, 1)$, $C = (0, -1)$, and $D = (-1, 2)$.

23. Consider the linear system

$$\frac{d\mathbf{Y}}{dt} = \begin{pmatrix} -2 & 1 \\ 0 & 2 \end{pmatrix} \mathbf{Y}.$$

(a) Show that $(0, 0)$ is a saddle.

(b) Find the eigenvalues and eigenvectors and sketch the phase plane.

(c) Sketch on the phase plane the solution curves with initial conditions $(1, 0.01)$ and $(1, -0.01)$.

(d) Estimate the time t at which the solutions with initial conditions $(1, 0.01)$ and $(1, -0.01)$ will be 1 unit apart.

3.4 COMPLEX EIGENVALUES

The techniques of the previous sections were based on the geometric observation that, for some linear systems, certain solution curves lie on straight lines in the phase plane. This geometric observation led to the algebraic notions of eigenvalues and eigenvectors. These, in turn, gave us the formulas for the general solution.

Unfortunately, these ideas do not work for all linear systems. Geometrically, we hit a road block when we encounter linear systems whose direction fields do not show any straight lines of solutions, as in the example below (see Figure 3.21). In this case it is the algebra of eigenvalues and eigenvectors which leads to an understanding of the system. Even though the method is different, the goals are the same: Starting with the entries in the coefficient matrix, understand the geometry of the phase plane, the $x(t)$- and $y(t)$-graphs, and find the general solution.

Complex numbers

In this section and the rest of the book, we use complex numbers extensively. Complex numbers are "numbers" of the form $x + iy$ where x and y are real numbers and i is the "imaginary" number $\sqrt{-1}$.

One word of caution: All mathematicians and almost everyone else denote the imaginary number $\sqrt{-1}$ by the letter i. Electrical engineers use the letter i for the current (because "current" starts with "c"), so they use the letter j for $\sqrt{-1}$.

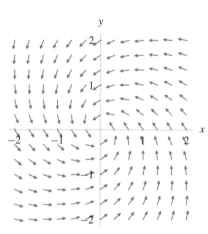

Figure 3.21
The direction field for

$$\frac{d\mathbf{Y}}{dt} = \begin{pmatrix} -2 & -3 \\ 3 & -2 \end{pmatrix} \mathbf{Y}.$$

Apparently there are no straight-line solutions.

A Linear System Without Straight-Line Solutions

Consider the system

$$\frac{d\mathbf{Y}}{dt} = \mathbf{AY} = \begin{pmatrix} -2 & -3 \\ 3 & -2 \end{pmatrix} \mathbf{Y}.$$

From the direction field for this system (see Figure 3.21), we see that there are no solution curves that lie on straight lines. Instead, solutions spiral around the origin.

The characteristic polynomial of $\mathbf{A}$ is

$$\det(\mathbf{A} - \lambda \mathbf{I}) = (-2 - \lambda)(-2 - \lambda) + 9,$$

which simplifies to

$$\lambda^2 + 4\lambda + 13.$$

The eigenvalues are the roots of the characteristic polynomial, that is, the solutions of the equation

$$\lambda^2 + 4\lambda + 13 = 0.$$

Hence, for this system, the eigenvalues are the complex numbers $\lambda_1 = -2 + 3i$ and $\lambda_2 = -2 - 3i$. So how are we going to find solutions and what information do complex eigenvalues give us?

General Solutions For Systems With Complex Eigenvalues

The most important thing to remember now is: Don't panic. Things are not nearly as complicated as they might seem. We cannot use the geometric ideas of Sections 3.2 and 3.3 to find solution curves that are straight lines because there aren't any straight-line solutions. However, the algebraic techniques we used in the those sections work the same way for complex numbers as they do for real numbers. The rules of arithmetic for complex numbers are exactly the same as for real numbers, so all of the computations we did in the previous sections are still valid even if the eigenvalues are complex. Consequently, our main observation about solutions of linear systems still holds even if the eigenvalues are complex. That is, given the linear system $d\mathbf{Y}/dt = \mathbf{AY}$, if λ is an eigenvalue for $\mathbf{A}$ and $\mathbf{Y}_0 = (x_0, y_0)$ is an eigenvector for the eigenvalue λ, then

$$\mathbf{Y}(t) = e^{\lambda t}\mathbf{Y}_0 = e^{\lambda t}\begin{pmatrix} x_0 \\ y_0 \end{pmatrix} = \begin{pmatrix} e^{\lambda t}x_0 \\ e^{\lambda t}y_0 \end{pmatrix}$$

is a solution. We can easily check this fact by differentiation, as we did before. We have

$$\frac{d\mathbf{Y}}{dt} = \frac{d}{dt}(e^{\lambda t}\mathbf{Y}_0) = \lambda e^{\lambda t}\mathbf{Y}_0 = e^{\lambda t}(\lambda\mathbf{Y}_0) = e^{\lambda t}\mathbf{AY}_0 = \mathbf{A}(e^{\lambda t}\mathbf{Y}_0) = \mathbf{AY}.$$

because $\mathbf{Y}_0$ is an eigenvector with eigenvalue λ, so $\mathbf{Y}(t)$ is a solution. Of course, we need to make sense of the fact that the exponential is now a complex function and the fact that the eigenvector may contain complex entries, but this is no real problem. The important thing to notice here is that this computation is exactly the same if the numbers are real or complex.

Example revisited

For the system

$$\frac{d\mathbf{Y}}{dt} = \mathbf{AY} \quad \text{where} \quad \mathbf{A} = \begin{pmatrix} -2 & -3 \\ 3 & -2 \end{pmatrix},$$

we already know that the eigenvalues are $\lambda_1 = -2 + 3i$ and $\lambda_2 = -2 - 3i$. We now find the eigenvector for $\lambda_1 = -2 + 3i$ just as we would if λ_1 were real, by solving for $\mathbf{Y}_0$ in the system of equations

$$\mathbf{AY}_0 = \lambda_1\mathbf{Y}_0.$$

That is we must find $\mathbf{Y}_0 = (x_0, y_0)$ such that

$$\begin{cases} -2x_0 - 3y_0 = (-2 + 3i)x_0 \\ 3x_0 - 2y_0 = (-2 + 3i)y_0 \end{cases}$$

or

$$\begin{cases} -3ix_0 - 3y_0 = 0 \\ 3x_0 - 3iy_0 = 0. \end{cases}$$

Just as in the case of real eigenvalues, these equations are not independent (multiply both sides of the first equation by i to obtain the second equation and recall that $i^2 = -1$). Thus, the solutions of these equations are all pairs of complex numbers (x_0, y_0) that satisfy $-3ix_0 - 3y_0 = 0$, or $x_0 = iy_0$. If we let $y_0 = 1$, then $x_0 = i$. In other words, the vector $(i, 1)$ is an eigenvector for eigenvalue $\lambda_1 = -2 + 3i$. We can double check this by computing

$$\mathbf{A}\begin{pmatrix} i \\ 1 \end{pmatrix} = \begin{pmatrix} -2 & -3 \\ 3 & -2 \end{pmatrix}\begin{pmatrix} i \\ 1 \end{pmatrix} = \begin{pmatrix} -2i - 3 \\ 3i - 2 \end{pmatrix} = (-2 + 3i)\begin{pmatrix} i \\ 1 \end{pmatrix}.$$

Thus we know that

$$\mathbf{Y}(t) = e^{(-2+3i)t}\begin{pmatrix} i \\ 1 \end{pmatrix} = \begin{pmatrix} ie^{(-2+3i)t} \\ e^{(-2+3i)t} \end{pmatrix}$$

is a solution of the system.

Obtaining Real-Valued Solutions
From Complex Solutions

So this is both good news and bad news. The good news is that we can find solutions to linear systems with complex eigenvalues. The bad news is that these solutions involve complex numbers. If this system is a model of populations, profits, or the position of a mechanical device, then only real numbers make sense. It is hard to imagine "$5i$" predators or a position of "$2 + 3i$" units from the rest position. In other words, the physical meaning of complex numbers is not readily apparent. We have to find a way of producing real solutions from complex solutions.

The key to getting real solutions from complex solutions is **Euler's formula**

$$e^{a+ib} = e^a e^{ib} = e^a(\cos b + i \sin b) = e^a \cos b + ie^a \sin b$$

for all real numbers a and b. (Using power series, one can verify that

$$e^{ib} = \cos b + i \sin b,$$

and Euler's formula follows using the laws of exponents.) In general, Euler's formula is the way that we exponentiate with complex exponents.

Euler's formula is one of the most remarkable identities in all of mathematics. It allows us to relate some of the most important functions and constants in a most intriguing way. For example, if we let $a = 0$ and $b = \pi$, then we obtain

$$e^{\pi i} = e^0 \cos \pi + ie^0 \sin \pi,$$

so

$$e^{\pi i} = -1.$$

We use Euler's formula to define a complex-valued exponential function. We have

$$e^{(\alpha+i\beta)t} = e^{\alpha t} e^{i\beta t}$$

$$= e^{\alpha t}(\cos \beta t + i \sin \beta t)$$

$$= e^{\alpha t} \cos \beta t + i e^{\alpha t} \sin \beta t.$$

For the example above, this gives

$$e^{(-2+3i)t} = e^{-2t} \cos 3t + i e^{-2t} \sin 3t.$$

We can now rewrite the solution $\mathbf{Y}(t)$ as

$$\mathbf{Y}(t) = (e^{-2t} \cos 3t + i e^{-2t} \sin 3t) \begin{pmatrix} i \\ 1 \end{pmatrix}$$

$$= \begin{pmatrix} (e^{-2t} \cos 3t + i e^{-2t} \sin 3t)i \\ e^{-2t} \cos 3t + i e^{-2t} \sin 3t \end{pmatrix}$$

$$= \begin{pmatrix} i e^{-2t} \cos 3t - e^{-2t} \sin 3t \\ e^{-2t} \cos 3t + i e^{-2t} \sin 3t \end{pmatrix},$$

which in turn can be broken into

$$\mathbf{Y}(t) = \begin{pmatrix} -e^{-2t} \sin 3t \\ e^{-2t} \cos 3t \end{pmatrix} + i \begin{pmatrix} e^{-2t} \cos 3t \\ e^{-2t} \sin 3t \end{pmatrix}.$$

So far we have only rearranged the solution $\mathbf{Y}(t)$ to isolate the part that involves the number i. We now use the fact that we are dealing with a *linear* system to find the required real solutions.

Theorem   *Suppose $\mathbf{Y}(t)$ is a complex-valued solution to a linear system*

$$\frac{d\mathbf{Y}}{dt} = \mathbf{AY} = \begin{pmatrix} a & b \\ c & d \end{pmatrix} \mathbf{Y},$$

where the coefficient matrix $\mathbf{A}$ has all real entries (a, b, c and d are real numbers). Suppose

$$\mathbf{Y}(t) = \mathbf{Y}_{\text{re}}(t) + i\,\mathbf{Y}_{\text{im}}(t)$$

where $\mathbf{Y}_{\text{re}}(t)$ and $\mathbf{Y}_{\text{im}}(t)$ are real-valued functions of t. Then $\mathbf{Y}_{\text{re}}(t)$ and $\mathbf{Y}_{\text{im}}(t)$ are both solutions of the original system $d\mathbf{Y}/dt = \mathbf{AY}$.

It is important to note that there are no i's in the expression $\mathbf{Y}_{\text{im}}(t)$. We have factored out the i from this expression.

To verify this theorem, we use the fact that $\mathbf{Y}(t)$ is a solution. In other words,

$$\frac{d\mathbf{Y}}{dt} = \mathbf{AY} \text{ for all } t.$$

Now we replace $\mathbf{Y}(t)$ with $\mathbf{Y}_{\text{re}}(t) + i\,\mathbf{Y}_{\text{im}}(t)$ on both sides of the equation. On the left-hand side, the usual rules of differentiation give,

$$\frac{d\mathbf{Y}}{dt} = \frac{d(\mathbf{Y}_{\text{re}} + i\mathbf{Y}_{\text{im}})}{dt}$$

$$= \frac{d\mathbf{Y}_{\text{re}}}{dt} + i\frac{d\mathbf{Y}_{\text{im}}}{dt}.$$

On the right-hand side, we use the fact that this is a linear system to obtain that

$$\mathbf{A}\mathbf{Y}(t) = \mathbf{A}(\mathbf{Y}_{\text{re}}(t) + i\,\mathbf{Y}_{\text{im}}(t))$$

$$= \mathbf{A}\,\mathbf{Y}_{\text{re}}(t) + i\mathbf{A}\,\mathbf{Y}_{\text{im}}(t).$$

So we have

$$\frac{d\mathbf{Y}_{\text{re}}}{dt} + i\frac{d\mathbf{Y}_{\text{im}}}{dt} = \mathbf{A}\mathbf{Y}_{\text{re}} + i\mathbf{A}\,\mathbf{Y}_{\text{im}}$$

for all t. Two complex numbers are equal only if both their real parts and their imaginary parts are equal. Hence the only way the above equation can hold is if

$$\frac{d\mathbf{Y}_{\text{re}}}{dt} = \mathbf{A}\mathbf{Y}_{\text{re}} \quad \text{and} \quad \frac{d\mathbf{Y}_{\text{im}}}{dt} = \mathbf{A}\mathbf{Y}_{\text{im}},$$

and this is exactly what it means to say $\mathbf{Y}_{\text{re}}(t)$ and $\mathbf{Y}_{\text{im}}(t)$ are solutions of $d\mathbf{Y}/dt = \mathbf{A}\mathbf{Y}$.

Completion of the first example

Recall that, for the system

$$\frac{d\mathbf{Y}}{dt} = \begin{pmatrix} -2 & -3 \\ 3 & -2 \end{pmatrix} \mathbf{Y},$$

we have shown that the complex vector-valued function

$$\mathbf{Y}(t) = \begin{pmatrix} -e^{-2t}\sin 3t \\ e^{-2t}\cos 3t \end{pmatrix} + i \begin{pmatrix} e^{-2t}\cos 3t \\ e^{-2t}\sin 3t \end{pmatrix}$$

is a solution. By taking real and imaginary parts, we know that both the real part

$$\mathbf{Y}_{\text{re}}(t) = \begin{pmatrix} -e^{-2t}\sin 3t \\ e^{-2t}\cos 3t \end{pmatrix}$$

and the imaginary part

$$\mathbf{Y}_{\text{im}}(t) = \begin{pmatrix} e^{-2t}\cos 3t \\ e^{-2t}\sin 3t \end{pmatrix}$$

are solutions of the original system. They are independent since their initial values $\mathbf{Y}_{\text{re}}(0) = (0, 1)$ and $\mathbf{Y}_{\text{im}}(0) = (1, 0)$ are independent. So the general solution of this system is

$$\mathbf{Y}(t) = k_1 \begin{pmatrix} -e^{-2t}\sin 3t \\ e^{-2t}\cos 3t \end{pmatrix} + k_2 \begin{pmatrix} e^{-2t}\cos 3t \\ e^{-2t}\sin 3t \end{pmatrix}$$

for constants k_1 and k_2. This can be rewritten in the form

$$\mathbf{Y}(t) = e^{-2t} \begin{pmatrix} -k_1\sin 3t + k_2\cos 3t \\ k_1\cos 3t + k_2\sin 3t \end{pmatrix}.$$

Qualitative analysis

The direction field for the system

$$\frac{d\mathbf{Y}}{dt} = \mathbf{AY} \quad \text{where} \quad \mathbf{A} = \begin{pmatrix} -2 & -3 \\ 3 & -2 \end{pmatrix}$$

indicates that solution curves spiral toward the origin (see Figure 3.22). The corresponding $x(t)$- and $y(t)$-graphs of solutions must alternate between positive and negative values with decreasing amplitude as the solution curve in the phase plane winds around the origin. The pictures of the solution curve and the $x(t)$- and $y(t)$-graphs confirm this, at least to some extent (see Figures 3.22 and 3.23). From these graphs, it appears that the solution winds only once around the origin before reaching $(0, 0)$. Actually, this solution spirals infinitely often, though these oscillations are difficult to detect. In Figure 3.24 we magnify a small portion of Figure 3.22. We see that, indeed, the solution does continue to spiral.

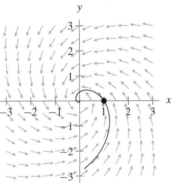

Figure 3.22
A solution curve in the phase plane for

$$\frac{d\mathbf{Y}}{dt} = \begin{pmatrix} -2 & -3 \\ 3 & -2 \end{pmatrix} \mathbf{Y}.$$

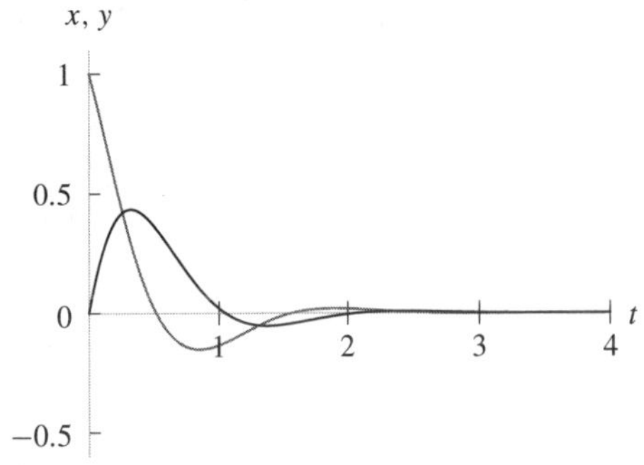

Figure 3.23
The $x(t)$- and $y(t)$-graphs of a solution to this differential equation

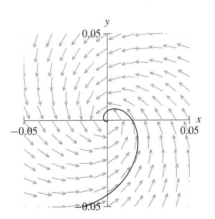

Figure 3.24
A magnification of Figure 3.22

The formula for the general solution of this system provides us with detailed behavior of this spiral. The oscillation in the x and y components of the solutions are caused by the sine and cosine terms. These trigonometric expressions are all of the form $\sin 3t$ and $\cos 3t$, so when t increases through $2\pi/3$, these terms return to their original values. Hence the period of the oscillation around the origin is always $2\pi/3$, no matter how large t is or how close the solution comes to the origin. Meanwhile, solutions must approach the origin because of the exponential term e^{-2t}. This term shows that the amplitude of the oscillations of the $x(t)$- and $y(t)$-graphs decreases at this very fast rate. This also explains why it is difficult to see these oscillations near the origin.

Happily, this sort of description of solutions can be accomplished without resorting to computing the general solution of the system. In fact, it can be obtained form the eigenvalues alone.

The Qualitative Behavior of Systems with Complex Eigenvalues

The discussion above can be generalized to any linear systems with complex eigenvalues. First find a complex solution by finding the complex eigenvalues and eigenvectors. Then take the real and imaginary parts of this solution to obtain two independent solutions (see Exercise 19 for the verification that the real and imaginary parts of a solution are independent solutions). Finally, form the general solution in the usual way as a linear combination of the two independent particular solutions. This is sometimes a very tedious process, but it works. If the general solution is what we need, then we can find it.

Just as in the case of real eigenvalues, we can tell a tremendous amount about the system from the complex eigenvalues without doing all the detailed computations to obtain the general solution. Suppose

$$\frac{d\mathbf{Y}}{dt} = \mathbf{AY}$$

is a linear system with complex eigenvalues $\lambda_1 = \alpha + i\beta$ and $\lambda_2 = \alpha - i\beta$, $\beta \neq 0$ (that complex eigenvalues always come in complex conjugate pairs is an interesting exercise, see Exercise 17). Then we know that the complex solutions have the form

$$\mathbf{Y}(t) = e^{(\alpha+i\beta)t}\mathbf{Y}_0,$$

where $\mathbf{Y}_0$ is a (complex) eigenvector of the matrix $\mathbf{A}$. We can rewrite this is

$$\mathbf{Y}(t) = e^{\alpha t}(\cos \beta t + i \sin \beta t)\mathbf{Y}_0.$$

Because $\mathbf{Y}_0$ is a constant, the real and imaginary parts of the solution $\mathbf{Y}(t)$ are a combination of two types of terms, the exponential and trigonometric terms. The effect of the exponential term on solutions depends on the sign of α. If $\alpha > 0$, then the $e^{\alpha t}$ term increases exponentially as $t \to \infty$, and the solution curve spirals off "toward infinity." If $\alpha < 0$, then the $e^{\alpha t}$ term tends to zero exponentially fast as t increases, so the solutions tend to the origin. If $\alpha = 0$, then the $e^{\alpha t}$ is identically one and the solutions oscillate with constant amplitude for all time. That is, the solutions are periodic.

The sine and cosine terms alternate from positive to negative and back again as t increases or decreases, so these terms make the x- and y-component functions oscillate. Hence, the solutions in the xy-phase plane spiral around $(0, 0)$. The period of this oscillation is the time it takes to go around once (say from one crossing of the positive x-axis to the next). The period is determined by β. The functions $\sin \beta t$ and $\cos \beta t$ satisfy the equations

$$\sin \beta(t + 2\pi/\beta) = \sin \beta t$$
$$\cos \beta(t + 2\pi/\beta) = \cos \beta t,$$

so increasing t by $2\pi/\beta$ returns $\sin \beta t$ and $\cos \beta t$ to their original values (see Figure 3.25).

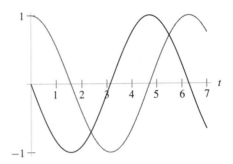

Figure 3.25
Graphs of $\cos \beta t$ and $\sin \beta t$ for $\beta = -1$. Note where the graphs cross the t axis.

We can summarize these observations with the following classification. Given a linear system

$$\frac{d\mathbf{Y}}{dt} = \mathbf{AY}$$

which has complex eigenvalues $\lambda_1 = \alpha + i\beta$ and $\lambda_2 = \alpha - i\beta$, $\beta \neq 0$, the solutions spiral around the origin with a period of $2\pi/\beta$. Moreover,

- If $\alpha < 0$, then the solutions spiral toward the origin. In this case, the origin is called a **spiral sink**.
- If $\alpha > 0$, then the solutions spiral away from the origin. In this case, the origin is called a **spiral source**.
- If $\alpha = 0$, then the solutions are *periodic*. They return exactly to their initial conditions in the phase plane and repeat the same closed curve over and over. In this case, the origin is called a **center** .

The question of "which way" the solutions spiral — clockwise or counterclockwise — can best be answered by looking at the direction field. Even one vector of the direction field is enough to tell which way a system with complex eigenvalues spirals. For example, if the direction field at $(1, 0)$ points down into the fourth quadrant, then the solutions must spiral in the clockwise direction. If the direction field at $(1, 0)$ points up into the first quadrant, then the solutions must spiral in the counterclockwise direction.

As we have seen, the $\sin \beta t$ and $\cos \beta t$ terms in the solutions make the solutions oscillate with a period of $2\pi/\beta$. This quantity is called the **natural period** of the system. Every solution takes this amount of time to wind once around the origin. The **natural frequency** is the reciprocal of the natural period, or $\beta/2\pi$.

A word of caution concerning terminology: In some subjects the terms natural frequency and natural period are reserved for linear systems with eigenvalues having zero real part (that is, only for centers, not spiral sinks or sources — that convention is reasonable because solutions of the system are periodic only for centers). We use the terms natural period and natural frequency for any linear system with complex eigenvalues.

A Spiral Source

Consider the initial-value problem

$$\frac{d\mathbf{Y}}{dt} = \mathbf{B}\mathbf{Y}, \quad \mathbf{Y}(0) = \begin{pmatrix} 1 \\ 1 \end{pmatrix} \quad \text{where} \quad \mathbf{B} = \begin{pmatrix} 0 & 2 \\ -3 & 2 \end{pmatrix}.$$

The eigenvalues are the roots of the characteristic polynomial

$$\det(\mathbf{B} - \lambda\mathbf{I}) = (0 - \lambda)(2 - \lambda) + 6 = \lambda^2 - 2\lambda + 6,$$

so the eigenvalues are $\lambda = 1 \pm i\sqrt{5}$. Since the real parts of the eigenvalues are positive, the origin is a spiral source, and the natural period of the system is $2\pi/\sqrt{5}$. Thus, the solution of the initial-value problem oscillates with increasing amplitude and constant period of $2\pi/\sqrt{5}$. The direction field (see Figure 3.26) shows that solutions spiral in the clockwise direction around the origin in the xy-phase plane.

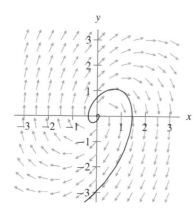

Figure 3.26
Direction field and the solution of the initial-value problem

$$\frac{d\mathbf{Y}}{dt} = \begin{pmatrix} 0 & 2 \\ -3 & 2 \end{pmatrix} \mathbf{Y} \quad \text{and} \quad \mathbf{Y}(0) = \begin{pmatrix} 1 \\ 1 \end{pmatrix}.$$

The solution spirals away from the origin.

To find the formula for the solution of the initial-value problem we must find an eigenvector (x_0, y_0) for one of the eigenvalues, say $1 + i\sqrt{5}$. In other words, we solve

$$\mathbf{B} \begin{pmatrix} x_0 \\ y_0 \end{pmatrix} = (1 + i\sqrt{5}) \begin{pmatrix} x_0 \\ y_0 \end{pmatrix}$$

or

$$\begin{cases} 2y_0 = (1 + i\sqrt{5})x_0 \\ -3x_0 + 2y_0 = (1 + i\sqrt{5})y_0. \end{cases}$$

Just as in the case of real eigenvalues, these two equations are dependent. (The second equation can be turned into the first by subtracting $2y_0$ from both sides, then multiplying both sides by $-(1+i\sqrt{5})/3$.) We only need one eigenvector, so we choose any convenient value for x_0 and solve for y_0. If we set $x_0 = 2$ then we must have $y_0 = 1+i\sqrt{5}$. Hence, for the eigenvalue $\lambda = 1 + i\sqrt{5}$, the vector $(2, 1 + i\sqrt{5})$ is an eigenvector.

The corresponding complex solution is

$$\mathbf{Y}_1(t) = e^{(1+i\sqrt{5})t} \begin{pmatrix} 2 \\ 1 + i\sqrt{5} \end{pmatrix}.$$

Rewriting this using Euler's formula we obtain

$$\mathbf{Y}_1(t) = e^t \begin{pmatrix} 2\cos(\sqrt{5}t) \\ \cos(\sqrt{5}t) - \sqrt{5}\sin(\sqrt{5}t) \end{pmatrix} + i e^t \begin{pmatrix} 2\sin(\sqrt{5}t) \\ \sqrt{5}\cos(\sqrt{5}t) + \sin(\sqrt{5}t) \end{pmatrix}.$$

The general solution is

$$\mathbf{Y}(t) = k_1 e^t \begin{pmatrix} 2\cos(\sqrt{5}t) \\ \cos(\sqrt{5}t) - \sqrt{5}\sin(\sqrt{5}t) \end{pmatrix} + k_2 e^t \begin{pmatrix} 2\sin(\sqrt{5}t) \\ \sqrt{5}\cos(\sqrt{5}t) + \sin(\sqrt{5}t) \end{pmatrix}.$$

where k_1 and k_2 are arbitrary constants. To solve the initial-value problem, we solve for k_1 and k_2 by setting the general solution at $t = 0$ equal to the initial condition $(1, 1)$ obtaining

$$k_1 \begin{pmatrix} 2 \\ 1 \end{pmatrix} + k_2 \begin{pmatrix} 0 \\ \sqrt{5} \end{pmatrix} = \begin{pmatrix} 1 \\ 1 \end{pmatrix}.$$

So $k_1 = 1/2$ and $k_2 = 1/(2\sqrt{5})$. The solution of the initial-value problem is

$$\mathbf{Y}(t) = \frac{1}{2} e^t \begin{pmatrix} 2\cos(\sqrt{5}t) \\ \cos(\sqrt{5}t) - \sqrt{5}\sin(\sqrt{5}t) \end{pmatrix} + \frac{1}{2\sqrt{5}} e^t \begin{pmatrix} 2\sin(\sqrt{5}t) \\ \sqrt{5}\cos(\sqrt{5}t) + \sin(\sqrt{5}t) \end{pmatrix}.$$

Example of an Undamped Harmonic Oscillator

Consider the harmonic oscillator with mass $m = 1$, spring constant $k_s = 2$ and with no damping, so $k_d = 0$. This is called an *undamped* harmonic oscillator. The second-order equation is

$$\frac{d^2y}{dt^2} = -2y$$

and the corresponding linear system is

$$\frac{d\mathbf{Y}}{dt} = \mathbf{CY} \quad \text{where} \quad \mathbf{C} = \begin{pmatrix} 0 & 1 \\ -2 & 0 \end{pmatrix} \quad \text{and} \quad \mathbf{Y} = \begin{pmatrix} y \\ v \end{pmatrix}.$$

The direction field for this system is given in Figure 3.27. We see from this picture that the solution curves encircle the origin. From this we predict that the eigenvalues for this system will be complex. It is difficult to determine from the direction field picture if solution curves are periodic or if the spiral very slowly toward or away from the origin. Since these equations model a mechanical system for which we have assumed there is no damping, we might suspect that the solution curves are periodic. We can verify this by computing the eigenvalues for the system. As a bonus, the eigenvalues give us the period of the oscillations.

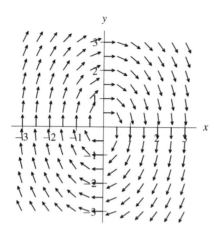

Figure 3.27
Direction field for the undamped harmonic oscillator system

$$\frac{d\mathbf{Y}}{dt} = \mathbf{CY} \quad \text{where} \quad \mathbf{C} = \begin{pmatrix} 0 & 1 \\ -2 & 0 \end{pmatrix}.$$

Although the direction field suggests that the eigenvalues of the system are complex, we cannot determine if the origin is a center, a spiral source, or a spiral sink.

The eigenvalues for the matrix $\mathbf{C}$ are the roots of its characteristic polynomial

$$\det(\mathbf{C} - \lambda\mathbf{I}) = (0 - \lambda)(0 - \lambda) + 2 = \lambda^2 + 2,$$

which are $\lambda = \pm i\sqrt{2}$. Hence, the origin is a center and all solutions are periodic. The imaginary part of the eigenvalue is $\sqrt{2}$, so the natural period of the system is $(2\pi)/\sqrt{2}$. This means that every solution completes one oscillation in $\sqrt{2}\pi$ time units regardless of its initial condition.

For the mass-spring system with these parameter values, the mass either stays at rest at the origin for all time or it oscillates back and forth about the rest position forever. The amplitude of these oscillations remains constant. This regular behavior is why watches are made with springs. Of course, real springs have some damping.

Paul's and Bob's CD Stores Revisited

Recall the model for Paul and Bob's compact disk stores from Section 3.1 using the linear system

$$\frac{d\mathbf{Y}}{dt} = \mathbf{AY}.$$

Now suppose that the coefficient matrix is

$$\mathbf{A} = \begin{pmatrix} 2 & 1 \\ -4 & -1 \end{pmatrix}.$$

We would like to predict the behavior of solutions to this system with as little computation as possible.

First we compute the eigenvalues from the characteristic polynomial

$$\det(\mathbf{A} - \lambda\mathbf{I}) = (2 - \lambda)(-1 - \lambda) + 4 = \lambda^2 - \lambda + 2 = 0$$

The roots are $(1 \pm i\sqrt{7})/2$, so we know that solutions will spiral around the equilibrium point at $(0, 0)$. Because the real part of the eigenvalues is $1/2$, we know that the origin is a spiral source. This information tells us that every solution (except the equilibrium point $(0, 0)$) spirals away from $(0, 0)$ in bigger and bigger loops as t increases. We can

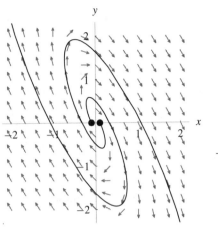

Figure 3.28
Phase plane for

$$\frac{d\mathbf{Y}}{dt} = \begin{pmatrix} 2 & 1 \\ -4 & -1 \end{pmatrix} \mathbf{Y}.$$

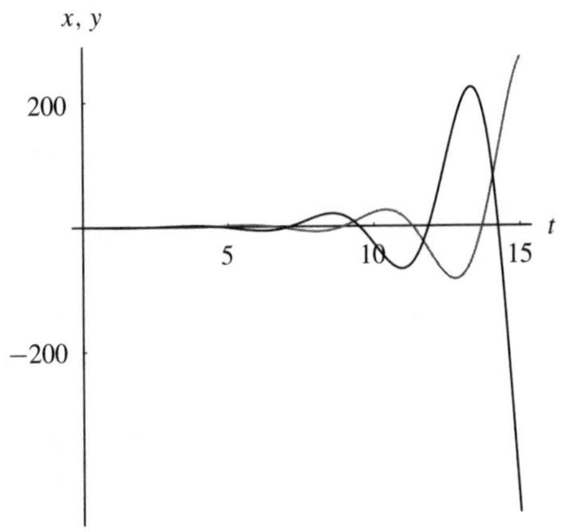

Figure 3.29
Graphs of a solution in the $x(t)$ and $y(t)$ for the system

$$\frac{d\mathbf{Y}}{dt} = \begin{pmatrix} 2 & 1 \\ -4 & -1 \end{pmatrix} \mathbf{Y}.$$

determine the direction (clockwise or counterclockwise) and approximate shape of the solution curves by sketching the phase plane (see Figure 3.28).

The $x(t)$- and $y(t)$-graphs of solutions oscillate with increasing amplitude. The period of these oscillations is $2\pi/(\sqrt{7}/2) = 4\pi/\sqrt{7} \approx 4.71$, and the amplitude increases like $e^{t/2}$. We sketch the qualitative behavior of the $x(t)$- and $y(t)$-graphs in Figure 3.29.

For Paul and Bob this means that either they will stay precisely at the "break even point" $x = y = 0$ or the profits and losses of their stores will go up and down with increasing amplitude (a boom to bust to boom business cycle). Also the equilibrium point at the origin is unstable, so even a tiny profit or loss by either store eventually leads to large oscillations in the profits of both stores. It would be very difficult to predict this behavior from just looking at the linear system without any computations.

Exercises for Section 3.4

1. Using real and imaginary parts, decompose the function

$$\mathbf{Y}(t) = e^{(1+3i)t} \begin{pmatrix} 2+i \\ 1 \end{pmatrix}$$

into two real-valued functions.

2. Using real and imaginary parts, decompose the function

$$\mathbf{Y}(t) = e^{(2+i)t} \begin{pmatrix} 1 \\ 4 \end{pmatrix}$$

into two real-valued functions.

In Exercises 3–8, each linear system has complex eigenvalues. For each of these systems:

   (a) Find the eigenvalues (if they are real then go on to the next problem).

   (b) Determine if the origin is a spiral sink, a spiral source, or a center.

   (c) Determine the natural period and natural frequency of the oscillations.

   (d) Determine the direction of the oscillations in the phase plane. (Do the solutions go clockwise or counterclockwise around the origin?)

   (e) Sketch the xy-phase plane and the $x(t)$- and $y(t)$-graphs for the solutions with the indicated initial conditions.

3. $\dfrac{d\mathbf{Y}}{dt} = \begin{pmatrix} 0 & 2 \\ -2 & 0 \end{pmatrix} \mathbf{Y}$ with initial condition $\mathbf{Y}_0 = (1, 0)$

4. $\dfrac{d\mathbf{Y}}{dt} = \begin{pmatrix} -1 & 2 \\ -1 & -1 \end{pmatrix} \mathbf{Y}$ with initial condition $\mathbf{Y}_0 = (0, 1)$

5. $\dfrac{d\mathbf{Y}}{dt} = \begin{pmatrix} 2 & 2 \\ -4 & 6 \end{pmatrix} \mathbf{Y}$ with initial condition $\mathbf{Y}_0 = (1, 1)$.

6. $\dfrac{d\mathbf{Y}}{dt} = \begin{pmatrix} 0 & 2 \\ -2 & -1 \end{pmatrix} \mathbf{Y}$ with initial condition $\mathbf{Y}_0 = (-1, 1)$

7. $\dfrac{d\mathbf{Y}}{dt} = \begin{pmatrix} 2 & -6 \\ 2 & 1 \end{pmatrix} \mathbf{Y}$ with initial condition $\mathbf{Y}_0 = (2, 1)$

8. $\dfrac{d\mathbf{Y}}{dt} = \begin{pmatrix} 1 & 4 \\ -3 & 2 \end{pmatrix} \mathbf{Y}$ with initial condition $\mathbf{Y}_0 = (1, -1)$

In Exercises 9–14, the linear systems are the same as in Exercises 3–8. For each of the linear systems

   (a) Find the general solution.

   (b) Find the particular solution with the given initial value.

   (c) Sketch the $x(t)$- and $y(t)$-graphs of the particular solution. (Compare these sketches with sketches you obtained in the problems above!)

9. $\dfrac{d\mathbf{Y}}{dt} = \begin{pmatrix} 0 & 2 \\ -2 & 0 \end{pmatrix} \mathbf{Y}$ with initial condition $\mathbf{Y}_0 = (1, 0)$

10. $\dfrac{d\mathbf{Y}}{dt} = \begin{pmatrix} -1 & 2 \\ -1 & -1 \end{pmatrix} \mathbf{Y}$ with initial condition $\mathbf{Y}_0 = (0, 1)$

11. $\dfrac{d\mathbf{Y}}{dt} = \begin{pmatrix} 2 & 2 \\ -4 & 6 \end{pmatrix} \mathbf{Y}$ with initial condition $\mathbf{Y}_0 = (1, 1)$.

12. $\dfrac{d\mathbf{Y}}{dt} = \begin{pmatrix} 0 & 2 \\ -2 & -1 \end{pmatrix} \mathbf{Y}$ with initial condition $\mathbf{Y}_0 = (-1, 1)$

13. $\dfrac{d\mathbf{Y}}{dt} = \begin{pmatrix} 2 & -6 \\ 2 & 1 \end{pmatrix} \mathbf{Y}$ with initial condition $\mathbf{Y}_0 = (2, 1)$

14. $\dfrac{d\mathbf{Y}}{dt} = \begin{pmatrix} 1 & 4 \\ -3 & 2 \end{pmatrix} \mathbf{Y}$ with initial condition $\mathbf{Y}_0 = (1, -1)$

15. The following six figures are graphs of functions $x(t)$. Two of them are graphs of the x-coordinate of a solution of a linear system with complex eigenvalues and the other four are not.

   (a) Identify which the graphs correspond to the graph of the x-coordinate of a solution of a linear system.

   (b) For these two graphs, give the natural period of the system and classify the equilibrium point at the origin as a spiral sink, a spiral source or a center.

   (c) For each of the other four graphs describe how you can be sure that they are not the graph of the x-coordinate of a linear system with complex eigenvalues. (Your answer should be a sentence or two describing what aspect of the graph makes it impossible for it to be the graph of the x-coordinate of a linear system with complex eigenvalues).

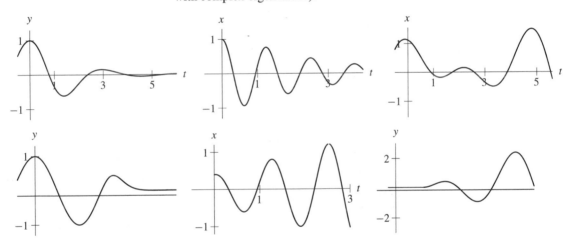

Figure 3.30
$x(t)$-graphs

16. Show that a matrix of the form

$$A = \begin{pmatrix} a & b \\ -b & a \end{pmatrix}$$

with $b \neq 0$ must have complex eigenvalues.

17. Suppose that a and b are real numbers and that the polynomial $\lambda^2 + a\lambda + b$ has $\lambda_1 = \alpha + i\beta$ as a root with $\beta \neq 0$. Show that $\lambda_2 = \alpha - i\beta$, the complex conjugate of λ_1, must also be a root. [*Hint*: There are (at least) two ways to attack this problem. Either look at the form of the quadratic formula for the roots, or notice that

$$(\alpha + i\beta)^2 + a(\alpha + i\beta) + b = 0$$

and take the complex conjugate of both sides of this equation.]

18. Suppose that the matrix $\mathbf{A}$ with real entries has complex eigenvalues $\lambda = \alpha + i\beta$ and $\bar{\lambda} = \alpha - i\beta$ with $\beta \neq 0$. Show that the eigenvectors of $\mathbf{A}$ must be complex, that is, show that if $\mathbf{Y}_0 = (x_0, y_0)$ is an eigenvector for $\mathbf{A}$, then either x_0 or y_0 or both have a non-zero imaginary part.

19. Suppose the matrix $\mathbf{A}$ with real entries has the complex eigenvalue $\lambda = \alpha + i\beta$, $\beta \neq 0$. Let $\mathbf{Y}_0$ be an eigenvector for λ and write $\mathbf{Y}_0 = \mathbf{Y}_1 + i\mathbf{Y}_2$ where $\mathbf{Y}_1 = (x_1, y_1)$ and $\mathbf{Y}_2 = (x_2, y_2)$ have real entries. Show that $\mathbf{Y}_1$ and $\mathbf{Y}_2$ are linearly independent. [*Hint:* Suppose they are not linearly independent. Then $(x_2, y_2) = k(x_1, y_1)$ for some constant k. Then $\mathbf{Y}_0 = (1 + ik)\mathbf{Y}_1$. Then use the fact that $\mathbf{Y}_0$ is an eigenvector of $\mathbf{A}$ and that $\mathbf{AY}_1$ contains no imaginary part.]

20. Suppose the matrix $\mathbf{A}$ with real entries has complex eigenvalues $\lambda = \alpha + i\beta$ and $\bar{\lambda} = \alpha - i\beta$. Suppose also that $\mathbf{Y}_0 = (x_1 + iy_1, x_2 + iy_2)$ is an eigenvector for the eigenvalue λ. Show that $\overline{\mathbf{Y}_0} = (x_1 - iy_1, x_2 - iy_2)$ is an eigenvector for the eigenvalue $\bar{\lambda}$. In other words, the complex conjugate of an eigenvector for λ is an eigenvector for $\bar{\lambda}$.

21. Consider the function $x(t) = e^{-\alpha t} \sin(\beta t)$ for $\alpha, \beta > 0$.

   (a) What is the distance between successive zeros of this function. More precisely, if $t_1 < t_2$ are such that $x(t_1) = x(t_2) = 0$ and $x(t) \neq 0$ for $t_1 < t < t_2$, then what is $t_2 - t_1$?

   (b) What is the distance between the first local maximum and the first local minimum of $x(t)$ for $t > 0$?

   (c) What is the distance between the first two local maxima of $x(t)$ for $t > 0$?

   (d) What is the distance between $t = 0$ and the first local maximum of $x(t)$ for $t > 0$?

22. Suppose we are given a function in the form $x(t) = e^{\alpha t}(k_1 \sin(\beta t) + k_2 \cos(\beta t))$ for some constants α, β, k_1, k_2. Show that this function can be rewritten in the form $x(t) = e^{\alpha t} K \cos(\beta t + \phi)$ for some constants K and ϕ. [*Hint:* Let $K = \sqrt{k_1^2 + k_2^2}$ and $\phi = \arctan(k_2/k_1)$ so $k_1 = K \cos(\phi)$ and $k_2 = K \sin(\phi)$. Remark: Sometimes solutions of linear systems with complex coefficients are written in this form. The K is interpreted as giving the amplitude and the ϕ as giving the "phase" of the solution.]

In Exercises 23–26, parameter values are given for the harmonic oscillator equations. For each set of parameter values:

   (a) Write the corresponding second order equation and first-order system.

   (b) Determine the eigenvalues.

   (c) Classify the oscillator as undamped, underdamped or overdamped.

   (d) Sketch the phase plane.

   (e) Give the general solution of the system.

23. Mass $m = 1$, spring constant $k_s = 2$ and coefficient of damping $k_d = 2$.

24. Mass $m = 1$, spring constant $k_s = 2$ and coefficient of damping $k_d = 0$.

25. Mass $m = 1$, spring constant $k_s = 4$ and coefficient of damping $k_d = 1$.

26. Mass $m = 0.02$, spring constant $k_s = 4$ and coefficient of damping $k_d = 1$.

27. Suppose we wish to make a clock with spring constant $k_s = 2$. If we assume there is no friction or damping ($k_d = 0$) then what mass m must be attached to the spring so that its natural period is one time unit?

28. Consider the harmonic oscillator with $m = 1$, $k_s = 2$ and $k_d = 1$.

(a) What is the natural period?

(b) If m is increased slightly, does the natural period increase or decrease? How fast does it increase or decrease?

(c) If k_s is increased slightly, does the natural period increase or decrease? How fast does it increase or decrease?

(d) If k_d is increased slightly, does the natural period increase or decrease? How fast does it increase or decrease?

29. As is pointed out in the text, an undamped or underdamped harmonic oscillator can be used to make a clock. If we arrange for the clock to "tick" off a time unit whenever the mass passes the rest position, then each tick of the clock is equal to half the natural period.

(a) If dirt increases the coefficient of damping slightly for the harmonic oscillator, will the clock run fast or slow?

(b) If as the spring ages it provides slightly less force for a given compression or extension, will this make the clock run fast or slow.

(c) If grime collects on the harmonic oscillator, slightly increasing the mass, will the clock run fast or slow?

(d) Suppose all of the above occur, the coefficient of damping increases slightly, the spring gets "tired" and the mass increases slightly, will the clock run fast or slow?

30. For the second-order, linear, homogeneous, constant coefficient equation

$$\frac{d^2y}{dt^2} + p\frac{dy}{dt} + qy = 0.$$

(a) Write this as a first-order system.

(b) What is the condition on p, and q which guarantee that the eigenvalues of the corresponding system are complex?

(c) What relationship between p and q guarantees that the origin is a spiral sink? What relationship guarantees that the origin is a spiral sink? What relationship guarantees that the origin is a center?

(d) If the eigenvalues are complex, what condition on p and q guarantees that solutions spiral around the origin in a clockwise direction?

31. Recall the "Method of the Lucky Guess" for finding solutions of second-order, linear, homogeneous constant coefficient equations (see Section 3.2). For the equation

$$\frac{d^2y}{dt^2} + p\frac{dy}{dt} + qy = 0$$

we guess that $y(t) = e^{st}$ is a solution and try to find s. Plugging y into the differential equation and dividing by e^{st} yields the quadratic equation

$$s^2 + ps + q = 0.$$

(a) If $s = \alpha + i\beta$ is a root, show that $y(t) = e^{(\alpha + i\beta)t}$ is a solution of the second-order differential equation.

(b) Show that $y(t) = e^{(\alpha + i\beta)t} = y_1(t) + iy_2(t)$ where $y_1(t) = e^{\alpha t}\cos(\beta t)$ and $y_2(t) = e^{\alpha t}\sin(\beta t)$.

(c) Show that y_1 and y_2 are solutions of the original second-order differential equation. [*Hint*: We know that $y(t) = y_1(t) + iy_2(t)$ is a solution from part (a), so plug this into the differential equation and remember that two complex numbers are equal if and only if their real and imaginary parts are equal.]

32. The slope field for the system

$$\frac{dx}{dt} = -0.9x - 2y$$
$$\frac{dy}{dt} = x + 1.1y$$

is given below. Plot the $x(t)$- and $y(t)$-graphs for the initial conditions A and B. What do these graphs have in common?

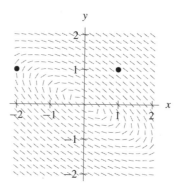

33. (Essay Question) We have seen that linear systems with real eigenvalues can be classified as sinks, sources or saddles depending on whether the eigenvalues are greater or less than zero. Linear systems with complex eigenvalues can be classified as spiral sources, spiral sinks or centers depending on the sign of the real part of the eigenvalue. Why is there not a type of linear system called a "spiral saddle"?

3.5 THE SPECIAL CASES: REPEATED AND ZERO EIGENVALUES

In the previous three sections we dealt with the linear systems

$$\frac{d\mathbf{Y}}{dt} = \mathbf{AY}$$

for which the 2×2 matrix $\mathbf{A}$ has either two distinct, real eigenvalues or complex eigenvalues. In these cases we were able to use the eigenvalues and eigenvectors to sketch the solutions in the xy-phase plane, to draw the $x(t)$-, and $y(t)$-graphs, and to derive an explicit formula for the general solution. We have not yet dealt with the case where the characteristic polynomial of **A** has only one root (a double root), so there is only one eigenvalue. In previous sections we also classified the equilibrium point at the origin as a sink, source, saddle, spiral sink, spiral source, or center depending on the sign of the eigenvalues (or the sign of their real parts) This classification scheme leaves out the cases where one or both eigenvalues are zero. In this section we will modify our methods to handle these remaining cases.

Most quadratic polynomials have two distinct, nonzero roots, so linear systems with only one eigenvalue or with an eigenvalue equal to zero are relatively rare. These systems are sometimes called "degenerate". Nevertheless, they are still important. These special systems form the "boundaries" between the most common types of linear systems. Whenever we study linear systems that depend on a parameter and the system changes behavior or bifurcates as the parameter changes, these special systems play a crucial role.

A System with Repeated Eigenvalues

Consider the linear system

$$\frac{d\mathbf{Y}}{dt} = \mathbf{A}\mathbf{Y} = \begin{pmatrix} -2 & 1 \\ 0 & -2 \end{pmatrix} \mathbf{Y}.$$

The direction field for this system looks somewhat different from the vector fields we have considered thus far in that there appears to be only one straight line of solutions (see Figure 3.31).

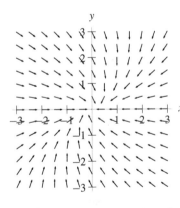

Figure 3.31

Direction field for the system

$$\frac{d\mathbf{Y}}{dt} = \begin{pmatrix} -2 & 1 \\ 0 & -2 \end{pmatrix} \mathbf{Y}.$$

The straight-line solutions all lie on the x-axis.

From an algebraic point of view, this matrix is also unusual. The eigenvalues of this system are the roots of the characteristic polynomial

$$\det(\mathbf{A} - \lambda\mathbf{I}) = (-2 - \lambda)(-2 - \lambda) - 0 = 0,$$

which has a double root at $\lambda = -2$. We say that **A** has a "repeated" eigenvalue $\lambda = -2$. We find the associated eigenvectors by solving $\mathbf{A}\mathbf{Y}_0 = -2\mathbf{Y}_0$ for $\mathbf{Y}_0 = (x_0, y_0)$. We have

$$\begin{cases} -2x_0 + y_0 = -2x_0 \\ -2y_0 = -2y_0, \end{cases}$$

which yields $y_0 = 0$. Therefore the eigenvectors corresponding to the eigenvalue $\lambda = -2$ all lie on the x-axis, so the only straight-line solutions for this system lie on this axis. The vector $(1, 0)$ is one eigenvector associated to $\lambda = -2$, so

$$\mathbf{Y}_1(t) = e^{-2t} \begin{pmatrix} 1 \\ 0 \end{pmatrix}$$

is a solution of this system. But this is only one solution, and as we know, we need two independent solutions to derive the general solution. Bummer.

On the other hand, this is not a complete catastrophe. Our goal is to understand the behavior of the solutions of this system. Writing a formula for the general solution certainly helps, but this is not the only option. We can always study the system using numerical and qualitative techniques.

To obtain a qualitative description of the solutions we start with the one straight line of solutions. Because the eigenvalue is negative, we know that solutions tend to the origin along this line as t increases. Looking at the direction field (or using Euler's method), we can sketch other solutions (see Figure 3.32). Every solution tends to the origin as t increases, so $(0, 0)$ is a sink. For initial conditions that do not lie on the straight-line solution, it appears that the corresponding solutions make a turn and arrive at the origin in a direction that is tangent to the straight-line solution. These solutions look as if they are "trying to spiral" around the origin, but the line of solutions somehow "gets in the way." We see in the next section that that linear systems with repeated eigenvalues form the "boundary" between linear systems that spiral and those with two independent lines of solutions.

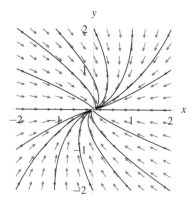

Figure 3.32
Phase plane for the system

$$\frac{d\mathbf{Y}}{dt} = \begin{pmatrix} -2 & 1 \\ 0 & -2 \end{pmatrix} \mathbf{Y}.$$

Note that all solution curves approach the sink at the origin.

The general solution for this example

In order to get some idea of how to compute the general solution of this system we go back to the original equations of the system and rewrite it in components,

$$\frac{dx}{dt} = -2x + y$$
$$\frac{dy}{dt} = -2y.$$

In these equations we spot a lucky break. The second equation for dy/dt does not contain any x's (the second equation is said to *decouple* from the system). So this is an equation that we can consider by itself,

$$\frac{dy}{dt} = -2y.$$

The general solution is $y(t) = k_2 e^{-2t}$ for some constant k_2. Now that we know what $y(t)$ is, we can substitute this back into the first equation, which becomes

$$\frac{dx}{dt} = -2x + k_2 e^{-2t}.$$

We must be living a good life (or have chosen this example very carefully) because this is a first-order, linear equation for $x(t)$, which we can also solve (see Section 1.8). The general solution is

$$x(t) = k_2 t e^{-2t} + k_1 e^{-2t},$$

where k_1 is another constant. (The reader who does not remember how to compute the formulas for solutions of these equations should take this as a warning to review the analytic techniques developed in Chapter 1.) We can write these solutions in vector form as

$$\begin{pmatrix} x(t) \\ y(t) \end{pmatrix} = \begin{pmatrix} k_2 t e^{-2t} + k_1 e^{-2t} \\ k_2 e^{-2t} \end{pmatrix}.$$

Now we study this solution closely to see if there is any pattern that can be generalized. Hidden inside this solution is the straight-line solution we found above. By regrouping terms we find

$$\mathbf{Y}(t) = \begin{pmatrix} x(t) \\ y(t) \end{pmatrix} = k_1 e^{-2t} \begin{pmatrix} 1 \\ 0 \end{pmatrix} + k_2 e^{-2t} \begin{pmatrix} t \\ 1 \end{pmatrix}.$$

The first term consists of the straight-line solutions. The second term is new and different from any other solution to a linear system we have seen so far because of the t in the first component. By taking $k_1 = 1$ and $k_2 = 0$, we get the solution

$$\mathbf{Y}_1(t) = e^{-2t} \begin{pmatrix} 1 \\ 0 \end{pmatrix}$$

and by taking $k_1 = 0$ and $k_2 = 1$, we get

$$\mathbf{Y}_2(t) = e^{-2t} \begin{pmatrix} t \\ 1 \end{pmatrix} = t e^{-2t} \begin{pmatrix} 1 \\ 0 \end{pmatrix} + e^{-2t} \begin{pmatrix} 0 \\ 1 \end{pmatrix}.$$

These two solutions have linearly independent initial conditions, $\mathbf{Y}_1(0) = (1, 0)$ and $\mathbf{Y}_2(0) = (0, 1)$. So the solution $\mathbf{Y}(t)$ above is the general solution of the linear system.

The Form of the General Solution

The solution above provides some hints on how to find a second independent solution for systems with repeated eigenvalues but only one straight-line solution. The solution $\mathbf{Y}_2(t)$ consists of 2 terms. One term is $t\mathbf{Y}_1(t)$, the product of t with our known straight-line solution. The other term is the product of e^{-2t} and the vector $(0, 1)$. From the phase portrait, it is difficult to tell why $(0, 1)$ is significant. We will examine the situation algebraically.

Consider a system

$$\frac{d\mathbf{Y}}{dt} = \mathbf{AY} = \begin{pmatrix} a & b \\ c & d \end{pmatrix} \mathbf{Y},$$

where $\mathbf{A}$ has a repeated eigenvalue λ but only one line of eigenvectors. If $\mathbf{V}_1$ is an eigenvector, then

$$\mathbf{Y}_1(t) = e^{\lambda t} \mathbf{V}_1$$

is a straight-line solution. Following the pattern from the example above, we guess that there is another solution of the form

$$\mathbf{Y}_2(t) = e^{\lambda t}(t\mathbf{V}_1 + \mathbf{V}_2)$$

where $\mathbf{V}_2$ is some vector yet to be determined.

We know how to find the eigenvector $\mathbf{V}_1$. To find the vector $\mathbf{V}_2$, we must use the differential equation. If the curve $\mathbf{Y}_2(t)$ is a solution of the system, it must satisfy the differential equation

$$\frac{d\mathbf{Y}_2}{dt} = \mathbf{AY}_2 \quad \text{for all } t.$$

Using the formula for $\mathbf{Y}_2(t)$, we expand the left-hand side, obtaining

$$\frac{d\mathbf{Y}_2}{dt} = \frac{d(e^{\lambda t}(t\mathbf{V}_1 + \mathbf{V}_2))}{dt}$$

$$= \frac{d(e^{\lambda t}t\mathbf{V}_1)}{dt} + \frac{d(e^{\lambda t}\mathbf{V}_2)}{dt}$$

$$= \lambda e^{\lambda t}t\mathbf{V}_1 + e^{\lambda t}\mathbf{V}_1 + \lambda e^{\lambda t}\mathbf{V}_2.$$

Expanding the right-hand side, we obtain

$$\mathbf{AY}_2 = \mathbf{A}(e^{\lambda t}(t\mathbf{V}_1 + \mathbf{V}_2))$$

$$= e^{\lambda t}t\mathbf{AV}_1 + e^{\lambda t}\mathbf{AV}_2$$

$$= e^{\lambda t}t(\lambda\mathbf{V}_1) + e^{\lambda t}\mathbf{AV}_2$$

$$= \lambda e^{\lambda t}t\mathbf{V}_1 + e^{\lambda t}\mathbf{AV}_2,$$

because $\mathbf{V}_1$ is an eigenvector for $\mathbf{A}$ with eigenvalue λ.

If $\mathbf{Y}_2(t)$ is a solution, then the two sides of the differential equation must be equal, so

$$\lambda e^{\lambda t}t\mathbf{V}_1 + e^{\lambda t}\mathbf{V}_1 + \lambda e^{\lambda t}\mathbf{V}_2 = \lambda e^{\lambda t}t\mathbf{V}_1 + e^{\lambda t}\mathbf{AV}_2,$$

for all t. Simplifying, we obtain

$$e^{\lambda t}(\mathbf{V}_1 + \lambda\mathbf{V}_2) = e^{\lambda t}(\mathbf{AV}_2)$$

or

$$\mathbf{V}_1 + \lambda\mathbf{V}_2 = \mathbf{AV}_2,$$

where $\mathbf{V}_1$ is our known eigenvector for $\mathbf{A}$. This equation tells us how to determine which vector we should use for $\mathbf{V}_2$, namely, any solution of the equation

$$\mathbf{AV}_2 - \lambda \mathbf{V}_2 = (\mathbf{A} - \lambda \mathbf{I})\mathbf{V}_2 = \mathbf{V}_1.$$

Since we know $\mathbf{V}_1 = (x_1, y_1)$, the equation above can be solved for the unknown vector $\mathbf{V}_2 = (x_2, y_2)$. Writing it out in components, this equation is

$$\begin{cases} (a - \lambda)x_2 + by_2 = x_1 \\ cx_2 + (d - \lambda)y_2 = y_1. \end{cases}$$

We summarize this calculation in the following theorem.

Theorem

Suppose $d\mathbf{Y}/dt = \mathbf{AY}$ is a linear system in which the 2×2 matrix $\mathbf{A}$ has a repeated real eigenvalue λ but only one line of eigenvectors. If $\mathbf{V}_1$ is an eigenvector, then the function $\mathbf{Y}_1(t) = e^{\lambda t}\mathbf{V}_1$ is a straight-line solution. There is another solution of the form

$$\mathbf{Y}_2(t) = e^{\lambda t}(t\mathbf{V}_1 + \mathbf{V}_2),$$

where $\mathbf{V}_2$ satisfies the equation $\mathbf{AV}_2 - \lambda \mathbf{V}_2 = \mathbf{V}_1$. The solutions $\mathbf{Y}_1(t)$ and $\mathbf{Y}_2(t)$ are independent. Consequently, the general solution has the form

$$\mathbf{Y}(t) = k_1 e^{\lambda t}\mathbf{V}_1 + k_2 e^{\lambda t}(t\mathbf{V}_1 + \mathbf{V}_2),$$

where k_1 and k_2 are arbitrary constants.

This is a rather lengthy process. In fact, there can be a lot of arithmetic involved in finding this general solution. Luckily, finding the vector $\mathbf{V}_2$ is not so hard in most interesting cases.

Instant replay

In the example that we did "by hand" above, namely the system

$$\frac{d\mathbf{Y}}{dt} = \mathbf{AY} = \begin{pmatrix} -2 & 1 \\ 0 & -2 \end{pmatrix} \mathbf{Y},$$

we computed that $\lambda = -2$ is an eigenvalue, that $\mathbf{V}_1 = (1, 0)$ is an eigenvector, and that there is only one line of eigenvectors. So we have the straight-line solution

$$\mathbf{Y}_1(t) = e^{-2t} \begin{pmatrix} 1 \\ 0 \end{pmatrix}.$$

To find the vector $\mathbf{V}_2$, we must solve the system of equations

$$\begin{cases} (-2 + 2)x_2 + 1y_2 = 1 \\ 0x_2 + (-2 + 2)y_2 = 0. \end{cases}$$

The second equation is $0 = 0$, but the first is $y_2 = 1$. Thus *any* vector whose second component is 1 can serve as $\mathbf{V}_2$. Usually we choose $\mathbf{V}_2$ as simple as possible, say $\mathbf{V}_2 = (0, 1)$, and we obtain the general solution

$$\mathbf{Y}(t) = k_1 e^{-2t} \begin{pmatrix} 1 \\ 0 \end{pmatrix} + k_2 e^{-2t} \left(t \begin{pmatrix} 1 \\ 0 \end{pmatrix} + \begin{pmatrix} 0 \\ 1 \end{pmatrix} \right) = \begin{pmatrix} k_1 e^{-2t} + k_2 t e^{-2t} \\ k_2 e^{-2t} \end{pmatrix},$$

which is exactly what we determined above.

Qualitative Analysis of Systems with Repeated Eigenvalues

Before discussing additional examples, we consider what the general solution tells us about the qualitative behavior of solutions. The form of the general solution is

$$\mathbf{Y}(t) = k_1 e^{\lambda t} \mathbf{V}_1 + k_2 e^{\lambda t}(t\mathbf{V}_1 + \mathbf{V}_2) = e^{\lambda t}(k_1\mathbf{V}_1 + k_2\mathbf{V}_2) + te^{\lambda t}k_2\mathbf{V}_1.$$

The dependence on t comes in two terms, the $e^{\lambda t}$ term and the $te^{\lambda t}$ term. If $\lambda < 0$ then both of these tend to zero as t increases. Consequently, the equilibrium point $(0, 0)$ is a sink. The $te^{\lambda t}$ term is much larger than $e^{\lambda t}$ if t is large. Consequently,

$$\mathbf{Y}(t) \approx te^{\lambda t}k_2\mathbf{V}_1$$

when t is large. Thus the solution tends to $(0, 0)$ in a direction tangent to the line of eigenvectors. If $\lambda > 0$, then all solutions (except the equilibrium solution) tend to infinity as t increases, so $(0, 0)$ is a source. Again the $te^{\lambda t}$ term dominates for large t if $k_2 \neq 0$ (see Figure 3.33 for typical examples of phase planes for these systems).

If we use the formula for the general solution to draw the phase plane, we see again that it looks as if the solutions (other than the one straight-line solution) are trying to spiral around $(0, 0)$. But they cannot spiral since the one straight-line solution gets in the way. The strange (and complicated) form of the general solution of systems with one straight-line of solutions is another symptom of the fact that these systems lie "between" systems with complex eigenvalues and those with real, distinct eigenvalues.

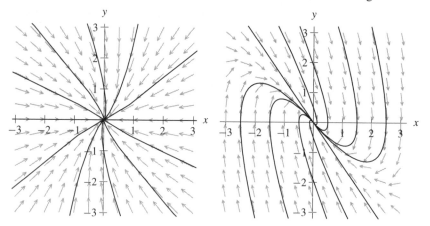

Figure 3.33
Typical phase planes for systems with repeated eigenvalues.

Harmonic Oscillator with Repeated Eigenvalues

We have seen that the harmonic oscillator system can have complex eigenvalues or two real eigenvalues. The complex eigenvalue case is called an **underdamped** oscillator while the two real distinct eigenvalues case is said to be **overdamped**. The solutions of the oscillator system in the underdamped case oscillate about the rest position. In the overdamped case solutions do not oscillate and in fact they can cross the rest position at most once. If the mass m and the spring constant k_s are held constant, then the type of system (underdamped or overdamped) is determined by the size of the coefficient of damping k_d. Small damping corresponds to the underdamped case and large damping to the overdamped case.

A natural question is: What is the dividing line between overdamped and underdamped cases, and what happens to solutions in that special case? What sort of harmonic oscillator system will we have? A bifurcation must occur at this parameter value. We would like to find this bifurcation value and study the corresponding system.

For example, suppose we fix the mass $m = 1$ and the spring constant $k_s = 2$ and let the coefficient of damping be a parameter. Then the second-order harmonic oscillator equation is

$$\frac{d^2 y}{dt^2} + k_d \frac{dy}{dt} + 2y = 0,$$

and the system is

$$\frac{d\mathbf{Y}}{dt} = \mathbf{CY} \quad \text{where} \quad \mathbf{C} = \begin{pmatrix} 0 & 1 \\ -2 & -k_d \end{pmatrix}, \quad \mathbf{Y} = \begin{pmatrix} y \\ v \end{pmatrix},$$

and $v = dy/dt$. The eigenvalues of this system are the roots of the characteristic polynomial

$$\det(\mathbf{C} - \lambda \mathbf{I}) = (0 - \lambda)(-k_d - \lambda) + 2 = \lambda^2 + k_d \lambda + 2.$$

Using the quadratic formula, we have

$$\lambda = \frac{-k_d \pm \sqrt{k_d^2 - 8}}{2},$$

and consequently, the eigenvalues are complex if $k_d^2 < 8$. Since the damping coefficient is always nonnegative, the dividing line between the underdamped and overdamped cases occurs at the bifurcation value $k_d = 2\sqrt{2}$. At this value, the system has repeated eigenvalues $\lambda = -\sqrt{2}$, and the system is

$$\frac{d\mathbf{Y}}{dt} = \begin{pmatrix} 0 & 1 \\ -2 & -2\sqrt{2} \end{pmatrix} \mathbf{Y}.$$

We say that this harmonic oscillator is **critically damped**.

Given the eigenvalue $\lambda = -\sqrt{2}$, we find the eigenvectors $\mathbf{V}_1$ by solving $\mathbf{CV}_1 = \lambda \mathbf{V}_1$, which is

$$\begin{pmatrix} 0 & 1 \\ -2 & -2\sqrt{2} \end{pmatrix} \begin{pmatrix} y_0 \\ v_0 \end{pmatrix} = -\sqrt{2} \begin{pmatrix} y_0 \\ v_0 \end{pmatrix}.$$

So the eigenvectors lie on the line $v_0 = -\sqrt{2} y_0$. For instance, one convenient eigenvector is $\mathbf{V}_1 = (1, -\sqrt{2})$. The phase plane for this system is shown in Figure 3.34. All solutions approach $(0, 0)$ as t increases, and all solutions are tangent to the line of eigenvectors as they approach the origin.

The eigenvalue and eigenvector computed above give us a straight-line solution

$$\mathbf{Y}_1(t) = e^{-\sqrt{2}t} \begin{pmatrix} 1 \\ -\sqrt{2} \end{pmatrix}.$$

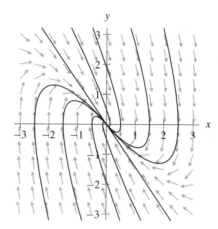

Figure 3.34

Direction field and solution curves for

$$\frac{d\mathbf{Y}}{dt} = \begin{pmatrix} 0 & 1 \\ -2 & -2\sqrt{2} \end{pmatrix} \mathbf{Y}.$$

Note that the solution curves approach the origin tangent to the straight-line solution.

We find a second independent solution by first finding a vector $\mathbf{V}_2$, that satisfies $(\mathbf{C} - \lambda\mathbf{I})\mathbf{V}_2 = \mathbf{V}_1$ or

$$\begin{pmatrix} 0 & 1 \\ -2 & -2\sqrt{2} \end{pmatrix} \mathbf{V}_2 + \sqrt{2}\mathbf{V}_2 = \begin{pmatrix} 1 \\ -\sqrt{2} \end{pmatrix}.$$

One such vector is $\mathbf{V}_2 = (\sqrt{2}/2, 0)$. Hence a second solution of the system is given by

$$\mathbf{Y}_2(t) = e^{-\sqrt{2}t}\left(t\begin{pmatrix} 1 \\ -\sqrt{2} \end{pmatrix} + \begin{pmatrix} \sqrt{2}/2 \\ 0 \end{pmatrix}\right).$$

We can form the general solution

$$\mathbf{Y}(t) = k_1\mathbf{Y}_1(t) + k_2\mathbf{Y}_2(t) = k_1\begin{pmatrix} e^{-\sqrt{2}t} \\ -\sqrt{2}e^{-\sqrt{2}t} \end{pmatrix} + k_2\begin{pmatrix} (\sqrt{2}/2 + t)e^{-\sqrt{2}t} \\ -\sqrt{2}te^{-\sqrt{2}t} \end{pmatrix}.$$

Paul's and Bob's CD Stores Revisited

Recall the model of Paul's and Bob's compact disk stores from Section 3.1. We suppose that the model has the form

$$\frac{d\mathbf{Y}}{dt} = \mathbf{AY} \quad \text{where} \quad \mathbf{A} = \begin{pmatrix} -5 & 1 \\ -1 & -3 \end{pmatrix} \mathbf{Y}.$$

The coefficients imply that if Bob is making money, then this will increase Paul's profits. But if Paul is making money, then this hurts Bob. In this model, rock 'n rollers don't like opera.

The eigenvalues for this system are the roots of the characteristic polynomial

$$\det(\mathbf{A} - \lambda\mathbf{I}) = \lambda^2 + 8\lambda + 16.$$

There is only one eigenvalue, $\lambda = -4$. Hence the origin is a sink and all solutions tend to $(0, 0)$ as t increases. The profits of both Paul's and Bob's stores tend toward zero no matter what the initial condition.

To find the eigenvectors, we solve

$$\begin{cases} -5x_1 + y_1 = -4x_1 \\ -1x_1 - 3y_1 = -4y_1, \end{cases}$$

which has as its solution set the line $x_1 = y_1$. From this information we conclude that as t increases, every solution tends to $(0, 0)$ tangent to the line $x = y$. As the values of x and y tend to zero, they are almost equal.

By looking at the direction field we can see which initial conditions lead to solutions that tend to zero along $x = y$ in the first quadrant and which initial conditions end up in the fourth quadrant, along with the corresponding $x(t)$- and $y(t)$-graphs (see Figures 3.35 and 3.36).

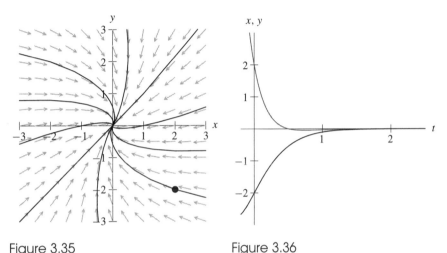

Figure 3.35
Phase plane for the system

$$\frac{d\mathbf{Y}}{dt} = \begin{pmatrix} -5 & 1 \\ -1 & -3 \end{pmatrix} \mathbf{Y}$$

Figure 3.36
The $x(t)$-, $y(t)$-graphs for the indicated solution.

The fact that the solutions tend to the line $x = y$ means that the two stores asymptotically have essentially the same profits or losses. This is a little surprising because the coefficients in the model imply that the profits of the two stores react in very different ways to the profits of the other store.

Effect of a small change in the coefficients

We might wonder what would happen if the coefficients were changed just a little. Suppose Bob decides that he will help Paul out, so he puts up a tasteful sign saying

Open your mind to Rock and Roll
(But do it someplace else, like Paul's Rock and Roll CD's)

The more people that come to Bob's store (the larger y is) the more this sign will help Paul. So the parameter b increases from 1 to, say, 1.1. The new system is

$$\frac{d\mathbf{Y}}{dt} = \mathbf{BY} = \begin{pmatrix} -5 & 1.1 \\ -1 & -3 \end{pmatrix} \mathbf{Y}.$$

We can compute that the eigenvalues of this matrix are complex and have negative real part. Hence, the origin is a spiral sink and all solutions still tend to $(0, 0)$. However, when the solutions are close to $(0, 0)$, they will oscillate instead of tending toward $(0, 0)$ along the line $y = x$. This small change in the system has made a distinct change in the qualitative behavior of the solutions. However, the oscillations are very subtle In fact, the phase plane and $x(t)$-, $y(t)$-graphs for $d\mathbf{Y}/dt = \mathbf{BY}$ are indistinguishable from those of $d\mathbf{Y}/dt = \mathbf{AY}$ (compare Figures 3.35 and 3.36 to Figures 3.37 and 3.38).

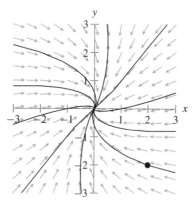

Figure 3.37
Phase plane for

$$\frac{d\mathbf{Y}}{dt} = \begin{pmatrix} -5 & 1.1 \\ -1 & -3 \end{pmatrix} \mathbf{Y}.$$

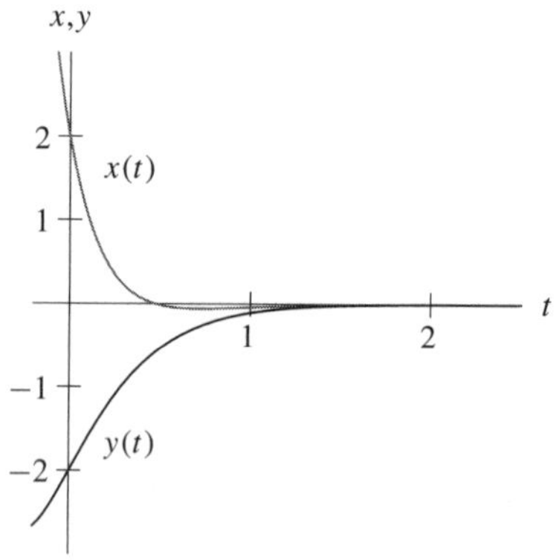

Figure 3.38
The $x(t)$-, $y(t)$-graphs for the indicated solution.

Systems For Which Every Vector is an Eigenvector

At this point we can deal with linear systems that have repeated real eigenvalues but only one line of eigenvectors. An example of a system with repeated eigenvalues and more than one line of eigenvectors is given by

$$\frac{d\mathbf{Y}}{dt} = \begin{pmatrix} a & 0 \\ 0 & a \end{pmatrix} \mathbf{Y}.$$

This system has a repeated eigenvalue $\lambda = a$, and *any* nonzero vector is an eigenvector for $\lambda = a$. In this case finding the general solution is equivalent to finding the general solutions of the two equations $dx/dt = ax$ and $dy/dt = ay$, which can be handled using the methods of Chapter 1. In the exercises we see that the only systems with one eigenvalue having more than one line of eigenvectors have coefficient matrix equal to $\lambda\mathbf{I}$, where λ is the eigenvalue (see Exercises 13 and 14). This is a very special case.

Systems with Zero as an Eigenvalue

We are close to having a complete understanding of linear systems and their phase planes. We can classify and sketch the behavior at $(0, 0)$ for the cases of real, complex, and repeated eigenvalues. The only case we have not yet explicitly considered is the case where one or both of the eigenvalues is zero. This case is important because it divides the linear systems with strictly positive eigenvalues (sources) and strictly negative eigenvalues (sinks) from those with one positive and one negative eigenvalue (saddles).

Suppose we have a linear system

$$\frac{d\mathbf{Y}}{dt} = \mathbf{A}\mathbf{Y}$$

and the matrix $\mathbf{A}$ has eigenvalues $\lambda_1 = 0$ and $\lambda_2 \neq 0$. Suppose $\mathbf{V}_1$ is an eigenvector for λ_1 and $\mathbf{V}_2$ is an eigenvector of λ_2.

This is the case of two real, distinct eigenvalues, and all of the algebra we did in Section 3.3 applies. Hence we can write down the general solution

$$\mathbf{Y}(t) = k_1 e^{\lambda_1 t}\mathbf{V}_1 + k_2 e^{\lambda_2 t}\mathbf{V}_2,$$

for k_1 and k_2 constants. But $\lambda_1 = 0$, so $e^{\lambda_1 t} = e^{0t} = 1$ for all t. Hence

$$\mathbf{Y}(t) = k_1 \mathbf{V}_1 + k_2 e^{\lambda_2 t}\mathbf{V}_2,$$

and the general solution $\mathbf{Y}(t)$ depends on t only through the $\mathbf{V}_2$ term. If we take $k_2 = 0$, then we have $\mathbf{Y}(t) = k_1 \mathbf{V}_1$, which does not depend on t. So all the points $k_1 \mathbf{V}_1$, for any k_1, are equilibrium points. Every point on the line of eigenvectors for the eigenvalue $\lambda_1 = 0$ is an equilibrium point. If $\lambda_2 < 0$, then the second term in the general solution tends to zero as t increases, so the solution

$$\mathbf{Y}(t) = k_1 \mathbf{V}_1 + k_2 e^{\lambda_2 t}\mathbf{V}_2,$$

tends to the equilibrium point $k_1 \mathbf{V}_1$ along a line parallel to $\mathbf{V}_2$. If $\lambda_2 > 0$, then the solution above tends away from the line of equilibrium points as t increases. This is enough information to sketch the phase planes.

An example with an eigenvalue equal to zero

Consider the system

$$\frac{d\mathbf{Y}}{dt} = \mathbf{A}\mathbf{Y} \quad \text{where} \quad \mathbf{A} = \begin{pmatrix} -3 & 1 \\ 3 & -1 \end{pmatrix}.$$

We compute the eigenvalues from the characteristic polynomial by solving

$$\det(\mathbf{A} - \lambda\mathbf{I}) = (-3 - \lambda)(-1 - \lambda) - 3 = 0,$$

or

$$\lambda^2 + 4\lambda = 0.$$

The eigenvalues are $\lambda_1 = 0$ and $\lambda_2 = -4$.

The eigenvectors $\mathbf{V}_1 = (x_1, y_1)$ for $\lambda_1 = 0$ satisfy

$$\begin{cases} -3x_1 + y_1 = 0x_1 \\ 3x_1 - y_1 = 0y_1. \end{cases}$$

They are on the line $y_1 = 3x_1$. For instance, $\mathbf{V}_1 = (1, 3)$ is an eigenvector for $\lambda_1 = 0$. Similarly, the eigenvectors $\mathbf{V}_2 = (x_2, y_2)$ for $\lambda_2 = -4$ satisfy

$$\begin{cases} -3x_2 + y_2 = -4x_2 \\ 3x_2 - y_2 = 4y_2, \end{cases}$$

and these equations can be simplified to $x_2 + y_2 = 0$, so the solutions of these equations are on the line $x_2 = -y_2$. So $\mathbf{V}_2 = (-1, 1)$ is an eigenvector for $\lambda_2 = -4$.

From this information we can draw the phase plane. There is a line of equilibrium points given by $y = 3x$ and every other solution approaches an equilibrium point on this line by following a line parallel to the line $y = -x$ (see Figure 3.39). The $x(t)$- and $y(t)$-graphs for the solution with initial condition $\mathbf{Y}(0) = (5, 2)$ and $\mathbf{Y}(0) = (1, 0)$ are given in Figure 3.40.

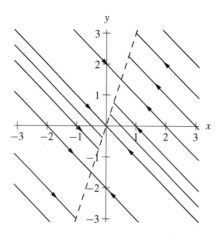

Figure 3.39
Phase plane for the system

$$\frac{d\mathbf{Y}}{dt} = \begin{pmatrix} -3 & 1 \\ 3 & -1 \end{pmatrix} \mathbf{Y}.$$

Solutions tend toward the line of equilibrium points.

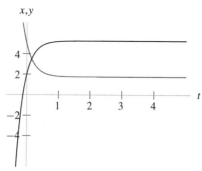

Figure 3.40
Graphs of $x(t)$ and $y(t)$ for solutions of

$$\frac{d\mathbf{Y}}{dt} = \begin{pmatrix} -3 & 1 \\ 3 & -1 \end{pmatrix} \mathbf{Y}$$

with initial positions $(5, 2)$ (left graphs) and $(1, 0)$ (right graphs).

Exercises for Section 3.5

In Exercises 1–4, each of the linear systems has one eigenvalue and one line of eigenvectors. For each of these linear systems:

(a) Find the eigenvalue.

(b) Find an eigenvector.

(c) Sketch the direction field.

(d) Sketch the phase plane, including the solution with the given initial condition.

(e) Sketch the $x(t)$- and $y(t)$-graphs of the solution with the given initial condition.

1. $\dfrac{d\mathbf{Y}}{dt} = \begin{pmatrix} -3 & 0 \\ 1 & -3 \end{pmatrix} \mathbf{Y}$ with initial condition $\mathbf{Y}_0 = (1, 0)$.

2. $\dfrac{d\mathbf{Y}}{dt} = \begin{pmatrix} 2 & 1 \\ -1 & 4 \end{pmatrix} \mathbf{Y}$ with initial condition $\mathbf{Y}_0 = (1, 0)$.

3. $\dfrac{d\mathbf{Y}}{dt} = \begin{pmatrix} -2 & -1 \\ 1 & -4 \end{pmatrix} \mathbf{Y}$ with initial condition $\mathbf{Y}_0 = (1, 0)$.

4. $\dfrac{d\mathbf{Y}}{dt} = \begin{pmatrix} 0 & 1 \\ -1 & -2 \end{pmatrix} \mathbf{Y}$ with initial condition $\mathbf{Y}_0 = (1, 0)$.

In Exercises 5–8, for each of the given systems:

(a) Find the general solution.

(b) Find the particular solution for the given initial condition.

(c) Sketch the $x(t)$- and $y(t)$-graphs of the solution and compare your sketches with those you made in the previous problem.

5. $\dfrac{d\mathbf{Y}}{dt} = \begin{pmatrix} -3 & 0 \\ 1 & -3 \end{pmatrix} \mathbf{Y}$ with initial condition $\mathbf{Y}_0 = (1, 0)$.

6. $\dfrac{d\mathbf{Y}}{dt} = \begin{pmatrix} 2 & 1 \\ -1 & 4 \end{pmatrix} \mathbf{Y}$ with initial condition $\mathbf{Y}_0 = (1, 0)$.

7. $\dfrac{d\mathbf{Y}}{dt} = \begin{pmatrix} -2 & -1 \\ 1 & -4 \end{pmatrix} \mathbf{Y}$ with initial condition $\mathbf{Y}_0 = (1, 0)$.

8. $\dfrac{d\mathbf{Y}}{dt} = \begin{pmatrix} 0 & 1 \\ -1 & -2 \end{pmatrix} \mathbf{Y}$ with initial condition $\mathbf{Y}_0 = (1, 0)$.

9. Given a quadratic $\lambda^2 + \alpha\lambda + \beta$, what condition on α and β guarantees

(a) that the quadratic has a double root?

(b) that the quadratic has zero as a root?

10. Evaluate the limit of $te^{\lambda t}$ as $t \to \infty$ if

(a) $\lambda > 0$ (b) $\lambda < 0$

Be sure and justify your answer.

11. Consider the matrix

$$A = \begin{pmatrix} 0 & 1 \\ q & p \end{pmatrix},$$

where $q, p \geq 0$. What condition on q and p guarantees

(a) that A has two real eigenvalues?

(b) that A has complex eigenvalues?

(c) that A has one eigenvalue and one line of eigenvectors?

12. Let

$$\mathbf{A} = \begin{pmatrix} a & b \\ c & d \end{pmatrix}.$$

Define the trace of $\mathbf{A}$ to be $\text{tr}(\mathbf{A}) = a + d$. Show that $\mathbf{A}$ has only one eigenvalue if and only if $(\text{tr}(\mathbf{A}))^2 - 4 \det(\mathbf{A}) = 0$.

13. Suppose

$$\mathbf{A} = \begin{pmatrix} a & b \\ c & d \end{pmatrix}$$

is a matrix with eigenvalue λ such that every nonzero vector is an eigenvector with eigenvalue λ, that is, $\mathbf{AY} = \lambda \mathbf{Y}$ for every vector $\mathbf{Y}$. Show that $a = d = \lambda$ and $b = c = 0$. [*Hint*: Since $\mathbf{AY} = \lambda \mathbf{Y}$ for every $\mathbf{Y}$, try $\mathbf{Y} = (1, 0)$ and $\mathbf{Y} = (0, 1)$.]

14. Suppose the matrix

$$\mathbf{A} = \begin{pmatrix} a & b \\ c & d \end{pmatrix}$$

has eigenvalue λ. Suppose that there are two linearly independent vectors $\mathbf{Y}_1$ and $\mathbf{Y}_2$ that are both eigenvectors for the eigenvalue λ. Show that every nonzero vector is an eigenvector with eigenvalue λ. What does this imply about a, b, c and d?

15. Consider a harmonic oscillator system with spring constant $k_s = 3$, damping coefficient $k_d = \sqrt{12}$, and mass $m = 1$. Find the solution with initial position $y(0) = -1$ and velocity $v(0) = 3$.

16. Consider a harmonic oscillator with spring constant $k_s = 2$, damping coefficient $k_d = 3$, and mass m.

(a) Find the value of m such that the resulting oscillator is critically damped.

(b) Find the eigenvalue and eigenvectors for this critically damped system.

(c) Sketch the phase plane for this critically damped system.

(d) Find the general solution for this critically damped system.

(e) Find the solution with initial position $y(0) = 0$ and initial velocity $v(0) = -2$.

17. Consider a harmonic oscillator with spring constant k_s, damping coefficient $k_d = 3$, and mass $m = 1$.

(a) Find the value of k_s such that the resulting oscillator is critically damped.

(b) Find the eigenvalue and eigenvectors for this critically damped system.

(c) Sketch the phase plane for this critically damped system.

(d) Find the general solution for this critically damped system.

(e) Find the solution with initial position $y(0) = 2$ and initial velocity $v(0) = 0$.

18. For the second-order, linear, homogeneous, constant coefficient equation

$$\frac{d^2y}{dt^2} + p\frac{dy}{dt} + qy = 0,$$

what condition on p and q guarantees that the corresponding system has only one eigenvalue and only one line of eigenvectors?

19. Recall the "Method of the Lucky Guess" for finding solutions of second-order, linear, homogeneous, constant coefficient equations from Section 3.2. For the equation

$$\frac{d^2y}{dt^2} + p\frac{dy}{dt} + qy = 0,$$

we guess that $y(t) = e^{st}$ is a solution and try to find s. Plugging this into the equation and dividing by e^{st} yields a quadratic equation for s

$$s^2 + ps + q = 0.$$

(a) What is the condition on p and q that guarantees that this quadratic has only one root?

(b) If $s = \lambda$ is the root, show that $y_1(t) = e^{\lambda t}$ and $y_2(t) = te^{\lambda t}$ are solutions of the second-order equation. (So, in the Method of the Lucky Guess, if the guess $y(t) = e^{st}$ yields only one solution, our next guess is $y(t) = te^{st}$.)

In Exercises 20–22, each of the given linear systems has zero as an eigenvalue. For each of these systems:

(a) Find the eigenvalues.

(b) Find the eigenvectors.

(c) Sketch the phase plane.

(d) Sketch the $x(t)$- and $y(t)$-graphs of the solution with initial condition $Y_0 = (1, 0)$.

(e) Find the general solution.

(f) Find the particular solution for the initial condition $Y_0 = (1, 0)$ and compare it with the sketch you made.

20. $\dfrac{d\mathbf{Y}}{dt} = \begin{pmatrix} 0 & 2 \\ 0 & -1 \end{pmatrix} \mathbf{Y}$ **21.** $\dfrac{d\mathbf{Y}}{dt} = \begin{pmatrix} 2 & 4 \\ 3 & 6 \end{pmatrix} \mathbf{Y}$ **22.** $\dfrac{d\mathbf{Y}}{dt} = \begin{pmatrix} 4 & 2 \\ 2 & 1 \end{pmatrix} \mathbf{Y}$

23. Let $\mathbf{A} = \begin{pmatrix} a & b \\ c & d \end{pmatrix}$.

(a) Show that, if one or both of the eigenvalues of $\mathbf{A}$ is zero, then the determinant of $\mathbf{A}$ is zero.

(b) Show that, if $\det \mathbf{A} = 0$, then at least one of the eigenvalues of $\mathbf{A}$ is zero.

24. Find the eigenvalues and sketch the phase planes for the following linear systems:

(a) $\dfrac{d\mathbf{Y}}{dt} = \begin{pmatrix} 0 & 2 \\ 0 & 0 \end{pmatrix} \mathbf{Y}$ (b) $\dfrac{d\mathbf{Y}}{dt} = \begin{pmatrix} 0 & -2 \\ 0 & 0 \end{pmatrix} \mathbf{Y}$

25. Find the general solution for the following linear systems:

(a) $\dfrac{d\mathbf{Y}}{dt} = \begin{pmatrix} 0 & 2 \\ 0 & 0 \end{pmatrix} \mathbf{Y}$ (b) $\dfrac{d\mathbf{Y}}{dt} = \begin{pmatrix} 0 & -2 \\ 0 & 0 \end{pmatrix} \mathbf{Y}$

26. Consider the linear system

$$\frac{d\mathbf{Y}}{dt} = \begin{pmatrix} a & 0 \\ 0 & d \end{pmatrix} \mathbf{Y}.$$

(a) Find the eigenvalues.

(b) Find the eigenvectors.

(c) Suppose $a = d < 0$. Sketch the phase plane and compute the general solution. (What are the eigenvectors in this case?)

(d) Suppose $a = d > 0$. Sketch the phase plane and compute the general solution.

27. Recall the model of Paul's and Bob's CD stores discussed in this section. Suppose Paul expands his store and that this means that an increase in his profits has an even more negative effect on Bob's profits, that is, the system becomes

$$\frac{d\mathbf{Y}}{dt} = \begin{pmatrix} -3 & 1 \\ -\frac{11}{10} & \frac{1}{3} \end{pmatrix} \mathbf{Y}.$$

(a) Sketch the phase plane for this system.

(b) What type of equilibrium point is the origin?

(c) Sketch and describe in a brief paragraph the behavior of the solution with initial condition $(2, 1)$.

3.6 ONE-PARAMETER FAMILIES OF LINEAR SYSTEMS

In the previous sections we have dealt with different types of linear systems

$$\frac{d\mathbf{Y}}{dt} = \mathbf{AY}, \quad \text{where} \quad \mathbf{A} = \begin{pmatrix} a & b \\ c & d \end{pmatrix}$$

is a 2×2 matrix and

$$\mathbf{Y}(t) = \begin{pmatrix} x(t) \\ y(t) \end{pmatrix}$$

is a vector of dependent variables. The qualitative behavior of solutions and the phase-plane diagram can be determined from the eigenvalues of $\mathbf{A}$. It is possible to create a table where one column is the nature of the eigenvalues (real or complex, positive or negative, and so forth), and the other column contains a sketch of the possible phase-plane pictures for that case. Creating this table is a very instructive exercise (see Exercises 1 and 2).

So far we have considered linear systems one at a time. In this section we consider how the different types of linear systems relate to one another by studying one-parameter families of systems. As a parameter in the coefficient matrix varies, the eigenvalues change, and this leads to a corresponding change in the phase plane of the system. If the eigenvalues change dramatically, for example, from positive to negative or from real to complex, then the phase plane will also change qualitatively. These changes correspond to bifurcations of the system.

An Example

Consider the linear system

$$\frac{d\mathbf{Y}}{dt} = \mathbf{AY}, \quad \text{where} \quad \mathbf{A} = \begin{pmatrix} -2 & a \\ -2 & 0 \end{pmatrix}.$$

This is a one-parameter family of linear systems with parameter a. The eigenvalues and eigenvectors depend on the value of a. We study this system by considering different values of a ranging from large positive to large negative. The parameter a may take on any real value (not just integer values).

The eigenvalues are the roots of the characteristic polynomial, that is, solutions of

$$\det(\mathbf{A} - \lambda\mathbf{I}) = \lambda^2 + 2\lambda + 2a = 0.$$

These eigenvalues are

$$\lambda = \frac{-2 \pm \sqrt{4 - 8a}}{2} = -1 \pm \sqrt{1 - 2a}.$$

If $a > 1/2$, then $1 - 2a < 0$ and the eigenvalues are complex. The real part is -1 and the natural period is $2\pi/\sqrt{1 - 2a}$. Hence for large a, the origin is a spiral sink. Solutions approach the origin at the exponential rate of e^{-t} and with very small period. The phase plane for $a = 10$ is shown in Figure 3.41.

Bifurcation from spiral sink to sink

If a is just slightly larger than $1/2$, solutions still spiral toward the origin. The period of the oscillations, which is given by $2\pi/\sqrt{1 - 2a}$, is very large for a near $1/2$.

When $a = 1/2$, the system is

$$\frac{d\mathbf{Y}}{dt} = \begin{pmatrix} -2 & \frac{1}{2} \\ -2 & 0 \end{pmatrix} \mathbf{Y}.$$

The characteristic polynomial becomes

$$\det(\mathbf{A} - \lambda\mathbf{I}) = \lambda^2 + 2\lambda + 1,$$

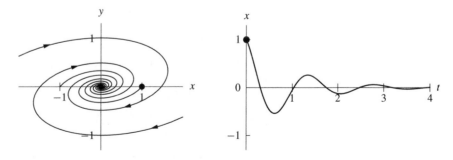

Figure 3.41
Phase plane and $x(t)$-graph for the indicated solution for the one-parameter family above with $a = 10$.

and hence the system has a repeated eigenvalue $\lambda = -1$. The solutions no longer spiral about the origin in the phase plane. There is a single line of solutions corresponding to the single line of eigenvectors. The phase plane and typical $x(t)$-graph is shown in Figure 3.42.

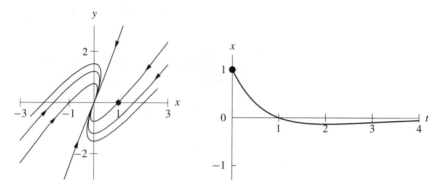

Figure 3.42
Phase plane and $x(t)$-graph for the indicated solution for the one-parameter family above with $a = 1/2$.

When $a = 1/2$ the origin is still a sink. However, a bifurcation occurs at this parameter value because the qualitative behavior of the solutions changes. As we have seen before, this is a somewhat subtle bifurcation. For values of a slightly larger than $1/2$, solutions oscillate with very large natural period. To observe one oscillation, we must watch a solution for a long time. Since the solutions are tending to the origin at an exponential rate, these oscillations may be very hard to detect (see Figure 3.43 which is almost indistinguishable from Figure 3.42). In applications there may be very little practical difference between a very slowly oscillating solution decaying toward the origin and a solution that does not oscillate.

For $a < 1/2$, the eigenvalues

$$\lambda = -1 \pm \sqrt{1 - 2a}$$

are both real. For $0 < a < 1/2$, the term $\sqrt{1 - 2a} < 1$, so both eigenvalues are negative. Hence the origin is a sink with two straight lines of solutions (see Figure 3.44).

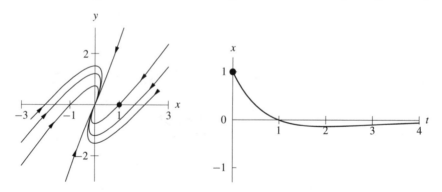

Figure 3.43
Phase plane and $x(t)$-graph for the indicated solution for the one-parameter family above with $a = 0.51$.

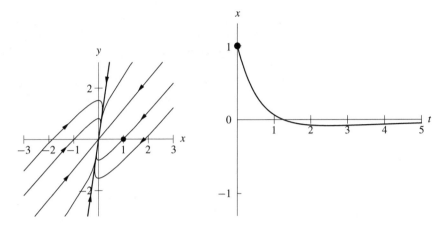

Figure 3.44
Phase plane and $x(t)$-graph for the indicated solution for the one-parameter family above with $a = 1/4$.

Bifurcation from sink to saddle

If a is reduced to zero, then the system again changes character, this time in a much more dramatic way. For $0 < a < 1/2$, both eigenvalues are negative, so the origin is a sink. When $a = 0$, the system becomes

$$\frac{d\mathbf{Y}}{dt} = \begin{pmatrix} -2 & 0 \\ -2 & 0 \end{pmatrix} \mathbf{Y}.$$

The eigenvalues are $\lambda_1 = -2$ and $\lambda_2 = 0$. Corresponding to $\lambda_1 = -2$, there is a straight-line solution that tends to the origin. The line of eigenvectors for $\lambda_2 = 0$ consists entirely of equilibrium points. Every solution of the system lies on a line that passes through one of these equilibrium points (see Figure 3.45). Therefore most solutions do not tend to the origin, and another bifurcation has taken place.

If a is just slightly greater than zero, all solutions tend toward the origin. In this case one of the eigenvalues will be just less than zero, so solutions tend relatively slowly

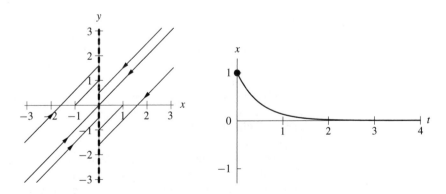

Figure 3.45
Phase plane and $x(t)$-graph for the indicated solution for the one-parameter family above with $a = 0$.

toward zero in the direction of the corresponding line of eigenvectors. Hence the short-term effect of changing a from slightly greater than zero to $a = 0$ is still small. The dramatic change comes in the long-term behavior of solutions.

For $a < 0$, the eigenvalue $-1 - \sqrt{1 - 2a}$ remains negative, but the other eigenvalue $-1 + \sqrt{1 - 2a}$ becomes positive. Hence the origin is a saddle, and most solutions eventually tend away from the origin as t increases (see Figure 3.46).

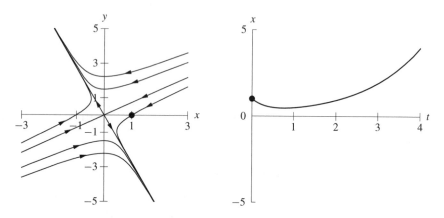

Figure 3.46
Phase plane and $x(t)$-graph for the indicated solution for the one-parameter family above with $a = -1$.

Classification

For the one-parameter family of systems

$$\frac{d\mathbf{Y}}{dt} = \begin{pmatrix} -2 & a \\ -2 & 0 \end{pmatrix} \mathbf{Y},$$

there are two bifurcation values of the parameter. At $a = 1/2$, the system changes from a spiral sink to a sink with two real eigenvalues and hence two lines of solutions. In the intermediate case, with $a = 1/2$, the system has a repeated eigenvalue and a single line of eigenvectors.

The other bifurcation occurs at $a = 0$. The system changes from a sink to a saddle. In the intermediate case, with $a = 0$, the system has a zero eigenvalue and the corresponding line of eigenvectors is a line of equilibrium points. We summarize this behavior in Figure 3.47.

For most a values, the origin is a sink or a saddle. The dividing lines between different types of behavior correspond to values of a for which the system is degenerate, either with repeated eigenvalues or with an eigenvalue equal to zero. This is typical of linear systems. For most linear systems, the origin is a (spiral) source, (spiral) sink, or saddle. The degenerate cases are important because they form the dividing line between the more common cases and hence indicate bifurcations.

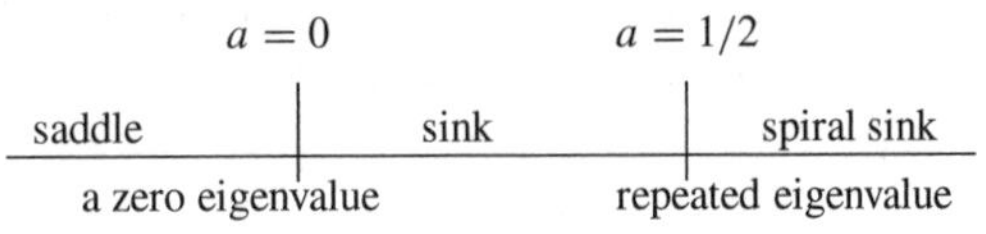

Figure 3.47
Classification of the behavior of
the one-parameter family of
systems for all possible a values.

$a = 0$ $a = 1/2$

saddle | sink | spiral sink

a zero eigenvalue   repeated eigenvalue

The Harmonic Oscillator

We can apply the same type of analysis to the harmonic oscillator equation

$$\frac{d^2 y}{dt^2} + \frac{k_d}{m}\frac{dy}{dt} + \frac{k_s}{m}y = 0,$$

where m is the mass, k_s is the spring constant, and k_d is the coefficient of damping, as usual. For convenience we will fix the mass $m = 1$ and the spring constant $k_s = 1$. This leaves only the coefficient of damping k_d to be determined. The second-order equation is then

$$\frac{d^2 y}{dt^2} + k_d\frac{dy}{dt} + y = 0,$$

which corresponds to the system

$$\frac{dy}{dt} = v$$

$$\frac{dv}{dt} = -y - k_d v$$

or the matrix system

$$\frac{d\mathbf{Y}}{dt} = \begin{pmatrix} 0 & 1 \\ -1 & -k_d \end{pmatrix}\mathbf{Y},$$

where $\mathbf{Y} = (y, v)$.

We summarize the possible behaviors of the harmonic oscillator by adjusting the parameter k_d from zero to very large (see the exercises for the "unphysical" case of $k_d < 0$).

Undamped oscillator

When $k_d = 0$, the eigenvalues of the system are $\pm i$ and the origin is a center. The solutions oscillate with constant amplitude and period 2π.

Underdamped oscillator

If k_d is just slightly greater than zero, the eigenvalues are equal to

$$-\frac{k_d}{2} \pm \frac{\sqrt{k_d^2 - 4}}{2}.$$

As long as $0 < k_d < 2$, the eigenvalues are complex with negative real part. Hence the origin is a spiral sink and solutions oscillate with decreasing amplitude. The amplitude of the oscillations decreases at the exponential rate of $e^{-tk_d/2}$, and the period of the oscillations is $4\pi/\sqrt{k_d^2 - 4}$.

Critically-damped oscillator

If $k_d = 2$, the system has a single eigenvalue equal to -1. Solutions no longer spiral in the phase plane. They tend to zero at the rate of te^{-t} or e^{-t}.

Overdamped oscillator

If $k_d > 2$, the system has two real, negative eigenvalues given by

$$-\frac{k_d}{2} \pm \frac{\sqrt{k_d^2 - 4}}{2}.$$

Again, no oscillation takes place. All solutions tend to the origin an an exponential rate, and most do so at the rate of

$$e^{(-k_d+\sqrt{k_d^2-4})t/2}.$$

However, a single line of solutions tends to zero at the rate determined by the other eigenvalue.

Conclusion

We have accomplished the goal we set for ourselves at the beginning of Section 3.1. Given a linear system, by doing arithmetic with the coefficients we can:

1. Get a good idea of the possible phase plane pictures for the system.

2. Classify the equilibrium point at (0,0) and describe the possible long-term behaviors of the components of solutions of the system.

3. Describe the qualitative behavior of solutions to initial-value problems.

4. Give explicit formulas for the general solution and solutions to initial-value problems.

The arithmetic that we have to do to accomplish all this is just the computation of the eigenvalues and eigenvectors.

This is a great and useful victory. Many very important "real-world" systems can be modeled very effectively with linear systems. Even when linear systems are not sufficiently accurate and we must use a nonlinear system, linear systems play an important role. It turns out that small parts of nonlinear systems can themselves be modeled by linear systems (that is, a model of a model). This process, known as **linearization**, is one of the most powerful tools of mathematics (see Chapter 4).

We end this section with an application of linear systems to economics (that has nothing to do with Paul and Bob).

A Model of Income and Interest Rates

We consider a simplified "Keynesian" model of the relationship between income and interest rates.* Let

$I(t) =$ national income at time t (total of all salaries for everybody in the country),

$r(t) =$ interest rate at time t (interest charged on loans and mortgages).

The model for the rate of change of the interest rate is

$$\frac{dr}{dt} = kI - \beta r - M,$$

where k, β, and M are parameters and are all positive. The first term, kI, represents the effect on interest rates of the demand for loans. The assumption is that the more money people make, the more they are willing to borrow. Hence an increase in income makes the demand for loans go up, so the interest rate rises. The next term, $-\beta r$, reflects the fact that the higher the interest rate, the less people will want to borrow, and the demand for loans will decrease. Finally, the parameter M represents the "money supply," that is, the amount of money in circulation. The larger the money supply, the more money there is in circulation, the less people need to borrow. The less people borrow, the less demand there is for loans which results in a decrease in the interest rate. Hence, an increase in M also causes dr/dt to decrease.

For the rate of change of income, we use

$$\frac{dI}{dt} = -s(I - T) + (G - T) - \alpha r,$$

where s, T, G, and α are positive parameters. The parameter s gives the fraction of after-tax income, which people save. The more people save, the less money they have available to buy things. This means that consumption goes down, so fewer people work overtime and income decreases. Hence the term $s(I - T)$ has a minus sign. The parameter T is the total amount paid in taxes while G is the total amount spent by the government. The term $G - T$ gives government spending minus taxes. If $G - T > 0$, then the government is spending more than it collects in taxes. Presumably this extra spending appears in people's paychecks and in benefits that the government provides. If $G - T > 0$, then income increases; hence the $G - T$ term is positive. Finally, the term $-\alpha r$ reflects the fact that the larger the interest rate, the less borrowing takes place. The less people borrow, the less they buy, and fewer people are employed making things. So a large interest rate causes a decrease in employment and hence in income.

Putting this together and rearranging terms we get

$$\frac{dI}{dt} = -sI - \alpha r - ((1 - s)T - G)$$

$$\frac{dr}{dt} = kI - \beta r - M,$$

which in matrix notation is

$$\frac{d\mathbf{Y}}{dt} = \mathbf{AY} - \begin{pmatrix} (1 - s)T - G \\ M \end{pmatrix},$$

---

*See Pierre N.V. Tu, *Dynamical Systems: An Introduction with Applications in Economics and Biology, Second Edition*, (Springer-Verlag, Berlin) 1992: p. 107.

where $\mathbf{Y} = (I, r)$ and $\mathbf{A} = \begin{pmatrix} -s & -\alpha \\ k & -\beta \end{pmatrix}$. Provided the determinant of $\mathbf{A}$ is nonzero, this system has only one equilibrium point, (I_0, r_0), given by

$$I_0 = \frac{\alpha M - \beta((1 - s)T - G)}{\beta s + \alpha k}$$

$$r_0 = \frac{-k((1 - s)T - G) - sM}{\beta s + \alpha k}$$

If we define new dependent variables x and y by

$$x = I - I_0$$
$$y = r - r_0,$$

then the new system is

$$\frac{dx}{dt} = -sx - \alpha y$$
$$\frac{dy}{dt} = kx - \beta y.$$

The eigenvalues of this system are $-(s + \beta)/2 \pm \sqrt{(s - \beta)^2 - 4\alpha k}/2$. Hence either both eigenvalues are real and negative or the eigenvalues are complex and the real part is negative. In either case all solutions tend to $x = y = 0$. For the original system, this means that every solution tends to $I = I_0$ and $r = r_0$.

If the eigenvalues are complex, then income and interest rates will oscillate around the equilibrium values with decreasing amplitude. If the eigenvalues are real, then income and interest rates approach equilibrium from a definite direction (see Figures 3.48 and 3.49).

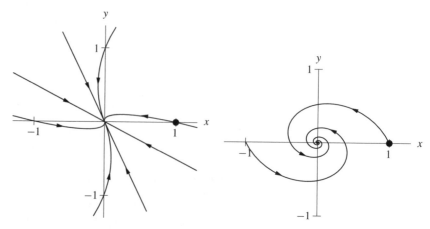

Figure 3.48
Phase planes for cases with real and complex eigenvalues, respectively.

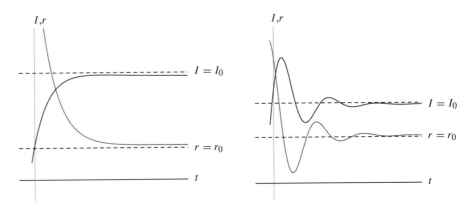

Figure 3.49
Graphs of $I(t)$ and $r(t)$ for the solutions indicated in the phase planes above for the two cases.

Reality check

The model predicts that every solution tend to the equilibrium point. This prediction means that both total income and interest rates always tend toward a fixed constant value as t increases. This is, perhaps unfortunately, not what really happens. Hence there must be other important factors determining interest and income rates that we have not taken into account.

This does not mean that the model is useless. It is possible that the model does reflect the behavior of income and interest during certain periods when these values don't change much. During these periods, the model may help predict how income and interest rates behave near the equilibrium value (see the exercises).

Exercises for Section 3.6

1. Construct a table of the possible linear systems as follows:
   (i) The leftmost column contains the system name (that is, sink, spiral sink, source, and so on) when it has one.
   (ii) The second column contains the condition on the eigenvalues that correspond to this case.
   (iii) The third column contains a small picture of two or more possible phase planes for this system.
   (iv) The fourth column contains $x(t)$- and $y(t)$-graphs of typical solutions indicated in your phase planes.

   [*Hint*: The most complete table will contain 14 cases. Don't forget the double eigenvalue and zero eigenvalue cases.]

2. Construct a table of all the possible harmonic oscillator systems.
   (i) The left-hand column contains the name of the system.
   (ii) The second column contains the condition on the eigenvalues corresponding the system.
   (iii) The third column contains the condition on the parameters m, k_s, and k_d for which the system satisfies the condition on the eigenvalues.

(iv) The fourth column contains the rate that solutions approach the origin and the natural period (if applicable).

(v) The fifth column contains sample phase-plane diagrams.

(vi) The sixth column contains typical $y(t)$- and $v(t)$-graphs for solutions indicated in the phase planes.

In Exercises 3–7, for each linear system:

(a) Compute the eigenvalues and eigenvectors.

(b) Classify the equilibrium point at $(0, 0)$ (that is, say whether it is a sink, source, and so on).

(c) Sketch the phase plane for the system.

(d) Sketch the $x(t)$- and $y(t)$-graphs for the solution with the given initial condition.

3. $\dfrac{d\mathbf{Y}}{dt} = \begin{pmatrix} -5 & 2 \\ 2 & 2 \end{pmatrix} \mathbf{Y}$, with initial condition $\mathbf{Y}_0 = (1, 0)$

4. $\dfrac{d\mathbf{Y}}{dt} = \begin{pmatrix} -1 & 4 \\ -5 & -2 \end{pmatrix} \mathbf{Y}$, with initial condition $\mathbf{Y}_0 = (2, 1)$

5. $\dfrac{d\mathbf{Y}}{dt} = \begin{pmatrix} -4 & 1 \\ 1 & 3 \end{pmatrix} \mathbf{Y}$, with initial conditions $\mathbf{Y}_0 = (0, -2)$

6. $\dfrac{d\mathbf{Y}}{dt} = \begin{pmatrix} 0 & 1 \\ -1 & -4 \end{pmatrix} \mathbf{Y}$, with initial condition $\mathbf{Y}_0 = (2, 0)$

7. $\dfrac{d\mathbf{Y}}{dt} = \begin{pmatrix} 0 & 1 \\ -4 & -1 \end{pmatrix} \mathbf{Y}$, with initial condition $\mathbf{Y}_0 = (0, 1)$

In Exercises 8–12, for each linear system: (which are the same as in Exercises 3–7)

(a) Compute the general solution.

(b) Compute the particular solution with the given initial condition.

(c) Sketch the $x(t)$- and $y(t)$-graphs of the particular solution and compare them with the sketches obtained in the exercises above.

8. $\dfrac{d\mathbf{Y}}{dt} = \begin{pmatrix} -5 & 2 \\ 2 & 2 \end{pmatrix} \mathbf{Y}$, with initial condition $\mathbf{Y}_0 = (1, 0)$

9. $\dfrac{d\mathbf{Y}}{dt} = \begin{pmatrix} -1 & 4 \\ -5 & -2 \end{pmatrix} \mathbf{Y}$, with initial condition $\mathbf{Y}_0 = (2, 1)$

10. $\dfrac{d\mathbf{Y}}{dt} = \begin{pmatrix} -4 & 1 \\ 1 & 3 \end{pmatrix} \mathbf{Y}$, with initial conditions $\mathbf{Y}_0 = (0, -2)$

11. $\dfrac{d\mathbf{Y}}{dt} = \begin{pmatrix} 0 & 1 \\ -1 & -4 \end{pmatrix} \mathbf{Y}$, with initial condition $\mathbf{Y}_0 = (2, 0)$

12. $\dfrac{d\mathbf{Y}}{dt} = \begin{pmatrix} 0 & 1 \\ -4 & -1 \end{pmatrix} \mathbf{Y}$, with initial condition $\mathbf{Y}_0 = (0, 1)$

In Exercises 13–20, consider the harmonic oscillators for each of the choices of values for the parameters m (mass), k_s (spring constant), and k_d (coefficient of damping).

(a) Write the corresponding second-order differential equation and the first-order system for these parameter values.

(b) Find the eigenvalues and eigenvectors of the linear system.

(c) Classify the oscillator (underdamped, overdamped, critically damped, or undamped) and, when appropriate, give the natural period.

(d) Sketch the phase plane of the linear system, including the solution with the given initial conditions.

(e) Sketch the $y(t)$- and $v(t)$-graphs of the solution with the given initial condition.

13. $m = 1$, $k_s = 2$, $k_d = 1$, with initial conditions $y(0) = 2$, $v(0) = 0$

14. $m = 1$, $k_s = 1$, $k_d = 4$, with initial condition $y(0) = 3$, $v(0) = 0$

15. $m = 10$, $k_s = 1$, $k_d = 4$, with initial condition $y(0) = 3$, $v(0) = 0$

16. $m = 2$, $k_s = 2$, $k_d = 3$, with initial condition $y(0) = -2$, $v(0) = 2$

17. $m = 2$, $k_s = 1$, $k_d = 3$, with initial conditions $y(0) = 0$, $v(0) = 3$

18. $m = 9$, $k_s = 1$, $k_d = 6$, with initial conditions $y(0) = 1$, $v(0) = 1$

19. $m = 2$, $k_s = 3$, $k_d = 0$, with initial conditions $y(0) = 2$, $v(0) = -3$

20. $m = 2$, $k_s = 3$, $k_d = 1$, with initial conditions $y(0) = 0$, $v(0) = -3$

In Exercises 21–28, consider the parameter values for harmonic oscillators given below (which are the same as those in Exercises 13–20).

(a) Find the general solution of the corresponding harmonic oscillator system.

(b) Find the particular solution for the given initial condition.

(c) Use the equations for the solution of the initial-value problem to draw the $y(t)$- and $v(t)$-graphs. Compare these graphs to your sketches for the corresponding problem above.

21. $m = 1$, $k_s = 2$, $k_d = 1$, with initial conditions $y(0) = 2$, $v(0) = 0$

22. $m = 1$, $k_s = 1$, $k_d = 4$, with initial condition $y(0) = 3$, $v(0) = 0$

23. $m = 10$, $k_s = 1$, $k_d = 4$, with initial condition $y(0) = 3$, $v(0) = 0$

24. $m = 2$, $k_s = 2$, $k_d = 3$, with initial condition $y(0) = -2$, $v(0) = 2$

25. $m = 2$, $k_s = 1$, $k_d = 3$, with initial conditions $y(0) = 0$, $v(0) = 3$

26. $m = 9$, $k_s = 1$, $k_d = 6$, with initial conditions $y(0) = 1$, $v(0) = 1$

27. $m = 2$, $k_s = 3$, $k_d = 0$, with initial conditions $y(0) = 2$, $v(0) = -3$

28. $m = 2$, $k_s = 3$, $k_d = 1$, with initial conditions $y(0) = 0$, $v(0) = -3$

29. Consider the one-parameter family of linear systems given by

$$\frac{d\mathbf{Y}}{dt} = \begin{pmatrix} a & 3 \\ -3 & 0 \end{pmatrix} \mathbf{Y}.$$

(a) Compute the eigenvalues for this system (they depend on a).

(b) Determine the bifurcation value(s) of a.

(c) Sketch the phase plane of the system for a at each of the bifurcation values and slightly on each side of the bifurcation values.

(d) Write a brief paragraph describing the transition that takes place in the system at each of the bifurcation values. Include in your description any change in the long-term behavior of solutions.

30. Consider the linear system

$$\frac{d\mathbf{Y}}{dt} = \begin{pmatrix} a & 1 \\ 0 & -1 \end{pmatrix} \mathbf{Y}$$

where a is a parameter that varies between -10 and 10.

(a) What are all the possible types of phase planes this system can have as a is varied?

(b) What are the different $x(t)$- and $y(t)$-graphs for the solutions with initial condition $(2, 1)$ as a is varied?

31. Repeat Exercise 30 for the system

$$\frac{d\mathbf{Y}}{dt} = \begin{pmatrix} a & 1 \\ -1 & 2 \end{pmatrix} \mathbf{Y}.$$

32. Repeat Exercise 30 for the system

$$\frac{d\mathbf{Y}}{dt} = \begin{pmatrix} a & -1 \\ 1 & 1 \end{pmatrix} \mathbf{Y}.$$

33. (a) Check that the matrix $\begin{pmatrix} 0 & 1 \\ -1 & 0 \end{pmatrix}$ has purely imaginary complex eigenvalues.

(b) Show that $\begin{pmatrix} \epsilon & 1 \\ -1 & 0 \end{pmatrix}$ has complex eigenvalues with nonzero real part for any sufficiently small number ϵ.

(c) Show that $\begin{pmatrix} 0 & 1+\epsilon \\ -1 & 0 \end{pmatrix}$ has complex eigenvalues with zero real part for any sufficiently small number ϵ.

34. Consider the matrix $\begin{pmatrix} 6 & 3 \\ 2 & 1 \end{pmatrix}$.

(a) Show that this matrix has a zero eigenvalue.

(b) Show that adding a small quantity ϵ to any one of the entries of this matrix gives a matrix that does not have zero as an eigenvalue.

35. Suppose material scientists discover a new type of fluid, called "magic-finger fluid", that has the following property: If an object is moving through magic-finger fluid, then instead of applying a force that slows the object (damping), the object is accelerated in the direction it is traveling! Moreover, suppose the amount of force the object feels accelerating it is proportional to its velocity (with proportionality constant k_{mf}), so the faster an object is going, the faster the magic-finger fluid accelerates the object in the direction it is moving.

(a) Write a second-order differential equation for a mass-spring system moving in magic-finger fluid. Write this second-order equation as a first-order system.

(b) Classify the possible behaviors of the linear system you constructed above.

36. An automobile's suspension system is essentially large springs with damping. When the car hits a bump, the springs are compressed. It is not unreasonable to use the harmonic oscillator system of differential equations to model the up and down motion, where $y(t)$ measures the amount the springs are stretched or compressed and $v(t)$ is the vertical velocity of the bouncing car.

Suppose that you are working for a company that designs suspension systems for cars. One day, your boss comes to you with the results of a market research survey indicating that most people want shock absorbers that "bounce twice" when compressed, then gradually return to their equilibrium position from above. That is, when the car hits a bump, the springs are compressed. Ideally they should expand, compress and expand, then settle back to the rest position. After the initial bump, the spring would pass through its rest position three times and approach the rest position from the expanded state.

(a) Sketch a graph of the position of the spring after hitting a bump, where $y(t)$ denotes the state of the spring at time t, $y > 0$ corresponds to the spring being stretched, and $y < 0$ corresponds to the spring being compressed.

(b) Explain (politely) why the behavior pictured in the figure is impossible with standard suspension systems that are accurately modeled by the harmonic oscillator system.

(c) What is your suggestion for a choice of a harmonic oscillator system that most closely approximates the desired behavior? Justify your answer with an essay.

37. Let $x(t)$ and $y(t)$ denote the populations of two interacting species of fish living in a small lake. Suppose we have determined that a good model for how these populations evolve is given by the linear system

$$\frac{dx}{dt} = 2x - y$$

$$\frac{dy}{dt} = 6x - 3y.$$

(A population of less than zero is not physically meaningful, so if either population ever reaches zero, we say that the corresponding species has become extinct.)

(a) If the initial populations are $x(0) = 200$ and $y(0) = 300$, what will happen to the populations as t increases?

(b) If the initial populations are $x(0) = 500$ and $y(0) = 100$, what will happen to the populations as t increases?

(c) Suppose the system is altered just a little so that the x species reproduces just a little bit faster (that is, the equation $dx/dt = 2x - y$ becomes, for example

$dx/dt = 2.01x - y$), but no changes occur in the y species or in the way the species interact. For the initial condition $x(0) = 200$ and $y(0) = 300$, what will happen to the population as t increases?

(d) Suppose the original model is altered so that the y species hinders the growth of the x species just a little less (that is, the equation $dx/dt = 2x - y$ becomes, for example, $dx/dt = 2x - .99y$), but no changes occur in the y species. For the initial condition $x(0) = 200$ and $y(0) = 300$, what will happen to the population as t increases?

38. Recall the model of Paul's and Bob's profits from their CD stores:

$$\frac{d\mathbf{Y}}{dt} = \mathbf{AY}.$$

Suppose $\mathbf{A}$ is the matrix $\begin{pmatrix} 1 & -2 \\ 3 & -1 \end{pmatrix}$.

(a) Sketch the phase plane for this system and describe how the profits of the two stores behave as t increases.

(b) If Bob or Paul make a small change in how they do business, the parameters in the model change just a little. What are all the possible types of behavior for the new system? (Draw the phase planes and describe how the profits of the stores change as t increases for each possible behavior.)

Exercises 39–42 refer to the model of income and interest rates on page 317, repeated here for convenience,

$$\frac{dI}{dt} = -sI - \alpha r - ((1 - s)T - G)$$

$$\frac{dr}{dt} = kI - \beta r - M.$$

39. Compute the equilibrium point for the model above. If all of s, α, k, and β are positive, verify that this is the only equilibrium point.

40.   **(a)** Compute the eigenvalues and eigenvectors for the system

$$\frac{dx}{dt} = -sx - \alpha y$$
$$\frac{dy}{dt} = kx - \beta y.$$

Note that your answer will be in terms of the parameters s, α, k, and β.

(b) Show that either the eigenvalues are both real and negative or that they are complex with negative real part [*Hint*: Remember, s, α, k and β, are all positive.]

41.   **(a)** What does the model predict will happen *over the long term* to income and interest rates if the parameters s (fraction of after-tax income saved), α (effect of interest rates on income), k ("transaction demand"), or β (effect of interest rates on interest rates) are increased or decreased? [*Hint*: Make a table with rows T, G, and M, and columns "increase" and "decrease". Enter the effect on I and r for each case.]

(b) What does the model predict will happen to the long income and interest rates if the parameters T (taxes), G (government spending) or M (money supply) are increased or decreased. [*Hint*: Make a table with rows T, G and M and columns "increase" and "decrease." Enter the effect on I and r for each case.]

(c) Suppose that income is very low and interest rates are very high. *If* you believe this model and you are the president (in control of taxes and government spending), what parameters would you adjust and why?

(d) Suppose that income is very low but interest rates are very high. *If* you believe this model and you are chairman of the Federal Reserve Bank (in control of the money supply), what would you do?

Remark: Actually the chairman of the Federal Reserve Bank also has considerable direct control of interest rates as well as the money supply.

42. Suppose the income and interest rates are not at the equilibrium values. If all the other parameters are fixed, is it possible to adjust the savings rate s so that the system does not oscillate as it approaches equilibrium?

3.7 LINEAR SYSTEMS IN THREE DIMENSIONS

So far we have studied linear systems with two dependent variables. For these systems, the behavior of solutions and the nature of the phase plane can be determined by computing the eigenvalues and eigenvectors of the 2×2 coefficient matrix. Once we have found two solutions with linearly independent initial conditions we can give the general solution.

In this section we show that the same is true for linear systems with three dependent variables. The eigenvalues and eigenvectors of the 3×3 coefficient matrix determine the behavior of solutions and the general solution. Three-dimensional linear systems have three eigenvalues, so the list of possible qualitatively distinct phase spaces is longer than for planar systems. Since we must deal with three scalar equations rather than two, the arithmetic can quickly become much more involved. The reader may wish to seek out software or a calculator capable of handling 3×3 matrices.

Independence and the Linearity Principle

The general form of a linear system with three dependent variables is

$$\frac{dx}{dt} = a_{11}x + a_{12}y + a_{13}z$$

$$\frac{dy}{dt} = a_{21}x + a_{22}y + a_{23}z$$

$$\frac{dz}{dt} = a_{31}x + a_{32}y + a_{33}z$$

where x, y and z are the dependent variables and the coefficients a_{ij}, $(i, j = 1, 2, 3)$, are constants. We can write this system in matrix form as

$$\frac{d\mathbf{Y}}{dt} = \mathbf{AY}$$

where $\mathbf{A}$ is the coefficient matrix

$$\mathbf{A} = \begin{pmatrix} a_{11} & a_{12} & a_{13} \\ a_{21} & a_{22} & a_{23} \\ a_{31} & a_{32} & a_{33} \end{pmatrix}$$

and $\mathbf{Y}$ is the vector of dependent variables,

$$\mathbf{Y} = \begin{pmatrix} x \\ y \\ z \end{pmatrix}.$$

To specify an initial condition for such a system we must give three numbers, x_0, y_0 and z_0.

The Linearity Principle holds for linear systems in all dimensions, so if $\mathbf{Y}_1(t)$ and $\mathbf{Y}_2(t)$ are solutions, then $k_1 \mathbf{Y}_1(t) + k_2 \mathbf{Y}_2(t)$ is also a solution for any constants k_1 and k_2.

Suppose $\mathbf{Y}_1(t)$, $\mathbf{Y}_2(t)$ and $\mathbf{Y}_3(t)$ are three solutions of the linear system

$$\frac{d\mathbf{Y}}{dt} = \mathbf{AY}.$$

If, for any point (x_0, y_0, z_0), there exist constants k_1, k_2 and k_3 such that

$$k_1 \mathbf{Y}_1(0) + k_2 \mathbf{Y}_2(0) + k_3 \mathbf{Y}_3(0) = (x_0, y_0, z_0),$$

then the general solution of the system is

$$\mathbf{Y}(t) = k_1 \mathbf{Y}_1(t) + k_2 \mathbf{Y}_2(t) + k_3 \mathbf{Y}_3(t).$$

In order for three solutions $\mathbf{Y}_1(t)$, $\mathbf{Y}_2(t)$ and $\mathbf{Y}_3(t)$ to give the general solution, the three vectors $\mathbf{Y}_1(0)$, $\mathbf{Y}_2(0)$ and $\mathbf{Y}_3(0)$ must point in "different directions", that is, no one of them can be in the plane through the origin and the other two. In this case the vectors $\mathbf{Y}_1(0)$, $\mathbf{Y}_2(0)$ and $\mathbf{Y}_3(0)$ (and the corresponding solutions) are said to be *linearly independent*. We present an algebraic technique for checking linear independence in the exercises (see Exercises 2 and 3).

An example

Consider the linear system

$$\frac{d\mathbf{Y}}{dt} = \mathbf{AY} = \begin{pmatrix} 0 & 0.1 & 0 \\ 0 & 0 & 0.2 \\ 0.4 & 0 & 0 \end{pmatrix} \begin{pmatrix} x \\ y \\ z \end{pmatrix}.$$

We can check that the curves

$$\mathbf{Y}_1(t) = e^{0.2t} \begin{pmatrix} 1 \\ 2 \\ 2 \end{pmatrix}$$

$$\mathbf{Y}_2(t) = e^{-0.1t} \begin{pmatrix} -\cos(\sqrt{0.03}t) - \sqrt{3}\sin(\sqrt{0.03}t) \\ -2\cos(\sqrt{0.03}t) + 2\sqrt{3}\sin(\sqrt{0.03}t) \\ 4\cos(\sqrt{0.03}t) \end{pmatrix}$$

$$\mathbf{Y}_3(t) = e^{-0.1t} \begin{pmatrix} -\sin(\sqrt{0.03}t) + \sqrt{3}\cos(\sqrt{0.03}t) \\ -2\sin(\sqrt{0.03}t) - 2\sqrt{3}\cos(\sqrt{0.03}t) \\ 4\sin(\sqrt{0.03}t) \end{pmatrix}$$

are solutions by substituting them into the differential equation. For example,

$$\frac{d\mathbf{Y}_1}{dt} = e^{0.2t} \begin{pmatrix} 0.2 \\ 0.4 \\ 0.4 \end{pmatrix}$$

and

$$\mathbf{A}\mathbf{Y}_1(t) = \begin{pmatrix} 0 & 0.1 & 0 \\ 0 & 0 & 0.2 \\ 0.4 & 0 & 0 \end{pmatrix} e^{0.2t} \begin{pmatrix} 1 \\ 2 \\ 2 \end{pmatrix} = e^{0.2t} \begin{pmatrix} 0.2 \\ 0.4 \\ 0.4 \end{pmatrix},$$

so $\mathbf{Y}_1(t)$ is a solution. The other two curves can be checked similarly (see Exercise 1). We can sketch these solution curves in the three-dimensional phase space (see Figure 3.50).

The initial conditions of these three solutions are $\mathbf{Y}_1(0) = (1, 2, 2)$, $\mathbf{Y}_2(0) = (-1, -2, 4)$, and $\mathbf{Y}_3(0) = (\sqrt{3}, -2\sqrt{3}, 0)$. These vectors are shown in Figure 3.51 where we can see that none of them is in the plane determined by the other two, hence they are linearly independent.

To find, for example, the solution $\mathbf{Y}(t)$ with initial position $\mathbf{Y}(0) = (2, 1, 3)$, we must solve

$$k_1\mathbf{Y}_1(0) + k_2\mathbf{Y}_2(0) + k_3\mathbf{Y}_3(0) = (2, 1, 3)$$

or

$$k_1 - k_2 + \sqrt{3}k_3 = 2$$
$$2k_1 - 2k_2 - 2\sqrt{3}k_3 = 1$$
$$2k_1 + 4k_2 = 3$$

This gives $k_1 = 4/3$, $k_2 = 1/12$ and $k_3 = \sqrt{3}/4$, so the solution we seek is

$$\mathbf{Y}(t) = (4/3)\mathbf{Y}_1(t) + (1/12)\mathbf{Y}_2(t) + (\sqrt{3}/4)\mathbf{Y}_3(t).$$

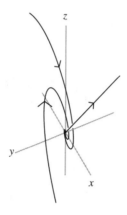

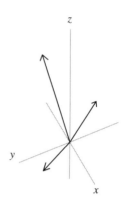

Figure 3.50
Solutions $\mathbf{Y}_1(t)$, $\mathbf{Y}_2(t)$ and $\mathbf{Y}_3(t)$.

Figure 3.51
Vectors $\mathbf{Y}_1(0) = (1, 2, 2)$,
$\mathbf{Y}_2(0) = (-1, -2, 4)$ and
$\mathbf{Y}_3(0) = (\sqrt{3}, -2\sqrt{3}, 0)$ in xyz-space.

Eigenvalues and Eigenvectors

Finding solutions of systems with three dependent variables is the same as for two variables. We begin by finding eigenvalues and eigenvectors. Suppose we are given a linear system

$$\frac{d\mathbf{Y}}{dt} = \mathbf{A}\mathbf{Y}$$

where $\mathbf{A}$ is a 3×3 matrix of coefficients and $\mathbf{Y} = (x, y, z)$. An *eigenvector* for the matrix $\mathbf{A}$ is a nonzero vector $\mathbf{Y}_1$ such that

$$\mathbf{A}\mathbf{Y}_1 = \lambda_1 \mathbf{Y}_1,$$

where λ_1 is the *eigenvalue* for $\mathbf{Y}_1$. If $\mathbf{Y}_1$ is an eigenvector for $\mathbf{A}$ with eigenvalue λ_1, then

$$\mathbf{Y}_1(t) = e^{\lambda_1 t} \mathbf{Y}_1$$

is a solution of the linear system.

Finding eigenvalues and eigenvectors for a 3×3 matrix

$$\mathbf{A} = \begin{pmatrix} a_{11} & a_{12} & a_{13} \\ a_{21} & a_{22} & a_{23} \\ a_{31} & a_{32} & a_{33} \end{pmatrix}$$

is very similar to two-dimensional systems, only requiring more arithmetic. In particular, we need the formula for the determinant of the 3×3 matrix.

Definition: The *determinant* of the matrix $\mathbf{A}$ is

$$\det \mathbf{A} = a_{11}(a_{22}a_{33} - a_{23}a_{32}) - a_{12}(a_{21}a_{33} - a_{23}a_{31}) + a_{13}(a_{21}a_{32} - a_{22}a_{32}).$$

Using the 3×3 identity matrix

$$\mathbf{I} = \begin{pmatrix} 1 & 0 & 0 \\ 0 & 1 & 0 \\ 0 & 0 & 1 \end{pmatrix},$$

we obtain the *characteristic polynomial* of $\mathbf{A}$ as

$$\det(\mathbf{A} - \lambda \mathbf{I}) = \det \begin{pmatrix} a_{11} - \lambda & a_{12} & a_{13} \\ a_{21} & a_{22} - \lambda & a_{23} \\ a_{31} & a_{32} & a_{33} - \lambda \end{pmatrix}.$$

As in the two-dimensional case we have,

Theorem *The eigenvalues of a 3×3 matrix $\mathbf{A}$ are the roots of its characteristic polynomial.*

To find the eigenvalues of a 3×3 matrix, we must find the roots of a cubic polynomial. This is not as easy as finding the roots of a quadratic. While there is a "cubic equation" analogous to the quadratic equation for finding the roots of a cubic, it is quite complicated. (It is used by computer algebra packages to give exact values of roots of cubics.) However, in cases where the cubic does not easily factor, we frequently turn to numerical techniques such as Newton's method for finding roots.

To find the corresponding eigenvectors, we must solve a system of three linear equations with three unknowns. Luckily, there are many examples of systems that illustrate the possible behaviors in three dimensions and for which the arithmetic is manageable.

An example

The simplest type of 3×3 matrix is a "diagonal" matrix — the only nonzero terms lie on the diagonal. For example, consider the system

$$\frac{d\mathbf{Y}}{dt} = \mathbf{A}\mathbf{Y} = \begin{pmatrix} -3 & 0 & 0 \\ 0 & -1 & 0 \\ 0 & 0 & -2 \end{pmatrix} \begin{pmatrix} x \\ y \\ z \end{pmatrix}.$$

The characteristic polynomial of $\mathbf{A}$ is $(-3 - \lambda)(-1 - \lambda)(-2 - \lambda)$, which is simple because so many of the coefficients of $\mathbf{A}$ are zero. The eigenvalues are the roots of this polynomial, that is, the solutions of

$$(-3 - \lambda)(-1 - \lambda)(-2 - \lambda) = 0.$$

Thus, the eigenvalues are $\lambda_1 = -3$, $\lambda_2 = -1$ and $\lambda_3 = -2$.

Finding the corresponding eigenvectors is also not too hard. For $\lambda_1 = -3$, we must solve

$$\mathbf{A}\mathbf{Y}_1 = -3\mathbf{Y}_1$$

for $\mathbf{Y}_1 = (x_1, y_1, z_1)$. That is,

$$
\begin{pmatrix} -3 & 0 & 0 \\ 0 & -1 & 0 \\ 0 & 0 & -2 \end{pmatrix} \begin{pmatrix} x_1 \\ y_1 \\ z_1 \end{pmatrix} = \begin{pmatrix} -3x_1 \\ -y_1 \\ -2z_1 \end{pmatrix} = -3 \begin{pmatrix} x_1 \\ y_1 \\ z_1 \end{pmatrix}.
$$

Solutions of this system of three equations with three unknowns are $y_1 = z_1 = 0$ and x_1 may have any (nonzero) value. So, in particular, $(1, 0, 0)$ is an eigenvector for $\lambda_1 = -3$. Similarly, we find that $(0, 1, 0)$ and $(0, 0, 1)$ are eigenvalues for $\lambda_2 = -1$ and $\lambda_3 = -2$, respectively. Note that $(1, 0, 0)$, $(0, 1, 0)$ and $(0, 0, 1)$ are linearly independent.

From these eigenvalues and eigenvectors we can construct solutions of the system

$$
\mathbf{Y}_1(t) = e^{-3t} \begin{pmatrix} 1 \\ 0 \\ 0 \end{pmatrix} = \begin{pmatrix} e^{-3t} \\ 0 \\ 0 \end{pmatrix},
$$

$$
\mathbf{Y}_2(t) = e^{-t} \begin{pmatrix} 0 \\ 1 \\ 0 \end{pmatrix} = \begin{pmatrix} 0 \\ e^{-t} \\ 0 \end{pmatrix},
$$

and

$$
\mathbf{Y}_3(t) = e^{-2t} \begin{pmatrix} 0 \\ 0 \\ 1 \end{pmatrix} = \begin{pmatrix} 0 \\ 0 \\ e^{-2t} \end{pmatrix}.
$$

Because this system is diagonal, we could have gotten this far "by inspection." If we write the system in components

$$
\frac{dx}{dt} = -3x
$$
$$
\frac{dy}{dt} = -y
$$
$$
\frac{dz}{dt} = -2z,
$$

we see that dx/dt depends only on x, dy/dt depends only on y and dz/dt depends only on z. We say that the system "decouples" since each coordinate may be dealt with independently. The solutions of these differential equations (by now) are easy to find.

Now that we have three independent solutions, we can solve any initial-value problem for this system. For example, to find the solution $\mathbf{Y}(t)$ with $\mathbf{Y}(0) = (2, 1, 2)$, we must find constants k_1, k_2 and k_3 such that

$$(2, 1, 2) = k_1 \mathbf{Y}_1(0) + k_2 \mathbf{Y}_2(0) + k_3 \mathbf{Y}_3(0).$$

So $k_1 = 2, k_2 = 1$ and $k_3 = 2$, and $\mathbf{Y}(t) = (2e^{-3t}, e^{-t}, 2e^{-2t})$ is the required solution.

Figure 3.52 is a sketch of the phase space. Note that the coordinate axes are lines of eigenvectors, so they form straight-line solutions. Since all three of the eigenvalues are negative, solutions along all three of the axes tend toward the origin. Since every other solution can be made up as a linear combination of the solutions on the axes, all solutions must tend to the origin and it is natural to call the origin a sink.

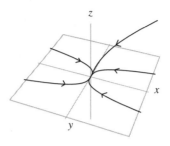

Figure 3.52
Phase space for $d\mathbf{Y}/dt = \mathbf{A}\mathbf{Y}$ for the
diagonal matrix $\mathbf{A}$ above.

Another example: three-dimensional behavior

Before giving the classification of linear systems in three dimensions, we give one more example which has behavior very different from what is possible in two dimensions.

Consider the system

$$\frac{d\mathbf{Y}}{dt} = \mathbf{C}\mathbf{Y} = \begin{pmatrix} 0.1 & -1 & 0 \\ 1 & 0.1 & 0 \\ 0 & 0 & -0.2 \end{pmatrix} \begin{pmatrix} x \\ y \\ z \end{pmatrix}.$$

The characteristic polynomial of $\mathbf{C}$ is

$$((0.1 - \lambda)(0.1 - \lambda) + 1)(-0.2 - \lambda) = (\lambda^2 - 0.2\lambda + 1.01)(-0.2 - \lambda),$$

so the eigenvalues are $\lambda_1 = -0.2$, $\lambda_2 = 0.1 + i$ and $\lambda_3 = 0.1 - i$. Corresponding to the real negative eigenvalue λ_1 we expect to see a line of solutions that approach the origin in the phase space. By analogy to the two-dimensional case, the complex eigenvalues with positive real part correspond to solutions that spiral away from the origin. This is a "spiral saddle," which is not possible in two dimensions.

We could find the eigenvectors associated with each eigenvalue as above and find the general solution. The eigenvectors for the complex eigenvalues are complex and to find the real solutions we would have to take real and imaginary parts just like in two dimensions (see Exercises). However, we are lucky again and this system also decouples into

$$\frac{dx}{dt} = x - 2y$$
$$\frac{dy}{dt} = 2x + y$$

and

$$\frac{dz}{dt} = -z.$$

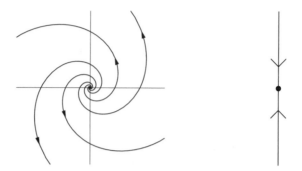

Figure 3.53
Phase plane for xy-system and phase line for z.

In the xy-plane, the eigenvalues are $1 \pm 2i$ so the origin is a spiral source. Along the z axis, all solutions tend toward zero as time increases (see Figure 3.53).

Combining these pictures we obtain a sketch of the three-dimensional phase space. Note that the z-coordinate of each solution decreases toward zero while, in the xy-plane, solutions spiral away from the origin (see Figure 3.54).

Figure 3.54
Phase space for $d\mathbf{Y}/dt = \mathbf{CY}$.

Classification of Three-Dimensional Linear Systems

While there are more possible types of phase space pictures for three-dimensional linear systems than for two dimensions, the list is still finite. Just as for two dimensions, the nature of the system is determined by the eigenvalues. Real eigenvalues correspond to straight-line solutions that tend toward the origin if the eigenvalue is negative and away from the origin if the eigenvalue is positive. Complex eigenvalues correspond to spiraling. Negative real parts indicate spiraling toward the origin while positive real parts indicate spiraling away from the origin.

Since the characteristic polynomial is a cubic, there are three eigenvalues (which might not all be distinct if there are repeated roots). It is always the case that at least one of the eigenvalues is real. The other two may be real or a complex conjugate pair (see exercises).

The most important types of three-dimensional linear systems can be divided into three categories: sinks, sources and saddles. Examples of the other cases (which include systems with double eigenvalues and zero eigenvalues) are given in the exercises.

Sinks

We call the equilibrium point at the origin a sink if all solutions tend toward it as time increases. If all three eigenvalues are real and negative, then there are three straight lines of solutions, all of which tend toward the origin. Since every other solution is a linear combination of these solutions, all solutions tend to the origin as time increase (see Figure 3.52).

The other possibility for a sink is to have one real negative eigenvalue and two complex eigenvalues with negative real parts. This means that there is one straight line of solutions tending to the origin and a plane of solutions which spiral toward the origin. All other solutions exhibit both of these behaviors (see Figure 3.55).

Sources

There are two possibilities for sources as well. We can either have three real and positive eigenvalues or one real positive eigenvalue and a complex conjugate pair with positive real parts. An example of such a phase space is given in Figure 3.56. Note that this system looks just like the sink in Figure 3.55 except the directions of the arrows have been reversed, so solutions move away form the origin as time increases.

Saddles

The equilibrium point at the origin is a saddle if, as time increases to infinity, some solutions tend toward it while other solutions move away from it. This can occur in

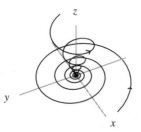

Figure 3.55
Example phase space for spiral sink.

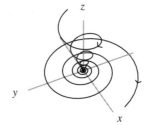

Figure 3.56
Example phase space for spiral source.

four different ways. If all the eigenvalues are real then we could have one positive and two negative or two positive and one negative. In the first case, one positive and two negative, there is one straight line of solutions which tend away from the origin as time increases and a plane of solutions which tend toward the origin as time increases. In the other case, two positive and one negative, there is a plane of solutions which tend away from the origin as time increases and a line of solutions which tend toward the origin as time increases. In both cases, all other solutions will eventually move away from the origin as time increases or decreases (see Figure 3.57).

The other two cases occur if there is only one real eigenvalue and the other two are a complex conjugate pair. If the real eigenvalue is negative and the real parts of the complex eigenvalues are positive, then as time increases, there is a straight line of solutions that tend toward the origin and a plane of solutions that tend away from it. All other solutions are a combination of these behaviors, so as time increases, they spiral around the straight line of solutions in ever widening loops (see Figure 3.54). The other possibility is that the real eigenvalue is positive and the complex eigenvalues have negative real part. In this case there is a straight line of solutions that tend away from the origin as time increases and a plane of solutions that spiral toward the origin as time increases. Every other solution spirals around the straight line of solutions while moving away from the origin (see Figure 3.58).

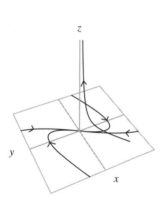

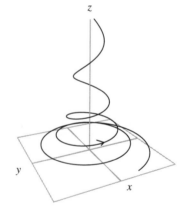

Figure 3.57
Example of a saddle with one positive and two negative eigenvalues .

Figure 3.58
Example of a saddle with one real eigenvalue and a complex conjugate pair of eigenvalues.

One more example

We end this section with one more example. All of the examples so far have been systems that decouple into systems of smaller dimension. Sadly, the general case is not so simple. The system below doesn't look too complicated because the coefficient matrix has many zero entries, however, it does not immediately decouple into lower dimensional systems.

Consider the system

$$\frac{d\mathbf{Y}}{dt} = \mathbf{AY} = \begin{pmatrix} 0 & 0.1 & 0 \\ 0 & 0 & 0.2 \\ 0.4 & 0 & 0 \end{pmatrix} \begin{pmatrix} x \\ y \\ z \end{pmatrix}.$$

The characteristic polynomial for $\mathbf{A}$ is $-\lambda^3 + 0.008$, so the eigenvalues are the solutions of

$$-\lambda^3 + 0.008 = 0.$$

That is, the eigenvalues are the cube roots of 0.008. Every number has three cube roots if we consider complex as well as real roots. The cube roots of 0.008 are $\lambda_1 = 0.2$, $\lambda_2 = 0.2e^{i2\pi/3}$ and $\lambda_3 = 0.2e^{-i2\pi/3}$. The last two eigenvalues may be written as $\lambda_2 = -0.1 + i\sqrt{0.03}$ and $\lambda_3 = -0.1 - i\sqrt{0.03}$.

This system is a saddle with one positive real eigenvalue and a complex conjugate pair of eigenvalues with negative real parts. Solutions spiral tightly around the line of eigenvectors associated to the eigenvalue $\lambda_1 = 2$. In order to sketch the phase plane, we must find the eigenvectors for these eigenvalues.

For $\lambda_1 = 2$, the eigenvalues are solutions of

$$\mathbf{AY}_1 = 2\mathbf{Y}_1$$

or

$$0.1y_1 = 2x_1$$
$$0.2z_1 = 2y_1$$
$$0.4x_1 = 2z_1.$$

In particular, $\mathbf{Y}_1 = (1/2, 1, 1)$ is one such eigenvector. This gives the direction of the straight line of solutions.

To find the plane of solutions that spiral toward the origin, we must find the eigenvectors for $\lambda_2 = -0.1 + i\sqrt{0.03}$. In other words, we must solve

$$\mathbf{AY}_2 = (-0.1 + i\sqrt{0.03})\mathbf{Y}_2,$$

for $\mathbf{Y}_2$. In other words,

$$y_2 = (-0.1 + i\sqrt{0.03})x_2$$
$$2z_2 = (-0.1 + i\sqrt{0.03})y_2$$
$$4x_2 = (-0.1 + i\sqrt{0.03})z_2.$$

This gives $\mathbf{Y}_2 = (-1 + i\sqrt{3}, -2 - i2\sqrt{3}, 4)$ as an eigenvector for λ_2. The corresponding solution to the system is

$$\mathbf{Y}_2(t) = e^{(-0.1 + i\sqrt{0.03})t}(-1 + i\sqrt{3}, -2 - i2\sqrt{3}, 4).$$

We can convert this into two real-valued solutions by taking real and imaginary parts. Since our goal is to find the plane on which solutions spiral, we need only look at the initial point $\mathbf{Y}_2(0) = (-1 + i\sqrt{3}, -2 - i2\sqrt{3}, 4)$. The initial points of the real and imaginary parts are $(-1, -2, 4)$ and $(\sqrt{3}, -2\sqrt{3}, 0)$ respectively. The plane on which solutions spiral toward the origin is the plane made up of all linear combinations of these two vectors. We can use this information to give a fairly accurate sketch of the phase space of this system (see Figure 3.50). We also sketch the graphs of the coordinate functions for one solution (see Figures 3.59 and 3.60). Note that for the example solution shown, all three coordinates tend to infinity as t increases because the eigenvector for the eigenvalue λ_1 has nonzero components for all three variables.

Three linearly independent solutions of this system are given above in the first example of this section (see page 327). We can see from this example that linear systems in three dimensions can be quite complicated (even when many of the coefficients are zero). However, the qualitative behavior is still determined by the eigenvalues, so it is possible to classify these systems without completely solving them.

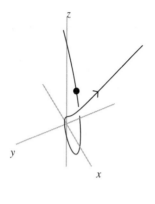

Figure 3.59
Phase space for system $d\mathbf{Y}/dt = \mathbf{AY}$.

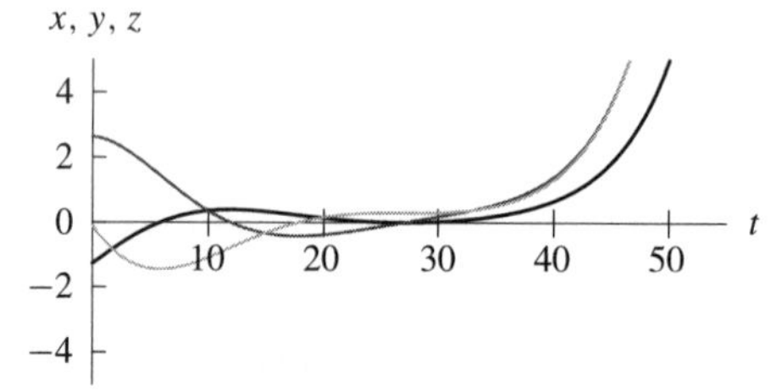

Figure 3.60
Graphs of $x(t)$, $y(t)$ and $z(t)$ for the indicated solution in Figure 3.59.

Exercises for Section 3.7

1. Consider the linear system

$$\frac{d\mathbf{Y}}{dt} = \mathbf{AY} = \begin{pmatrix} 0 & 0.1 & 0 \\ 0 & 0 & 0.2 \\ 0.4 & 0 & 0 \end{pmatrix} \begin{pmatrix} x \\ y \\ z \end{pmatrix}.$$

Check that the functions

$$\mathbf{Y}_2(t) = e^{-0.1t} \begin{pmatrix} -\cos(\sqrt{0.03}t) - \sqrt{3}\sin(\sqrt{0.03}t) \\ -2\cos(\sqrt{0.03}t) + 2\sqrt{3}\sin(\sqrt{0.03}t) \\ 4\cos(\sqrt{0.03}t) \end{pmatrix}$$

and

$$\mathbf{Y}_3(t) = e^{-0.1t} \begin{pmatrix} -\sin(\sqrt{0.03}t) + \sqrt{3}\cos(\sqrt{0.03}t) \\ -2\sin(\sqrt{0.03}t) - 2\sqrt{3}\cos(\sqrt{0.03}t) \\ 4\sin(\sqrt{0.03}t) \end{pmatrix}$$

are solutions to the system.

2. If a vector $\mathbf{Y}_3$ lies in the plane determined by the two vectors $\mathbf{Y}_1$ and $\mathbf{Y}_2$, then we can write $\mathbf{Y}_3$ as a linear combination of $\mathbf{Y}_1$ and $\mathbf{Y}_2$. That is,

$$\mathbf{Y}_3 = k_1\mathbf{Y}_1 + k_2\mathbf{Y}_2$$

for some constants k_1 and k_2. But then

$$k_1 \mathbf{Y}_1 + k_2 \mathbf{Y}_2 - \mathbf{Y}_3 = (0, 0, 0).$$

Show that, if

$$k_1 \mathbf{Y}_1 + k_2 \mathbf{Y}_2 - k_3 \mathbf{Y}_3 = (0, 0, 0)$$

with not all of k_1, k_2 and $k_3 = 0$, then the vectors are not linearly independent. [*Hint:* Start by assuming that $k_3 \neq 0$ and show that $\mathbf{Y}_3$ is in the plane determined by $\mathbf{Y}_1$ and $\mathbf{Y}_2$. Then treat the other cases.]

Remark: This computation leads to the theorem that three vectors $\mathbf{Y}_1$, $\mathbf{Y}_2$ and $\mathbf{Y}_3$ are linearly independent if and only if the only solution of

$$k_1 \mathbf{Y}_1 + k_2 \mathbf{Y}_2 + k_3 \mathbf{Y}_3 = (0, 0, 0)$$

is $k_1 = k_2 = k_3 = 0$.

3. Using the technique of the previous problem, determine whether or not the following sets of three vectors are linearly independent.

(a) $(1, 2, 1)$, $(1, 3, 1)$, $(1, 4, 1)$.

(b) $(2, 0, -1)$, $(3, 2, 2)$, $(1, -2, -3)$.

(c) $(1, 2, 0)$, $(0, 1, 2)$, $(2, 0, 1)$.

(d) $(-3, \pi, 1)$, $(0, 1, 0)$, $(-2, -2, -2)$.

In Exercises 4–7, we consider the linear system

$$\frac{d\mathbf{Y}}{dt} = \mathbf{AY}$$

with the given coefficient matrix $\mathbf{A}$. Each of these system decouples into a two-dimensional system and a one-dimensional system. For each exercise,

(a) compute the eigenvalues;

(b) determine how the system decouples;

(c) sketch the two-dimensional phase plane and one-dimensional phase line for the decoupled systems; and

(d) sketch the full phase space.

4. $\mathbf{A} = \begin{pmatrix} 0 & 1 & 0 \\ -1 & 0 & 0 \\ 0 & 0 & 2 \end{pmatrix}$ **5.** $\mathbf{A} = \begin{pmatrix} -2 & 3 & 0 \\ 3 & -2 & 0 \\ 0 & 0 & -1 \end{pmatrix}$

6. $\mathbf{A} = \begin{pmatrix} 1 & 0 & 3 \\ 0 & -1 & 0 \\ -3 & 0 & 1 \end{pmatrix}$ **7.** $\mathbf{A} = \begin{pmatrix} 1 & 0 & 0 \\ 0 & 2 & -1 \\ 0 & -1 & 2 \end{pmatrix}$

Exercises 8–9 consider the properties of the cubic polynomial

$$p(\lambda) = \alpha \lambda^3 + \beta \lambda^2 + \gamma \lambda + \delta,$$

where α, β, γ and δ are real numbers.

8. **(a)** Show that, if α is positive, then the limit of $p(\lambda)$ as $\lambda \to \infty$ is ∞ and the limit of $p(\lambda)$ as $\lambda \to -\infty$ is $-\infty$.

(b) Show that, if α is negative, then the limit of $p(\lambda)$ as $\lambda \to \infty$ is $-\infty$ and the limit of $p(\lambda)$ as $\lambda \to -\infty$ is ∞.

(c) Using the above, show that $p(\lambda)$ must have at least one real root (that is, at least one real number λ_0 such that $p(\lambda_0)=0$). [*Hint*: Look at the graph of $p(\lambda)$.]

9. Suppose $a + ib$ is a root of $p(\lambda)$ (so $p(a + ib) = 0$). Show that $a - ib$ is also a root. [Hint: Remember that a complex number is zero if and only if both its real and imaginary parts are zero and compute $p(a + ib)$ and $p(a - ib)$.]

In Exercises 10–13, we consider the linear system

$$\frac{d\mathbf{Y}}{dt} = \mathbf{BY}$$

with the given coefficient matrix. These systems do not fit into the classification of the most common types of systems given in the text. However, the equations for dx/dt and dy/dt decouple from the dz/dt. For each of these systems,

(a) compute the eigenvalues;

(b) sketch the xy phase plane and the z phase line; and

(c) sketch the phase space.

10. B $= \begin{pmatrix} -2 & 1 & 0 \\ 0 & -2 & 0 \\ 0 & 0 & -1 \end{pmatrix}$ **11. B** $= \begin{pmatrix} -2 & 1 & 0 \\ 0 & -2 & 0 \\ 0 & 0 & 1 \end{pmatrix}$

12. B $= \begin{pmatrix} -1 & 2 & 0 \\ 2 & -4 & 0 \\ 0 & 0 & -1 \end{pmatrix}$ **13. B** $= \begin{pmatrix} -1 & 2 & 0 \\ 2 & -4 & 0 \\ 0 & 0 & 0 \end{pmatrix}$

In Exercises 14–15, we consider two additional cases of the linear system

$$\frac{d\mathbf{Y}}{dt} = \mathbf{CY}$$

which do not fit into the classification in the text. These systems do not decouple into lower dimensional systems. For each of these systems,

(a) compute the eigenvalues;

(b) compute the eigenvectors;

(c) sketch (as best you can) the phase portrait of the system. [*Hint*: Use the eigenvalues and eigenvectors and also vectors in the vector field.]

14. C $= \begin{pmatrix} -2 & 1 & 0 \\ 0 & -2 & 1 \\ 0 & 0 & -2 \end{pmatrix}$ **15. C** $= \begin{pmatrix} 0 & 1 & 0 \\ 0 & 0 & 1 \\ 0 & 0 & 0 \end{pmatrix}$

16. For the linear system

$$\frac{d\mathbf{Y}}{dt} = \mathbf{AY} = \begin{pmatrix} 2 & -1 & 0 \\ 0 & -2 & 3 \\ -1 & 3 & -1 \end{pmatrix} \begin{pmatrix} x \\ y \\ z \end{pmatrix},$$

(a) Show that $\mathbf{Y}_0 = (1, 1, 1)$ is an eigenvector of the coefficient matrix by computing $\mathbf{AY}_0$. What is the eigenvalue for this eigenvector?

(b) Find the other two eigenvalues for the matrix $\mathbf{A}$.

(c) Classify the system (source, sink, ...).

(d) Sketch the phase portrait [*Hint*: Use the other eigenvalues and find the other eigenvectors.]

17. For the linear system

$$\frac{d\mathbf{Y}}{dt} = \mathbf{AY} = \begin{pmatrix} -4 & 3 & 0 \\ 0 & -1 & 1 \\ 5 & -5 & 0 \end{pmatrix} \begin{pmatrix} x \\ y \\ z \end{pmatrix},$$

(a) Show that $\mathbf{Y}_0 = (1, 1, 0)$ is an eigenvector of the coefficient matrix by computing $\mathbf{AY}_0$. What is the eigenvalue for this eigenvector?

(b) Find the other two eigenvalues for the matrix $\mathbf{A}$.

(c) Classify the system (source, sink, ...).

(d) Sketch the phase portrait [*Hint*: Use the other eigenvalues and find the other eigenvectors.]

18. Consider the linear system

$$\frac{d\mathbf{Y}}{dt} = \mathbf{BY} = \begin{pmatrix} -10 & 10 & 0 \\ 28 & -1 & 0 \\ 0 & 0 & -8/3 \end{pmatrix} \begin{pmatrix} x \\ y \\ z \end{pmatrix}.$$

This system is related to the Lorenz system studied in Section 2.7. See also Section 4.4.

(a) Find the characteristic polynomial and the eigenvalues.

(b) Find the eigenvectors.

(c) Sketch the phase space.

(d) Comment on how the fact that the system "decouples" helps in the computations and in sketching the phase space.

Many years later, when Glen finally retires from writing math texts, he decides to join his friends and former collaborators Paul and Bob in the Compact Disc business. He opens a store specializing in New Age music located in between Paul's and Bob's stores. Let $z(t)$ be Glen's profits at time t (with $x(t)$ and $y(t)$ representing Paul's and Bob's profits at time t, respectively). Suppose the three stores effect each other in such a way that

$$\frac{dx}{dt} = -y + z$$

$$\frac{dy}{dt} = -x + z$$

$$\frac{dz}{dt} = z$$

19. (a) If Glen makes a profit, does this help or hurt Paul and Bob's profits?

(b) If Paul and Bob are making profits, does this help or hurt Glen's profits?

20. Write this system in matrix form and find the eigenvalues. Use them to classify the system.

21. Suppose, at time $t = 0$, both Paul and Bob are making (equal) small profits, but Glen is just breaking even ($x(0) = y(0)$ are small and positive, but $z(0) = 0$).

(a) Sketch the solution curve in the xyz-phase space.

(b) Sketch the $x(t)$-, $y(t)$- and $z(t)$-graphs of the solution.

(c) Describe what happens to the profits of each store.

22. Suppose at time $t = 0$, both Paul and Bob are just breaking even, but Glen is making a small profit ($x(0)$ and $y(0)$ are zero, but $z(0)$ is small and positive).

(a) Sketch the solution curve in the xyz-phase space.

(b) Sketch the $x(t)$-, $y(t)$- and $z(t)$-graphs of the solution.

(c) Describe what happens to the profits of each store.

Lab 3.1   Bifurcations in Linear Systems

In Chapter 3 we have studied techniques for solving linear systems. Given the coefficient matrix for the system, we can use these techniques to classify the system, describe the qualitative behavior of solutions, and give a formula for the general solution. In this lab we consider a two parameter family of linear systems. The goal is to better understand how different linear systems are related to each other. In other words, what bifurcations occur in parameterized families of linear systems.

Consider the linear system

$$\frac{dx}{dt} = ax + by$$
$$\frac{dy}{dt} = -x - y$$

where a and b are parameters which can take on any real value. In your report, address the following items:

1. For each value of a and b, classify the linear system as source, sink, center, spiral sink, and so forth. Draw a picture of the ab-plane and indicate the values of a and b for which the system is of each type (e.g., shade the values of a and b for which the system is a sink red, for which it is a source blue, and so forth). Be sure to describe all of the computations involved in creating this picture.

2. As the values of a and b are changed so that the point (a, b) moves from one region to another, the type of the linear system changes, that is, a bifurcation occurs. Which of these bifurcations is important for the long term behavior of solutions? Which of these bifurcations corresponds to a dramatic change in the phase plane or the $x(t)$- and $y(t)$-graphs?

Your report: Address the items above in the form of a short essay. Include any computations necessary to produce the picture in part 1. You may include phase planes and/or graphs of solutions to illustrate your essay, but your answer should be complete and understandable without the pictures.

Lab 3.2   RLC Circuits

We have already seen examples of differential equations that serve as models of simple electrical circuits involving only a resistor, a capacitor and a voltage source In this lab we consider slightly more complicated circuits consisting of a resistor, a capacitor, an inductor and a voltage source (see Figure 3.61). The behavior of the system can be described by specifying the current moving around the circuit and the changes in voltages across each component of the circuit. In this lab we take an axiomatic approach to the relationship between the current and the voltages. Readers interested in the more information on the derivation of these laws are referred to texts in electric circuit theory.

In the tradition of electrical engineering, we let i denote the current moving around the circuit. We let v_T, v_C and v_L denote the voltages across the voltage source, the capacitor and the inductor, respectively. Also, we let R denote the resistance, C the capacitance and L the inductance of the associated components of the circuit. These are indicated in Figure 3.61. We think of v_T, R, C and L as parameters set by the person building the circuit. The quantities i, v_C and v_L depend on time.

We need the following basic relationships between the quantities above. First, Kirchhoff's Voltage Law states that the sum of the voltage changes around a closed

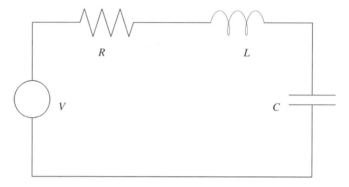

Figure 3.61
An RLC circuit.

loop must be zero. For our circuit, this gives

$$v_T - Ri = v_C + v_L.$$

Next, we need the relationship between current and voltage in the capacitor and the inductor. In a capacitor, the current is proportional to the rate of change of the voltage. The proportionality constant is the capacitance C. Hence, we have

$$C\frac{dv_C}{dt} = i.$$

In an inductor, the voltage is proportional to the rate of change of the current. The proportionality constant is the inductance L. Hence, we have

$$L\frac{di}{dt} = v_L.$$

In this lab we consider the possible behavior of the circuit above for several different input voltages.

In your report, address the following questions:

1. First, set the input voltage to zero, that is, assume $v_T = 0$. Using the three equations above, write a first-order system of differential equations with dependent variables i and v_C. [Hint: Use the first equation to eliminate v_L from the third equation. You should have R, C and L as parameters in your system.]

2. Find the eigenvalues of the resulting system in terms of the parameters R, C and L. What are the possible phase planes for your system given that R, C and L are always non-negative? Sketch the phase plane and the $v_C(t)$- and $i(t)$-graphs for each case.

3. Convert the first-order system of equations from part 1 into a second-order differential equation involving only v_C (and not i). (This is the form of the equation that you will typically find in electric circuit theory texts.)

4. Repeat part 1 assuming that v_T is nonzero. The resulting system will have R, C, L and v_T as parameters.

5. The units used in applications are volts and amps for voltages and currents, ohms for resistors, farads for capacitors and Henrys for inductors. A typical, off-the-shelf circuit, might have parameter values $R = 2000$ ohms (or 2 kilo-ohms), $C = 2 \cdot 10^{-7}$

farads (or 0.2 micro farads) and $L = 1.5$ Henrys. Assuming zero-input, $v_T = 0$, and the initial values of the current and voltage are $i(0) = 0$, $v_C(0) = 10$, describe the behavior of the current and voltage for this circuit.

6. Repeat part 5 using a voltage source of $v_T = 10$ volts.

Your report: In your report, address the items above. Show all algebra and justify all steps. In parts 5 and 6 you may work either analytically or numerically. Give phase planes and graphs of solutions as appropriate.

Lab 3.3   Measuring Mass in Space

The effects on the human body of prolonged weightlessness during space flights is not completely understood. One important variable which must be monitored is the astronaut's "weight." However, weight refers to the force of gravity on a body. What actually must be measured is body mass. To perform this measurement, a mass-spring system is used where the mass is the body of the astronaut. The astronaut sits in a special chair attached to springs. The frequency of the oscillation of the astronaut in the chair is measured and from this the weight is computed.*

In your report address the following items:.

1. Suppose the chair has a mass of 20kg. The system is initially calibrated by placing a known mass in the chair and measuring the period of oscillations. Suppose that a 25kg mass placed in the chair results in an oscillation with period of 1.3 seconds per oscillation. We assume that the coefficient of damping of the apparatus is very small (so as a first approximation we assume that there is no damping). What will be the period of oscillations of an astronaut with mass 60kg? What would your frequency of oscillation be?

2. Does it matter whether or not the calibration is done on the earth or in space? (It would be much better if it could be done on the earth since it is expensive to launch 25kg masses into space.)

3. Suppose an error is made during the calibration and the actual frequency resulting when a 25kg mass is placed in the chair is 1.31 seconds instead of 1.3 seconds. How much error then results in the measurement of the mass of astronaut with mass 60kg? With mass 80kg?

4. Suppose a small amount of damping develops in the chair. How seriously does this affect the measurements? How could you determine if damping were present (that is, what measurements would you perform during the calibration phase)?

Your Report: In your report you should address each of the items above. Show all algebraic computations which you perform and justify all assertions. While this lab does not require numerical approximations of solutions (since we can explicitly solve all the equations involved) you may include sketches of solutions and/or computer generated graphs **if appropriate**.

_____

*This information comes from the "Q& A" column of the *The New York Times*, Science Section, August 1, 1995, page C6.

Lab 3.4   Find Your Own Harmonic Oscillator

In the text we claimed that the harmonic oscillator could be used as a simple model for many different situations. The key ingredients are a restoring force which pushes the system back toward a rest position and (perhaps) a damping force.

For this lab you are to find such a system. It may be mechanical, biological, psychological, political, financial or whatever. You will have to do some analysis on your system, so you will need to be able to obtain some data on its "motion." You can either record the data yourself or use published data (with proper references).

In your report, address the following items:

1. In a short essay, carefully describe the system you propose to model with a harmonic oscillator equation. State the dependent variable, describe the rest position, the restoring force and the damping.

2. Collect data on the behavior of your system either by observing your system or from published sources (taking care to give proper references).

3. Use your data to estimate the parameter values appropriate for your system. Classify your system (e.g., overdamped, critically damped, underdamped). Compare solutions of the harmonic oscillator equation with the data you have collected. Describe how well the solutions of the model fit the data and discuss any discrepancies.

Your report: In your report you should address the three items above in the form of a short essay. In part 2 describe how you obtained your data (either citing appropriate references or giving details of your experimental procedure). In part 3, be sure to address any problems that arose in using the harmonic oscillator as a model. (If your data differs significantly from your solutions, then you may want to adjust the parameter values that you are using or search for another situation which is better modeled by the harmonic oscillator equations.)

Lab 3.5   A Baby Bottle Harmonic Oscillator

If you performed the previous lab experiment, you undoubtedly saw that harmonic oscillators are everywhere. Here is a harmonic oscillator that one of the authors discovered late one night.

Consider a "U" shaped tube or pipe partially filled with fluid. (For example, a common design of baby bottle has this shape, see Figure below.)

If held upright, the level of fluid will come to rest at an equal height on each side

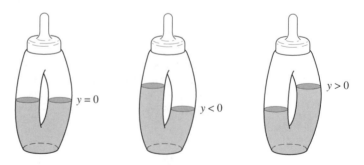

Figure 3.62
A U-shaped baby bottle.

of the bottle. If we tip the bottle quickly to one side, then return it to level, there will be more fluid in one side than in the other. Gravity pulling on this extra fluid will provide a restoring force, pushing the system back toward the rest position where both sides are at an equal level.

In your report, address the following items:

1. Let $y(t)$ denote the height of the fluid in one side of bottle with $y = 0$ corresponding to the rest position. Assume that the restoring force is proportional to the difference in the mass in the left and right-hand sides of the bottle. Also assume, as a first approximation, that the damping force is proportional to the velocity. Use this to give a differential equation model of this system based on the harmonic oscillator equation.

2. Buy or borrow such a baby bottle and observe it. Estimate the natural period of the oscillations and the rate of decay of the amplitude of the oscillations. Use the observations to estimate the parameter values for the harmonic oscillator equation model.

3. Compare solutions of the harmonic oscillator equation in part 2 to your observations. In particular, if you increase the total mass in the bottle (but leave the size of the initial displacement unchanged) does the model make the correct prediction of the change of behavior of the fluid in the bottle (e.g., does the natural period increase or decrease in this situation)?

Your report: Address each of the three questions above in the form of a short essay. In part 1, carefully describe the relationship between the forces and the differential equation model. In part 2, describe the experiments you performed to obtain data and the computations necessary for computing the coefficients in the harmonic oscillator equation from your data. In part 3, discuss any discrepancies between your data and the predictions of the model.

NONLINEAR SYSTEMS

4

In Chapter 3 we saw that, using a combination of analytic and geometric techniques, we can understand linear systems completely. In this chapter we combine these linear techniques with some additional qualitative methods to tackle nonlinear systems. While these techniques do not allow us to determine the phase portrait of all nonlinear systems, we see that we are able to handle some important nonlinear systems.

We first show how a nonlinear system can be approximated near an equilibrium point by a linear system. This process, known as "linearization," is one of the most frequently used techniques in applications. By studying the linear approximation, we can surmise the behavior of solutions of the nonlinear system, at least near the equilibrium point.

In the remainder of the chapter, we study special types of models and the nonlinear systems associated with them. These special nonlinear systems are important both because they arise in applications and because the techniques involved in their analysis are delightful.

4.1 EQUILIBRIUM POINT ANALYSIS

By the techniques used in Chapter 2 to analyze a given nonlinear system, we are able to get a good idea of the behavior of its solutions over most of the phase plane. However, near an equilibrium point the behavior of solutions is more difficult to determine. Numerical analysis of solutions near equilibrium points is difficult because the vector field is close to zero. This means that many time steps are necessary to follow a numerical approximation of a solution even a short distance. Qualitative analysis is difficult near an equilibrium point because the vector field can change direction radically near the fixed point. In this section we introduce a technique for using our knowledge of linear systems to study nonlinear systems near equilibrium points.

The Van der Pol Equation

To illustrate how to analyze the behavior of solutions near an equilibrium point we begin with a simple but important example of a nonlinear system, the Van der Pol system. Recall that the Van der Pol system is

$$\frac{dx}{dt} = y$$
$$\frac{dy}{dt} = -x + (1 - x^2)y \qquad -x + y - yx^2$$

This system was studied numerically in Section 2.5, page 188. The direction field and phase plane are shown in Figure 4.1.

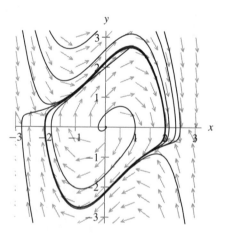

Figure 4.1
Direction field and phase plane for the Van der Pol system.

The only equilibrium point of this system is at the origin. We wish to understand how solutions near the origin behave. The x-nullcline for this system is the x-axis and the y-nullcline is the curve $y = x/(1 - x^2)$. In Figure 4.2 we sketch these two nullclines in the vicinity of the origin. Note that the arrangement of these curves tell us that

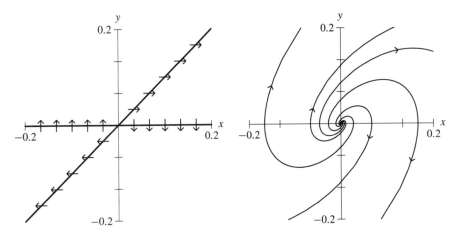

Figure 4.2

Nullclines near the origin for the
Van der Pol equation

Figure 4.3

Phase plane for the Van der Pol
system near the origin

solutions circle around the origin. But do these solutions spiral in, spiral away, or lie on periodic solutions? We have no way of telling, at least not by using only the nullclines.

We can use numerical techniques to resolve this issue. In Figure 4.3 we use Euler's method to plot solutions near the origin. Note that they do seem to spiral away.

We can understand why solutions spiral away from the origin by approximating the Van der Pol system with another system whose solutions we know. The Van der Pol equation is nonlinear, but there is only one nonlinear term, namely the x^2y term in the equation for dy/dt. If x and y are small, then this term is much smaller than any of the other terms in the equation. For example, if $x = 0.1 = y$, then the x^2y term is 0.001, which is significantly smaller. If $x = 0.01 = y$, then $x^2y = 10^{-6}$, which is again much smaller than either x or y. This suggests that we can approximate the Van der Pol system by one in which we simply neglect the x^2y term, at least if both x and y are close to 0. If we drop this term from the system, then we are left with

$$\frac{dx}{dt} = y$$
$$\frac{dy}{dt} = -x + y,$$

which is a linear system. So unlike the Van der Pol system, we can solve this system. Using techniques from the last chapter, we find that the eigenvalues of this system are $(1 \pm \sqrt{3}i)/2$. Since these eigenvalues are complex with positive real part, we know that solutions spiral away from the origin for the linear system.

The linear system and the Van der Pol system have vector fields that are very close to each other near the equilibrium point at the origin. Since solutions of the linear system spiral away from the origin, it is not surprising that solutions of the Van der Pol system that start near the origin also spiral away.

The technique we applied above is called **linearization**. Near the equilibrium point, we approximate the nonlinear system by a linear system. For initial conditions near the equilibrium point, the solutions of the nonlinear system and the linear approximation stay close together at least for an interval of time.

The Competing Species Model

Now let's return to the system of differential equations governing a pair of competing species that we began to analyze in Section 2.6. The differential equations are

$$\frac{dx}{dt} = 2x\left(1 - \frac{x}{2}\right) - xy$$

$$\frac{dy}{dt} = 3y\left(1 - \frac{y}{3}\right) - 2xy$$

The phase plane for this system is sketched in Figure 4.4.

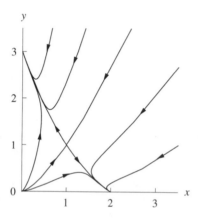

Figure 4.4
Phase plane for the system

$$\frac{dx}{dt} = 2x\left(1 - \frac{x}{2}\right) - xy$$

$$\frac{dy}{dt} = 3y\left(1 - \frac{y}{3}\right) - 2xy.$$

There are several equilibrium points for this system, but we will concentrate only on the one at $(1, 1)$. The linear systems we studied in Chapter 3 all have equilibrium points at the origin. Hence the first step in comparing the nonlinear system near the equilibrium point $(1, 1)$ to a linear system is to move the equilibrium point to the origin via a change of variables. Once the equilibrium point is at the origin, we can use the same ideas as the example above to identify the linear approximation.

To move the equilibrium point to the origin, we introduce two new variables u and v by the formulas $u = x - 1$ and $v = y - 1$. Note that u and v are close to 0 when (x, y) is close to $(1, 1)$. To obtain the system in the new variables, we first compute

$$\frac{du}{dt} = \frac{d(x - 1)}{dt} = \frac{dx}{dt}$$

$$\frac{dv}{dt} = \frac{d(y - 1)}{dt} = \frac{dy}{dt}$$

The right-hand sides of the system in the new variables are given by

$$\frac{dx}{dt} = 2x\left(1 - \frac{x}{2}\right) - xy$$

$$= 2(u + 1)\left(1 - \frac{u + 1}{2}\right) - (u + 1)(v + 1)$$

$$= -u - v - u^2 - uv,$$

and

$$\frac{dy}{dt} = 3y\left(1 - \frac{y}{3}\right) - 2xy$$
$$= 3(v+1)\left(1 - \frac{v+1}{3}\right) - 2(u+1)(v+1)$$
$$= -2u - v - 2uv - v^2.$$

Putting these together we obtain

$$\frac{du}{dt} = -u - v - u^2 - uv$$
$$\frac{dv}{dt} = -2u - v - 2uv - v^2.$$

As we expect, the origin is an equilibrium point for this system. The expression for du/dt involves some linear terms $(-u - v)$ and some nonlinear terms $(-u^2 - uv)$. For dv/dt the linear terms are $(-2u - v)$ and the nonlinear terms are $(-2uv - v^2)$. As above, near the origin the nonlinear terms are much smaller than the linear terms. So we can approximate the nonlinear system near $(u, v) = (0, 0)$ with the linear system

$$\frac{du}{dt} = -u - v$$
$$\frac{dv}{dt} = -2u - v.$$

The eigenvalues of this system are $-1 \pm \sqrt{2}$. One of these numbers is positive and the other is negative, so the origin is a saddle point for the linear system (see Figure 4.5). Since the linear and nonlinear systems are approximately the same, the phase plane for the nonlinear system near the equilibrium point $x = y = 1$ looks like the uv-phase plane of the linear system. We can use this information to improve our sketch of the nonlinear system in the xy-phase plane. In particular, this allows us to conclude that there are only two curves of solutions in the xy-plane that tend toward the equilibrium point $(1, 1)$ as t increases.

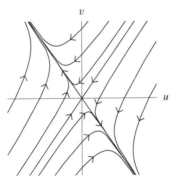

Figure 4.5
Phase plane for

$$\frac{du}{dt} = -u - v$$
$$\frac{dv}{dt} = -2u - v,$$

the linear approximation of the competitive system near $x = y = 1$.

A Nonpolynomial Example

The two examples above have vector fields that are polynomials. When the vector field is a polynomial and the equilibrium point under consideration is the origin, it is very easy to identify which terms are linear and which are nonlinear. However, not all vector fields are polynomials.

Consider the system

$$\frac{dx}{dt} = y$$

$$\frac{dy}{dt} = -y - \sin x$$

This system will be studied extensively in Sections 4.2 and 4.3. (It is a model for the swinging motion of a damped pendulum.) The equilibria of this system occur at the points $x = 0, \pm\pi, \pm2\pi, \ldots, y = 0$. Suppose we want to study the solutions that are close to the equilibrium point at $x = y = 0$. Because of the $\sin x$ term, it is not clear what the nonlinear terms of this system are. However, from calculus we know that the power series expansion of $\sin x$ is

$$\sin x = x - \frac{x^3}{3!} + \frac{x^5}{5!} - \cdots.$$

Therefore we can write

$$\frac{dx}{dt} = y$$

$$\frac{dy}{dt} = -y - \left(x - \frac{x^3}{3!} + \frac{x^4}{5!} \cdots \right).$$

Again, near the origin we drop the nonlinear terms, and we are left with the linear system

$$\frac{dx}{dt} = y$$

$$\frac{dy}{dt} = -y - x$$

The eigenvalues of this system are easily computed to be $(-1 \pm \sqrt{3}i)/2$. Since these eigenvalues are complex with negative real parts, we expect the corresponding equilibrium point for the nonlinear system to be a spiral sink (see Figures 4.6 and 4.7).

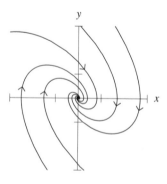

Figure 4.6
Phase plane for the system

$$\frac{dx}{dt} = y$$

$$\frac{dy}{dt} = -y - x.$$

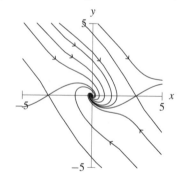

Figure 4.7
Phase plane for the system

$$\frac{dx}{dt} = y$$

$$\frac{dy}{dt} = -y - \sin x.$$

Linearization

The next step is to make the process of linearization more orderly. In the preceding examples, we first move the equilibrium point to the origin (if necessary) by changing variables. Then we identify the linear and nonlinear terms, using calculus if necessary.

Consider the general form of a nonlinear system

$$\frac{dx}{dt} = f(x, y)$$
$$\frac{dy}{dt} = g(x, y).$$

Suppose that (x_0, y_0) is an equilibrium solution for this system. This means that

$$f(x_0, y_0) = 0 = g(x_0, y_0).$$

We wish to understand what happens to solutions near (x_0, y_0) — that is, to linearize the system near (x_0, y_0). We introduce new variables to move the equilibrium point to the origin,

$$u = x - x_0$$
$$v = y - y_0.$$

If x and y are close to the equilibrium point (x_0, y_0), then u and v are close to 0.

Since x_0 and y_0 are constants, the differential equations for u and v are

$$\frac{du}{dt} = \frac{d(x - x_0)}{dt} = \frac{dx}{dt} = f(x, y) = f(x_0 + u, y_0 + v)$$
$$\frac{dv}{dt} = \frac{d(y - y_0)}{dt} = \frac{dy}{dt} = g(x, y) = g(x_0 + u, y_0 + v)$$

or

$$\frac{du}{dt} = f(x_0 + u, y_0 + v)$$
$$\frac{dv}{dt} = g(x_0 + u, y_0 + v).$$

When $u = v = 0$, the right-hand side of this equation vanishes, so we have moved the equilibrium point to the origin (of the uv plane).

Now we would like to be able to eliminate the "higher-order" or nonlinear terms in the expressions for f and g. Since f and g can include exponentials, logarithms, and trigonometric functions, it is not always clear which terms are higher order and which are linear. In this case it is necessary to study f and g more closely.

The basic idea of differential calculus is that the derivative of a function provides the "best linear approximation" of the function. For functions of two variables, the best linear approximation at a particular point is given by the tangent plane. Hence we have

$$f(x_0 + u, y_0 + v) \approx f(x_0, y_0) + u\frac{\partial f}{\partial x}(x_0, y_0) + v\frac{\partial f}{\partial y}(x_0, y_0).$$

The right-hand side is the equation for the tangent plane of the graph of f at (x_0, y_0). (We can also think of the right-hand side as the first-degree Taylor polynomial approximation of f.)

Thus we can rewrite the system for du/dt and dv/dt as

$$\frac{du}{dt} = f(x_0, y_0) + u\frac{\partial f}{\partial x}(x_0, y_0) + v\frac{\partial f}{\partial y}(x_0, y_0) + \dots$$

and

$$\frac{dv}{dt} = g(x_0, y_0) + u\frac{\partial g}{\partial x}(x_0, y_0) + v\frac{\partial g}{\partial y}(x_0, y_0) + \dots,$$

where "..." stands for the terms making up the difference between the tangent plane and the function. These are precisely the terms we wish to ignore when forming the linear approximation of the system. Since $f(x_0, y_0) = 0 = g(x_0, y_0)$, we can write this more succinctly as

$$\begin{pmatrix} \dfrac{du}{dt} \\ \dfrac{dv}{dt} \end{pmatrix} = \begin{pmatrix} \dfrac{\partial f}{\partial x}(x_0, y_0) & \dfrac{\partial f}{\partial y}(x_0, y_0) \\ \dfrac{\partial g}{\partial x}(x_0, y_0) & \dfrac{\partial g}{\partial y}(x_0, y_0) \end{pmatrix} \begin{pmatrix} u \\ v \end{pmatrix} + \dots.$$

The 2×2 matrix of partial derivatives in this expression is called the **Jacobian matrix** of the system at (x_0, y_0).

Hence the **linearized system** at the equilibrium point (x_0, y_0) is

$$\begin{pmatrix} \dfrac{du}{dt} \\ \dfrac{dv}{dt} \end{pmatrix} = \begin{pmatrix} \dfrac{\partial f}{\partial x}(x_0, y_0) & \dfrac{\partial f}{\partial y}(x_0, y_0) \\ \dfrac{\partial g}{\partial x}(x_0, y_0) & \dfrac{\partial g}{\partial y}(x_0, y_0) \end{pmatrix} \begin{pmatrix} u \\ v \end{pmatrix}.$$

This is the system we use to study the behavior of solutions of the nonlinear system near the equilibrium point (x_0, y_0). Note that to create this system, we need only know the partial derivatives of the components of the vector field at the equilibrium point. We do not have to explicitly compute the change of variables moving the equilibrium point to the origin.

As always with differential calculus, the derivative of a nonlinear function provides only a "local" approximation. Hence the solutions of the linearized system are close to solutions of the nonlinear system only near the equilibrium point. How close to the equilibrium point we must be for the linear approximation to be any good depends on the size of the nonlinear terms.

More Examples of Linearization

Consider the nonlinear system

$$\frac{dx}{dt} = -2x + 2x^2 = f(x, y)$$

$$\frac{dy}{dt} = -3x + y + 3x^2 = g(x, y).$$

There are two equilibrium points for this system, at $(0, 0)$ and $(1, 0)$. To understand nearby solution curves, we first compute the Jacobian matrix

$$\begin{pmatrix} \dfrac{\partial f}{\partial x}(x, y) & \dfrac{\partial f}{\partial y}(x, y) \\ \dfrac{\partial g}{\partial x}(x, y) & \dfrac{\partial g}{\partial y}(x, y) \end{pmatrix} = \begin{pmatrix} -2 + 4x & 0 \\ -3 + 6x & 1 \end{pmatrix}.$$

At the two equilibrium points, we have

$$
\begin{pmatrix} \dfrac{\partial f}{\partial x}(0,0) & \dfrac{\partial f}{\partial y}(0,0) \\[2mm] \dfrac{\partial g}{\partial x}(0,0) & \dfrac{\partial g}{\partial y}(0,0) \end{pmatrix} = \begin{pmatrix} -2 & 0 \\ -3 & 1 \end{pmatrix}
$$

and

$$
\begin{pmatrix} \dfrac{\partial f}{\partial x}(1,0) & \dfrac{\partial f}{\partial y}(1,0) \\[2mm] \dfrac{\partial g}{\partial x}(1,0) & \dfrac{\partial g}{\partial y}(1,0) \end{pmatrix} = \begin{pmatrix} 2 & 0 \\ 3 & 1 \end{pmatrix}.
$$

Near $(0,0)$ the phase plane for the nonlinear system should resemble that of the linearized system

$$
\frac{d\mathbf{Y}}{dt} = \begin{pmatrix} -2 & 0 \\ -3 & 1 \end{pmatrix} \mathbf{Y}.
$$

The eigenvalues of this system are -2 and 1, so the origin is a saddle. We can compute (good review) that $(0,1)$ is an eigenvector for eigenvalue 1 and $(1,1)$ is an eigenvector for eigenvalue -2. Using this information we can sketch the phase plane for the linear system (see Figure 4.8).

At the other equilibrium point, $(1,0)$, the linearized system is

$$
\frac{d\mathbf{Y}}{dt} = \begin{pmatrix} 2 & 0 \\ 3 & 1 \end{pmatrix} \mathbf{Y}.
$$

Here the eigenvalues are 2 and 1, so for this system the origin is a source. Using the fact that $(0,1)$ is an eigenvector for eigenvalue 1 and $(1,3)$ is an eigenvector for eigenvalue 2, we can sketch the phase plane (see Figure 4.9).

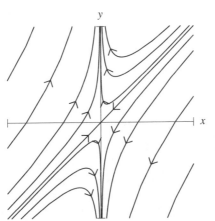

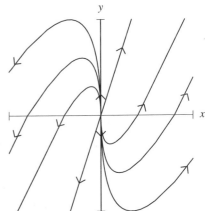

Figure 4.8
Phase plane for the system

$$
\frac{d\mathbf{Y}}{dt} = \begin{pmatrix} -2 & 0 \\ -3 & 1 \end{pmatrix} \mathbf{Y}.
$$

Figure 4.9
Phase plane for the system

$$
\frac{d\mathbf{Y}}{dt} = \begin{pmatrix} 2 & 0 \\ 3 & 1 \end{pmatrix} \mathbf{Y}.
$$

At least near the two equilibrium points, the phase plane for the nonlinear system resembles that of the linearized systems. Numerically generated approximate solutions of the system are shown in Figure 4.10. If we magnify the phase plane near $(0, 0)$ and $(1, 0)$, we see that the solution curves do indeed look like those of the corresponding linearized systems (see Figure 4.10).

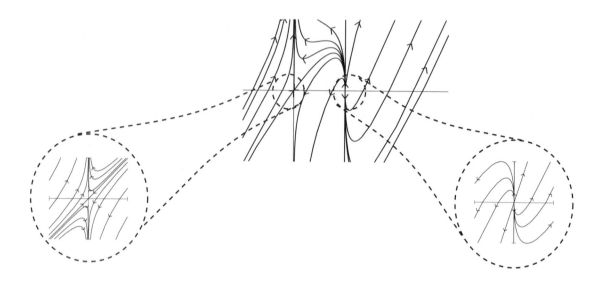

Figure 4.10
Solution curves and magnifications near the fixed points for the system

$$\frac{dx}{dt} = -2x + 2x^2$$

$$\frac{dy}{dt} = -3x + y + 3x^2.$$

Classification of Equilibrium Points

The idea of linearization is to use a linear system to approximate the behavior of solutions of a nonlinear system near an equilibrium point. The solutions of the nonlinear system near the equilibrium point are close to solutions of the approximating linear system, at least for a short time interval. For most systems, the information gained by studying the linearization is enough to determine the long-term behavior of solutions of the nonlinear system near the equilibrium point.

For example, consider the system

$$\frac{dx}{dt} = f(x, y)$$

$$\frac{dy}{dt} = g(x, y).$$

Suppose (x_0, y_0) is an equilibrium point and let

$$\mathbf{J} = \begin{pmatrix} \dfrac{\partial f}{\partial x}(x_0, y_0) & \dfrac{\partial f}{\partial y}(x_0, y_0) \\ \dfrac{\partial g}{\partial x}(x_0, y_0) & \dfrac{\partial g}{\partial y}(x_0, y_0) \end{pmatrix}$$

be the Jacobian matrix at (x_0, y_0). The linearized system is

$$\begin{pmatrix} \dfrac{du}{dt} \\ \dfrac{dv}{dt} \end{pmatrix} = \mathbf{J} \begin{pmatrix} u \\ v \end{pmatrix}.$$

If the Jacobian matrix has only negative eigenvalues or complex eigenvalues with negative real part, then $(u, v) = (0, 0)$ is a sink and all solutions of the linearized system approach $(u, v) = (0, 0)$ as t tends to infinity. For the nonlinear system, solutions that start near $(x, y) = (x_0, y_0)$ approach $(x, y) = (x_0, y_0)$ as t tends to infinity. Hence we call (x_0, y_0) a sink for the nonlinear system. If the eigenvalues are complex, then (x_0, y_0) is called a spiral sink. We reiterate that this says nothing about solutions of the nonlinear system with initial positions far from (x_0, y_0).

Similarly, if the Jacobian matrix has only positive eigenvalues or complex eigenvalues with positive real part, then solutions with initial position near (x_0, y_0) move away from (x_0, y_0) as t increases. The equilibrium point (x_0, y_0) of the nonlinear system is called a source. If the eigenvalues are complex, then (x_0, y_0) is called a spiral source.

If the Jacobian matrix has one positive and one negative eigenvalue, then (x_0, y_0) is called a saddle equilibrium point for the nonlinear system. Just as for a linear system, there will be exactly two curves of solutions that approach the equilibrium point as t increases and exactly two curves of solutions that approach the equilibrium point as t decreases (see Figures 4.8 and 4.10). For the nonlinear system, these curves of solutions need not be straight lines. All other solutions with initial position near (x_0, y_0) move away as t increases and as t decreases.

Separatrices

The four special curves of solutions that tend toward a saddle equilibrium point as t increases or decreases are called **separatrices**. (The singular is *separatrix*.) These curves are of special importance because they separate solutions with different behavior (hence the name). The two separatrices on which solutions tend toward the saddle as t increases are called **stable separatrices**, while those on which solutions tend toward the saddle as t decreases are called **unstable separatrices**.

In the system

$$\frac{dx}{dt} = -2x + 2x^2$$

$$\frac{dy}{dt} = -3x + y + 3x^2$$

studied above, the origin is a saddle (see Figures 4.8 and 4.10). The stable separatrix that tends toward the origin as t increases separates the strip between $x = 0$ and $x = 1$ into two parts. Initial conditions in this strip above the stable separatrix give solutions that tend to $y = +\infty$ as t increases. Initial conditions in the strip below the stable separatrix give solutions that tend to $y = -\infty$ as t increases (see Figure 4.11).

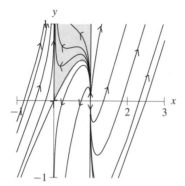

Figure 4.11
Separatrices of $(0, 0)$ for the system

$$\frac{dx}{dt} = 2x + 2x^2$$

$$\frac{dy}{dt} = -3x + y + 3x^2$$

and regions of the strip between $x = 0$ and $x = 1$ with different long term behavior.

When Linearization Fails

Unfortunately, in some cases the information given by the linearized system is not enough to completely determine the behavior of solutions of the nonlinear system near the equilibrium point.

For example, consider the system

$$\frac{dx}{dt} = y - (x^2 + y^2)x$$

$$\frac{dy}{dt} = -x - (x^2 + y^2)y.$$

The origin is the only equilibrium point for this system. The linearized system at the origin is

$$\begin{pmatrix} \dfrac{du}{dt} \\ \dfrac{dv}{dt} \end{pmatrix} = \begin{pmatrix} 0 & 1 \\ -1 & 0 \end{pmatrix} \begin{pmatrix} u \\ v \end{pmatrix}.$$

The eigenvalues of this system are $\pm i$ and hence it is a center. All nonzero solutions of the linearized system lie on periodic solutions that wind around $(0, 0)$ in the clockwise direction. But there are no periodic solutions for the nonlinear system. To see this, we regard the vector field as a sum of two vector fields, the linear field $(y, -x)$ and the nonlinear vector field $(-(x^2 + y^2)x, -(x^2 + y^2)y)$. The linear vector field is always tangent to circles centered at the origin. But the vector field $(-(x^2+y^2)x, -(x^2+y^2)y)$ always points directly toward $(0, 0)$, since this vector field is basically a multiple of the field$(-x, -y)$ by the positive number $x^2 + y^2$. The net effect of adding these two vector fields is a vector field that always points "into" a circle centered at the origin. This forces solutions to spiral into $(0, 0)$ rather than to circulate on periodic orbits (see Figure 4.12).

Note that if we merely change the signs of the higher-order terms in the above system, then the resulting system,

$$\frac{dx}{dt} = y + (x^2 + y^2)x$$

$$\frac{dy}{dt} = -x + (x^2 + y^2)y,$$

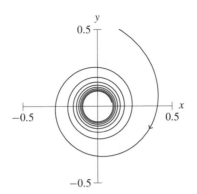

Figure 4.12
The solution curves for the system

$$\frac{dx}{dt} = y - (x^2 + y^2)x$$

$$\frac{dy}{dt} = -x - (x^2 + y^2)y$$

spiral slowly toward the equilibrium point at the origin even though its linearization is a center.

has the same linearization near $(0, 0)$, but now solutions spiral away from the origin.

In this example, the solutions of the nonlinear system near the origin and the solutions of the linearized system are still approximately the same, at least for a short amount of time. The problem is that because the linearized system is a center, any small perturbation can change the long-term behavior of the solutions. Even the very small perturbation caused by the inclusion of the nonlinear terms can turn the center into a spiral sink or a spiral source.

Fortunately there are only two situations in which the long-term behavior of solutions near an equilibrium point of the nonlinear system and its linearization can differ. One is when the linearized system is a center; the other is when the linearized system has zero as an eigenvalue (see the exercises for examples). In every other case, the local picture of the nonlinear system near an equilibrium point looks like its linearization.

Exercises for Section 4.1

1. Consider the following three systems:

(i)
$$\frac{dx}{dt} = 2x + y$$
$$\frac{dy}{dt} = -y + x^2$$

(ii)
$$\frac{dx}{dt} = 2x + y$$
$$\frac{dy}{dt} = y + x^2$$

(iii)
$$\frac{dx}{dt} = 2x + y$$
$$\frac{dy}{dt} = -y - x^2$$

All three have an equilibrium point at $(0, 0)$. Which two systems have phase planes with the same "local picture" near $(0, 0)$? Justify your answer. [*Hint*: Very little computation is required for this exercise, but be sure to give complete justification.]

2. Consider the following three systems:

(i)
$$\frac{dx}{dt} = 3 \sin x + y$$
$$\frac{dy}{dt} = 4x + \cos y - 1$$

(ii)
$$\frac{dx}{dt} = -3 \sin x + y$$
$$\frac{dy}{dt} = 4x + \cos y - 1$$

(iii)
$$\frac{dx}{dt} = -3 \sin x + y$$
$$\frac{dy}{dt} = 4x - 3 \cos y - 3$$

All three have an equilibrium point at $(0, 0)$. Which two systems have phase planes with the same "local picture" near $(0, 0)$? Justify your answer. [*Hint*: Very little computation is required for this exercise, but be sure to give complete justification.]

3. Consider the system

$$\frac{dx}{dt} = -2x + y$$

$$\frac{dy}{dt} = -y + x^2.$$

For the equilibrium point at $(0, 0)$,

    (a) find the linearized system at $(0, 0)$;

    (b) classify the equilibrium point $(0, 0)$ as source, sink, center, and so on; and

    (c) sketch the phase plane near $(0, 0)$.

4. Consider the system

$$\frac{dx}{dt} = -2x + y$$

$$\frac{dy}{dt} = -y + x^2.$$

For the equilibrium point at $(2, 4)$,

    (a) find the linearized system at $(2, 4)$;

    (b) classify the equilibrium point $(2, 4)$ as source, sink, center, and so on;

    (c) sketch the phase plane near $(2, 4)$.

5. Consider the system

$$\frac{dx}{dt} = -x$$

$$\frac{dy}{dt} = -4x^3 + y.$$

    (a) Show that the origin is the only equilibrium point.

    (b) Find the linearized system at the origin.

    (c) Classify the linearized system and sketch its phase plane.

6. Consider the system

$$\frac{dx}{dt} = -x$$

$$\frac{dy}{dt} = -4x^3 + y.$$

    (a) Find the general solution of the equation

$$\frac{dx}{dt} = -x.$$

    [*Hint*: This is as easy as it looks.]

    (b) Using this general solution in place of x, find the general solution of the equation

$$\frac{dy}{dt} = -4x^3 + y.$$

    [*Hint*: This provides some review of linear equations in Section 1.8.]

    (c) Use the information above to form the general solution of the system.

    (d) Find the solution curves of the system that tend toward the origin as t increases and as t decreases.

(e) Sketch the solution curves in the xy-phase plane corresponding to these solutions. These are the separatrices.

(f) Compare the sketch of the linearized system in the previous problem with the sketch of the separatrix solutions. Comment on why the two pictures differ.

In Exercises 7–16, we restrict attention to the first quadrant, $x \geq 0$, $y \geq 0$. For each system,

(a) find all equilibria (in the first quadrant) and determine which are sinks, saddles, sources, and so on;

(b) sketch the x- and y-nullclines;

(c) indicate the direction of the solution curve in any regions bounded by the nullclines; and

(d) sketch the phase plane of the system.

7.
$$\frac{dx}{dt} = x(-x - 3y + 150)$$
$$\frac{dy}{dt} = y(-2x - y + 100)$$

8.
$$\frac{dx}{dt} = x(10 - x - y)$$
$$\frac{dy}{dt} = y(30 - 2x - y)$$

9.
$$\frac{dx}{dt} = x(100 - x - 2y)$$
$$\frac{dy}{dt} = y(150 - x - 6y)$$

10.
$$\frac{dx}{dt} = x(-x - y + 100)$$
$$\frac{dy}{dt} = y(-x^2 - y^2 + 2500)$$

11.
$$\frac{dx}{dt} = x(-x - y + 40)$$
$$\frac{dy}{dt} = y(-x^2 - y^2 + 2500)$$

12.
$$\frac{dx}{dt} = x(-4x - y + 160)$$
$$\frac{dy}{dt} = y(-x^2 - y^2 + 2500)$$

13.
$$\frac{dx}{dt} = x(-8x - 6y + 480)$$
$$\frac{dy}{dt} = y(-x^2 - y^2 + 2500)$$

14.
$$\frac{dx}{dt} = x(2 - x - y)$$
$$\frac{dy}{dt} = y(y - x^2)$$

15.
$$\frac{dx}{dt} = x(2 - x - y)$$
$$\frac{dy}{dt} = y(y - x)$$

16.
$$\frac{dx}{dt} = x(x - 1)$$
$$\frac{dy}{dt} = y(x^2 - y)$$

For each system in Exercises 17–20,

(a) find all equilibria and determine whether they are sinks, sources, saddles, and so on;

(b) sketch the x- and y-nullclines;

(c) indicate the direction of the solution curve in any regions bounded by the nullclines; and

(d) sketch the entire phase plane.

17.
$$\frac{dx}{dt} = (x - y)(1 - x - y)$$
$$\frac{dy}{dt} = x(2 - y)$$

18.
$$\frac{dx}{dt} = (x + y)(1 - x - y)$$
$$\frac{dy}{dt} = x(2 - y)$$

19.
$$\frac{dx}{dt} = (x + y)(1 - x - y)$$
$$\frac{dy}{dt} = x(2 + y)$$

20.
$$\frac{dx}{dt} = (x + y)(-x - y)$$
$$\frac{dy}{dt} = x(2 - y)$$

21. Consider the system

$$\frac{dx}{dt} = -x^3$$
$$\frac{dy}{dt} = -y + y^2.$$

Note that this system has equilibrium points at $(x, y) = (0, 0)$ and $(x, y) = (0, 1)$.

(a) Find the linearized system at $(0, 0)$.

(b) Find the eigenvalues and eigenvectors and sketch the phase plane of the linearized system at $(0, 0)$.

(c) Find the linearized system at $(0, 1)$.

(d) Find the eigenvalues and eigenvectors and sketch the phase plane of the linearized system at $(0, 1)$.

(e) Sketch the phase plane of the nonlinear system. [*Hint*: The equations decouple, so first draw two phase lines.]

(f) Describe why the phase plane for the linearized systems and the phase plane for the nonlinear system near the equilibrium points look so different.

22. When a nonlinear system depends on a parameter, then as the parameter changes, the equilibrium points can change. That is, as the parameter changes, a bifurcation can occur. Consider the one-parameter system

$$\frac{dx}{dt} = x^2 - a$$
$$\frac{dy}{dt} = -y(x^2 + 1)$$

where a is a parameter.

(a) Show that for $a < 0$ the system has no equilibrium points.

(b) Show that for $a > 0$ the system has two equilibrium points.

(c) Show that when $a = 0$ the system has exactly one equilibrium point.

(d) Find the linearization of the equilibrium point when $a = 0$ and compute the eigenvalues for this equilibrium point.

Remark: The system changes from having no fixed points to having two fixed points when the parameter a is increased through $a = 0$. We say that the system has a bifurcation when $a = 0$ and that $a = 0$ is a bifurcation point.

23. Continuing the study of the system

$$\frac{dx}{dt} = x^2 - a$$
$$\frac{dy}{dt} = -y(x^2 + 1)$$

we started in Exercise 22,

- (a) use the nullclines and the direction field to sketch the phase plane for the system when $a = -1$;
- (b) use the nullclines, the direction field and the linearization at the equilibrium point to sketch the phase plane for $a = 0$; and
- (c) use the nullclines, the direction field and the linearization at the equilibrium points to sketch the phase plane for $a = 1$.

Remark: The transition from a system with no equilibrium points, to a system with one equilibrium point with zero as an eigenvalue, and then to a system with saddle and sink equilibrium points is typical of bifurcations which create fixed points. More examples follow.

In Exercises 24–29 each system depends on a parameter a. In each case,

- (a) find all equilibrium points;
- (b) determine all values of a at which a bifurcation occurs; and
- (c) in a short paragraph complete with pictures, describe the phase plane at, before, and after each bifurcation value.

24.
$$\frac{dx}{dt} = y - x^2$$
$$\frac{dy}{dt} = y - a$$

25.
$$\frac{dx}{dt} = y - x^2$$
$$\frac{dy}{dt} = a$$

26.
$$\frac{dx}{dt} = y - x^2$$
$$\frac{dy}{dt} = y - x - a$$

27.
$$\frac{dx}{dt} = y - ax^3$$
$$\frac{dy}{dt} = y - x$$

28.
$$\frac{dx}{dt} = y - x^2 + a$$
$$\frac{dy}{dt} = y + x^2 - a$$

29.
$$\frac{dx}{dt} = y - x^2 + a$$
$$\frac{dy}{dt} = y + x^2$$

30. The following is a model for a pair of competing species for which dy/dt depends on a parameter a.

$$\frac{dx}{dt} = x(-x - y + 70)$$
$$\frac{dy}{dt} = y(-2x - y + a)$$

Find the two values of a for which there are bifurcations. Describe the fate of the x and y populations before and after each bifurcation.

31. Suppose two species X and Y are to be introduced onto an island. It is known that the two species compete, but the precise nature of their interactions is unknown. We assume that the populations $x(t)$ and $y(t)$ of X and Y, respectively, at time t are modeled by a system

$$\frac{dx}{dt} = f(x, y)$$
$$\frac{dy}{dt} = g(x, y).$$

(a) Suppose $f(0, 0) = g(0, 0) = 0$, that is, $x = y = 0$ is an equilibrium point. What does this say about the ability of X and Y to migrate to the island?

(b) Suppose that a small population of just X or just Y will rapidly reproduce. What does this imply about $\partial f / \partial x(0, 0)$ and $\partial g / \partial y(0, 0)$?

(c) Since X and Y compete for resources, the presence of either of the species will decrease the rate of growth of the population of the other. What does this say about $\partial f / \partial y(0, 0)$ and $\partial g / \partial x(0, 0)$?

(d) Using the assumptions from parts (a) through (c), what type(s) of equilibrium point (sink, source, center, and so on) could $(0, 0)$ possibly be? [*Hint*: There may be more than one possibility; if so, list them all.]

(e) For each of the possibilities listed above, sketch a possible phase plane near $(0, 0)$.

Be sure to justify all answers.

32. Suppose two identical countries Y and Z are engaged in an arms race. Let $y(t)$ and $z(t)$ denote the size of the arms stockpiles of Y and Z, respectively, at time t. We model this situation, as usual, with a system of differential equations

$$\frac{dy}{dt} = h(y, z)$$
$$\frac{dz}{dt} = k(y, z).$$

Suppose that all we know about the functions h and k are the following:

(i) If country Z's arms stockpile is not changing, then any increase in size of Y's arms stockpile results in a decrease in the rate of arms building in country Y. The same is true for country Z.

(ii) If either country increases its arms, the other responds by increasing its rate of arms production.

(a) What do the above imply about $\partial h / \partial y$ and $\partial k / \partial z$?

(b) What do the above imply about $\partial h / \partial z$ and $\partial k / \partial y$?

(c) What types of equilibrium points (source, sink, spiral source, and so on) are possible for this system? Justify your answer. [*Hint*: Suppose you have an equilibrium point. What do your results in the previous two parts imply about the Jacobian matrix at that equilibrium point?]

33. For species X and Y as in the previous two exercises, suppose that species X reproduces very quickly if it is on the island without any Y's present, and that species Y reproduces slowly if there are no X's present. Also suppose that the growth rate of species X is decreased a relatively large amount by the presence of Y, but that species Y is indifferent to X's population.

(a) What can you say about $\partial f / \partial x(0, 0)$ and $\partial g / \partial y(0, 0)$?

(b) What can you say about $\partial f / \partial y(0, 0)$ and $\partial g / \partial x(0, 0)$?

(c) What are the possible type(s) (sink, source, center, and so on) for the equilibrium point at $(0, 0)$? [*Hint*: More than one type may be possible; if so, list them all.]

(d) For each of the types listed above, sketch the phase plane near $(0, 0)$.

Be sure to justify all answers.

34. For the two species X and Y of the previous exercise, suppose that both X and Y reproduce very slowly. Also suppose that competition between these two species is very intense.

(a) What can you conclude about $\partial f/\partial x(0, 0)$ and $\partial g/\partial y(0, 0)$?

(b) What can you conclude about $\partial f/\partial y(0, 0)$ and $\partial g/\partial x(0, 0)$?

(c) What type(s) of equilibrium point (sink, source, center, and so on) could $(0, 0)$ possibly be? [*Hint*: More than one type may be possible; if so, list all the possibilities.]

(d) For each of the possible types of equilibrium point, sketch a possible phase plane near $(0, 0)$.

Remember to justify all answers.

4.2 HAMILTONIAN SYSTEMS

As we have emphasized many times, nonlinear systems of differential equations are almost impossible to solve explicitly. We have also seen that solution curves of systems may behave in many different ways, and there are no qualitative techniques that are guaranteed to work in all cases. Fortunately there are certain special types of nonlinear systems that arise often in practice and for which there are special techniques that enable us to gain some understanding of the phase plane. In this and the next section we will discuss two of these special types of nonlinear systems. But first, we pause for a story.

How This Book Came To Be

Paul and Glen have been writing a differential equations textbook for the past ten years. They want their book to be filled with brilliant and witty new ideas about differential equations, but they are finding that new ideas are hard to come by.

More troubling to them is the following observation that they have both made over the years. Whenever Paul comes to the office in the morning with a witty new idea, they both work feverishly on his idea, and at first their creative juices flow. They work diligently, but eventually the idea doesn't pan out. So their enthusiasm wanes and with it, their creativity. On the other hand, whenever Glen comes in the morning with a new idea, no matter how trivial, something different happens. They again both work feverishly. Sometimes Paul's enthusiasm wanes, but Glen is always excited. There is good give and take; they never give up. At least something comes of the idea and the book progresses. Now this bothered Paul and Glen. Why should their creative energy depend so critically on who has the first idea? They decided to model their plight with a system of differential equations.

They let $x(t)$ denote Glen's and $y(t)$ Paul's level of enthusiasm at time t. Now $x(t)$ and $y(t)$ are difficult to measure because of a lack of standardized units of enthusiasm. However, it is clear that $y(t) > 0$ means that Paul is enthusiastic, whereas $y(t) < 0$ means that Paul is glum. When $y(t) = 0$, Paul just sits there, neither happy nor sad, and similarly for Glen.

Both Glen and Paul have observed that Glen's enthusiasm changes at a rate directly proportional to Paul's level of enthusiasm. When Paul is excited, Glen gets more enthusiastic, but when Paul is glum, Glen loses his verve. A simple equation catching this behavior is

$$\frac{dx}{dt} = y.$$

Paul is a bit more difficult to categorize. When Glen is mildly enthusiastic, Paul's enthusiasm goes up. But when Glen gets wildly excited, Paul starts to lose enthusiasm. Apparently, when the torrent of ideas spilling out of Glen becomes too great, Paul gets a headache and tunes Glen out. On the other hand, when Glen is down, Paul is really glum. For the rate of change of Paul's enthusiasm then, use

$$\frac{dy}{dt} = x - x^2.$$

The graph of $dy/dt = x - x^2$ as a function of x (Figure 4.13) captures exactly Paul's enthusiasm level. We can see that $dy/dt > 0$ if $0 < x < 1$, but $dy/dt < 0$ otherwise.

The system of differential equations they settle on is

$$\frac{dx}{dt} = y$$
$$\frac{dy}{dt} = x - x^2.$$

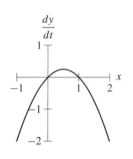

Figure 4.13
The graph of
$dy/dt = x - x^2$ as a
function of x.

The direction field for this system is shown in Figure 4.14. There are two equilibrium points, one at the origin and one at $x = 1$, $y = 0$. The technique of linearization from Section 4.1 can be used to study the solutions near the equilibrium points. At the origin the Jacobian matrix is

$$\begin{pmatrix} 0 & 1 \\ 1 & 0 \end{pmatrix},$$

which has eigenvalues ± 1. Hence the origin is a saddle.

The equilibrium point at $(x, y) = (1, 0)$ has Jacobian matrix

$$\begin{pmatrix} 0 & 1 \\ -1 & 0 \end{pmatrix},$$

which has eigenvalues $\pm i$. The linearized system is a center. As we saw in Section 4.1, this is one of the cases when the long-term behavior of solutions of the nonlinear system near the equilibrium point is not completely determined by the linearization. The

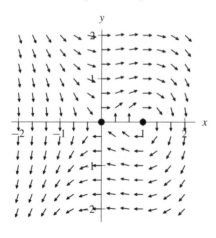

Figure 4.14
The direction field for

$$\frac{dx}{dt} = y$$
$$\frac{dy}{dt} = x - x^2.$$

behavior of the nonlinear system near the equilibrium point $(1, 0)$ could be that of a spiral sink, a spiral source, or a center.

Numerical approximations of solutions give the phase plane shown in Figure 4.15. This figure implies that solutions behave in a very regular way. The solutions with initial conditions near $(1, 0)$ seem to form closed loops corresponding to periodic solutions. Also, the unstable and stable separatrices emerging from the saddle at the origin into the first and fourth quadrants seem to form a single loop.

The qualitative techniques we have studied so far do not allow us to predict that this system will have such regular solution curves. Also, because the numerical approximations of solutions are only approximations, we should be cautious. Distinguishing solution curves that form closed loops from those that spiral very slowly can be very difficult. We would like to have techniques that can be used to verify the special behavior of this system.

Conserved quantities

Paul and Glen were so mystified by the behavior of their system that they decided to show it to their friend Bob. Bob exclaimed that he had seen this system before and that it had a conserved quantity. Noting the confused looks he got from Paul and Glen, he explained:

Definition: Suppose

$$\frac{dx}{dt} = f(x, y)$$

$$\frac{dy}{dt} = g(x, y)$$

is a system of differential equations. A real-valued function $H(x, y)$ of the two variables x and y is a **conserved quantity** if H is constant along all solution curves of the system. That is, if $(x(t), y(t))$ is a solution of the system, then $H(x(t), y(t))$ is constant, or

$$\frac{d}{dt} H(x(t), y(t)) = 0.$$

Bob remembered that

$$H(x, y) = \frac{1}{2}y^2 - \frac{1}{2}x^2 + \frac{1}{3}x^3$$

is a conserved quantity for the system in question. To check this, suppose $(x(t), y(t))$ is a solution of the system. Then we compute

$$\begin{aligned}
\frac{d}{dt} H(x(t), y(t)) &= \frac{\partial H}{\partial x} \cdot \frac{dx}{dt} + \frac{\partial H}{\partial y} \cdot \frac{dy}{dt} \\
&= (-x + x^2) \cdot y + y \cdot (x - x^2) \\
&= 0,
\end{aligned}$$

where the first equality follows from the chain rule and the second equality uses the fact that $(x(t), y(t))$ is a solution to replace dx/dt with y and dy/dt with $x - x^2$.

This means that the solution curves always lie along the level curves of H. The level curves of H are shown in Figure 4.16. We see in this picture that around the equilibrium point $(1, 0)$, the level curves of H form closed circles and the branches of

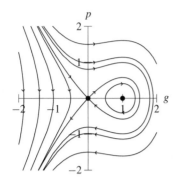

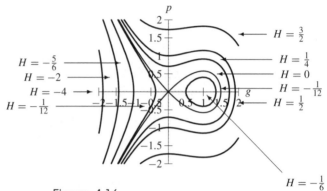

Figure 4.15
The phase plane of the system

Figure 4.16
The level curves of H.

$$\frac{dx}{dt} = y$$

$$\frac{dy}{dt} = x - x^2.$$

the level curve emanating from the origin toward the right connect into a loop. This agrees with our analysis of the phase plane for the system (see Figure 4.15).

The fact that the right-hand branches of the stable and unstable separatrices of the origin form a single loop is very special. This type of solution curve is a **saddle connection**. Inside the saddle connection, all solution curves are periodic, whereas outside the saddle connection, all solutions have the property that both $x(t)$ and $y(t)$ tend to $-\infty$ as $t \to \infty$. This explains, to Paul and Glen's great relief, why their daily productivity depends so crucially on who has the first idea. If $y(0) > 0$ but $x(0) = 0$, the solution curve eventually tends to $x = y = -\infty$. But if $x(0) > 0$, $y(0) = 0$ [with $x(0)$ not too large], then both $x(t)$ and $y(t)$ are periodic, with $x(t) > 0$ for all t. We sketch graphs of $x(t)$ and $y(t)$ for initial conditions near $(0, 0)$ in Figure 4.17.

This was a source of great satisfaction to both Paul and Glen. So much was their happiness that they invited Bob to become a coauthor of their book. Bob reluctantly agreed, but only after Glen and Paul promised to listen to *The Marriage of Figaro* in its entirety.

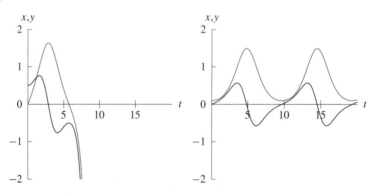

Figure 4.17
Graphs of $x(t)$ and $y(t)$ for solutions with initial conditions $y(0) > 0$, $x(0) = 0$ and $y(0) = 0$, $0 < x(0) < 1$.

Hamiltonian Systems

What makes the preceding analysis work is the fact that the system of differential equations modeling Glen and Paul's enthusiasm levels is a special kind of system called a Hamiltonian system. [named for William Rowan Hamilton (1805–1865), an Irish mathematician].

Definition: A system of differential equations

$$\frac{dx}{dt} = f(x, y)$$
$$\frac{dy}{dt} = g(x, y)$$

is called a **Hamiltonian system** if there exists a real-valued function $H(x, y)$ such that for all x and y,

$$f(x, y) = \frac{\partial H}{\partial y}(x, y)$$

$$g(x, y) = -\frac{\partial H}{\partial x}(x, y).$$

The function H is called the *Hamiltonian function* for the system.

Note that H is always a conserved quantity for such a system. We can verify this by letting $(x(t), y(t))$ be any solution of the system; then

$$\frac{d}{dt} H(x(t), y(t)) = \frac{\partial H}{\partial x} \cdot \frac{dx}{dt} + \frac{\partial H}{\partial y} \cdot \frac{dy}{dt}$$
$$= \frac{\partial H}{\partial x} \left(\frac{\partial H}{\partial y} \right) + \frac{\partial H}{\partial y} \left(-\frac{\partial H}{\partial x} \right)$$
$$= 0.$$

The first equality is the chain rule and the second equality uses the fact that the system is Hamiltonian and that $(x(t), y(t))$ is a solution to replace dx/dt with $\partial H/\partial y$ and dy/dt with $-\partial H/\partial x$.

So, as above, solution curves of the system lie along the level curves of H; that is, the curves $H(x, y) = $ constant. Sketching the phase plane for a Hamiltonian system is the same as sketching the level sets of the Hamiltonian function.

Examples of Hamiltonian Systems:
The Harmonic Oscillator

Recall that the undamped harmonic oscillator system is

$$\frac{dy}{dt} = v$$
$$\frac{dv}{dt} = -qy,$$

where q is a positive constant. If we let

$$H(y, v) = \frac{1}{2}v^2 + \frac{q}{2}y^2,$$

then

$$\frac{dy}{dt} = v = \frac{\partial H}{\partial v}$$

and

$$\frac{dv}{dt} = -qy = -\frac{\partial H}{\partial y}.$$

Hence the undamped harmonic oscillator system is a Hamiltonian system. The level sets of the function H are ellipses in the yv-plane that correspond to the phase-plane picture of the solutions of the undamped harmonic oscillator (see Section 3.4). This Hamiltonian function is sometimes called the **energy function** for the oscillator.

The Nonlinear Pendulum

Consider a pendulum made of a light rigid rod of length l with a ball at one end of mass m. The ball is called the *bob* and the rigid rod the *arm* of the pendulum. We assume that all the mass of the pendulum is in the bob, neglecting the mass of the rod. The other end of the rigid rod is attached to the wall in such a way that it can turn through an entire circle in a plane perpendicular to the ground. The position of the bob at time t is given by an angle $\theta(t)$, which we choose to measure in the counterclockwise direction with $\theta = 0$ corresponding to the downward vertical axis (see Figure 4.18).

We assume that there are only two forces acting on the pendulum; gravity and friction. The constant gravitational force equal to mg is in the downward direction, where g is the acceleration of gravity near earth ($g \approx 9.8\text{m/s}^2$). Only the component of this force tangent to the circle of motion affects the motion of the pendulum. This component is $-mg \sin\theta$ (see Figure 4.19). There is also a force due to friction, which we assume to be proportional to the velocity of the bob.

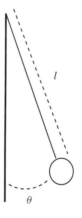

Figure 4.18
Picture of a pendulum with
rod length l and angle θ.

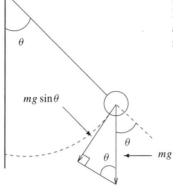

Figure 4.19
Decomposition of the force of gravity into components along the pendulum arm and tangent to the circle of motion of the pendulum bob.

The position of the bob at time t is given by the point $(l \sin \theta(t), -l \cos \theta(t))$ on the circle of radius l (remember that $\theta = 0$ corresponds to straight down). The speed of the bob is the length of the velocity vector, which is $l d\theta/dt$. The component of the acceleration that points along the direction of the motion of the bob has length $l d^2\theta/dt^2$. We take the force due to friction to be proportional to the velocity, so this force is $-k_d(l d\theta/dt)$, where $k_d > 0$ is a parameter that corresponds to the *coefficient of damping*.

Putting all this together with Newton's law "force equals mass times acceleration," we find the equations of motion,

$$ml \frac{d^2\theta}{dt^2} = -k_d l \frac{d\theta}{dt} - mg \sin \theta$$

or

$$\frac{d^2\theta}{dt} + \frac{k_d}{m} \frac{d\theta}{dt} + \frac{g}{l} \sin \theta = 0.$$

We can write this as a nonlinear system in the usual manner by letting $v = d\theta/dt$ be the velocity. Then the equations become

$$\frac{d\theta}{dt} = v$$
$$\frac{dv}{dt} = -\frac{k_d}{m} v - \frac{g}{l} \sin \theta.$$

Remember that θ is an angular variable, so that $\theta = 0$, $\theta = 2\pi$, $\theta = 4\pi$, and so forth all measure the same position: the pendulum passing through its lowest point. The values $\theta = \pi$, 3π, 5π, ... occur when the pendulum is at the highest point of its circular motion.

The ideal pendulum

In practice, there is always a force due to friction acting on a pendulum. For the moment, however, suppose there is no friction in this model. This is an "ideal" case that does not occur in the real world. However, it is not an unreasonable model for a very well-built and well-lubricated pendulum.

When no friction is present, the coefficient k_d vanishes. For convenience we suppose that the pendulum arm has unit length; that is, we suppose $l = 1$. (We consider the effect of adjusting the arm length in the exercises.) The equations of motion of the ideal pendulum with unit length arm are

$$\frac{d\theta}{dt} = v$$

$$\frac{dv}{dt} = -g \sin \theta.$$

The first step is to find the equilibrium points. We must have $v = 0$ from the first equation and $\sin \theta = 0$ from the second, so that $v = 0$, $\theta = n\pi$, where n is any integer, yield all the equilibrium points. When θ is an even multiple of π and $v = 0$, the bob hangs motionless in a downward position, an obvious equilibrium position for the pendulum. When θ is an odd multiple of π and $v = 0$, there is also a rest position corresponding to the bob balanced perfectly motionless in an upright position. This kind of equilibrium is hard to see in practice: Just when you manage to balance the pendulum perfectly, someone twitches, and the resulting current of air moves the pendulum out of equilibrium and into motion in one direction or the other.

We give a method below for determining whether a given system is a Hamiltonian system and, if so, how to compute the Hamiltonian. For the moment we use the rabbit-out-of-the-hat method and claim that the differential equations for the ideal pendulum form a Hamiltonian system, with the Hamiltonian function given by

$$H(\theta, v) = \frac{1}{2}v^2 - g \cos \theta.$$

This follows from the computation of partial derivatives of H,

$$\frac{\partial H}{\partial \theta} = g \sin \theta = -\frac{dv}{dt}, \quad \frac{\partial H}{\partial v} = v = \frac{d\theta}{dt}.$$

We can describe the graph of the function H as follows: For each fixed θ, the $v^2/2$ term implies that the graph of H is a parabola in the v direction. The critical points occur at $v = 0$, $g \sin \theta = 0$ or $\theta = 0, \pm\pi, \pm 2\pi, \ldots$. The points $v = 0$, $\theta = 0, \pm 2\pi, \pm 4\pi, \ldots$ are local minima, and the points $v = 0$, $\theta = \pm\pi, \pm 3\pi, \ldots$ are saddle points of the graph.

The level curves of H are plotted in Figure 4.20. We know that the vector field is always tangent to these curves. The θ component of the vector field equals v, so the vector field points to the right when $v > 0$ and to the left when $v < 0$. Using this fact, we can assign directions along these curves and thereby determine the phase portrait. This is shown in Figure 4.21.

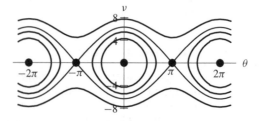

Figure 4.20
Level curves for the ideal pendulum.

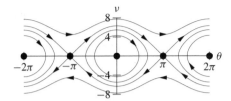

Figure 4.21
Phase plane for the ideal pendulum.

Figure 4.22
Special solutions for the pendulum
equations.

There are three different types of solution curves present in the phase-plane picture. These three types are shown schematically in Figure 4.22 and labeled A, B, and C.
Around the equilibrium points at $v = 0$, $\theta = 0$, $\pm 2\pi$, $\pm 4\pi$, ... we find periodic solutions traversed in the clockwise direction. These are the type A solutions in Figure 4.22.
Since the θ-value along these solution curves never reaches π, 3π, ..., it follows that
the pendulum never passes through the upright position. Consequently, the pendulum
simply oscillates back and forth periodically with the maximum and minimum θ-values
determined by where the solution curve crosses the θ-axis. This is the usual swinging
motion we associate with a pendulum. The graph of $y(t)$ for such a solution is shown
in Figure 4.23.

On the other hand, a solution curve such as B in Figure 4.22 corresponds to the
pendulum rotating forever in a counterclockwise direction. Note that $v \neq 0$ along such
a solution curve, so the pendulum never reaches a point where its velocity is 0. The
graph of $y(t)$ for such a solution is increasing for all t if $v > 0$ and decreasing for all t
if $v < 0$ (see Figure 4.23).

The intermediate types of solutions (see C in Figure 4.22) are separatrices of saddle
equilibrium points forming saddle connections. These solutions tend toward and come
from the upright equilibrium position as time tends to $\pm\infty$. The $y(t)$-graph of such a
solution is shown in Figure 4.23. In order to be on such a solution, we must choose
the initial angle and velocity perfectly. With slightly too high an initial velocity, the
pendulum swings past the vertical position; with slightly to low an initial velocity, the
pendulum falls back. The separatrices of the saddle equilibrium points separate the
"oscillating" solutions, which oscillate back and forth, from the "rotating" solutions,
which swing around and around.

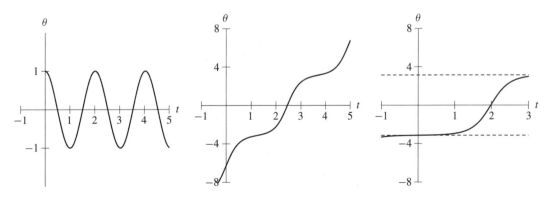

Figure 4.23
Graphs of $y(t)$ for the three different types of solutions represented in Figure 4.22.

The value of the Hamiltonian function for the ideal pendulum at a particular point (y, v) is called the **energy**. The physical principle of conservation of energy applies to the ideal pendulum in the same way that it does to the undamped harmonic oscillator. This is one way we could predict that the undamped harmonic oscillator and the ideal pendulum are Hamiltonian systems. A more mathematical approach is given next.

Finding Hamiltonian Systems

Hamiltonian systems are special kinds of systems of differential equations in several senses. As we have seen, we can "solve" a Hamiltonian system in the plane in a qualitative sense once we know the Hamiltonian function. All we need to do is plot the level curves of H and sketch in the directions to determine the phase portrait. Unfortunately Hamiltonian systems are fairly rare. Given a system, we would like to be able to determine whether or not it is a Hamiltonian system and, if it is, to determine the Hamiltonian function.

Suppose we have a system of equations

$$\frac{dx}{dt} = f(x, y)$$

$$\frac{dy}{dt} = g(x, y)$$

and wish to check whether it is Hamiltonian. We ask whether there exists a function $H(x, y)$ such that for all (x, y),

$$f(x, y) = \frac{\partial H}{\partial y}(x, y)$$

and

$$g(x, y) = -\frac{\partial H}{\partial x}(x, y).$$

If such a function H exists (and has continuous second partial derivatives), then

$$\frac{\partial^2 H}{\partial x \partial y} = \frac{\partial^2 H}{\partial y \partial x}.$$

Therefore if the system is Hamiltonian, then

$$\frac{\partial f}{\partial x} = \frac{\partial^2 H}{\partial x \partial y} = \frac{\partial^2 H}{\partial y \partial x} = -\frac{\partial g}{\partial y}.$$

That is, to check whether a system may be Hamiltonian, we compute $\partial f/\partial x$ and $\partial g/\partial y$ and check if $\partial f/\partial x = -\partial g/\partial y$. If this equation does not hold for all (x, y) then the system is not Hamiltonian.

For example, consider the nonlinear system

$$\frac{dx}{dt} = f(x, y) = x + y^2$$

$$\frac{dy}{dt} = g(x, y) = y^2 - x.$$

We compute

$$\frac{\partial f}{\partial x} = 1 \neq -2y = -\frac{\partial g}{\partial y}.$$

Therefore this system is not a Hamiltonian system.

Constructing Hamiltonian functions

For the system

$$\frac{dx}{dt} = f(x, y)$$
$$\frac{dy}{dt} = g(x, y),$$

if

$$\frac{\partial f}{\partial x} = -\frac{\partial g}{\partial y},$$

then the system is Hamiltonian. We can verify this by actually constructing the Hamiltonian function as follows: If we are to have

$$f(x, y) = \frac{\partial H}{\partial y}(x, y),$$

then integrating both sides of this equation with respect to y, we must have

$$H(x, y) = \int f(x, y) dy$$

up to a "constant of integration" that may depend on x. That is, we can write

$$H(x, y) = \int f(x, y) dy + \phi(x),$$

where ϕ is some function of x alone that is to be determined. To find ϕ, we simply differentiate $H(x, y)$ with respect to x and equate the result with $-g(x, y)$. We find

$$\frac{\partial H}{\partial x} = \frac{\partial}{\partial x} \int f(x, y) dy + \phi'(x) = -g(x, y).$$

That is,

$$\phi'(x) = -g(x, y) - \frac{\partial}{\partial x} \int f(x, y) dy.$$

Integration of the right-hand side with respect to x then determines ϕ and consequently H.

Two examples

Recall the system of differential equations governing Glen and Paul's book-writing dilemma:

$$\frac{dx}{dt} = f(x, y) = y$$
$$\frac{dy}{dt} = g(x, y) = x - x^2.$$

Since

$$\frac{\partial f}{\partial x} = 0 = -\frac{\partial g}{\partial y},$$

we know this system is Hamiltonian. To find H, we first integrate f with respect to y, finding

$$H(x, y) = \int y \, dy = \frac{1}{2}y^2 + \phi(x).$$

Next we must have

$$x - x^2 = -\frac{\partial H}{\partial x}$$

$$= -\frac{\partial}{\partial x}\left(\frac{1}{2}y^2 + \phi(x)\right)$$

$$= -\phi'(x).$$

Integrating $\phi'(x) = -x + x^2$ with respect to the variable x, we find

$$\phi(x) = -\frac{1}{2}x^2 + \frac{1}{3}x^3,$$

so

$$H(x, y) = \frac{1}{2}y^2 - \frac{1}{2}x^2 + \frac{1}{3}x^3.$$

This is exactly the function we found earlier. Of course, we could add an arbitrary constant to H, but this would not change the shape of the level curves of H.

As another example, consider the system

$$\frac{dx}{dt} = f(x, y) = -x \sin y + 2y$$

$$\frac{dy}{dt} = g(x, y) = -\cos y.$$

This system is Hamiltonian since

$$\frac{\partial f}{\partial x} = -\sin y = -\frac{\partial g}{\partial y}$$

for all x and y. So we first integrate f with respect to y, finding

$$H(x, y) = x \cos y + y^2 + \phi(x).$$

Then we must have

$$-\cos y = -\frac{\partial H}{\partial x}$$

$$= -\frac{\partial}{\partial x}(x \cos(y) + y^2 + \phi(x))$$

$$= -\cos y + \phi'(x).$$

Therefore we can choose ϕ to be any constant, say 0, and our Hamiltonian function is

$$H(x, y) = x \cos y + y^2.$$

We study this system more closely in the exercises.

Equilibrium Points of Hamiltonian Systems

Hamiltonian systems have a number of special properties not shared by general systems of differential equations. For example, suppose (x_0, y_0) is our equilibrium point for the Hamiltonian system

$$\frac{dx}{dt} = \frac{\partial H}{\partial y}(x, y)$$

$$\frac{dy}{dt} = -\frac{\partial H}{\partial x}(x, y).$$

The Jacobian matrix at this equilibrium point is given by

$$\begin{pmatrix} \dfrac{\partial^2 H}{\partial x \partial y} & \dfrac{\partial^2 H}{\partial y^2} \\ -\dfrac{\partial^2 H}{\partial x^2} & -\dfrac{\partial^2 H}{\partial y \partial x} \end{pmatrix},$$

where each of these partial derivatives is evaluated at (x_0, y_0). Since

$$\frac{\partial^2 H}{\partial x \partial y} = \frac{\partial^2 H}{\partial y \partial x},$$

the Jacobian matrix assumes the form

$$\begin{pmatrix} \alpha & \beta \\ \gamma & -\alpha \end{pmatrix},$$

where $\alpha = \partial^2 H / \partial x \partial y$, $\beta = \partial^2 H / \partial y^2$, and $\gamma = -\partial^2 H / \partial x^2$.

The characteristic polynomial of this matrix is

$$(\alpha - \lambda)(-\alpha - \lambda) - \beta\gamma = \lambda^2 - \alpha^2 - \beta\gamma.$$

This means that the eigenvalues are given by the roots of

$$\lambda^2 = \alpha^2 + \beta\gamma,$$

which are

$$\lambda = \pm\sqrt{\alpha^2 + \beta\gamma}.$$

Thus we see that there are only three possibilities for the eigenvalues.

1. If $\alpha^2 + \beta\gamma > 0$, both eigenvalues are real and have opposite signs.

2. If $\alpha^2 + \beta\gamma < 0$, both eigenvalues are imaginary with real part equal to zero.

3. If $\alpha^2 + \beta\gamma = 0$, then 0 is the only eigenvalue.

In particular, it is not possible to have eigenvalues that are complex and have nonzero real parts for a Hamiltonian system in the plane. So solutions cannot spiral into or away from an equilibrium point. Similarly, we cannot have two positive or two negative eigenvalues. Thus Hamiltonian systems cannot have equilibrium points that are sinks or sources. This gives us another preliminary indication whether a given system is Hamiltonian: If the direction field indicates the presence of sinks or sources, it is not Hamiltonian. Two phase planes are pictured in Figure 4.24. We can say for sure that the phase plane on the left does not come from a Hamiltonian system because it has an equilibrium point that is a sink. The phase plane on the right might be that of a Hamiltonian system. The only way to determine whether it is Hamiltonian is to study the formulas for the system as we have done above.

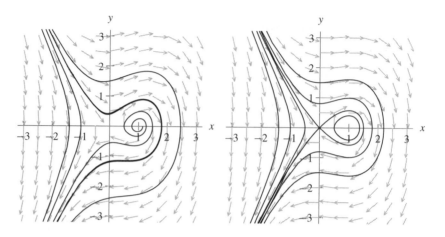

Figure 4.24
The phase plane on the left cannot be a Hamiltonian system, whereas the phase plane on the right might be Hamiltonian.

Exercises for Section 4.2

1. For the system

$$\frac{dx}{dt} = y$$

$$\frac{dy}{dt} = x^3 - x,$$

(a) show that the system is Hamiltonian with Hamiltonian function

$$H(x, y) = \frac{y^2}{2} + \frac{x^2}{2} - \frac{x^4}{4};$$

(b) sketch the level sets of H; and

(c) sketch the phase plane for the system. Include a description of all equilibrium points and any saddle connections.

2. For the system

$$\frac{dx}{dt} = x \cos(xy)$$

$$\frac{dy}{dt} = -y \cos(xy);$$

(a) show that the system is a Hamiltonian system with Hamiltonian function

$$H(x, y) = \sin(xy);$$

(b) sketch the level sets of H; and

(c) sketch the phase plane of the system. Include a description of all equilibrium points and any saddle connections.

3. For the system

$$\frac{dx}{dt} = -x \sin y + 2y$$

$$\frac{dy}{dt} = -\cos y;$$

(a) check that the system is a Hamiltonian system with Hamiltonian function

$$H(x, y) = x \cos y + y^2;$$

(b) sketch the level sets of H; and

(c) sketch the phase plane of the system. Include a description of all equilibrium points and any saddle connections.

[*Hint*: Take advantage of whatever technology you have available for drawing level sets and phase planes. Then interpret the pictures you obtain.]

In Exercises 4–8, we continue the study of the ideal pendulum system with bob mass m and arm length l given by

$$\frac{d\theta}{dt} = v$$

$$\frac{dv}{dt} = -\frac{g}{l} \sin \theta.$$

4. (a) What is the linearization of the ideal pendulum system above at the equilibrium point $(0, 0)$?

(b) Using $g = 9.8$ meters/sec^2, how should l and m be chosen so that small swings of the pendulum have period 1 second?

5. For the linearization of the ideal pendulum above at $(0, 0)$, the period of the oscillation is independent of the amplitude. Does the same statement hold for the ideal pendulum itself? Is the period of oscillation the same no matter how high the ideal pendulum swings? If not, will the period be shorter or longer for high swings?

6. An ideal pendulum clock — a clock containing an ideal pendulum which "ticks" once for each swing of the pendulum arm — keeps perfect time when the pendulum makes very high swings. Will the clock run slow or fast if the amplitude of the swings is very small?

7. (a) If the arm length of the ideal pendulum is doubled from l to $2l$, what is the effect on the period of small amplitude swinging solutions?

(b) What is the rate of change of the period of small amplitude swings as l is varied?

8. Will an ideal pendulum clock that keeps perfect time on earth run slow or fast on the moon?

In Exercises 9–12, determine whether the given system is Hamiltonian. If so, find a Hamiltonian function.

9. $\dfrac{dx}{dt} = x - 3y^2$

$\dfrac{dy}{dt} = -y$

10. $\dfrac{dx}{dt} = \sin x \cos y$

$\dfrac{dy}{dt} = -\cos x \sin y$

11.
$$\frac{dx}{dt} = x \cos y$$
$$\frac{dy}{dt} = -y \cos x$$

12.
$$\frac{dx}{dt} = 1$$
$$\frac{dy}{dt} = y$$

13. Which of the following phase planes could possibly correspond to that of a Hamiltonian system? If not, explain why.

a.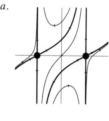

b.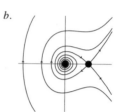

c.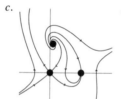

d.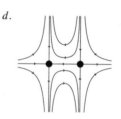

14. Consider the system of differential equations of the form

$$\frac{dx}{dt} = F(y)$$
$$\frac{dy}{dt} = G(x).$$

That is, dx/dt depends only on y, and dy/dt depends only on x. Show that this system is Hamiltonian. What is the Hamiltonian function?

15. Consider the nonlinear system

$$\frac{dx}{dt} = 1 - y^2$$
$$\frac{dy}{dt} = x(1 + y^2).$$

(a) Is this a Hamiltonian system?

(b) Use a computer to plot the phase portrait of this system. Do you think that there is a constant of motion for this system? Can you find one?

(c) Suppose we multiply this vector field by the function

$$f(x, y) = \frac{1}{1 + y^2}.$$

The new vector field is

$$\frac{dx}{dt} = \frac{1 - y^2}{1 + y^2}$$

$$\frac{dy}{dt} = x.$$

Is this system Hamiltonian?

(d) Can you now find a constant of motion for the original system?

(e) Why does this method work?

16. Consider the nonlinear system

$$\frac{dx}{dt} = -yx^2$$

$$\frac{dy}{dt} = x + 1.$$

(a) Is this a Hamiltonian system?

(b) Can you find a constant of motion? *Hint*: Using the technique of the previous exercise, multiply this system by a function that puts the new system in the form

$$\frac{dx}{dt} = F(y)$$

$$\frac{dy}{dt} = G(x).$$

17. Consider the system of differential equations

$$\frac{dx}{dt} = 1 - y^2$$

$$\frac{dy}{dt} = x(2 - y).$$

Find a constant of motion for this system.

In Exercises 18–20, we study bifurcations of parameterized families of Hamiltonian systems. Because the equilibrium points of Hamiltonian systems assume a special form, their bifurcations do also.

18. For the system

$$\frac{dx}{dt} = y$$

$$\frac{dy}{dt} = x^2 - a,$$

where a is a parameter, do the following.

(a) Verify that this system is Hamiltonian and with Hamiltonian function

$$H(x, y) = \frac{y^2}{2} - \frac{x^3}{3} + ax.$$

(b) Show that for $a \geq 0$, the system has equilibrium points at $x = \pm\sqrt{a}$, $y = 0$ and no equilibrium points when $a < 0$ (so $a = 0$ is a bifurcation value of the parameter).

(c) Linearize the system at each of the equilibrium points and determine the behavior of solutions near the equilibrium points.

(d) Sketch the level curves of H (and hence the phase plane of the system) for $a = -1$, $a = -0.5$, $a = 0$, $a = 0.5$, and $a = 1$. [*Hint:* Think about the effect of changing a on the graph of H. It is worthwhile to employ whatever graphing technology you have available.]

(e) Describe in a brief paragraph the bifurcation that takes place at $a = 0$.

19. Consider the system of differential equations depending on a parameter a,

$$\frac{dx}{dt} = x^2 + 2xy$$
$$\frac{dy}{dt} = a - 2xy - y^2.$$

Describe in detail the bifurcation that occurs when $a = 0$.

20. Suppose Glen and Paul decide to modify the model of their enthusiasm levels because Paul notes that his enthusiasm level also goes down at a constant rate over time. Therefore the new equations are

$$\frac{dx}{dt} = y$$
$$\frac{dy}{dt} = x - x^2 - a.$$

(a) Describe in detail any major bifurcations that occur in this family.

(b) Suppose $a = 1$. What happens to Glen and Paul's book no matter how enthusiastic both authors start out?

4.3 DISSIPATIVE SYSTEMS

The Hamiltonian systems discussed in the previous chapter are obviously idealized systems. In the real world, pendulums do not swing forever in a periodic motion. Rather they eventually wind down; energy is dissipated. In this section we will discuss in detail these types of dissipative systems. We begin by modifying the ideal pendulum by considering a small damping term.

The Nonlinear Pendulum with Friction

Recall that the second-order equation governing the motion of the pendulum is

$$\frac{d^2\theta}{dt^2} + \frac{k_d}{m}\frac{d\theta}{dt} + \frac{g}{l}\sin\theta = 0,$$

where k_d is the coefficient of damping , m is the mass of the pendulum bob, l is the length of the pendulum arm, and g is the acceleration of gravity ($g \approx 9.8\text{m/s}^2$). All the parameters are positive. The term

$$\frac{k_d}{m}\frac{d\theta}{dt}$$

results from a damping force due, for example, to air resistance or friction at the pivot point of the pendulum arm. We assumed that $k_d = 0$ in our idealized system. Here we do not make this assumption.

Introducing the velocity $v = d\theta/dt$, this second-order differential equation can be written as a nonlinear system in the usual manner,

$$\frac{d\theta}{dt} = v$$

$$\frac{dv}{dt} = -\frac{k_d}{m}v - \frac{g}{l}\sin\theta.$$

This system is no longer Hamiltonian, since our test for existence of a Hamiltonian function fails:

$$\frac{\partial}{\partial\theta}(v) = 0 \neq -\frac{k_d}{m} = -\frac{\partial}{\partial v}\left(-\frac{k_d}{m}v - \frac{g}{l}\sin\theta\right).$$

The equilibrium points and nullclines

We begin the study of the damped pendulum system

$$\frac{d\theta}{dt} = v$$

$$\frac{dv}{dt} = -\frac{k_d}{m}v - \frac{g}{l}\sin\theta$$

by finding the equilibrium points. As for the ideal pendulum, the equilibrium points occur at $\theta = 0, \pm\pi, \pm 2\pi, \ldots, v = 0$.

The Jacobian matrix of the vector field at (θ, v) is

$$\begin{pmatrix} 0 & 1 \\ -\frac{g}{l}\cos\theta & -\frac{k_d}{m} \end{pmatrix}.$$

When $\theta = 0, \pm 2\pi, \pm 4\pi, \ldots$ and $v = 0$ (pendulum hanging downward), this matrix becomes

$$\begin{pmatrix} 0 & 1 \\ -\frac{g}{l} & -\frac{k_d}{m} \end{pmatrix}.$$

The linearized system at these equilibrium points has this matrix as its coefficient matrix. The eigenvalues of this matrix are roots of the characteristic equation

$$-\lambda\left(-\frac{k_d}{m} - \lambda\right) + \frac{g}{l} = 0,$$

or

$$\lambda^2 + \frac{k_d}{m}\lambda + \frac{g}{l} = 0.$$

Hence the eigenvalues are

$$-\frac{k_d}{2m} \pm \sqrt{\left(\frac{k_d}{2m}\right)^2 - \frac{g}{l}}.$$

There are three different cases, depending on the sign of $(k_d/(2m))^2 - g/l$. If this quantity is negative, then the eigenvalues of the Jacobian matrix are complex. The real part is $-k_d/(2m)$, which is negative, so in this case the equilibrium is a spiral sink. When $(k_d/(2m))^2 - g/l > 0$, we find two real distinct eigenvalues. Since

$$0 < \left(\frac{k_d}{2m}\right)^2 - \frac{g}{l} < \left(\frac{k_d}{2m}\right)^2,$$

we have that

$$\frac{k_d}{2m} > \sqrt{\left(\frac{k_d}{2m}\right)^2 - \frac{g}{l}},$$

and both eigenvalues are negative. Finally, when $(k_d/(2m))^2 - g/l = 0$, we have a repeated eigenvalue $-k_d/(2m)$, which is again negative.

Thus in each case we find that the equilibrium points at $\theta = 0, \pm 2\pi, \pm 4\pi, \ldots, v = 0$ are sinks. If we focus attention on the case of a pendulum with only a small amount of friction, then k_d is very small. For k_d sufficiently small, we have that $(k_d/(2m))^2 - g/l$ is negative and the points $\theta = 0, \pm 2\pi, \pm 4\pi, \ldots, v = 0$ are spiral sinks. We deal only with this case here (see the exercises for the other cases).

At the other equilibrium points, $\theta = \pm \pi, \pm 3\pi, \ldots, v = 0$, which correspond to the pendulum standing upright, the Jacobian matrix is

$$\begin{pmatrix} 0 & 1 \\ \dfrac{g}{l} & -\dfrac{k_d}{m} \end{pmatrix},$$

and this is the coefficient matrix for the linearized system at these points. The eigenvalues are

$$-\frac{k_d}{2m} \pm \sqrt{\left(\frac{k_d}{2m}\right)^2 + \frac{g}{l}}.$$

These eigenvalues are always real because k_d, m, l, and g are all positive. The eigenvalue

$$-\frac{k_d}{2m} + \sqrt{\left(\frac{k_d}{2m}\right)^2 + \frac{g}{l}}$$

is positive, whereas the other eigenvalue,

$$-\frac{k_d}{2m} - \sqrt{\left(\frac{k_d}{2m}\right)^2 + \frac{g}{l}},$$

is negative. Hence the equilibrium points $\theta = \pm \pi, \pm 3\pi, \ldots, v = 0$ are saddles.

To find the θ- and v-nullclines, note that $d\theta/dt = 0$ when $v = 0$, so the vector field points up or down along the θ-axis. The v-nullcline is where $dv/dt = 0$, which occurs on the curve

$$v = -\frac{mg}{k_d l} \sin\theta;$$

so the vector field points in a horizontal direction along this curve. We can determine the direction of the vector field in a region between two nullclines by checking the direction at a single point (see Figure 4.25).

Coupled with our analysis at the equilibrium points, we can now describe the behavior of solution curves near the equilibrium points as shown in Figure 4.26. Note how the solution curves circulate about the equilibrium points at $\theta = 0, \pm 2\pi, \pm 4\pi, \ldots,$ $v = 0$ in a clockwise direction. We can verify that the stable and unstable separatrices at $\theta = \pm \pi, \pm 3\pi, \ldots, v = 0$ point in the directions indicated in Figure 4.26 by finding the linearization at these points or by using the direction field.

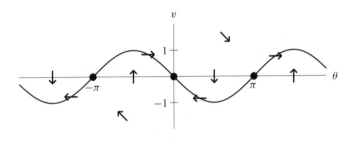

Figure 4.25
Nullclines for the system

$$\frac{d\theta}{dt} = v$$

$$\frac{dv}{dt} = \frac{k_d}{m}v - \frac{g}{l}\sin\theta.$$

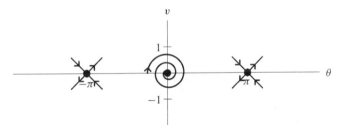

Figure 4.26
Phase plane near the equilibrium points for the system

$$\frac{d\theta}{dt} = v$$

$$\frac{dv}{dt} = \frac{k_d}{m}v - \frac{g}{l}\sin\theta.$$

The Effects of Dissipation

At this point, Figure 4.26 contains all the information about the phase plane of the pendulum with small damping available by qualitative techniques that we have at our disposal. We do not have the benefit of a constant of motion as we did in the frictionless case, so we cannot determine the entire phase plane. However, using the Hamiltonian function

$$H(\theta, v) = \frac{1}{2}v^2 - \frac{g}{l}\cos\theta$$

that we developed in the last section for the ideal pendulum, we can complete the picture. If $(\theta(t), v(t))$ is a solution of the ideal pendulum system (with $k_d = 0$), then

$$\frac{d}{dt}H(\theta(t), v(t)) = 0.$$

When $k_d \neq 0$, this is no longer true. However, if $(\theta(t), v(t))$ is a solution of the pendulum system with $k_d \neq 0$, we have

$$\frac{dH}{dt} = \frac{\partial H}{\partial \theta}\frac{d\theta}{dt} + \frac{\partial H}{\partial v}\frac{dv}{dt}$$

$$= (\frac{g}{l}\sin\theta)v + v\left(-\frac{k_d}{m}v - \frac{g}{l}\sin\theta\right)$$

$$= -\frac{k_d}{m}v^2$$

$$\leq 0.$$

That is, we no longer have that H is constant along solution curves of the system, but rather $H(\theta(t), v(t))$ is decreasing whenever $v(t) \neq 0$. Hence solution curves in the θv-phase plane cross level sets of H moving from larger to smaller H values.

This provides us with a strategy for determining the phase plane. First, we draw the level curves of H as we did before (see Figure 4.27). When the value of k_d/m is small and the velocity v is small, the value of H decreases slowly along solutions. We can sketch the phase plane by sketching curves that flow in the same general direction as the solution curves of the ideal pendulum but that move from level curves of higher H value toward lower H value (see Figure 4.28).

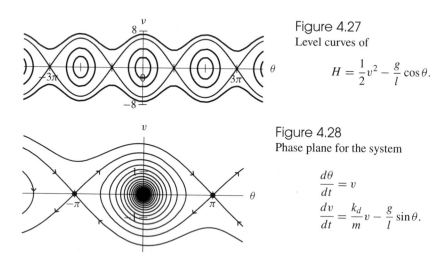

Figure 4.27
Level curves of

$$H = \frac{1}{2}v^2 - \frac{g}{l}\cos\theta.$$

Figure 4.28
Phase plane for the system

$$\frac{d\theta}{dt} = v$$

$$\frac{dv}{dt} = \frac{k_d}{m}v - \frac{g}{l}\sin\theta.$$

The fact that H is decreasing along solution curves allows us to determine the fate of stable and unstable separatrices emanating from saddle points: The unstable separatrices must fall into adjacent sinks whereas the stable separatrices come from "infinity" (see Figure 4.28). Also, the equilibrium point analysis we performed earlier agrees with what we have just done. The whole picture begins to fit together in a neat package, with the local analysis near equilibrium points fully complementing the global structure provided by all the level curves of H. Finally, if we plot θ versus t for a typical solution curve, we see that the pendulum eventually tends to oscillate about the downward-pointing rest position, as it should (see Figure 4.29).

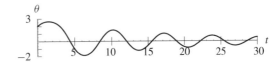

Figure 4.29
A θt-plot for a typical solution to the system

$$\frac{d\theta}{dt} = v$$

$$\frac{dv}{dt} = \frac{k_d}{m}v - \frac{g}{l}\sin\theta.$$

Liapounov Functions

The function H above plays a very different role for the damped pendulum than it does for the ideal pendulum. Solutions of the damped pendulum system move across level sets of H from higher to lower values. The function H is called a **Liapounov function**, after the Russian mathematician Aleksandr Mikhailovich Liapounov (1857–1918). This idea can be generalized as follows:

Definition: Suppose

$$\frac{dx}{dt} = f(x, y)$$

$$\frac{dy}{dt} = g(x, y)$$

is a system of differential equations. A function $L(x, y)$ is called a **Liapounov function** if, for every solution $(x(t), y(t))$ that is not an equilibrium point,

$$\frac{d}{dt}L(x(t), y(t)) \leq 0$$

for all t with strict inequality except for a discrete set of t's.

So the value of a Liapounov function never increases and usually decreases along a nonequilibrium solution. A Liapounov function can be a great help in drawing the phase plane for a system. Solution curves must cross level sets of the Liapounov function from higher to lower values.

The Damped Harmonic Oscillator

Consider the damped harmonic oscillator system

$$\frac{dy}{dt} = v$$

$$\frac{dv}{dt} = -qy - pv,$$

where q and p are positive constants. When $p = 0$, we saw in Section 4.2, page 369, that the system is Hamiltonian, with Hamiltonian function

$$H(y, v) = \frac{1}{2}v^2 + \frac{q}{2}y^2.$$

If $p > 0$ and $(y(t), v(t))$ is a solution of the system, then we can compute the rate of change of H along the solution by

$$\frac{d}{dt}H(y(t), v(t)) = \frac{\partial H}{\partial y}\frac{dy}{dt} + \frac{\partial H}{\partial v}\frac{dv}{dt}$$

$$= qy \cdot v + v(-qy - pv)$$

$$= -pv^2$$

$$\leq 0.$$

Hence $H(y(t), v(t))$ decreases at a nonzero rate (except when $v = 0$). This implies that H is a Liapounov function for the damped harmonic oscillator system. The level sets of H are ellipses in the yv-plane, so solutions of the damped harmonic oscillator system (with $p > 0$) cross these level sets from larger to smaller values of H. Since H has one global minimum at the origin and no other critical points, all solutions must tend toward the origin as time increases (see Figures 4.30 and 4.31). This agrees with our analysis of the damped harmonic oscillator from the point of view of linear systems and will be useful in Section 5.3.

As we mentioned before (see page 370), the value of H at a point (y, v) is called the energy of the harmonic oscillator in position y with velocity v. Saying that H decreases along solutions of the damped harmonic oscillator system is a precise way of saying that the damping dissipates energy. The same holds for the pendulum with friction.

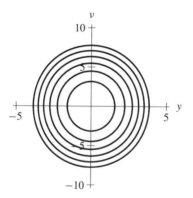

Figure 4.30
Level sets for

$$H(y, v) = \frac{1}{2}v^2 + \frac{q}{2}y^2.$$

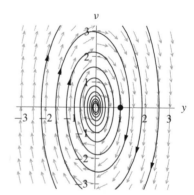

Figure 4.31
Phase plane for the damped harmonic oscillator

$$\frac{dy}{dt} = v$$

$$\frac{dv}{dt} = -qy - pv$$

with p small.

Gradient Systems

Not all systems have Liapounov functions. Finding a Liapounov function for a system, even if we somehow know that the system has a Liapounov function, can be extremely challenging. However, for some modeling problems the physical system being modeled provides a motivation for a "lucky guess" of a Liapounov function. This is the case for the damped pendulum and the damped harmonic oscillator. The Hamiltonian function of the undamped system becomes a Liapounov function when damping is added.

In some cases the Liapounov function is an integral part of the construction of the system. Such systems are called **gradient systems**. We begin with an example.

Navigation by Smell

Many animal species use smell to navigate through their environment. A lobster, for example, can use its antennae to detect very small concentrations of chemicals in the water around it (that is, it "smells" the water). With this ability the lobster can determine which direction a smell is coming from and hence find food in the murky deep. Although it is not clear how lobsters accomplish this, one possible method is to determine local variations in the concentration and move in the direction in which the concentration increases fastest. *

To describe how we can design a mechanical lobster that navigates by smell, we begin by assuming our lobster can only move on a flat two-dimensional plane. Let $S(x, y)$ equal the concentration at (x, y) of the chemicals making up the smell of a dead fish. We know from vector calculus that at (x, y) the direction in which S increases the fastest is given by the gradient vector

$$\nabla S(x, y) = \left(\frac{\partial S}{\partial x}(x, y), \frac{\partial S}{\partial y}(x, y) \right).$$

This defines a vector field on the xy-plane. We assume that our model lobster always moves in the direction that increases the smell the fastest. Hence the velocity of our lobster's motion points in the same direction as the gradient of S. If we let $(x(t), y(t))$ denote the location of the lobster at time t, then the velocity vector is given by

$$\left(\frac{dx}{dt}(t), \frac{dy}{dt}(t) \right) = \nabla S(x(t), y(t))$$

or

$$\frac{dx}{dt} = \frac{\partial S}{\partial x}(x, y)$$
$$\frac{dy}{dt} = \frac{\partial S}{\partial y}(x, y).$$

We use this system of equations to determine our lobster's motion. If the initial position is (x_0, y_0) at time $t = 0$, then the solution curve in the phase plane satisfying this initial condition is our prediction of a lobster's path.

---

*The ability of lobsters to do this "computation" in turbulent fluids is only beginning to be duplicated by technology: See Lipkin, R, "Tracking Undersea Scent" *Science News* 147, 5, (1994): p. 78, for a discussion of a "robo-lobster."

An example: Two dead fish

Suppose $S(x, y)$ is defined for $-2 \leq x \leq 2$ and $-2 \leq y \leq 2$ and is given by the formula

$$S(x, y) = \frac{x^2}{2} - \frac{x^4}{4} - \frac{y^2}{2} + 8.$$

The graph of S is shown in Figure 4.32. We restrict attention to the region of the xy-plane with x and y between 2 and -2 because this function S takes on negative values when x or y is large, and negative concentrations are not physically meaningful. More realistic, but more algebraically complicated choices for S are considered in the exercises.

We can see (or compute by the techniques of vector calculus) that S has a local maximum at each of the points $(-1, 0)$ and $(1, 0)$ and a saddle point at $(0, 0)$. A function of this sort might arise if there were a smelly dead fish at each of the points $(-1, 0)$ and $(1, 0)$.

The gradient of S gives rise to a vector field and, as above, this specifies a system of differential equations

$$\frac{dx}{dt} = \frac{\partial S}{\partial x} = x - x^3$$
$$\frac{dy}{dt} = \frac{\partial S}{\partial y} = -y.$$

By construction, the velocity vector of a solution of this system is equal to the gradient of S and hence points in the direction for which S increases the fastest. If $(x(t), y(t))$ is a solution, we compute the rate of change of S along the solution by

$$\frac{d}{dt} S(x(t), y(t)) = \frac{\partial S}{\partial x} \frac{dx}{dt} + \frac{\partial S}{\partial y} \frac{dy}{dt}$$

$$= (x - x^3)(x - x^3) + (-y)(-y)$$

$$= (x - x^3)^2 + y^2$$

$$\geq 0.$$

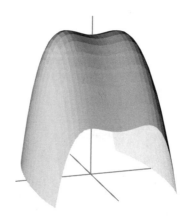

Figure 4.32
Graph of the function
$S(x, y) = x^2/2 - x^4/4 - y^2/2 + 8.$

Hence S increases at every point where $\nabla S \neq 0$. These are precisely the points where the vector field of the system is zero — that is, the equilibrium points of the system.

To sketch the phase plane of the system

$$\frac{dx}{dt} = \frac{\partial S}{\partial x} = x - x^3$$
$$\frac{dy}{dt} = \frac{\partial S}{\partial y} = -y,$$

we begin by sketching the level sets of S (see Figure 4.33). The direction field points in the direction of the gradient of S, which is perpendicular to the level sets. To sketch the solution curves, we draw curves always perpendicular to the level sets of S moving in the direction of increasing S (see Figure 4.34).

We can verify by linearization that the equilibrium points $(\pm 1, 0)$ are sinks while the point $(0, 0)$ is a saddle (see the exercises). Lobsters that start with $y(0) < 0$ tend toward the equilibrium point $(-1, 0)$ whereas those with $y(0) > 0$ tend toward $(1, 0)$. If a lobster starts precisely on the y-axis [$y(0) = 0$], then it will tend toward the saddle equilibrium point at the origin, unable to decide between the two fish.

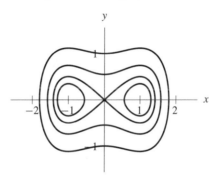

Figure 4.33
Level sets of

$$S(x, y) = \frac{1}{2}x^2 - \frac{1}{4}x^4 - \frac{1}{2}y^2 + 8.$$

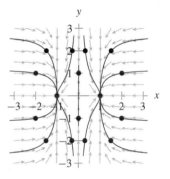

Figure 4.34
Phase plane for the system

$$\frac{dx}{dt} = \frac{\partial S}{\partial x} = x - x^3$$
$$\frac{dy}{dt} = \frac{\partial S}{\partial y} = -y.$$

General Form of Gradient Systems

The vector field for the system in the preceding example is the gradient of the function S, hence it is not unreasonable to call it a gradient system.

Definition: A system of differential equations

$$\frac{dx}{dt} = f(x, y)$$
$$\frac{dy}{dt} = g(x, y)$$

is called a **gradient system** if there is a function G such that

$$f(x, y) = \frac{\partial G}{\partial x}(x, y)$$

$$g(x, y) = \frac{\partial G}{\partial y}(x, y)$$

for all (x, y).

As in the example, if $(x(t), y(t))$ is a solution for a gradient system with vector field given by the gradient of $G(x, y)$, then

$$\frac{d}{dt}G(x(t), y(t)) = \frac{\partial G}{\partial x}\frac{dx}{dt} + \frac{\partial G}{\partial y}\frac{dy}{dt}$$

$$= \left(\frac{\partial G}{\partial x}\right)^2 + \left(\frac{\partial G}{\partial y}\right)^2$$

$$\geq 0.$$

That is, the value of G increases along every orbit except at critical points of G. The function $-G$ decreases along orbits and hence is a Liapounov function for this system.

Properties of Gradient Systems

Just as in the case of Hamiltonian systems, gradient systems enjoy several special properties that make their phase planes relatively simple. For one thing, gradient systems cannot possess periodic solutions. Since the gradient function must increase along all non-equilibrium solutions, a solution curve can never return to its starting place.

There are also restrictions on the types of equilibrium points that occur in a gradient system. Suppose

$$\frac{dx}{dt} = \frac{\partial G}{\partial x}(x, y)$$

$$\frac{dy}{dt} = \frac{\partial G}{\partial y}(x, y)$$

is a gradient system. Then the Jacobian matrix associated with this vector field is

$$\begin{pmatrix} \dfrac{\partial^2 G}{\partial x^2} & \dfrac{\partial^2 G}{\partial y \partial x} \\ \dfrac{\partial^2 G}{\partial x \partial y} & \dfrac{\partial^2 G}{\partial y^2} \end{pmatrix}.$$

Because the mixed partial derivatives of G are equal, this matrix assumes the form

$$\begin{pmatrix} \alpha & \beta \\ \beta & \gamma \end{pmatrix},$$

where $\alpha = \partial^2 G/\partial x^2$, $\beta = \partial^2 G/\partial y \partial x$, and $\gamma = \partial^2 G/\partial y^2$. The eigenvalues are roots of the equation

$$\lambda^2 - (\alpha + \gamma)\lambda + \alpha\gamma - \beta^2 = 0,$$

which are

$$\frac{\alpha + \gamma \pm \sqrt{(\alpha + \gamma)^2 - 4\alpha\gamma + 4\beta^2}}{2}.$$

Simplifying, we see that the eigenvalues equal

$$\frac{1}{2}(\alpha + \gamma) \pm \frac{1}{2}\sqrt{(\alpha - \gamma)^2 + 4\beta^2}.$$

Since the term $(\alpha - \gamma)^2 + 4\beta^2 \geq 0$, it follows that these eigenvalues never have an imaginary part; that is, the eigenvalues are always real. Gradient systems do not have spiral sinks, spiral sources or centers.

It is important to note that not all systems of differential equations that possess Liapounov functions are gradient systems. For example, the linear system

$$\frac{dx}{dt} = -x + y$$

$$\frac{dy}{dt} = -x - y$$

has a Liapounov function given by $L(x, y) = x^2 + y^2$. We check this by computing the derivative of L along a solution $(x(t), y(t))$:

$$\begin{aligned}
\frac{d}{dt}L(x(t), y(t)) &= 2x\frac{dx}{dt} + 2y\frac{dy}{dt} \\
&= 2x(-x + y) + 2y(-x - y) \\
&= -2(x^2 + y^2) \\
&\leq 0.
\end{aligned}$$

Since $dL/dt = 0$ only at the origin, this function decreases along all nonequilibrium solution curves. But L cannot be a gradient system, since the eigenvalues at $(0,0)$ are complex $(-1 \pm i)$; so the origin is a spiral sink, which cannot occur in gradient systems.

Exercises for Section 4.3

1. Consider the system

$$\frac{dx}{dt} = -x^3$$

$$\frac{dy}{dt} = -y^3.$$

(a) Verify that

$$L(x, y) = \frac{x^2}{2} + \frac{y^2}{2}$$

is a Liapounov function for the system.

(b) Sketch the level sets of L.

(c) What can you conclude about the phase plane of the system from the information in parts (a) and (b) above? (Sketch the phase plane and write a short essay describing what you know about the phase plane and how you know it.)

2. Consider the system

$$\frac{dx}{dt} = y$$

$$\frac{dy}{dt} = -x - \frac{y}{4} + x^2.$$

(a) Verify that the function

$$L(x, y) = \frac{y^2}{2} + \frac{x^2}{2} - \frac{x^3}{3}$$

is a Liapounov function for the system.
(b) Sketch the level sets of L.
(c) What can you conclude about the phase plane of the system from the information in parts (a) and (b)? (Sketch the phase plane and write a short essay describing what you know about the phase plane and how you know it.)

3. Consider the system

$$\frac{dx}{dt} = y$$

$$\frac{dy}{dt} = -4x - 0.1y.$$

(a) Verify that all solutions tend toward the origin as t increases, and sketch the phase plane. [*Hint*: The system is linear.]
(b) Verify that

$$L(x, y) = x^2 + y^2$$

is *not* a Liapounov function for the system.
(c) Verify that

$$K(x, y) = 2x^2 + \frac{y^2}{2}$$

is a Liapounov function for the system.

In Exercises 4–11, we consider the damped pendulum system

$$\frac{d\theta}{dt} = v$$

$$\frac{dv}{dt} = -\frac{g}{l}\sin\theta - \frac{k_d}{m}v,$$

where k is the damping coefficient, m is the mass of the bob, l is the length of the arm, and g is the acceleration of gravity ($g \approx 9.8\text{m/s}^2$).

4. What relationship must hold between the parameters k_d, m, and l for the period of a small swing back and forth of the damped pendulum to be one second.

5. Suppose we have a pendulum clock that uses a slightly damped pendulum to keep time (that is, k_d is positive but $k_d \approx 0$). The clock "ticks" each time the pendulum arm crosses $\theta = 0$. If the mass of the pendulum bob is increased, does the clock run fast or slow?

6. Suppose we have a pendulum clock that uses only a slightly damped pendulum to keep time. The clock "ticks" each time the pendulum arm crosses $\theta = 0$.

(a) As the clock "winds down" (so the amplitude of the swings decreases), does the clock run slower or faster?

(b) If the initial push of the pendulum is large so that the pendulum swings very close to the vertical, will the clock run too fast or too slow?

7. For fixed values of k_d and l, for what values of the mass m will the pendulum be usable as a clock?

8. Suppose we take $l = 9.8$m (so $g/l = 1$), $m = 1$, and k_d large, say $k_d = 4$. For the damped pendulum system above and with this choice of parameter values, do the following.

(a) Find the eigenvalues and eigenvectors of the linearized system at the equilibrium point $(0, 0)$.

(b) Find the eigenvalues and eigenvectors of the linearized system at the equilibrium point $(\pi, 0)$.

(c) Sketch the phase plane near the equilibrium points.

(d) Sketch the entire phase plane. [*Hint*: Begin by sketching the level sets of H as in the text.]

9. Suppose we have a pendulum clock that uses a slightly damped oscillator to keep time. Suppose that the clock "ticks" each time the pendulum arm crosses $\theta = 0$, but the arm must reach a height of $\theta = \pm 0.1$ to record the swing (that is, if the entire swing takes place with $-0.1 < \theta < 0.1$, then the clock doesn't tick). Suppose one tick is one second. In terms of the parameters k_d, m, and l, give a rough estimate of how long the clock can keep accurate time. Comment on why pendulum clocks must be wound.

10. (a) For the slightly damped pendulum ($k_d > 0$ but k_d close to zero) find the set of all initial conditions $(\theta(0), v(0))$ that execute exactly two complete revolutions for $t > 0$ (that is, pass the vertical position exactly twice) before settling into a back and forth swinging motion. Sketch the phase plane for the slightly damped pendulum and shade these initial conditions.

(b) Repeat part (a) for solutions that execute exactly five complete revolutions for $t > 0$ before settling into back and forth swinging motion.

11. Suppose that instead of adding damping to the ideal pendulum, we add a small amount of "antidamping"; that is, we take k_d slightly negative in the damped pendulum system. Physically this would mean that whenever the velocity is nonzero, the pendulum is accelerated in the direction of motion.

(a) Linearize and classify the equilibrium points in this situation.

(b) Sketch the phase plane for this system.

(c) Describe in a brief paragraph the behavior of a solution with initial condition near $\theta = v = 0$.

12. Let
$$G(x, y) = x^3 - 3xy^2.$$

(a) What is the gradient system with vector field given by the gradient of G?

(b) Sketch the graph of G and the level sets of G.

(c) Sketch the phase plane of the gradient system in part (a).

13. Let
$$G(x, y) = x^2 - y^2.$$

(a) What is the gradient system with vector field given by the gradient of G?

(b) Classify the equilibrium point at the origin. [*Hint*: This system is linear.]

(c) Sketch the graph of G and the level sets of G.

(d) Sketch the phase plane of the gradient system in part (a).

Remark: This is why saddle equilibrium points have the name *saddle*.

14. Let
$$G(x, y) = x^2 + y^2.$$

(a) What is the gradient system with vector field given by the gradient of G?

(b) Classify the equilibrium point at the origin. [*Hint*: The system is linear.]

(c) Sketch the graph of G and the level sets of G.

(d) Sketch the phase plane of the gradient system in part (a).

15. For the two dead fish example given by the system
$$\frac{dx}{dt} = x - x^3$$
$$\frac{dy}{dt} = -y,$$

(a) find the linearized system for the equilibrium point at the origin and verify that the origin is a saddle;

(b) find the linearized system for the equilibrium point at $(1, 0)$ and verify that this point is a sink;

(c) from the eigenvalues and eigenvectors of the system in part (b), determine from which direction the model lobster will approach the equilibrium point $(1, 0)$; and

(d) check that the linearized system at the equilibrium point $(-1, 0)$ is the same as that at $(1, 0)$.

16. The system for the two dead fish example
$$\frac{dx}{dt} = x - x^3$$
$$\frac{dy}{dt} = y,$$

has the special property that the equations "decouple"; that is, the equation for dx/dt depends only on x, and the equation for dy/dt depends only on y.

(a) Sketch the phase lines for the dx/dt and dy/dt equations.

(b) Using these phase lines, sketch the phase plane of the system.

17. Suppose the smell of a bunch of dead fish in the region $-2 \leq x \leq 2, -2 \leq y \leq 2$ is given by the function
$$S(x, y) = x^2 + y^2 - \frac{x^4 + y^4}{4} - 3x^2 y^2 + 100.$$

(a) What is the gradient system whose vector field is the gradient of S?

(b) (Using technology to numerically approximate solutions) Sketch the phase plane for this system.

(c) How many dead fish are there, and where are they?

(d) Using the results from part (b), sketch the level sets of S.

(e) Why is the model not realistic for large values of x or y?

18. A reasonable model for the smell at (x, y) of a dead fish located at (x_1, y_1) is given by

$$S_1(x, y) = \frac{1}{(x - x_1)^2 + (y - y_1)^2 + 1}.$$

That is, S_1 is given by 1 over the distance to the dead fish squared plus 1.

(a) Form the function S giving the total smell from three dead fish located at $(1, 0)$, $(-1, 0)$, and $(0, 2)$.

(b) Sketch the level sets of S.

(c) Sketch the phase plane of the gradient system

$$\frac{dx}{dt} = \frac{\partial S}{\partial x}(x, y)$$
$$\frac{dy}{dt} = \frac{\partial S}{\partial y}(x, y).$$

(d) Write out explicitly the formulas for the right-hand sides of the equations in part (c).

(e) Why did we use distance squared plus 1 instead of just distance squared in the definition of S_1? Why did we use distance squared plus 1 instead of just distance plus 1 in the definition of S_1?

19. Suppose

$$\frac{dx}{dt} = f(x, y)$$
$$\frac{dy}{dt} = g(x, y)$$

is a gradient system. That is, there exists a function $G(x, y)$ such that $f = \partial G / \partial x$ and $g = \partial G / \partial y$.

(a) Verify that if f and g have continuous partial derivatives, then

$$\frac{\partial f}{\partial y}(x, y) = \frac{\partial g}{\partial x}(x, y)$$

for all (x, y).

(b) Use this to show that the system

$$\frac{dx}{dt} = x^2 + 3xy$$
$$\frac{dy}{dt} = 2x + y^3$$

is *not* a gradient system.

20. The following portrait cannot occur for a gradient system. Explain why.

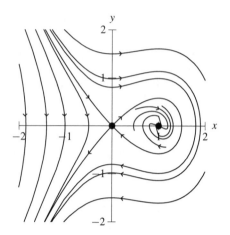

21. Let

$$H(y, v) = \frac{1}{2}v^2 + V(y)$$

for some function V, and consider the Hamiltonian system

$$\frac{dy}{dt} = \frac{\partial H}{\partial v}(y, v) = v$$

$$\frac{dv}{dt} = -\frac{\partial H}{\partial y}(y, v) = -\frac{dV}{dy}(y).$$

Show that H is a Liapounov function for the system

$$\frac{dy}{dt} = v$$

$$\frac{dv}{dt} = -\frac{dV}{dy}(y) - kv$$

for any value of $k > 0$.

22. Consider the Hamiltonian system

$$\frac{dx}{dt} = \frac{\partial H}{\partial y}$$

$$\frac{dy}{dt} = -\frac{\partial H}{\partial x}$$

and the gradient system

$$\frac{dx}{dt} = \frac{\partial H}{\partial x}$$

$$\frac{dy}{dt} = \frac{\partial H}{\partial y},$$

where H is the same function in each case. What can you say about the relationship between the two phase portraits of these two systems?

4.4 NONLINEAR SYSTEMS IN THREE DIMENSIONS

We saw in Section 2.7 that solutions of differential equations with three dependent variables are curves in three-dimensional space. These curves can loop around each other in very complicated ways. In Section 3.7 we studied the behavior of linear systems with three dependent variables. The behavior of linear systems can be determined by the eigenvalues and eigenvectors, however, the list of possible behaviors is much longer than for planar systems.

In this section we consider two examples of nonlinear systems in three dimensions. The first is a population model with three species forming a food chain. For the second example, we return to the Lorenz equations studied in Section 2.7. We can use the technique of linearization of equilibrium points and numerical approximation (Euler's method) to obtain a little bit of information about these systems. As we have emphasized in the previous sections, there are very few all-purpose tools for differential equations in three and higher dimensions, and these systems are an area of active research in mathematics.

A Food Chain Model

We have studied the behavior of populations living in isolation and of pairs of species interacting in predator-prey, cooperative, and competitive systems. These systems only scratch the surface of the types of complicated interactions found in nature. One possibility is the formation of a food chain of three or more species.   A single species can be both predator and prey. An example that has been studied recently involves the Balsam fir tree, the moose, and the wolf. The fir trees are eaten by the moose, and the moose (particularly young and infirm individuals) are eaten by wolves. A natural question is whether or not changes in the wolf population affects the tree population. Recent studies of the tree/moose/wolf population in Isle Royal National Park * indicate that changes in the wolf population may affect the trees.

We form a model of a system with three species, each one eaten by the next. For convenience, we call these species trees, moose, and wolves in analogy to the example above. Let

$$x(t) = \text{population of trees at time } t,$$
$$y(t) = \text{population of moose at time } t,$$
$$z(t) = \text{population of wolves at time } t.$$

We assume that each of the populations in isolation can be modeled with a logistic equation and that the effect of interaction between species is proportional to the product of the populations. The behavior of solutions will depend on the parameters chosen for the growth rate, carrying capacity, and affect of interaction. In order to get an idea of how solutions of systems of this form behave, we begin by taking all of the parameters to be 1. (This certainly is not the case for trees, moose, and wolves, but it makes the arithmetic which follows much more pleasant.) Our model is

---

*See McLaren, B.E. and Peterson, R.O., "Wolves, Moose and Tree Rings on Isle Royale," *Science* 266 (1994): p. 1555.

$$\frac{dx}{dt} = x(1 - x) - xy$$

$$\frac{dy}{dt} = y(1 - y) + xy - yz$$

$$\frac{dz}{dt} = z(1 - z) + yz.$$

Note that the growth of the tree population is decreased by the presence of moose (the $-xy$ term), the growth of moose population is increased by the trees (the $+xy$ term) but decreased by the wolves (the $-yz$ term), and the growth of wolf population is increased by the moose (the $+yz$ term).

   We can find the equilibrium points of this system by setting the right-hand sides of the equations to zero and solving for x, y, and z. The equilibrium points are $(0, 0, 0)$, $(1, 0, 0)$, $(0, 1, 0)$, $(0, 0, 1)$, $(1, 0, 1)$, and $(2/3, 1/3, 4/3)$. Of these, only the point $(2/3, 1/3, 4/3)$ has all three coordinates nonzero, so the three species can coexist in equilibrium at these populations. The Jacobian matrix for this system at (x, y, z) is

$$\begin{pmatrix} 1 - 2x - y & -x & 0 \\ y & 1 - 2y + x - z & -y \\ 0 & z & 1 - 2z + y \end{pmatrix}.$$

So at the equilibrium point $(2/3, 1/3, 4/3)$, the Jacobian is

$$\begin{pmatrix} -2/3 & -2/3 & 0 \\ 1/3 & -1/3 & -1/3 \\ 0 & 4/3 & -4/3. \end{pmatrix}$$

The characteristic polynomial of the matrix is

$$-(\lambda^3 + (7/3)\lambda^2 + (20/9)\lambda + 8/27).$$

We can use Newton's method to show that the eigenvalues are approximately $\lambda_1 \approx -1$, $\lambda_2 \approx -0.67 + 0.67i$, and $\lambda_3 \approx -0.67 - 0.67i$. Since λ_1 is negative and λ_2 and λ_3 have negative real part, this equilibrium point is a sink, and all solutions with initial conditions sufficiently close to this point will tend toward it as t increases. In fact, numerical simulations show that every solution with initial conditions in the first octant (where x, y, and z are all positive) tend to this equilibrium point. Hence we expect that over the long run the populations will settle at these values and the species will coexist in equilibrium.

Effect of a disease among the top predator

Suppose that a disease affects the population of wolves, the predator at the top of the food chain. Suppose this disease kills a certain small fraction of the population each year. We can adjust our model to include this element by adding the term $-\gamma z$ to the dz/dt equation, where γ is a parameter indicating the fraction of the population killed by the disease per unit time. The new model is

$$\frac{dx}{dt} = x(1 - x) - xy$$

$$\frac{dy}{dt} = y(1 - y) + xy - yz$$

$$\frac{dz}{dt} = z(1 - z) - \gamma z + yz.$$

We can again compute the equilibrium points for this system, which will now depend on the value of γ. The only equilibrium point with all three coordinates nonzero is $((2 - \gamma)/3, (1 + \gamma)/3, (4 - 2\gamma)/3))$. Assuming that γ is small, this point will still be a sink. We can ask what effect adjusting γ has on the populations. To do this we compute the derivative with respect to γ of each of the coordinates of the equilibrium point. For $z = (4 - 2\gamma)/3$, the derivative with respect to γ is

$$\frac{d}{d\gamma}\left(\frac{4 - 2\gamma}{3}\right) = -\frac{2}{3},$$

which is negative. This is as we expect, since an increase in the effect of the disease should decrease the number of wolves. For $y = (1 + \gamma)/3$, the derivative with respect to γ is

$$\frac{d}{d\gamma}\left(\frac{1 + \gamma}{3}\right) = \frac{1}{3},$$

which is positive. So an increase in γ is good for the moose population, again as we would expect. Finally, for $x = (2 - \gamma)/3$, the derivative with respect to γ is

$$\frac{d}{d\gamma}\left(\frac{2 - \gamma}{3}\right) = -\frac{1}{3},$$

which is negative. An increase in γ decreases the equilibrium population of trees.

There are two important remarks to be made. First, although it is easy to make simple models of systems representing long food chains, it is not so easy to find such systems in nature. Most predators have a variety of prey species to choose from. When one prey species is in short supply, the predator changes its diet.

Second, care must be taken when modifying a system to account for changing situations. Although the discussion above might seem obvious from common sense, if we had chosen to modify the equations to take into account the wolf disease by reducing the growth-rate parameter of the wolf population, then we would obtain different results (see the exercises).

Lorenz Equations

The Lorenz equations are

$$\frac{dx}{dt} = \sigma(y - x)$$

$$\frac{dy}{dt} = \rho x - y - xz$$

$$\frac{dz}{dt} = -\beta z + xy,$$

where σ, ρ, and β are parameters. In Section 2.7, we fixed the parameter values to $\sigma = 10$, $\beta = 8/3$, and $\rho = 28$ to obtain the system studied by Lorenz:

$$\frac{dx}{dt} = 10(y - x)$$
$$\frac{dy}{dt} = 28x - y - xz$$
$$\frac{dz}{dt} = \frac{-8z}{3} + xy.$$

Numerical approximations that we studied in that section showed that solutions looped around the phase space in a very complicated way (see Figure 4.35). With the aid of the technique of linearization near the equilibrium points, we can gain a little more insight into the behavior of these orbits. (However, the full story will have to wait for Chapter 6.)

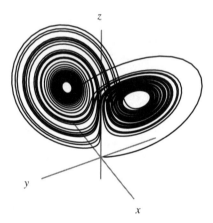

Figure 4.35
One solution curve of the Lorenz system in xyz-phase space.

The equilibrium points are $(0, 0, 0)$, $(6\sqrt{2}, 6\sqrt{2}, 27)$, and $(-6\sqrt{2}, -6\sqrt{2}, 27)$. We will linearize around each of these points to determine the local phase-space picture. The Jacobian matrix at (x, y, z) is

$$\begin{pmatrix} -10 & 10 & 0 \\ 28 - z & -1 & -x \\ y & x & -8/3 \end{pmatrix}.$$

At the origin the Jacobian is

$$\begin{pmatrix} -10 & 10 & 0 \\ 28 & -1 & 0 \\ 0 & 0 & -8/3 \end{pmatrix},$$

so the linear system that approximates the Lorenz system near the origin is

$$\frac{dx}{dt} = -10x + 10y)$$
$$\frac{dy}{dt} = 28x - y$$
$$\frac{dz}{dt} = -8/3z.$$

This system decouples since the dx/dt and dy/dt equations do not depend on z and the dz/dt equation does not depend on x or y. We have already studied this system in Section 3.7 and found that the origin is a sink in the z direction and a saddle in the xy-plane. In the full three-dimensional picture, the origin is a saddle with a plane of initial conditions tending toward the origin as t increases and a line of initial conditions tending toward the origin as t decreases. The picture of the phase space is reproduced in Figure 4.36.

Next we consider the equilibrium point $(6\sqrt{2}, 6\sqrt{2}, 27)$. The Jacobian at this point is

$$\begin{pmatrix} -10 & 10 & 0 \\ 1 & -1 & -6\sqrt{2} \\ 6\sqrt{2} & 6\sqrt{2} & -8/3 \end{pmatrix}.$$

The eigenvalues are $\lambda_1 \approx -13.8$, $\lambda_2 \approx 0.094 + 10.2i$, and $\lambda_3 \approx 0.094 - 10.2i$, so this point is a spiral saddle that has a line of solutions that tend toward the equilibrium point as t increases and a plane of solutions that spiral toward the equilibrium point as t decreases. Solutions approach the equilibrium point quickly along the direction of the eigenvector of the negative eigenvalue, then spiral slowly away along the plane corresponding to the complex eigenvalues. By computing the eigenvectors, we can obtain the orientation of the straight line of solutions and the plane of spiraling solutions. This yields a good representation of the phase space near the equilibrium point, which we sketch in Figure 4.36.

We can compute the linearization for the other equilibrium point $(-6\sqrt{2}, -6\sqrt{2}, 27)$ in the same manner, and we find that it is also a spiral saddle. These computations give us the "local" picture of the phase space near the equilibrium points (see Figure 4.36). Placing this next to a picture of a solution of the full nonlinear Lorenz system, we start to see some order in the way the solution behaves (see Figure 4.37). The solution approaches one of the equilibrium points $(\pm 6\sqrt{2}, \pm 6\sqrt{2}, 27)$ along the straight line corresponding to the negative eigenvalue. When it gets close to the equilibrium point, it begins to spiral away. When the spiral is large enough, the solution becomes involved with the saddle at the origin and either returns to repeat its previous pattern or goes to the equilibrium point "on the other side."

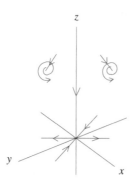

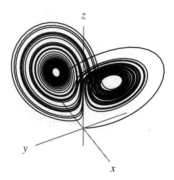

Figure 4.36
Local phase-space pictures
around the three
equilibrium points.

Figure 4.37
Solution of the nonlinear Lorenz
system.

Reality Check

The study of the local behavior near the equilibrium points has given us a bit more
insight into the complicated behavior of solutions of the Lorenz system. However, the
local pictures do not tell the whole story. This is what makes the study of this and
other three-dimensional systems so hard. Calculus and linearization techniques tell us
a great deal about small parts of the phase space, but the behavior of solutions also
depends on what is going on "globally." The development of the tools needed to study
these systems as well as understand individual models is an active area of mathematical
research.

Exercises for Section 4.4

Exercises 1–4, refer to the three-species food chain model from the text:

$$\frac{dx}{dt} = x(1 - x) - xy$$
$$\frac{dy}{dt} = y(1 - y) + xy - yz$$
$$\frac{dz}{dt} = z(1 - z) + yz.$$

In the text we studied the linearization for the equilibrium point where all three species
coexisted. The other equilibrium points are listed below. For each,

(a) find the Jacobian at this point;

(b) find the eigenvalues and eigenvectors;

(c) classify the equilibrium point;

(d) sketch the phase space of the linearized system; and

(e) discuss in a few sentences what happens to a solution with initial value near the
equilibrium point. [*Hint*: There will be several cases depending on which populations are nonzero. You need not consider negative populations, but consider all
combinations of zero and positive populations that are near the equilibrium point.]

1. $(1, 0, 0)$ **2.** $(0, 1, 0)$

3. $(0, 0, 1)$ **4.** $(1, 0, 1)$

5. Discuss why there is an equilibrium point at $(1, 0, 1)$ with trees and wolves coexisting at their carrying capacities, but there are not equilibrium points at $(0, 1, 1)$ and $(1, 1, 0)$.

   (a) First give the "mathematical" reason by studying the differential equations.

   (b) Relate this to the model to see if your conclusions are "physically reasonable."

6. Suppose that a moose disease enters the region of study and that βy moose per time unit are killed by the disease (where β is small).

   (a) Modify the model to include the moose disease.

   (b) What are the equilibrium points for the new model (recall β is small)?

   (c) How does an increase in β effect the equilibrium populations of trees, moose, and wolves where all three species coexist?

7. Suppose that instead of a disease, the growth rate of the wolves changes. The equation governing the wolf population becomes

$$\frac{dz}{dt} = \gamma z(1 - z) + yz,$$

where γ is a constant near 1. The equations for dx/dt and dy/dt remain unchanged.

   (a) Find the equilibrium points of this new system [*Hint*: The equilibrium points will depend on the parameter γ. Remember, γ is close to 1.]

   (b) What effect does a decrease in γ have on the equilibrium population where all three species coexist?

   (c) Is the change in the equilibrium point what you expected? Why or why not?

Lab 4.1   Hard and Soft Springs

In this lab we continue our study of second-order equations by considering "nonlinear springs." In Section 2.4 we developed the model of a spring based on Hooke's Law. Hooke's Law asserts that the restoring force of a spring is proportional to its displacement, and this assumption leads to the second-order equation

$$m\frac{d^2 y}{dt^2} + k_s\, y = 0.$$

Since the resulting differential equation is linear, we say that the spring is linear. In this case, the restoring force is $-k_s\, y$. In addition, we assume that the friction or damping force is proportional to the velocity. The resulting second-order equation is

$$m\frac{d^2 y}{dt^2} + k_d\frac{dy}{dt} + k_s\, y = 0.$$

Hooke's Law is an idealized model that works well for small oscillations. In fact, the restoring force of a spring is roughly linear if the displacement of the spring from its equilibrium position is small, but it is generally more accurate to model the restoring force by a cubic of the form $-ky+by^3$ where b is small relative to k. If b is negative, the spring is said to be hard, and if b is positive, the spring is soft. In this lab we consider the behavior of hard and soft springs for particular values of the parameters. A table of choices for the parameters is given below (your instructor will tell you which parameter value(s) to consider).

In your report, you should analyze the phase planes and $y(t)$- and $v(t)$-graphs to describe the long-term behavior of the solutions to the equations:

1. (Hard spring with no damping) The first equation that you should study is the hard spring with no damping. That is, $k_d = 0$ and $b = b_1$. Examine solutions using both their graphs and the phase plane. Consider the periods of the periodic solutions that have the initial condition $v(0) = 0$. Sketch the graph of the period as a function of the initial condition $y(0)$. Is there a minimum period? A maximum period? If so, how do you interpret these extrema?

2. (Hard spring with damping) Now use the given value of k_d and $b = b_1$ to introduce damping into the discussion. What happens to the long-term behavior of solutions in this case? Determine the value of the damping parameter that separates the underdamped case from the overdamped case.

3. (Soft spring with no damping) Consider the soft spring that corresponds to the positive value b_2 of b. Over what range of y-values is this model reasonable? Consider the periods of the periodic solutions that have the initial condition $v(0) = 0$. Sketch the graph of the period as a function of the initial condition $y(0)$. Is there a minimum period? A maximum period? Use the phase portrait to help justify your answer.

4. (Soft spring with damping) Using the given values of k_d and $b = b_2$, what happens to the long-term behavior of solutions in this case? Determine the value of the damping parameter that separates the underdamped case from the overdamped case.

5. From a physical point of view, what's the difference between a hard spring and a soft spring?

Your report: Address each of the items above in the form of a short essay. You may illustrate your essay with pictures of phase planes and graphs of solutions, however, your essay should be complete and understandable without the pictures. Make sure you relate the behavior of the solutions to the motion of the associated mass and spring systems.

Table 4.1 Choices for the parameter values. Assume the mass $m = 1$ unless told otherwise by your instructor.

Choice	k_s	k_d	b_1	b_2
1	$k_s = 0.1$	$k_d = 0.15$	$b_1 = -0.005$	$b_2 = 0.005$
2	$k_s = 0.2$	$k_d = 0.20$	$b_1 = -0.008$	$b_2 = 0.008$
3	$k_s = 0.3$	$k_d = 0.20$	$b_1 = -0.009$	$b_2 = 0.009$
4	$k_s = 0.2$	$k_d = 0.20$	$b_1 = -0.005$	$b_2 = 0.005$
5	$k_s = 0.1$	$k_d = 0.10$	$b_1 = -0.005$	$b_2 = 0.005$
6	$k_s = 0.3$	$k_d = 0.20$	$b_1 = -0.007$	$b_2 = 0.007$
7	$k_s = 0.3$	$k_d = 0.15$	$b_1 = -0.007$	$b_2 = 0.007$
8	$k_s = 0.1$	$k_d = 0.15$	$b_1 = -0.004$	$b_2 = 0.004$
9	$k_s = 0.2$	$k_d = 0.15$	$b_1 = -0.005$	$b_2 = 0.005$
10	$k_s = 0.3$	$k_d = 0.20$	$b_1 = -0.008$	$b_2 = 0.008$

Lab 4.2   Higher Order Approximations of the Pendulum

In previous chapters, we studied the behavior of second-order, homogeneous linear equations (like the harmonic oscillator) by reducing them to first-order linear systems. This "reduction" technique can be applied to nonlinear equations as well, and in this lab we study the ideal pendulum and approximations to the pendulum using this technique.

In the text, we modeled the ideal pendulum by the second-order, nonlinear equation

$$\frac{d^2\theta}{dt^2} + \frac{g}{l}\sin\theta = 0,$$

where θ is the angle from the vertical, g is the gravitational constant ($g = 32$ ft/s^2), and l is the length of the rod of the pendulum, that is, the radius of the circle on which the mass travels. In this lab, we are going to compare the results of numerical simulation of this model with the results obtained from two approximations to this model. The first approximation is a linear approximation given by

$$\frac{d^2\theta}{dt^2} + \frac{g}{l}\theta = 0.$$

The second approximation is a cubic approximation

$$\frac{d^2\theta}{dt^2} + \frac{g}{l}\left(\theta - \frac{\theta^3}{6}\right) = 0.$$

Recall from calculus that the expression $\theta - \theta^3/6$ represents the first two terms in the power series expansion of $\sin\theta$ about $\theta = 0$. We are especially interested in how close the solutions of the approximations of the ideal pendulum equation are to the original ideal pendulum equation. In particular, we are interested in how closely the periods of the periodic orbits of the approximations of the pendulum equation relate to the periods of the periodic orbits of original equation. Your instructor will tell you what value of the parameter l (the length of the pendulum arm) you should use.

Your report should include:

1. A phase portrait analysis for all three equations. Compare and contrast these phase portraits from the point of view of how well the linear and cubic equations approximate the ideal pendulum.

2. In order to study how the periods of the periodic orbits are related, consider the one-parameter family of initial conditions parameterized by θ_0 where $\theta(0) = \theta_0$ and $\theta'(0) = 0$ (no initial velocity). In other words, you should study the various solutions that begin at a given angle with zero velocity. For what intervals of initial conditions do the periods of the periodic orbits of

$$\frac{d^2\theta}{dt^2} + \frac{g}{l}\theta = 0.$$

and

$$\frac{d^2\theta}{dt^2} + \frac{g}{l}\left(\theta - \frac{\theta^3}{6}\right) = 0.$$

closely approximate the periods of the periodic orbits of the ideal pendulum? (The computation of the periods in the linear approximation can be done exactly using the techniques of Chapter 3. Analytic techniques exist for computing the periods of the periodic orbits of the other two equations, but in this lab you should work numerically.) You should plot graphs of the period as a function of θ_0 using a relatively small table (5,10, or 15 entries) of periods obtained using direct numerical simulation of the model.

3. Another family of initial conditions is $\theta(0) = 0$ and $\theta'(0) = v_0$. In this family, the initial velocity is the parameter. Initially, the pendulum points straight down with a given velocity v_0. What changes from your results in part 2 above?

4. Suppose you are a clock maker who makes clocks based on the motion of a pendulum. For each of the three equations, what would you do to double the period of the oscillation?

Your report: Address each of the items above in the form of a short essay. Be as systematic as possible when collecting data and present this data in a concise and clear format. You may illustrate your essay with pictures of phase planes and graphs of solutions or of the data that you collect. However, your essay should be complete and understandable without the pictures.

Lab 4.3   Off Line Springs

Consider a mass m restricted to sliding back and forth along a rod as shown below. We attach the mass to a spring whose other end is a fixed distance a from the rod. Suppose the rest length of the spring (the length at which it applies no force) is L and the spring constant is k_s. Below we will consider the cases where the quantity L is larger than, equal to, and smaller than the distance a

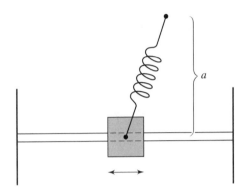

Figure 4.38
Mass sliding on a rod attached to a spring.

We let $y(t)$ denote the position of the mass on the rod at time t with the point on the rod where the spring is shortest corresponding to $y = 0$. When the mass is at position y, the spring will have length $\sqrt{y^2 + a^2}$. Since the spring has rest length L, when the mass is at position y, the amount the spring is stretched or compressed is given by

$$\sqrt{y^2 + a^2} - L$$

(with positive values corresponding to stretched and negative corresponding to compressed). The force exerted by the spring is given by

$$k_s \left(\sqrt{y^2 + a^2} - L \right).$$

Only the component of this force along the rod acts to move the mass. The fraction of this force along the rod is given by the cosine of the angle between the rod and the spring, which is

$$\frac{y}{\sqrt{y^2 + a^2}}.$$

Hence, the equation of motion of the mass is given by

$$m \frac{d^2 y}{dt^2} = k_s \left(\sqrt{y^2 + a^2} - L \right) \frac{y}{\sqrt{y^2 + a^2}}$$

$$= k_s y \left(1 - \frac{L}{\sqrt{y^2 + a^2}} \right).$$

In your report, address the following items:

1. Convert the second-order equation above into a first-order system. Locate the equilibrium points and classify them. [Hint: There will be three cases, $a > L$, $a = L$ and $a < L$, and each case has a different number of equilibrium points.]

2. Give a sketch of the phase plane for each case and write a brief paragraph describing all of the different possible qualitative behaviors of the mass and spring for each of the three cases.

3. Add the effect of a small amount of damping to the system above and repeat parts 1 and 2.

4. Add another spring to the system as shown below. Derive the equations governing the motion of the mass along the rod if the second spring has the same spring constant and the same rest length as the first spring. Repeat parts 1 and 2 (assuming no damping) for this system.

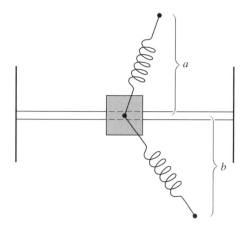

Figure 4.39
Mass sliding on a rod with two springs.

Your report: Address each of the items above in the form of a short essay. Include all relevant calculations. While this lab does not explicitly require numerical work, you may find it helpful to choose values of the parameters and sketch some phase planes. You may include some of these pictures in your report. Be sure to specify the specific parameter values you chose for each picture and remember that the pictures do not replace the words.

FORCING AND RESONANCE

5

In this chapter we consider some particularly important nonautonomous, second-order equations and systems along with the mathematical techniques used to study them. These examples are important because they occur in numerous applications and because the mathematical ideas developed to attack them are quite appealing.

We begin with a system called the forced harmonic oscillator. The harmonic oscillator models (among other things) a mass attached to a spring sliding on a table. The only forces included in the model are the restoring force of the spring and the damping. Any other force affecting the motion of the mass is referred to as an external force. A harmonic oscillator with external force that depends only on time is a forced harmonic oscillator.

In the first three sections, we give techniques for finding solutions of forced harmonic oscillator equations for typical forcing functions. Using these special equations as a guide, we discuss the phenomenon of resonance. In the final two sections, we consider some nonlinear forced equations. In this context, we discuss recent work modeling the collapse of the Tacoma Narrows Bridge (see the cover of this book) and certain so-called chaotic differential equations.

5.1 FORCED HARMONIC OSCILLATORS

Recall that the harmonic oscillator equation is a model for a physical system in which a quantity is subject both to a restoring force and to damping. The restoring force is assumed to be proportional to the displacement and the damping is assumed to be proportional to the velocity. This yields a linear, second-order, constant coefficient equation that can be converted into a first-order system.

A device that can be very informally modeled using the harmonic oscillator system is the motion of a crystal glass tapped (gently) with a fork. The position variable y measures the amount the shape of the glass is deformed from its rest position. The restoring force is the force that pushes the glass back toward its rest shape. After the tap by the fork, the ringing sound is caused by oscillations of the glass around the rest position, so this is an underdamped oscillator. The frequency of the sound is the natural frequency of the oscillations. As an external force, we imagine an opera singer standing near the glass singing a fixed note. The sound waves push against the glass and deform it slightly. The size and direction of this external force depend only on time.

Using a linear system to model something as complicated as the deformation of a glass is a vast simplification. The goal of this model is not to give precise numerical predictions but rather to provide a heuristic framework to think about the mathematics.

In the first two sections of this chapter we consider how external forces affect the motion of a harmonic oscillator. In particular, how can an opera singer break the glass by singing a particular note?

Equations for the Forced Harmonic Oscillator

Recall that the equations for the harmonic oscillator come from Newton's law

$$\text{mass} \times \text{acceleration} = \text{force},$$

which gives

$$m\frac{d^2y}{dt^2} = -k_s\,y - k_d\frac{dy}{dt} \quad \text{or} \quad \frac{d^2y}{dt^2} + \frac{k_d}{m}\frac{dy}{dt} + \frac{k_s}{m}y = 0,$$

where m is the mass, $-k_s\,y$ is the spring force, and $-k_d\,(dy/dt)$ is the force from damping.

To include an "external" force, we must add another term to the right-hand side of this equation. This forcing term can be any function. We consider external forcing that depends only on time t, so the external force is given by a function $g(t)$. Typical examples of forcing functions include $g(t) = $ constant for a force pushing at a constant strength against the mass, and $g(t) = e^{-at}$ for a force whose strength decreases exponentially with time. A particularly important example is $g(t) = \sin(\omega t)$, called sinusoidal forcing with period $2\pi/\omega$ (or frequency $\omega/(2\pi)$). This corresponds to a force that alternately pushes and pulls the mass back and forth in a periodic fashion (like sound waves pushing against a glass).

The new equation is

$$m\frac{d^2y}{dt^2} = -k_s y - k_d \frac{dy}{dt} + g(t),$$

which is more frequently written

$$\frac{d^2y}{dt^2} + \frac{k_d}{m}\frac{dy}{dt} + \frac{k_s}{m}y = \frac{g(t)}{m}.$$

To simplify the notation, we set the mass $m = 1$ and let $p = k_d$, $q = k_s$. The general form of the equation is then

$$\frac{d^2y}{dt^2} + p\frac{dy}{dt} + qy = g(t).$$

This is a second-order, linear, constant coefficient, **nonhomogeneous**, nonautonomous equation. The new adjective nonhomogeneous refers to the fact that the right-hand side of the equation is nonzero. We also refer to this as a forced harmonic oscillator, or a forced equation, for short. An equation of this form with the right-hand side equal to zero is called a **homogeneous** or an **unforced equation**. Our goal for the forced harmonic oscillator equation is the same as for the harmonic oscillator. Given the parameters p and q and the forcing function $g(t)$, describe the solutions, give sketches of the solutions and, if possible, give formulas for solutions. In this section we give examples of a method for finding explicit solutions of the forced harmonic oscillator system for special types of forcing functions.

The method used here is a "lucky guess" method with the impressive name **The Method of Undetermined Coefficients**. An alternative, somewhat more systematic technique is the method of Laplace transforms given in Chapter 8. Again it must be emphasized that, as with nonlinear equations, there is no method that gives the general solution for arbitrary forcing terms. The equations we can solve explicitly are the special cases. The goal of this section is to obtain the general solution for a few special cases and to study the behavior of solutions in these cases.

Outline of the Technique

To find the general solution of a forced harmonic oscillator equation

$$\frac{d^2y}{dt^2} + p\frac{dy}{dt} + qy = g(t),$$

we take full advantage of the fact that we already know how to find the general solution of the autonomous equation

$$\frac{d^2y}{dt^2} + p\frac{dy}{dt} + qy = 0,$$

which corresponds to the system

$$\frac{dy}{dt} = v$$

$$\frac{dv}{dt} = -qy - pv.$$

Extended Linearity Principle

Suppose $y_p(t)$ is a particular solution of the nonhomogeneous equation (the forced equation)

$$\frac{d^2y}{dt^2} + p\frac{dy}{dt} + qy = g(t),$$

and $y_h(t)$ is a solution of the corresponding homogeneous equation (the unforced equation)

$$\frac{d^2y}{dt^2} + p\frac{dy}{dt} + qy = 0.$$

Then $y_p(t) + y_h(t)$ is also a solution of the forced equation. Hence, if $k_1 y_1(t) + k_2 y_2(t)$ is the general solution of the homogeneous equation, then

$$k_1 y_1(t) + k_2 y_2(t) + y_p(t)$$

is the general solution of the nonhomogeneous equation.

We verify this theorem by substituting $y_p(t) + y_h(t)$ into the nonhomogeneous equation. We get

$$\frac{d^2}{dt}(y_h + y_p) + p\frac{d}{dt}(y_h + y_p) + q(y_h + y_p)$$

$$= \left(\frac{d^2 y_h}{dt} + \frac{d^2 y_p}{dt}\right) + p\left(\frac{dy_h}{dt} + \frac{dy_p}{dt}\right) + q\left(y_h + y_p\right)$$

$$= \left(\frac{d^2 y_h}{dt} + p\frac{dy_h}{dt} + qy_h\right) + \left(\frac{d^2 y_p}{dt} + p\frac{dy_p}{dt} + qy_p\right)$$

$$= 0 + g(t),$$

so $y_h(t) + y_p(t)$ is a solution of the nonhomogeneous equation.

From this we obtain the following algorithm for finding explicit solutions of forced harmonic oscillator systems.

Step 1 Find the general solution of the corresponding unforced system or the homogeneous second-order equation.

Step 2 Find **one** particular solution of the forced system or nonhomogeneous second-order equation.

Step 3 Add the results of Steps 1 and 2 to obtain the general solution of the forced system.

We are already adept at solving the unforced harmonic oscillator equations, so Step 1 is doable. Step 2, finding a particular solution of the forced system, is more problematic.

A First-Order Example

The Extended Linearity principle and the three-step algorithm above also apply to first-order, linear, forced (or nonhomogeneous) equations of the form

$$\frac{dy}{dt} + qy = g(t),$$

where q is a constant. The corresponding unforced (or homogeneous) equation is

$$\frac{dy}{dt} + qy = 0$$

(see Exercises 1 and 2). We begin with two first-order examples to demonstrate the techniques involved in The Method of Undetermined Coefficients. Consider the first-order, linear equation

$$\frac{dy}{dt} + 2y = e^{-5t}.$$

(We could also solve this equation using integrating factors — see Section 1.8.)

The first step is to find the general solution of the related homogeneous equation

$$\frac{dy}{dt} + 2y = 0.$$

By separating variables (or by "inspection"), the general solution is

$$y(t) = ke^{-2t},$$

where k is a constant.

In the second step we must find a solution $y_p(t)$ of the nonhomogeneous equation

$$\frac{dy}{dt} + 2y = e^{-5t}.$$

We use guess-and-test, that is, we try to guess $y_p(t)$. Since the right-hand side of the equation is e^{-5t}, it is reasonable to assume that, whatever we guess for $y_p(t)$ should also contain an e^{-5t} term. If we try $y_p(t) = e^{-5t}$ and substitute into the left-hand side of the nonhomogeneous equation, we get

$$\frac{dy_p}{dt} + 2y_p = -5e^{-5t} + 2e^{-5t} = -3e^{-5t},$$

which is not equal to the right-hand side e^{-5t}. Consequently this is not a good guess. However, it is not so far off since the left-hand side and right-hand side of the equation evaluated at this guess differ only by a constant multiple.

We can improve the guess by making it more general. Our new guess is $y_p(t) = ae^{-5t}$, where a is a coefficient we must determine. If we substitute this guess into the equation, the left-hand side becomes

$$\frac{dy_p}{dt} + 2y_p = \frac{d(ae^{-5t})}{dt} + 2ae^{-5t} = -5ae^{-5t} + 2ae^{-5t} = -3ae^{-5t}.$$

We must have

$$-3ae^{-5t} = e^{-5t}$$

for $y_p(t)$ to be a solution. Canceling the e^{-5t}, we obtain

$$-3a = 1.$$

Hence if we choose $a = -1/3$, then $y_p(t) = -(1/3)e^{-5t}$ is a solution.

To form the general solution of the nonhomogeneous equation, we add the general solution of the homogeneous equation and the particular solution of the nonhomogeneous equation, obtaining

$$y(t) = ke^{-2t} - \tfrac{1}{3}e^{-5t},$$

where k is an arbitrary constant that we can adjust to satisfy initial conditions.

To summarize, the idea for finding a particular solution of the nonhomogeneous equation is to make a reasonable guess at the form of the solution but leave some coefficients to be adjusted when the guess is tested in the equation. This motivates the name Method of Undetermined Coefficients. Some coefficients are left undetermined so that they can be adjusted at the last moment.

Second guessing

As with any guess-and-test method, we run the risk that our first guess (no matter how reasonable) will not be correct. What we do in this case is guess again.

The equation

$$\frac{dy}{dt} + 2y = 3e^{-2t}$$

looks very similar to the example above. To find the general solution, we again start by finding the general solution of the homogeneous equation

$$\frac{dy}{dt} + 2y = 0.$$

This is the same homogeneous equation as above, so the general solution is again $y(t) = ke^{-2t}$.

To find a particular solution of the nonhomogeneous equation

$$\frac{dy}{dt} + 2y = 3e^{-2t}$$

(following the ideas above), we make the reasonable guess $y_p(t) = ae^{-2t}$, with a as the undetermined "coefficient." Substituting this guess into the nonhomogeneous equation, the left-hand side becomes

$$\frac{dy_p}{dt} + 2y_p = \frac{d(ae^{-2t})}{dt} + 2ae^{-2t} = -2ae^{-2t} + 2ae^{-2t} = 0.$$

This is very upsetting. No matter how we pick the coefficient a when we substitute $y_p(t)$ into the left-hand side of the equation, we get zero. There is not a particular solution of the nonhomogeneous equation of the form $y_p(t) = ae^{-2t}$.

The problem is that e^{-2t} is a solution of the homogeneous equation. When we substitute it into the left-hand side, we are guaranteed to get zero. On the other hand, our guess must contain the factor e^{-2t} in order to have any hope of having the left-hand side of the equation equal the right-hand side. However, this leaves a wide variety of choices of possible guesses.

We need a second guess for $y_p(t)$ which contains an e^{-2t} term, is not a solution of the homogeneous equation, and is no more complicated than it needs to be. So we try,

$$y_p(t) = ate^{-2t},$$

where a is a coefficient to be determined; that is, we multiply our first guess by t. (We could motivate this guess by finding the general solution using integrating factors, but then it would not be a guess anymore.)

Substituting this into the left-hand side of the nonhomogeneous equation

$$\frac{dy}{dt} + 2y = 3e^{-2t},$$

we obtain

$$\frac{dy_p}{dt} + 2y_p = (ae^{-2t} - 2ate^{-2t}) + 2ate^{-2t} = ae^{-2t}.$$

If we take $a = 3$, then $y_p(t) = 3te^{-2t}$ is a solution. Now we see why multiplying the first guess by t is a good idea. The derivative of $y_p(t) = ate^{-2t}$ via the Product Rule has one term that is a multiple of $y_p(t)$ and another that contains only a constant multiple of e^{-2t}.

Rule of thumb for second guessing

The above is a good example of what is so unsatisfying about lucky-guess (guess-and-test) methods. How did we know to make the second guess equal to t times the first guess? The answer is that we have either seen a similar problem before or we can figure out at least the form of the guess by another technique.

It turns out that for linear equations of any order, the examples above give a general rule of thumb for how to find particular solutions. The form of the guess depends on the right-hand side (or t term) of the equation. If the first guess doesn't yield a particular solution (which will occur if the right-hand side is a solution of the homogeneous equation), we try a second guess by multiplying the first guess by t.

A Second-Order Example

Consider the nonhomogeneous second-order equation

$$\frac{d^2y}{dt^2} + 5\frac{dy}{dt} + 6y = 2e^{-4t}.$$

Suppose we want to find the solution with initial conditions $y(0) = 2$ and $y'(0) = -1$. Finding the general solution of a second-order nonhomogeneous equation follows the same three steps as the first-order case: Find the general solution of the homogeneous equation, find a particular solution of the nonhomogeneous equation, and add.

Solving the homogeneous equation

The first step is to find the general solution of the homogeneous equation

$$\frac{d^2y}{dt^2} + 5\frac{dy}{dt} + 6y = 0.$$

We have a choice of methods. We can convert the equation to the system

$$\frac{dy}{dt} = v$$

$$\frac{dv}{dt} = -6y - 5v$$

and compute the eigenvalues and eigenvectors. Alternately, we can use the lucky guess method; that is, we can guess that the equation has a solution of the form $y(t) = e^{st}$ and solve for s. Substituting $y(t) = e^{st}$ into the equation and simplifying gives

$$s^2 + 5s + 6 = 0,$$

which is the characteristic polynomial. The roots of this quadratic are -2 and -3, and the general solution is

$$y(t) = k_1 e^{-2t} + k_2 e^{-3t}.$$

Since we already have a good feel for the geometry of solutions and the information contained in the eigenvalues, it doesn't matter which technique we choose to solve the homogeneous equation. (Since it doesn't matter, you should use the method that you find easiest.)

Finding one solution of the nonhomogeneous equation

To find a particular solution of the nonhomogeneous equation

$$\frac{d^2 y}{dt^2} + 5\frac{dy}{dt} + 6y = 2e^{-4t},$$

we guess

$$y_p(t) = ae^{-4t},$$

where a is the undetermined coefficient to be determined. Substituting this guess into the nonhomogeneous equation, we obtain

$$16ae^{-4t} - 20ae^{-4t} + 6ae^{-4t} = 2e^{-4t}.$$

Solving for a, we obtain $a = 1$. A particular solution is $y_p(t) = e^{-4t}$. Hence the general solution of the nonhomogeneous equation is

$$y(t) = k_1 e^{-2t} + k_2 e^{-3t} + e^{-4t}.$$

To find values of k_1 and k_2, we use the given initial conditions $y(0) = 2$ and $y'(0) = -1$. Taking the derivative of the general solution, we have

$$\frac{dy}{dt} = -2k_1 e^{-2t} - 3k_2 e^{-3t} - 4e^{-4t}.$$

To satisfy the initial conditions, we must have

$$2 = y(0) = k_1 + k_2 + 1$$

and

$$-1 = y'(0) = -2k_1 - 3k_2 - 4.$$

Solving this system of equations for k_1 and k_2 yields $k_1 = 6$ and $k_2 = -5$. The particular solution is

$$y(t) = 6e^{-2t} - 5e^{-3t} + e^{-4t}.$$

Just as above, if the first guess for the particular solution had not yielded a solution, we would have made a second guess by multiplying the first guess by t (see the exercises).

An Example from Electric Circuit Theory

We present an example of a nonhomogeneous second-order equation motivated by electric circuit theory (with realistic and rather messy numbers). (Nonexperts in electric

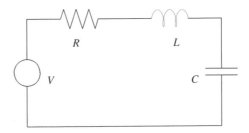

Figure 5.1
Schematic of the RLC circuit to be modeled.

circuits may skip to the differential equation.) Consider the resistor-inductor-capacitor (RLC) circuit pictured in Figure 5.1. Suppose $V_S(t) = 2e^{-t/3}$ volts at time t; that is, the strength of the voltage source is decaying at an exponential rate.

Suppose the $R = 4\text{k}\Omega = 4000\Omega$, $C = 0.25\mu\text{F} = 0.25 \times 10^{-6}\text{F}$, and $L = 1.6\text{H}$ (where Ω, F and H stand for ohms, farads and henrys, standard units in electric circuit theory). The voltage across the capacitor $v(t)$ is a solution of the second-order equation

$$LC\frac{d^2v}{dt^2} + RC\frac{dv}{dt} + v = V_S(t).$$

Using the values given above, we have

$$0.4 \times 10^{-6}\frac{d^2v}{dt^2} + 10^{-3}\frac{dv}{dt} + v = 2e^{-t/3}.$$

We can put this equation into a more familiar form by multiplying through by 2.5×10^6, obtaining

$$\frac{d^2v}{dt^2} + 2500\frac{dv}{dt} + 2.5 \times 10^6v = 5 \times 10^6e^{-t/3}.$$

These numbers are imposing but are realistic for the standard units of measure in electric circuits.

To find the general solution, we follow the procedure above. First, we find the general solution of the homogeneous equation. Then we find a particular solution of the nonhomogeneous equation.

To find the general solution of

$$\frac{d^2v}{dt^2} + 2500\frac{dv}{dt} + 2.5 \times 10^6v = 0$$

we can use the algebraic methods of Chapter 3, or the lucky-guess method. In either case, the eigenvalues are $(-1.25 \pm i\sqrt{0.9375}) \times 10^3$, so the origin is a spiral sink. Using the approximation 0.968 for $\sqrt{0.9375}$, the general solution is

$$v(t) = k_1e^{-1250t}\sin(968t) + k_2e^{-1250t}\cos(968t),$$

where k_1 and k_2 are constants. (Note that the solution decays incredibly quickly to zero, but oscillates very rapidly.)

Now to find a particular solution of the nonhomogeneous equation

$$\frac{d^2v}{dt^2} + 2500\frac{dv}{dt} + 2.5 \times 10^6v = 5 \times 10^6e^{-t/3},$$

we guess $v_p(t) = ae^{-t/3}$, where a is an undetermined coefficient. To find the appropriate a value, we substitute the guess back into the differential equation. The left-hand side is

$$\frac{d^2v_p}{dt} + 2500\frac{dv_p}{dt} + 2.5 \times 10^6 v_p = \frac{d^2(ae^{-t/3})}{dt} + 2500\frac{d(ae^{-t/3})}{dt} + 2.5 \times 10^6 ae^{-t/3}$$

$$= \frac{a}{9}e^{-t/3} - 2500\frac{a}{3}e^{-t/3} + 2.5 \times 10^6 ae^{-t/3}$$

$$= \left(\frac{1}{9} - \frac{2500}{3} + 2.5 \times 10^6\right)ae^{-t/3}.$$

Next, we set this equal to the right-hand side of the nonhomogeneous equation, which is $5 \times 10^6 e^{-t/3}$. Cancelling the $e^{-t/3}$ term and solving for a, we see that $a \approx 2$. So our particular solution is (approximately) $v_p(t) = 2e^{-t/3}$.

We can now form the general solution

$$v(t) = k_1 e^{-1250t}\sin(968t) + k_2 e^{-1250t}\cos(968t) + 2e^{-t/3},$$

where k_1 and k_2 are constants that we adjust according to the initial conditions. The first two terms of the general solution oscillate very quickly and decrease very quickly to zero. Hence for large t, every solution will be close to the particular solution. Graphs of the solution with initial conditions $v(0) = 1$, $v'(0) = 0$ are shown in Figures 5.2 and 5.3. To see the oscillations, we must look very close to $t = 0$ (see Figure 5.3). In circuit theory language, we would say that the natural response (the solution of the homogeneous equation) decays quickly and every solution approaches the **forced response** (the particular solution of the homogeneous equation).

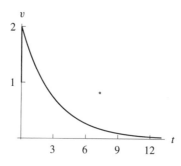

Figure 5.2
Solution of

$$\frac{d^2v}{dt^2} + 2500\frac{dv}{dt} + 2.5 \times 10^6 = 5 \times 10^6 e^{-t/3}$$

with initial condition $v(0) = 1$, $v'(0) = 0$.

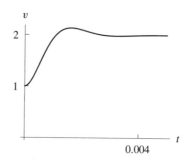

Figure 5.3
The same solution that is graphed in Figure 5.2. Here we show the graph only for t that are very close to 0 in order to illustrate the rapid increase from the initial value $v(0) = 1$.

Exercises for Section 5.1

1. Consider the first-order linear differential equation

$$\frac{dy}{dt} + qy = g(t)$$

where q is a constant.

(a) Verify that, if $y_1(t)$ is a solution of the homogeneous equation

$$\frac{dy}{dt} + qy = 0$$

and $y_p(t)$ is a solution of the nonhomogeneous equation

$$\frac{dy}{dt} + qy = g(t),$$

then $y_1(t) + y_p(t)$ is also a solution of the nonhomogeneous equation.

(b) Verify that, if $y_1(0) \neq 0$, then $ky_1(t) + y_p(t)$ is the general solution of the nonhomogeneous equation.

2. Suppose $y_p(t)$ is a solution of

$$\frac{dy}{dt} + 3y = 4\sin(2t).$$

Show that $2y_p(t)$ is *not* a solution of this equation.

In Exercises 3–12, find the general solution of the given differential equation. Then find the particular solution that corresponds to the initial condition specified.

3. $\dfrac{dy}{dt} + 5y = 2e^{-2t}, \quad y(0) = 2$

4. $\dfrac{dy}{dt} + 5y = 2e^{-5t}, \quad y(0) = 2$

5. $\dfrac{dy}{dt} - 2y = 2e^{-2t}, \quad y(0) = -2$

6. $\dfrac{d^2y}{dt^2} + 4y = 2e^{-3t}, \quad \begin{aligned} y(0) &= 2 \\ y'(0) &= 0 \end{aligned}$

7. $\dfrac{d^2y}{dt^2} + 2y = e^{-2t}, \quad \begin{aligned} y(0) &= 2 \\ y'(0) &= -1 \end{aligned}$

8. $\dfrac{d^2y}{dt^2} + 2\dfrac{dy}{dt} = -3e^{-2t}, \quad \begin{aligned} y(0) &= 1 \\ y'(0) &= 0 \end{aligned}$

9. $\dfrac{d^2y}{dt^2} + 2\dfrac{dy}{dt} + 5y = e^t, \quad \begin{aligned} y(0) &= 2 \\ y'(0) &= 0 \end{aligned}$

10. $\dfrac{d^2y}{dt^2} - \dfrac{dy}{dt} + y = e^{-t}, \quad \begin{aligned} y(0) &= 2 \\ y'(0) &= 0 \end{aligned}$

11. $\dfrac{d^2y}{dt^2} - \dfrac{dy}{dt} = 2e^t, \quad \begin{aligned} y(0) &= -2 \\ y'(0) &= 1 \end{aligned}$

12. $\dfrac{d^2y}{dt^2} - \dfrac{dy}{dt} - y = e^{-3t}, \quad \begin{aligned} y(0) &= 2 \\ y'(0) &= 0 \end{aligned}$

In Exercises 13–15, find the general solution of the differential equation.

13. $\dfrac{d^2y}{dt^2} + 5\dfrac{dy}{dt} + 6y = 2e^{-t}$

14. $\dfrac{d^2y}{dt^2} + 5\dfrac{dy}{dt} + 6y = 2e^{-2t}$

15. $\dfrac{d^2y}{dt^2} + 2\dfrac{dy}{dt} + y = 3e^{-t}$

16. The "lucky guess" method to find a particular solution of a forced equation says that, if the forcing is a trigonometric function, then you should guess a trigonometric function for the particular solution. If the forcing is some other type of function, then you should adjust your guess accordingly. For example, for a forced harmonic oscillator of the form

$$\frac{d^2y}{dt^2} + 4y = -3t^2 + 2t + 3,$$

the forcing function is the quadratic polynomial $g(t) = -3t^2 + 2t + 3$. It is reasonable to guess that a particular solution in this case will also be a quadratic polynomial. Hence, we guess $y_p(t) = at^2 + bt + c$. The constants a, b, and c are determined by substituting $y_p(t)$ into the equation.

    (a) Find the general solution of the forced harmonic oscillator specified above in this exercise.

    (b) Find the particular solution with the initial condition $y(0) = 2$, $y'(0) = 0$.

In Exercises 17–18, use the method of Exercise 16 to find the general solution of the specified equation.

17. $\dfrac{d^2y}{dt^2} + 4y = 3t + 2$ **18.** $\dfrac{d^2y}{dt^2} + 2\dfrac{dy}{dt} = 3t + 2$

19. In the exercises above we have seen how to use the "lucky-guess" method to find particular solutions of forced equations with exponential forcing. We can also use this approach to solve equations whose forcing functions are sums of these types of functions. More precisely, suppose that $y_1(t)$ is a solution of the equation

$$\frac{d^2y}{dt^2} + p\frac{dy}{dt} + qy = g(t)$$

and that $y_2(t)$ is a solution of the equation

$$\frac{d^2y}{dt^2} + p\frac{dy}{dt} + qy = h(t).$$

Show that $y_1(t) + y_2(t)$ is a solution of the equation

$$\frac{d^2y}{dt^2} + p\frac{dy}{dt} + qy = g(t) + h(t).$$

In Exercises 20–24, use the method of Exercise 19 to find the general solution of the differential equation.

20. $\dfrac{d^2y}{dt^2} + 4y = 3 + 2t + e^{-3t}$ **21.** $\dfrac{d^2y}{dt^2} + 4y = 4t^2 + e^{-2t}$

22. $\dfrac{dy}{dt} + 2y = 2t + 5e^{-3t}$ **23.** $\dfrac{d^2y}{dt^2} + 2y = 2t + 5e^{-3t}$

24. $\dfrac{d^2y}{dt^2} + \dfrac{dy}{dt} + 2y = e^{-3t} + 3e^{-4t}$

In Exercises 25–26, use the method of Exercise 19 to find the solution of the given initial-value problem.

25. $\dfrac{d^2 y}{dt^2} + 4y = 5t^3 + e^{-t}$,

$y(0) = 2, \; y'(0) = 0$

26. $\dfrac{d^2 y}{dt^2} + 2\dfrac{dy}{dt} + 4y = e^{-2t} + 1 + t$,

$y(0) = 0, \; y'(0) = 0$

27. Suppose the equation

$$\frac{d^2 v}{dt^2} + 2\frac{dv}{dt} + 6v = 2e^{-t/4}$$

is the equation for the voltage across the capacitor in an RLC circuit. (These numbers are not very realistic for actual circuits).

(a) Find the general solution of this equation.

(b) Suppose a constant voltage source is added to the circuit so that the equation becomes

$$\frac{d^2 v}{dt^2} + 2\frac{dv}{dt} + 6v = 4 + 2e^{-t/4}.$$

Find the general solution of this equation.

(c) In a paragraph, describe the long-term behaviors of the solutions of these two equations.

5.2 SINUSOIDAL FORCING AND RESONANCE

In this section we consider the forced harmonic oscillator equation

$$\frac{d^2 y}{dt^2} + \frac{k_d}{m}\frac{dy}{dt} + \frac{k_s}{m}y = g(t)$$

with a a particular type of forcing term $g(t)$. In the last section we considered this equation where $g(t)$ was given by exponential or polynomial functions. Here we consider the case where $g(t)$ is a trigonometric function.

Sinusoidal Forcing

One type of forcing term that frequently arises in applications is periodic forcing, where the oscillator is subjected to periodic shaking back and forth. Examples range from the shaking of a building in an earthquake to the pressure waves of a sound striking a crystal glass. To model such a force, we add a forcing term $g(t)$ that is periodic in t. The most familiar periodic functions are the trigonometric functions, sine and cosine. We focus on a sinusoidal forcing term $g(t) = \sin(\omega t)$ However, the same ideas apply to forcing terms involving other combinations of sines and cosines. The frequency of the forcing term is called the **forcing frequency**. In this case the forcing frequency is $\omega/(2\pi)$.

We begin with a forced, undamped oscillator

$$\frac{d^2y}{dt^2} + 2y = \sin(\omega t).$$

We can think of this differential equation as a model for a unit mass attached to a spring, with spring constant 2, sliding on a frictionless table. At the same time, the table is being rocked back and forth with period $2\pi/\omega$ in such a way that the mass is pulled or pushed with an extra force equal to $\sin(\omega t)$. We use the method of undetermined coefficients of the previous section to find the general solution of this equation. Of course, this method is valid for any choice of spring constant; we fix $q = 2$ simply to reduce the number of parameters.

General solution of the unforced system

The first step in the method of undetermined coefficients is to find the solution of the unforced equation

$$\frac{d^2y}{dt^2} + 2y = 0.$$

The general solution of this system is

$$y(t) = k_1 \cos(\sqrt{2}t) + k_2 \sin(\sqrt{2}t),$$

where k_1 and k_2 are constants. These solutions oscillate around $(0,0)$ with natural frequency $\sqrt{2}/2\pi$ (see Figure 5.4).

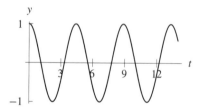

Figure 5.4

Solution of the unforced equation

$$\frac{d^2y}{dt^2} + 2y = 0$$

with initial conditions $y(0) = 1$, $y'(0) = 0$.

Particular solution of the forced system

The second step is to find one particular solution of the nonhomogeneous equation

$$\frac{d^2y}{dt^2} + 2y = \sin(\omega t).$$

As before, we use the lucky-guess method to find this solution. To make a lucky (that is, intelligent) guess, we note that the forcing term is $\sin(\omega t)$. From Euler's equation we know that

$$e^{i\omega t} = \cos(\omega t) + i\sin(\omega t),$$

so the forcing function is the imaginary part of the complex exponential.

Since we know how to solve nonhomogeneous equations whose forcing terms are exponentials, this suggests that we consider the new equation

$$\frac{d^2y}{dt^2} + 2y = e^{i\omega t}.$$

As we have seen in Chapter 3, a particular solution $y_p(t)$ of this complex differential equation will have both a real and an imaginary part; that is, we can write $y_p(t) = y_1(t) + iy_2(t)$. Just as we saw earlier, $y_1(t)$ is the solution of the nonhomogeneous equation whose forcing term is $\cos(\omega t)$, and $y_2(t)$ is the solution of the equation whose forcing term is $\sin(\omega t)$. Thus our particular solution is the imaginary part $y_2(t)$ of the particular solution that satisfies the equation

$$\frac{d^2 y}{dt^2} + 2y = e^{i\omega t}.$$

As in the previous section, to solve this complex differential equation, we guess $y_p(t) = ae^{i\omega t}$ where a is to be determined by the differential equation. Substituting this expression into the differential equation, we obtain

$$-a\omega^2 e^{i\omega t} + 2ae^{i\omega t} = e^{i\omega t}.$$

Dividing both sides by $e^{i\omega t}$ and solving for a yields

$$a = \frac{1}{2 - \omega^2}$$

as long as $2 - \omega^2 \neq 0$. Hence a particular solution of the complex forced equation is given by

$$y_1(t) + iy_2(t) = \frac{1}{2 - \omega^2} e^{i\omega t} = \frac{1}{2 - \omega^2} \left(\cos(\omega t) + i \sin(\omega t) \right).$$

Choosing the imaginary part of this solution then gives the particular solution

$$y_2(t) = \frac{1}{2 - \omega^2} \sin(\omega t)$$

of the sinusoidally forced equation. Consequently, the general solution of the nonhomogeneous equation is

$$y(t) = k_1 \cos(\sqrt{2}t) + k_2 \sin(\sqrt{2}t) + \frac{1}{2 - \omega^2} \sin(\omega t),$$

where k_1 and k_2 are constants determined by the initial conditions as usual.

One question that arises is why did we not simply guess $a \sin(\omega t)$ at the outset? After all, the particular solution we found does indeed only involve sines. However, as we will see below when we consider damped oscillators, it is very often the case that particular solutions of sinusoidally forced equations involve both sines and cosines. The guess of a complex exponential yields these solutions expeditiously. Incidentally, this complex guess is related to objects called **phasors** used in electric circuit theory (see the exercises).

The general solution obtained above is only valid if $2 - \omega^2 \neq 0$. This is the case where the forcing frequency is not equal to the natural frequency of the system. In this case the general solution is the sum of trigonometric functions with periods $2\pi/\sqrt{2}$ and $2\pi/\omega$. For some values of t, all of these terms have the same sign and the amplitude of the solution is large. For other values of t, certain terms have different signs and there is cancellation. This is particularly evident if the forcing frequency is close to the natural frequency. Examples of typical graphs of $y(t)$ for three different choices of ω are given in Figures 5.5–5.7.

Figure 5.5
Solution of $d^2y/dt^2 + 2y = \sin(\omega t)$ for $y(0) = 1$, $y'(0) = 0$, and $\omega = 0.5$.

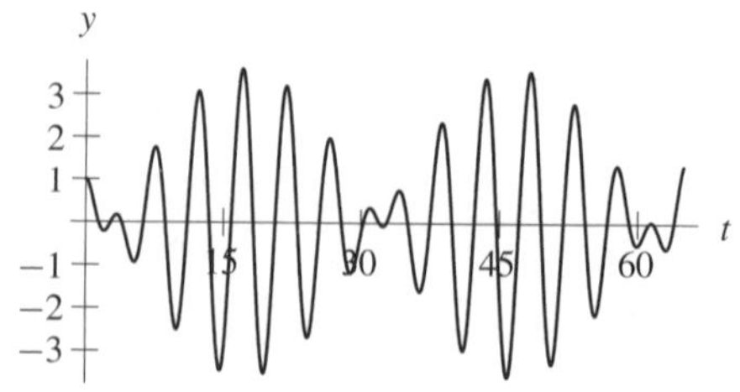

Figure 5.6
Solution of $d^2y/dt^2 + 2y = \sin(\omega t)$ for $y(0) = 1$, $y'(0) = 0$, and $\omega = 1.2$.

We can hear the phenomenon pictured in Figure 5.6 when listening to a slightly out of tune piano. The sound from two strings with almost the same pitch add to make a "wha-wha" sound called beating modes.

Adjusting the forcing frequency can have a large effect on the solutions (or response) of the system. For our example, when $\omega^2 \approx 2$, the coefficient $|1/(2 - \omega^2)|$ is large. The particular solution can oscillate with amplitude as large as $|1/(2 - \omega^2)|$. Hence if we "tune" the forcing frequency so that it is close to the natural frequency, we expect the solutions eventually to exhibit large oscillations even for small initial conditions (see Figure 5.7).

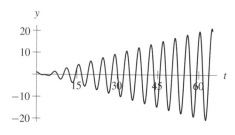

Figure 5.7
Solution of

$$\frac{d^2y}{dt^2} + 2y = \sin(\omega t)$$

for $y(0) = 1$ and $y'(0) = 0$ where the forcing is nearly resonant ($\omega = 1.4$). In other words, the forcing and natural frequencies are almost equal.

All our computations above rely on the crucial assumption that the forcing frequency $\omega/2\pi$ is not equal to the natural frequency $\sqrt{2}/2\pi$, or $\omega \neq \pm\sqrt{2}$. The case when the forcing frequency equals the natural frequency ($\omega = \pm\sqrt{2}$ in this example) is called **resonant forcing**. When the forcing frequency is close to the natural frequency, the system is said to be **near resonant**.

Resonance

The terms **resonant forcing** and **resonance** are used to describe any periodic forcing of a harmonic oscillator equation whose frequency is equal to the natural frequency. The differences between the solutions for resonant forcing and nonresonant forcing are both dramatic and dangerous. To study what happens, we consider the case of sinusoidal forcing in the example above.

Suppose $\omega = \sqrt{2}$, so

$$\frac{d^2y}{dt^2} + 2y = \sin(\sqrt{2}t)$$

and the forcing frequency and the natural frequency agree. The system is said to be in resonance. Since we can no longer divide by $2 - \omega^2$ as we did in the previous example, we must repeat all of the algebra above.

To find a particular solution of the forced system, we begin by studying the related equation with complex exponential forcing,

$$\frac{d^2 y}{dt^2} + 2y = e^{i\sqrt{2}t}.$$

Suppose we make the usual guess that the particular solution is $ae^{i\sqrt{2}t}$. Substituting this guess into the differential equation as above yields

$$-2a + 2a = 1.$$

No value of a satisfies this equation. Therefore, we must alter our guess.

This difficulty is exactly the same as we encountered in Section 5.1. The guess does not give a solution of the nonhomogeneous equation because it is a solution of the homogeneous equation. Using the rule of thumb for second-guessing in Section 5.1, we make the new guess $y_p(t) = ate^{i\sqrt{2}t}$. Using

$$\frac{dy_p}{dt} = ae^{i\sqrt{2}t} + ia\sqrt{2}te^{i\sqrt{2}t}$$

and

$$\frac{d^2 y_p}{dt^2} = 2ia\sqrt{2}e^{i\sqrt{2}t} - 2ate^{i\sqrt{2}t},$$

the equation

$$\frac{d^2 y_p}{dt^2} + 2y_p = e^{i\sqrt{2}t}$$

becomes

$$(2ia\sqrt{2}e^{i\sqrt{2}t} - 2ate^{i\sqrt{2}t}) + 2ate^{i\sqrt{2}t} = e^{i\sqrt{2}t}.$$

Dividing through by the exponential term and solving for the undetermined coefficient a gives

$$a = \frac{1}{2i\sqrt{2}}.$$

Since $1/i = -i$, we have $a = -i/(2\sqrt{2})$ and

$$y_p(t) = -\frac{i}{2\sqrt{2}}te^{i\sqrt{2}t}.$$

Using Euler's formula, we find

$$y_p(t) = y_1(t) + iy_2(t) = \frac{1}{2\sqrt{2}}t\sin(\sqrt{2}t) - \frac{i}{2\sqrt{2}}t\cos(\sqrt{2}t).$$

As before, the particular solution of the original equation

$$\frac{d^2 y}{dt^2} + 2y = \sin(\sqrt{2}t)$$

is the imaginary part of $y_p(t)$, or

$$y_2(t) = -\frac{t\cos(\sqrt{2}t)}{2\sqrt{2}}.$$

Therefore the general solution of this equation is

$$y(t) = k_1\cos(\sqrt{2}t) + k_2\sin(\sqrt{2}t) - \frac{t\cos(\sqrt{2}t)}{2\sqrt{2}}$$

for any constants k_1 and k_2.

Qualitative nature of resonance

From the formula for the general solution, we can determine that having the natural frequency equal to the forcing frequency has a qualitative effect on solutions. The particular solution $y(t) = -t\cos(\sqrt{2}t)/(2\sqrt{2})$ of the nonhomogeneous equation oscillates with larger and larger amplitude as t increases. The amplitude increases linearly with slope $1/(2\sqrt{2})$ (see Figure 5.8).

When t is large, the term $-t\cos(\sqrt{2}t)/(2\sqrt{2})$ oscillates with much larger amplitude than the other terms in the general solution. No matter what the initial condition (that is, choice of k_1 and k_2), the behavior of $y(t)$ is eventually dominated by the $t\cos(\sqrt{2}t)/(2\sqrt{2})$ term. The amplitude of the oscillation grows in an unbounded fashion.

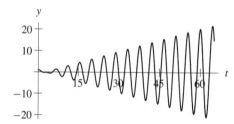

Figure 5.8
Graph of the solution to

$$\frac{d^2y}{dt^2} + 2y = \sin(\sqrt{2}t)$$

with $y(0) = 1$, $y'(0) = 0$.

Dependence of the amplitude of solutions on the forcing frequency

If we think of the forcing frequency as an adjustable parameter, then we can ask how the maximum amplitude of solutions changes as the forcing frequency approaches the natural frequency. At the resonant frequency, the amplitude of the solutions does not reach a maximum but continues to grow for all t.

In Figure 5.9, we plot the maximum possible amplitude of the solution of

$$\frac{d^2y}{dt^2} + 2y = \sin(\omega t)$$

with initial conditions $y(0) = 1$, $y'(0) = 0$ as a function of the forcing frequency ω. The forcing frequency and natural frequency must be very close in order to observe the growth of amplitude due to resonance.

The resonance effect occurs even if the amplitude of the forcing term is small. Adding a parameter ϵ to control the amplitude of the forcing, the equation becomes

$$\frac{d^2y}{dt^2} + 2y = \epsilon\sin(\omega t).$$

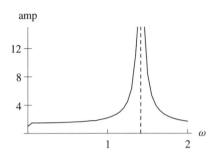

Figure 5.9

Maximum amplitude of the solutions of the equation

$$\frac{d^2y}{dt^2} + 2y = \sin(\omega t)$$

as a function of ω for the solution with the initial conditions $y(0) = 1$, $y'(0) = 0$.

A particular solution for the nonresonant case ($\omega \neq \pm\sqrt{2}$) is

$$y(t) = \frac{\epsilon}{2 - \omega^2} \sin(\omega t).$$

Even if ϵ is very small, if the forcing frequency is close enough to the natural frequency ($\omega \approx \pm\sqrt{2}$), then this solution will oscillate with large amplitude.

For resonant forcing

$$\frac{d^2y}{dt^2} + 2y = \epsilon \sin(\sqrt{2}t),$$

a particular solution is

$$y(t) = \frac{-\epsilon t \cos(\sqrt{2}t)}{2\sqrt{2}}.$$

So all solutions of this equation have oscillations that eventually grow at the rate $\epsilon t/(2\sqrt{2})$, which is large for large t. This is an important property of resonance. Informally we say that the effects of small forcing at the natural frequency add up to large effects over time. There is nothing special about the particular value of the spring constant we used in this example. Also, the phenomenon of resonance can occur for many different types of periodic forcing where the frequency of the forcing equals the natural frequency of the oscillator (see Section 5.3).

Damped Oscillator with Sinusoidal Forcing

We use the lucky guess technique in one more example, a damped harmonic oscillator with sinusoidal forcing. Consider the example above with the damping coefficient equal to 1 and the forcing frequency equal to $\omega/(2\pi)$. The second-order equation is

$$\frac{d^2y}{dt^2} + \frac{dy}{dt} + 2y = \sin(\omega t).$$

First we find the general solution for the unforced oscillator

$$\frac{d^2y}{dt^2} + \frac{dy}{dt} + 2y = 0,$$

which corresponds to the system

$$\frac{dy}{dt} = v$$

$$\frac{dv}{dt} = -2y - v.$$

The eigenvalues are $(-1 \pm i\sqrt{7})/2$ with eigenvectors $(1, (-1 \pm i\sqrt{7})/2)$, respectively. Thus this oscillator is underdamped with a natural period of $4\pi/\sqrt{7}$. The general solution of the homogeneous equation is

$$y(t) = k_1 e^{-t/2} \cos(\sqrt{7}\, t/2) + k_2 e^{-t/2} \sin(\sqrt{7}\, t/2).$$

Now to find a particular solution of the forced equation, we again note that the sinusoidal forcing term is the imaginary part of $e^{i\omega t}$. So we begin by studying

$$\frac{d^2 y}{dt^2} + \frac{dy}{dt} + 2y = e^{i\omega t}.$$

To find a particular solution of this equation, we guess

$$y_p(t) = a e^{i\omega t}$$

where a is the undetermined coefficient to be computed.

Differentiating $y_p(t)$ twice and substituting into the differential equation, we obtain

$$-a\omega^2 e^{i\omega t} + ia\omega e^{i\omega t} + 2a e^{i\omega t} = e^{i\omega t}.$$

Simplifying this equation yields

$$(2 - \omega^2)a + i\omega a = 1$$

or

$$a = \frac{1}{(2 - \omega^2) + i\omega} = \frac{(2 - \omega^2) - i\omega}{(2 - \omega^2)^2 + \omega^2}.$$

Therefore,

$$y_p(t) = \frac{(2 - \omega^2) - i\omega}{(2 - \omega^2)^2 + \omega^2} e^{i\omega t}$$

$$= \frac{(2 - \omega^2) - i\omega}{(2 - \omega^2)^2 + \omega^2} (\cos(\omega t) + i\sin(\omega t)).$$

If we write $y_p(t) = y_1(t) + iy_2(t)$ as above, the imaginary part

$$y_2(t) = \frac{(2 - \omega^2)}{(2 - \omega^2)^2 + \omega^2} \sin(\omega t) - \frac{\omega}{(2 - \omega^2)^2 + \omega^2} \cos(\omega t)$$

is a particular solution of the equation

$$\frac{d^2 y}{dt^2} + \frac{dy}{dt} + 2y = \sin(\omega t).$$

Hence the general solution of this equation is

$$y(t) = k_1 e^{-t/2} \cos(\sqrt{7}t/2) + k_2 e^{-t/2} \sin(\sqrt{7}t/2) +$$

$$\frac{(2 - \omega^2)}{(2 - \omega^2)^2 + \omega^2} \sin(\omega t) - \frac{\omega}{(2 - \omega^2)^2 + \omega^2} \cos(\omega t),$$

which is quite a mouthful. Note that the particular solution of the nonhomogeneous equation involves both sines and cosines even though the forcing function is $\sin(\omega t)$. This is the reason that the most obvious first guess $y_p(t) = a\sin(\omega t)$ does not always work.

In contrast to the undamped case, the denominators that appear in the particular solution are never zero, no matter what the value of ω. However, the amplitude of the oscillations of the particular solution does vary with ω.

Analysis of the Solution Formulas

The formula for the general solution of the sinusoidally forced, damped oscillator

$$\frac{d^2y}{dt^2} + \frac{dy}{dt} + 2y = \sin(\omega t)$$

is

$$y(t) = k_1 e^{-t/2} \cos(\sqrt{7}t/2) + k_2 e^{-t/2} \sin(\sqrt{7}t/2) +$$

$$\frac{(2 - \omega^2)}{(2 - \omega^2)^2 + \omega^2} \sin(\omega t) - \frac{\omega}{(2 - \omega^2)^2 + \omega^2} \cos(\omega t),$$

which is very complicated. Even though this formula allows us to compute the exact location of the solution at a given time, the qualitative behavior of the solutions in this case is still unclear. By studying these equations, we can read off certain aspects that are important to the qualitative behavior of solutions.

Long-term behavior of solutions

First, the $e^{-t/2}$ factor in the terms of the general solution of the unforced system means that these terms tend to zero as t increases. The choices of k_1 and k_2 (which we adjust to satisfy the initial condition) do not matter for large t. All solutions eventually approach $y_p(t)$, which is a periodic solution. In Figure 5.10 we give the graphs of three different solutions. In each case the long-term behavior is the same, and it resembles the behavior of the forcing function. We can describe this behavior efficiently with the following sentence: For a forced, damped oscillator, the natural response decays exponentially and the solutions approach the forced response. The **natural response** is the solution of the unforced or homogeneous equation. The **forced response** is the particular solution of the forced or nonhomogeneous equation. In electric circuit theory, the forced response is sometimes called the **steady state**, even though it is oscillating with the same frequency as the forcing function.

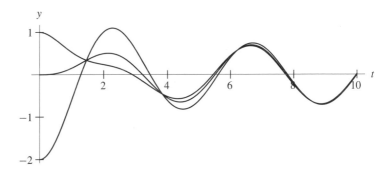

Figure 5.10
Three solutions of $d^2y/dt^2 + dy/dt + 2y = \sin(\omega t)$ for $\omega = \sqrt{2}$. All have initial velocity $y'(0) = 0$.

Effect of the forcing frequency

The maximum amplitude of solutions changes as the forcing frequency is adjusted. The maximum amplitude of solutions may sometimes be quite large (see Figure 5.11), but if damping is present, then solutions do not tend to infinity. Although this is mathematically reassuring, in practice we must remember that physical systems break down at large finite values. Even if some damping is present, the phenomenon of resonance can have dramatic (or disastrous) consequences.

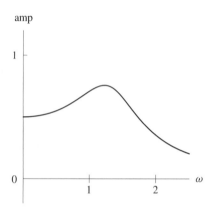

Figure 5.11
Graph of an upper bound for the amplitude of solutions at large time versus the forcing frequency ω for the forced damped oscillator

$$\frac{d^2y}{dt^2} + \frac{dy}{dt} + 2y = \sin(\omega t).$$

Exercises for Section 5.2

In Exercises 1–6, find the solution to the initial-value problem specified. Give an upper bound for the maximum of $|y(t)|$, and describe the long-term behavior of the solution $y(t)$.

1. $\dfrac{d^2y}{dt^2} + 4y = \cos(3t),$

   $y(0) = 1,\ y'(0) = 0$

2. $\dfrac{d^2y}{dt^2} + 4y = 0.5\sin(4t),$

   $y(0) = 0,\ y'(0) = 1$

3. $\dfrac{d^2y}{dt^2} - 4y = \sin(2t),$

   $y(0) = 1,\ y'(0) = -2$

4. $\dfrac{d^2y}{dt^2} + 4y = 0.2\cos(2t),$

   $y(0) = 1,\ y'(0) = 0$

5. $\dfrac{d^2y}{dt^2} + 2\dfrac{dy}{dt} + 4y = 0.4\cos(2t),$

   $y(0) = 1,\ y'(0) = 0$

6. $\dfrac{d^2y}{dt^2} + 10\dfrac{dy}{dt} + 4y = \cos(3t),$

   $y(0) = 2,\ y'(0) = 1$

7. For a forced harmonic oscillator equation of the form

$$\frac{d^2y}{dt^2} + p\frac{dy}{dt} + qy = \epsilon \cos(\omega t),$$

where p and q are positive, show that any solution $y(t)$ oscillates with frequency approximately equal to $\omega/(2\pi)$ for sufficiently large t.

8. The commercials for a maker of recording tape show a crystal glass being broken by the sound of a recording of an opera singers voice. Is this a good test of the quality of the tape? Why or why not?

9. Legend has it that the marching band of a major university is forbidden from playing the song "Hey Jude". The reason is that whenever the band played this song, the fans would stomp their feet in time to the music and after the second chorus, the stadium structure would begin to rock quite noticeably. The administration insisted that there was no danger that the stadium would collapse, but banned the song as a precaution. No other song has this effect on the stadium (although, presumedly, the fans will stomp to other songs). Give a possible explanation of why the stadium structure reacted so violently to the stomping accompanying the song "Hey Jude" and to no other song?

In Exercises 10–12, we explore the different types of notation used to express solutions of harmonic oscillator equations with trigonometric forcing functions. Sadly, the differences in terminology sometimes make it quite difficult for mathematicians and engineers to communicate, even though we are talking about the same things.

10. Given a function

$$g(t) = a \cos(\omega t) + b \sin(\omega t),$$

where a and b are constants, show that there exist constants k and ϕ such that

$$g(t) = k \cos(\omega t + \phi).$$

[*Hint*: Use the formula for the cosine of a sum to rewrite $\cos(\omega t + \phi)$ and solve for k and ϕ.]
Remark: Writing $g(t)$ in terms of just cosine instead of sine and cosine makes it easier to determine the amplitude of the oscillation of $g(t)$ (it is just k). The quantity ϕ is called the **phase** of the oscillation.

11. Given a function

$$g(t) = a \cos(\omega t) + b \sin(\omega t),$$

where a and b are constants, show that g can be written in the form

$$g(t) = \text{Re}(ke^{i\phi} e^{i\omega t}).$$

[*Hint*: Use Euler's formula and Exercise 10.]
Remark: The complex expression $ke^{i\phi}$ is called a **phasor**. If we know that $g(t)$ has the form $k \cos(\omega t + \phi)$, then we need know only the constants k and ϕ, the amplitude and the phase, to know the function g. Hence we can use the phasor $ke^{i\phi}$ as a notation for the function $g(t) = ke^{i\phi} e^{i\omega t}$.

12. Show that if $k_1 e^{i\phi_1}$ is the phasor representing $g_1(t)$ and $k_2 e^{i\phi_2}$ is the phasor representing $g_2(t)$ (both with frequency $\omega/(2\pi)$), then the phasor representing $g_1(t) + g_2(t)$ is $k_1 e^{i\phi_1} + k_2 e^{i\phi_2}$. [*Hint*: Write the function $g_1(t)$ and $g_2(t)$ as the real part of the phasor time $e^{i\omega t}$, add them, and simplify.]
Remark: Phasor terminology is widely used in electric circuit theory, where an algebra of phasors is developed suited to dealing with currents and voltages in circuits with periodic voltage sources.

5.3 QUALITATIVE ANALYSIS OF THE FORCED HARMONIC OSCILLATOR

In this section we consider the forced harmonic oscillator equation

$$\frac{d^2 y}{dt^2} + p\frac{dy}{dt} + qy = g(t)$$

or the corresponding system

$$\frac{dy}{dt} = v$$
$$\frac{dv}{dt} = -qy - pv + g(t)$$

from a qualitative point of view. Our main goal is to better understand the phenomenon of resonance.

The tools we use for qualitative analysis of the unforced harmonic oscillator — the vector field, the phase plane and the eigenvalues and eigenvectors — no longer tell the complete story. The right-hand side of the equation depends on time, so the vector field changes with time. Figure 5.12 shows two solutions for the harmonic oscillator system with $p = 0$, $q = 2$ and $g(t) = 0.2\sin(t)$ in the yv-plane. The Uniqueness Theorem says that two different solutions cannot be in the same place at the same time, but it does not prevent solutions from being at the same place at different times. Since the vector field changes with time, the solution curves can cross in the yv-plane, unlike the case of autonomous systems.

We use ideas from Hamiltonian systems (see Section 4.2) to give insight into how the forced harmonic oscillator works and how forcing at the natural frequency (resonant forcing) causes such a dramatic effect.

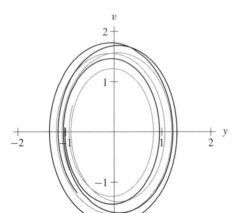

Figure 5.12
Two solutions of the system

$$\frac{dy}{dt} = v$$
$$\frac{dv}{dt} = -2y + 0.2\sin t$$

in the yv-plane.

Energy for the Harmonic Oscillator

To understand the qualitative behavior of the forced harmonic oscillator, we use the concept of energy for Hamiltonian systems. Consider the undamped harmonic oscillator with mass $m = 1$, spring constant $q = 2$, and damping coefficient $p = 0$

$$\frac{d^2 y}{dt^2} + 2y = 0,$$

which corresponds to the system

$$\frac{dy}{dt} = v$$
$$\frac{dv}{dt} = -2y.$$

In matrix form, this system is

$$\frac{d\mathbf{Y}}{dt} = \begin{pmatrix} 0 & 1 \\ -2 & 0 \end{pmatrix} \mathbf{Y},$$

where $\mathbf{Y} = (y, v)$.

To understand the qualitative behavior of solutions, we borrow an idea from physics. The energy of the harmonic oscillator system is defined to be

$$H(y, v) = \frac{v^2}{2} + y^2.$$

That is, if we think of this undamped harmonic oscillator system as a model for the motion of a unit mass attached to a spring with spring constant 2 and sliding on a frictionless table, then $H(y, v)$ is the energy when the mass is at position y with velocity v. The term $v^2/2$ is the kinetic energy, and the term y^2 is the potential energy of the system. This function is called the Hamiltonian or energy function for the system.

If $(y(t), v(t))$ is a solution, then the Hamiltonian or energy function is constant along this curve. (This is the mathematical version of the law of conservation of energy.) We can double-check this by computing

$$\frac{d}{dt} H(y(t), v(t)) = \frac{\partial H}{\partial y}(y(t), v(t))\frac{dy}{dt} + \frac{\partial H}{\partial v}(y(t), v(t))\frac{dv}{dt}$$

$$= 2(y(t))\frac{dy}{dt} + v(t)\frac{dv}{dt}$$

$$= 2y(t)v(t) + v(t)(-2y(t)) = 0.$$

Since $d(H(y(t), v(t))/dt = 0$ for all t, solutions remain on level curves of H. We can sketch the phase plane by sketching the level curves of H (see Figure 5.13).

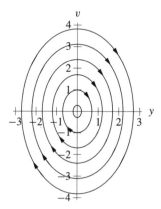

Figure 5.13
Level sets of

$$H(y, v) = \frac{v^2}{2} + y^2$$

and solutions of the system

$$\frac{dy}{dt} = v$$

$$\frac{dv}{dt} = -2y.$$

Energy and the Forced Harmonic Oscillator

Now we return to the undamped, forced harmonic oscillator equation

$$\frac{d^2y}{dt^2} + 2y = g(t)$$

or

$$\frac{dy}{dt} = v$$

$$\frac{dv}{dt} = -2y + g(t).$$

We again let $H(y, v) = v^2/2 + y^2$, which is the energy of the system. Because there is an external force, the energy is no longer constant. We obtain the change in energy along a solution $(y(t), v(t))$ by computing

$$\frac{d}{dt}(H(y(t), v(t))) = \frac{\partial H}{\partial y}(y(t), v(t))\frac{dy}{dt} + \frac{\partial H}{\partial v}(y(t), v(t))\frac{dv}{dt}$$

$$= 2y\frac{dy}{dt} + v\frac{dv}{dt}$$

$$= 2yv + v(-2y + g(t))$$

$$= v(t)g(t).$$

So the rate of change of the energy at time t depends on the velocity and on the forcing function.

If $v(t)$ and $g(t)$ have the same sign, then $d(H(y(t), v(t))/dt > 0$ and H is increasing. This makes sense because $v(t)$ and $g(t)$ having the same sign means the external force is pushing the mass in the direction it is going. The push increases the velocity of the mass and energy increases. If $g(t)$ and $v(t)$ have opposite signs, then the mass is being pushed against the direction of motion. The push decreases the velocity, the energy decreases, and $d(H(y(t), v(t))/dt < 0$.

Sinusoidal Forcing

Suppose we have an undamped, forced harmonic oscillator of the form

$$\frac{d^2y}{dt^2} + 2y = \epsilon \sin(\omega t),$$

which corresponds to the system

$$\frac{dy}{dt} = v$$

$$\frac{dv}{dt} = -2y + \epsilon \sin(\omega t).$$

So the forcing is a sine function with period $2\pi/\omega$. The parameter ϵ controls the strength or amplitude of the forcing. The forcing oscillates periodically, pushing and pulling the mass. We can think of this as a simple model of a building being shaken by and earthquake or a crystal glass being deformed by the sound of a note sung by an opera singer.

Suppose the external force is very small; that is, the parameter ϵ is very small. Because the forcing is very small, solutions of the forced oscillator system stay close to solutions of the unforced system, at least for a small amount of time. If $(y_1(t), v_1(t))$ is a solution of the forced system, then

$$v_1(t) \approx a \sin(\sqrt{2}\,t) + b \cos(\sqrt{2}\,t)$$

for some constants a and b for small t. But then

$$\frac{d}{dt}(H(y_1(t), v_1(t)) = v_1(t)\,\epsilon \sin(\omega t)$$

$$\approx (a \sin(\sqrt{2}\,t) + b \cos(\sqrt{2}\,t))\,\epsilon \sin(\omega t).$$

Both these terms oscillate between positive and negative values.

When the forcing is not resonant ($\omega \neq \pm\sqrt{2}$), the two terms in the rate of change of the energy oscillate with different frequencies. Sometimes they will be "in phase" and have the same sign. Later they will be "out of phase" and have opposite signs. The energy will increase in the first case and decrease in the second. So the energy of the system oscillates and the amplitude of solutions rises and falls. The external forcing alternately adds energy then takes it away. This explains the "beating modes" we observed in the solution in the previous section (see Figure 5.6 on page 426).

Resonance

In the case of resonant forcing, when $\omega = \pm\sqrt{2}$, the natural frequency and the forcing frequency are equal. The external forcing is always pushing or pulling in the direction in which the mass is moving and energy is only added to the system.

Resonance occurs for any sort of periodic forcing when the forcing frequency is equal to the natural frequency. In our simple model of a crystal glass and an opera singer, if the singer holds a note equal to the natural frequency of the oscillations of the glass, then we expect that the energy of these oscillations will continually build. Eventually the amplitude of the oscillations becomes so large that the glass breaks.

Energy, Forcing, and Damped Oscillators

We consider one more case, the forced, damped harmonic oscillator. Consider

$$\frac{d^2y}{dt^2} + \frac{dy}{dt} + 2y = g(t),$$

which corresponds to the system

$$\frac{dy}{dt} = v$$
$$\frac{dv}{dt} = -2y - v + g(t).$$

For the mass-spring model, this corresponds to taking $m = 1$, spring constant $k_s = q = 2$, and coefficient of damping $k_d = p = 1$. We can again compute that the change in the energy $H(y, v) = v^2/2 + y^2$ along a solution $(y(t), v(t))$ is

$$\frac{d}{dt} H(y(t), v(t)) = -(v(t))^2 + v(t)g(t).$$

Hence, damping removes energy from the system and forcing either adds or removes energy depending on whether the external force is pushing with or against the motion.

A useful observation is that, if $g(t)$ is bounded and the velocity $v(t)$ is increasing, then eventually v becomes so large that damping removes more energy than can be added by the external force. Hence a system with damping and bounded forcing will have bounded solutions. (However, these solutions could get quite large if the damping is small.)

The steady-state solution

Since the damping term implies that energy is continually being removed from the system, we can informally say that all the energy provided by the initial condition eventually dissipates. Hence for large t, the only energy available to the system is that provided by the external force. Every solution, no matter what the initial condition is, tends toward the same solution, which is called the **steady-state** solution. If the forcing term is periodic (for example, a sine or cosine), then the steady-state solution oscillates with the same frequency as the forcing function. (So the steady-state solution is not necessarily constant.) This is another way of saying that the natural response (the solution of the unforced or homogeneous equation) tends to zero, leaving the forced response (the particular solution of the nonhomogeneous equation).

Exercises for Section 5.3

1. Suppose an opera singer can break a glass by singing a particular note.

   (a) Will the singer have to sing a higher or a lower note to break the an identical glass that is half full of water?

   (b) Suppose both notes are within the singer's range. Will it be harder or easier to break the glass when it is half full of water?

2. The glass harmonica is a musical instrument invented by Ben Franklin. It consists of a collection of crystal glasses, one glass for each note. The musician rubs her slightly damp finger around the edge of the glass and the resulting vibrations of the glass make a very pure tone. The instrument is tuned by pouring water into the glasses. The harder the musician pushes against the glass, the louder the note (but the frequency stays the same). Explain how the glass harmonica works.

3. Suppose the suspension system of the average car can be fairly well modeled by a underdamped harmonic oscillator system with natural period of 2 seconds. How far apart should "speed bumps" be placed so that a car traveling at 10 miles per hour over several bumps will bounce more and more violently with each bump?

4. Consider a unit mass sliding on a frictionless table attached to a spring with spring constant $k_s = 16$. Suppose the mass is lightly tapped by a hammer every T seconds.

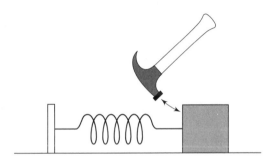

Mass tapped periodically with a hammer.

Suppose that the first tap occurs at time $t = 0$ and before that time the mass is at rest. Describe what happens to the energy of the system and the motion of the mass for the following choices of the tapping period T:

   (a) $T = 1$ (b) $T = 3/2$ (c) $T = 2$ (d) $T = 5/2$ (e) $T = 3$

5. Consider the forced harmonic oscillator

$$\frac{d^2 y}{dt^2} + qy = e^{-t},$$

which corresponds to the system

$$\frac{dy}{dt} = v$$

$$\frac{dv}{dt} = -qy + e^{-t}.$$

   (a) How does the energy $H(y, v) = (v^2 + qy^2)/2$ change along solutions of this system?

   (b) What is the long-term behavior of solutions of this system?

   (c) How does the long-term behavior of a solution of the forced system differ from the behavior of the solution of the unforced system with the same initial conditions?

6. Consider the forced damped harmonic oscillator

$$\frac{d^2y}{dt^2} + p\frac{dy}{dt} + qy = e^{-t},$$

which corresponds to the system

$$\frac{dy}{dt} = v$$

$$\frac{dv}{dt} = -qy - pv + e^{-t}.$$

(a) How does the energy $H(y, v) = (v^2 + qy^2)/2$ change along solutions of this system?

(b) What is the long-term behavior of solutions of this system?

(c) How does the long-term behavior of a solution of the forced system differ from the behavior of the solution of the unforced system with the same initial conditions?

7. Consider the forced harmonic oscillator equations

$$\frac{d^2y}{dt^2} + 4y = 0.2t$$

and

$$\frac{d^2y}{dt^2} + 4y = 0.2\cos(2t).$$

The energy for each of these systems is given by $H(y, v) = (1/2)v^2 + y^2$.

(a) Compute the rate of change of the energy along a solution $(y(t), v(t))$ for both equations.

(b) For which forced oscillator do you expect the energy to grow the fastest and why?

5.4 THE TACOMA NARROWS BRIDGE

On July 1, 1940, the $6 million Tacoma Narrows Bridge opened for traffic. On November 7, 1940, during a windstorm, the bridge broke apart and collapsed. During its short stand the bridge, a suspension bridge more than a mile long, was known as "Galloping Gertie" because the roadbed oscillated dramatically in the wind. The collapse of the bridge proved to be a scandal in more ways than one, including the fact that, because the insurance premiums had been embezzled, the bridge was uninsured.*

The roadbed of a suspension bridge hangs from vertical cables that are attached to cables strung between towers (see Figure 5.14 for a schematic picture). If we think of the vertical cables as long springs, then it is tempting to model the oscillations of the

---

*The story of the bridge and its collapse can be found in Martin Braun, *Differential Equations and Their Applications*, Springer-Verlag, 1993, p. 173, and Matthys Levy and Mario Salvadori, *Why Buildings Fall Down: How Structures Fail*, W. W. Norton Co., 1992, p. 109

Figure 5.14
Schematic of a suspension bridge

roadbed with a harmonic oscillator equation. We can think of the wind as somehow providing periodic forcing. It is very tempting to say, "Ah-ha; the collapse must be due to resonance."

It turns out that things are not quite so simple. We know that, to cause dramatic effects, the forcing frequency of a forced harmonic oscillator must be very close to its natural frequency. The wind seldom behaves in such a nice way for very long, and it would be very bad luck indeed if the oscillations caused by the wind happened to have a frequency almost exactly the same as the natural frequency of the bridge.

Recent research on the dynamics of suspension bridges (by two mathematicians, A. C. Lazer and P. J. McKenna[*]) indicates that the linear harmonic oscillator does not make an accurate model of the movement of a suspension bridge. The vertical cables do act like springs when they are stretched. That is, when the roadbed is below its rest position, the cables pull up. However, when the roadbed is significantly above its rest position, the cables are slack, so they do not push down. Hence the roadbed feels less force trying to pull it back into the rest position when it is pushed up than when it is pulled down (see Figure 5.15).

In this section we study a model for a system with these properties. The system of equations we study was developed by Lazer and McKenna from more complicated models of oscillations of suspension bridges. This system gives considerable insight into the possible behaviors of suspension bridges and even hints at how they can be made safer.

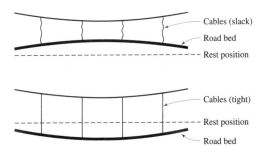

Figure 5.15
Close-up of the vertical cable when the roadbed is
above and below the rest position.

[*]See "Large-amplitude Periodic Oscillations in Suspension Bridges: Some New Connections with Nonlinear Analysis" by A. C. Lazer and P. J. McKenna, in SIAM Review, Vol. 32, No. 4, 1990, pp. 537-578.

Derivation of the Equations

The model we consider for the motion of the bridge uses only one variable to describe the position of the bridge. We assume that the bridge oscillates up and down as in Figure 5.15. We let $y(t)$ (measured in feet or meters) denote the vertical position of the center of the bridge, with $y = 0$ corresponding to the position where the cables are taut but not stretched. We let $y < 0$ correspond to the position in which the cables are stretched and $y > 0$ correspond to the position in which the cables are slack (see Figure 5.16). Of course, using one variable to study the motion of the bridge ignores many possible motions, and we comment on other models at the end of this section.

To develop a model for $y(t)$, we consider the forces that act on the center of the bridge. Gravity provides a constant force in the negative direction of y. We also assume that the cables provide a force that pulls the bridge up when $y < 0$ and which is proportional to y. On the other hand, when $y > 0$, the cable provides no force. When $y \neq 0$ there is also a restoring force that pulls y back toward $y = 0$ due to the stretching of roadbed. Finally, there will also be some damping, which is assumed to be proportional to dy/dt. We choose units so that the mass of the bridge is 1.

Based on these assumptions, the equation developed by Lazer and McKenna to model a suspension bridge on a calm day (no wind) is

$$\frac{d^2y}{dt^2} + \alpha\frac{dy}{dt} + \beta y + c(y) = -g.$$

The first term is the vertical acceleration. The second term, $\alpha(dy/dt)$, arises from the damping. Since suspension bridges are relatively flexible structures, we assume that α is small. The term βy accounts for the force provided by the material of the bridge pulling the bridge back toward $y = 0$. The function $c(y)$ accounts for the pull of the cable when $y < 0$ (and the lack thereof when $y \geq 0$), and therefore, it is given by

$$c(y) = \begin{cases} \gamma y & \text{if } y < 0; \\ 0 & \text{if } y \geq 0. \end{cases}$$

The constant g represents the force due to gravity.

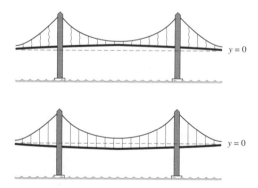

Figure 5.16
Positions of the bridge corresponding to $y > 0$ and $y < 0$.

We can convert this to a system in the usual way, obtaining

$$\frac{dy}{dt} = v$$

$$\frac{dv}{dt} = -\beta y - c(y) - \alpha v - g.$$

This is an autonomous system. Simplifying the right-hand side of this system by combining the $-\beta y$ term with the terms in $c(y)$, we obtain

$$\frac{dy}{dt} = v$$

$$\frac{dv}{dt} = -h(y) - \alpha v - g,$$

where $h(y)$ is the piecewise-defined function

$$h(y) = \begin{cases} ay & \text{if } y < 0; \\ by & \text{if } y \geq 0, \end{cases}$$

and $a = \beta + \gamma$ and $b = \beta$.

To study this example numerically, we choose particular values of the parameters. (These values are not motivated by any particular bridge.) Following Lazer and McKenna, we take $a = 17$, $b = 13$, $\alpha = 0.01$, and $g = 10$. So the system we study is

$$\frac{dy}{dt} = v$$

$$\frac{dv}{dt} = -h(y) - 0.01v - 10,$$

where

$$h(y) = \begin{cases} 17y & \text{if } y < 0; \\ 13y & \text{if } y \geq 0. \end{cases}$$

We can easily compute that this system has only one equilibrium point which is given by $(y, v) = (-10/17, 0)$. The y-coordinate of this equilibrium point is negative because gravity forces the bridge to sag a little, stretching the cables. The linearization around this point indicates that it is a spiral sink. This is what we expect because there is a small amount of damping present. The direction field and sketch of solution curves are shown in Figure 5.17 and 5.18. The solutions spiral toward the equilibrium point very slowly. The discontinuity of the direction field along the line $y = 0$ is difficult to see because the vector field is nearly vertical on both sides of the line.

The behavior of the solutions indicate that the bridge oscillates around the equilibrium position. The amplitude of the oscillation dies out slowly due to the small amount of damping. Because the forces controlling the motion change abruptly along $y = 0$, solutions also change direction when they cross from the left half-plane to the right half-plane.

The effect of wind

To add the effect of wind into the model, we add an extra term to the right-hand side of the equation. The effect of the wind is very difficult to quantify. Not only are there

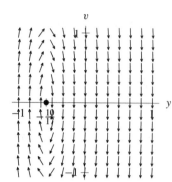

Figure 5.17
Direction field for the system

$$\frac{dy}{dt} = v$$

$$\frac{dv}{dt} = -h(y) - 0.01v - 10.$$

There is a spiral sink at $(y, v) = (-10/17, 0)$. In Figure 5.18, we see that solution curves for the system spiral slowly toward the equilibrium point.

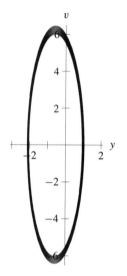

Figure 5.18
Solution curves spiral toward equilibrium very slowly.

gusts of more or less random duration and strength, but also the way in which the wind interacts with the bridge can be very complicated. Even if we assume that the wind has constant speed and direction, the effect on the bridge need not be constant. As air moves past the bridge, swirls or vortices (like those at the end of an oar in water) form above and below the roadbed. When they become large enough, these vortices "break off," causing the bridge to rebound. Hence even a constant wind can give a periodic push to the bridge.

Despite these complications, for simplicity, we assume that the wind provides a forcing term of the form $\lambda \sin(\mu t)$. Since it is unlikely that turbulent winds will give a forcing term with constant amplitude λ or constant frequency $\mu/(2\pi)$, we will look for behavior of solutions that persist for a range of λ and μ values.

The system with forcing is given by

$$\frac{dy}{dt} = v$$

$$\frac{dv}{dt} = -h(y) - 0.01v - 10 + \lambda \sin(\mu t).$$

This is a fairly simple model for the complicated behavior for a bridge moving in the wind, but we see below that even this simple model has solutions that behave in a surprising way.

Behavior of Solutions

We have seen that a linear system with damping and sinusoidal forcing will have one periodic solution to which every other orbit tends as time increases. In other words, no matter what the initial conditions, the long-term behavior of the system will be the same. The amplitude and frequency of this periodic solution are determined by the amplitude and frequency of the forcing term (see Section 5.2).

The behavior of the system

$$\frac{dy}{dt} = v$$

$$\frac{dv}{dt} = -h(y) - 0.01v - 10 + \lambda \sin(\mu t)$$

is quite different. Below we describe results of a numerical study of these solutions of this system carried out by Glover, Lazer, and McKenna.

If, for example, we choose $\mu = 4$ and λ very small ($\lambda < 0.05$), every solution tends toward a periodic solution with small magnitude near $y = -10/17$ (see Figure 5.19). For this periodic solution, $y(t)$ is negative for all t. Since this solution never crosses $y = 0$, it behaves just like the solution of the forced linear system

$$\frac{dy}{dt} = v$$

$$\frac{dv}{dt} = -17y - 0.01v - 10 + \lambda \sin(4t).$$

In terms of the behavior of the bridge, this means that, in light winds, we expect the bridge to oscillate with small amplitude. Gravity keeps the bridge sagging downward and the cables are always stretched somewhat (see Figure 5.20). In this range, modeling the cables as linear springs is reasonable.

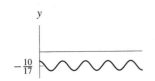

Figure 5.19
Solution of the system with small forcing.

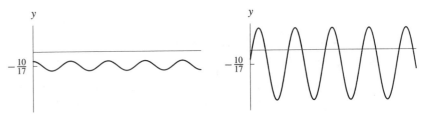

Figure 5.20
Schematic of the bridge oscillating in light winds.

As λ increases, a new phenomenon is observed. Initial conditions near $(y, v) = (-10/17, 0)$ still yield solutions that oscillate with small amplitude (see Figure 5.21). However, if $y(0) = -10/17$ but $v(0)$ is large, solutions can behave differently. There is another periodic solution that oscillates around $y = -10/17$, but with (relatively) large amplitude (see Figure 5.22).

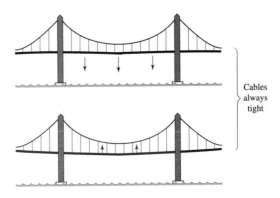

Figure 5.21
Solution of the forced system with larger forcing than in Figure 5.19 and initial conditions near the equilibrium.

Figure 5.22
Solution of the forced system with the same large forcing as in Figure 5.21, but with initial conditions further from the equilibrium.

This has dramatic implications for the behavior of the bridge. If the initial displacement is small, then we expect to see small oscillations as before. However, if a gust of wind gives the bridge a kick large enough to cause it to rise above $y = 0$, then the cables will go slack and the linear model will no longer be accurate (see Figure 5.23). In this situation the bridge can start oscillating with much larger amplitude and these oscillations *do not die out*. So, in a moderate wind ($\lambda > 0.06$), a single strong gust could suddenly cause the bridge to begin oscillating with much larger amplitude, perhaps with devastating consequences.

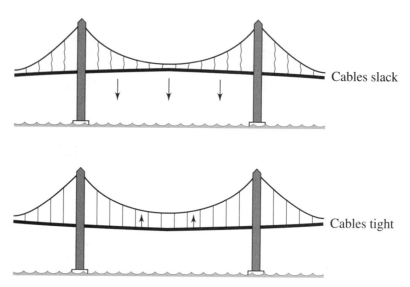

Figure 5.23
Schematic of large amplitude oscillations of the bridge.

Varying the parameters

As mentioned above, because the effects of the wind are not particularly regular, we should investigate the behavior of solutions as λ and μ are varied. It turns out that the large amplitude periodic solution persists for a fairly large range of λ and μ. This means that even in winds with uneven velocity and direction, the sudden jump in behavior to a persistent oscillation with large amplitude is possible.

Does This Explain the Tacoma Narrows Bridge Disaster?

As with any simple model of a complicated system, a note of caution is in order. To construct this model, we have made a number of simplifying assumptions. These include, but are not limited to, assuming that the bridge oscillates in one piece. The bridge can oscillate in two or more sections (see Figure 5.24). To include this in our model, we would have to include a new independent variable for the position along the bridge. The resulting model is a partial differential equation.

Another factor we have ignored is the fact that the roadbed of the bridge has width as well as length. The final collapse of the Tacoma Narrow Bridge was preceded by

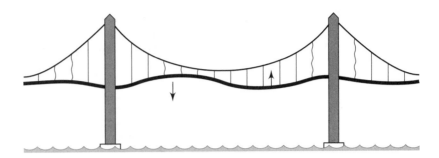

Figure 5.24
More complicated forms of oscillation of a suspension bridge.

violent twisting motions of the roadbed alternately stretching and loosening the cables on either side of the road (see the cover of this book). Analysis of a model including the width gives considerable insight into the final moments before the bridge's collapse (see Lazer and McKenna).

This being said, the simple model discussed above still helps a great deal in understanding the behavior of the bridge. If this simple system can feature the surprising appearance of large amplitude periodic solutions, then it is not at all unreasonable to expect that more complicated and more accurate models will also exhibit this behavior. So this model does what it is supposed to do; it tells us what to look for when studying the behavior of a flexible suspension bridge.

Exercises for Section 5.4

For Exercises 1–3, recall that our simple model of a suspension bridge is

$$\frac{d^2y}{dt^2} + \alpha\frac{dy}{dt} + \beta y + c(y) = -g,$$

where α is the coefficient of damping, β is a parameter corresponding to the stiffness of the roadbed, the function $c(y)$ accounts for the pull of the cables, and g is the gravitational constant. For each of the following modifications of bridge design,

(a) discuss which parameters are changed; and

(b) discuss how you expect a change in the parameter values to effect the solutions. (For example, does the modification make the system look more or less like a linear system?)

1. The "stiffness" of the roadbed is increased, for example, by reinforcing the concrete or adding extra material that makes it harder for the roadbed to bend.

2. The coefficient of damping is increased.

3. The strength of the cables is increased.

4. The figure below is a schematic for an alternate suspension bridge design called the Lazer-McKenna light flexible long span suspension bridge. Why does this design avoid the problems of the standard suspension bridge design? Discuss this in a paragraph and give model equations similar to those in the text for this design.

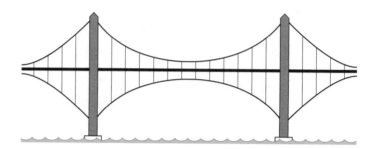

A schematic of the Lazer-McKenna light flexible long span suspension bridge.

In Exercises 5–8, we consider another application of the ideas in this section. Lazer and McKenna observe that the equations they study may also be used to model the up-and-down motion of an object floating in water, which can rise completely out of the water. This has serious implications for the behavior of a ship in heavy seas.

Suppose we have a cube made of a light substance floating in water. Gravity always pulls the cube downward. The cube floats at an equilibrium level where the mass of the water displaced equals the mass of the cube. If the cube is higher or lower than the equilibrium level, then there is a restoring force proportional to the size of the displacement. We assume that the bottom and top of the cube stay parallel to the surface of the water at all times and that the system has a small amount of damping.

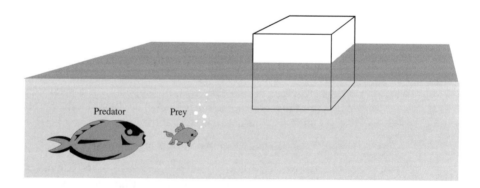

Cube floating in water.

5. Write a differential equation model for the up-and-down motion of the cube, assuming that it always stays in contact with the water and is never completely submerged.

6. Write a differential equation for the up-and-down motion of the cube, assuming that it always stays in contact with the water, but that it can be completely submerged.

7. Write a differential equation model for the up-and-down motion of the cube, assuming that it will never be completely submerged but can rise completely out of the water by some distance.

8. **(a)** Adjust each of the models in Exercises 5–7 to include the effect of waves on the motion of the cube (assuming the top and bottom remain parallel to the average water level).

   (b) Discuss the implications of the behavior of solutions of this system considered in the text for the motion of the cube.

5.5 PERIODIC FORCING OF NONLINEAR SYSTEMS AND CHAOS

In Section 5.2 we studied the forced harmonic oscillator with periodic (sinusoidal) forcing. We found that the periodic forcing affects the amplitude of the oscillating solutions and could even cause the system to "blow up" in the case of resonant forcing. However, just as in the case of the unforced harmonic oscillator, we saw that there was not much difference in the qualitative behavior of the solutions of the periodically forced oscillator when the initial conditions were changed.

In Section 5.4 we studied a particular nonlinear system with external periodic forcing and saw that it had some surprising behavior. In this section we continue the study of periodically forced nonlinear systems. These systems are examples of systems that have *chaotic* behavior. In this context, the word chaotic means that solutions exhibit infinitely many different qualitative behaviors. Moreover, these different solutions are packed closely together, so that any change in the initial conditions has a radical effect on the long-term behavior of the solution.

Our goal in this section is different from our goals in the previous sections. It is impossible to hope for analytic solutions of a chaotic system since there are infinitely many qualitatively different types of solutions. Also a complete qualitative analysis of solutions is impossible for the same reason. Instead we emphasize the startling behavior of these systems so that we can recognize "chaos" when we see it in other systems. We note that these systems are the subject of active research in mathematics. If our presentation seems incomplete, it is because the complete story has yet to be discovered.

A Periodically Forced Duffing's Equation

We begin with an example of a nonlinear system that we "completely understand," that is, we can draw the phase plane and describe the behavior of every solution. We then add a periodic forcing term to this system and develop techniques for gaining insight for the resulting nonautonomous system.

The nonlinear system we consider is a Duffing's equation of the form

$$\frac{dy}{dt} = v$$

$$\frac{dv}{dt} = y - y^3.$$

This is a Hamiltonian system with energy $H(y, v) = v^2/2 - y^2/2 + y^4/4$. Hence we can draw the phase plane for this system by drawing the level curves of H (see Figure 5.25). The system has two center equilibrium points at $(\pm 1, 0)$ and a saddle equilibrium point at $(0, 0)$. The solutions that tend to the saddle as $t \to -\infty$ also tend to it as $t \to +\infty$. All other solutions are periodic, either looping around one of the centers or looping around all three equilibrium points.

We modify this system by adding a sinusoidal forcing term to obtain the system

$$\frac{dy}{dt} = v$$

$$\frac{dv}{dt} = y - y^3 + \epsilon \sin t.$$

Note that the amplitude of the forcing is given by ϵ and the period is 2π. (See the exercises for the interpretation of this system as a second-order equation.)

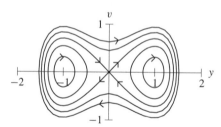

Figure 5.25
Phase plane for the system

$$\frac{dy}{dt} = v$$

$$\frac{dv}{dt} = y - y^3.$$

The Return Map

As we pointed out in Section 5.1, the usual tools for dealing with systems — the vector field, direction field, and phase plane — are not relevant to nonautonomous systems because the vector field changes with time. We really need a three-dimensional picture with y-, v-, and t-axes. We would then think of solutions as moving along curves in the three dimensional yvt-space. An attempt to render such a drawing is given in Figure 5.26. This figure contains only three solutions and yet it is fairly difficult to visualize. Also, to understand the long-term behavior of solutions, we would have to extend the t-axis quite far. We need to find a better way to depict solutions.

To represent the solutions of this system, we use an idea that is due to the person who first saw and understood the ramifications of chaotic systems, the mathematician Henri Poincaré. The goal is to replace the three-dimensional picture in Figure 5.26 by a two-dimensional picture. Fix an initial point $y(0) = y_0$, $v(0) = v_0$. In Figure 5.27 this point is represented by a point in the plane $t = 0$. If we draw the plane $t = T$ for some later time T in the yvt-space, we can ask where the solution that starts at (y_0, v_0)

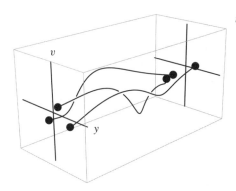

Figure 5.26
Three solutions of the system

$$\frac{dy}{dt} = v$$

$$\frac{dv}{dt} = y - y^3 + \epsilon \sin t$$

in the yvt-space.

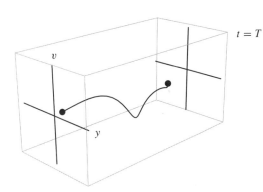

Figure 5.27
Initial point on the plane $t = 0$ and the intersection of the solution on the plane $t = T$.

at time $t = 0$ meets this plane. By following the solution curve until time $t = T$, we can determine where this meeting point is (see Figure 5.27).

Now we take advantage of the fact that the forcing term is periodic. Take $T = 2\pi$. The solution starting at $t = 0$ intersects the plane $t = 2\pi$ at some point $(y_1, v_1) = (y(2\pi), v(2\pi))$. One way to follow this solution further is to extend the t-axis. However, because the forcing is periodic, the differential equations are periodic with period 2π. Hence the curve that starts at (y_1, v_1) when $t = 2\pi$ is a translation by 2π time units of the curve that starts at (y_1, v_1) on the plane $t = 0$. So we can follow the solution for another 2π time units without adding to the t-axis. After 2π more time units, the solution hits the plane $t = 2\pi$ again, this time at a point (y_2, v_2), which is equal to $(y(4\pi), v(4\pi))$. We can repeat the process by moving the point (y_2, v_2) to the plane $t = 0$, and then following the solution for another 2π time units (See Figure 5.28).

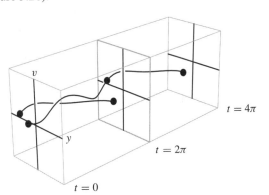

Figure 5.28
The phase space from $0 \leq t \leq 4\pi$. Note that the section from 2π to 4π is just a translation of the section from 0 to 2π.

This picture is still very complicated when t gets large since there are many strands of the solution running from $t = 0$ to $t = 2\pi$. However, the behavior of the solution can be recovered from a simpler picture. Suppose we turn this picture so that the t-axis goes directly into the paper. Then we see only the points where the solution pierces the plane $t = 0$, that is, the points $(y_0, v_0) = (y(0), v(0))$, $(y_1, v_1) = (y(2\pi), v(2\pi))$, $(y_2, v_2) = (y(4\pi), v(4\pi))$, ... This sequence of points is enough to give a good idea of the long-term behavior of the solution.

We thus replace the picture of the solution curves in the three-dimensional phase space with a function that takes a given initial value (y_0, v_0) in the plane $t = 0$ to its "first return" point (y_1, v_1), also in the plane $t = 0$. This function is called the **return map** or **Poincaré return map**; we think of points in the plane $t = 0$ as being taken or "mapped" to the next point of intersection with this plane by the return map. By repeatedly applying this function, we see the points where the solution pierces the plane in succession.

Return map for the unforced system

To illustrate this procedure, we first consider a simple case. Let $\epsilon = 0$, so that we have the unforced system

$$\frac{dy}{dt} = v$$

$$\frac{dv}{dt} = y - y^3.$$

This is autonomous system, but we use the ideas above to give a picture of the return map in this special case. If we start with an initial condition near the equilibrium point at $(1, 0)$, then the solution in yvt-space spirals around this equilibrium point (see Figure 5.29). The return map for this solution picks out points on this solution curve at 2π time intervals and thus yields a sequence of points on a loop around the $(1, 0)$. If we take a different initial condition farther from the $(1, 0)$, the solution spirals at a different rate around the $(1, 0)$, and the resulting return map yields a sequence of points that progress more slowly around the $(1, 0)$. The equilibrium points of the system give solutions which always return to the same point on the plane. Such points are called **fixed points** for the Poincaré return map. Finally, the solutions that tend from the saddle equilibrium point back to itself give a sequence of points that progress from the origin and back to the origin via the Poincaré return map (see Figure 5.30).

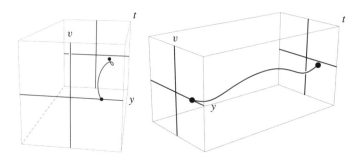

Figure 5.29
In the yvt-space, two views of a solution with initial point near $(1, 0)$ for the unforced system.

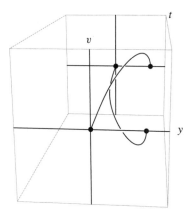

Figure 5.30
The yvt-space picture of a solution that tends from the saddle at $(0, 0)$ as $t \to -\infty$ back to $(0, 0)$ as $t \to +\infty$ and the corresponding Poincaré return map.

In Figure 5.31, we depict the results of applying the return map to several different initial conditions in the unforced system. In this case, the Poincaré return map picture looks very much like the phase plane. Initial conditions near the $(\pm 1, 0)$ give solutions that stay on closed loops around $(\pm 1, 0)$, so in the Poincaré map picture these solutions give sequences of points on loops around the $(\pm 1, 0)$. The distribution of points around these loops is different at different radii because the rate at which the solutions spiral around the origin decreases as the radius increases. The orbits connecting the saddle fixed point to itself give sequences running from the origin back to the origin.

Just as with the phase space, it is very instructive to watch these pictures as they are computed. The order in which the sequence of points appears for the return map gives much more information than the static picture. However, we can still use the completed Poincaré map picture to draw conclusions about solutions.

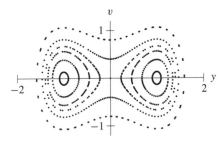

Figure 5.31
The Poincaré return map for many different solutions of the system

$$\frac{dy}{dt} = v$$
$$\frac{dv}{dt} = y - y^3.$$

Return Map for the Forced Nonlinear System

We now return to the forced nonlinear system

$$\frac{dy}{dt} = v$$
$$\frac{dv}{dt} = y - y^3 + \epsilon \sin t.$$

To compute solutions numerically we need to choose a value of ϵ. For the sake of illustration we take $\epsilon = 0.06$; other values of ϵ are considered in the exercises. Our goal is to use the Poincaré return map pictures to interpret the behavior of solutions of this system.

We fix an initial condition near the center $(1, 0)$, compute the solution in the yvt-space and then the Poincaré return map picture. The resulting image is given in Figure 5.32. The $y(t)$-graph of this solution is given in Figure 5.33. We can use this picture to predict the behavior of the solution: At least for times that are a multiples of 2π, the position of the solution stays relatively near $(1, 0)$.

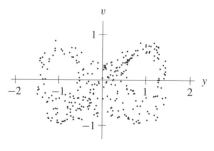

Figure 5.32
Poincaré return map for a solution with initial point near $(1, 0)$ for the forced Duffing system with $\epsilon = 0.06$.

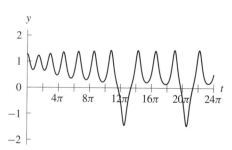

Figure 5.33
Graph of the solutions in Figure 5.32 above on the ty-plane.

The solutions that start near the saddle at the origin are much more interesting. The Poincaré return map for such a solution is given in Figure 5.34. Instead of a simple curve connecting the origin to itself all on one side of the phase plane, we see a cloud of points on both sides of the origin.

Figure 5.34 allows us to predict that a solution that starts near the origin behaves in a wild way. The y-coordinate takes on both positive and negative values. Moreover, because the points appear to follow no particular pattern, we can predict that the $y(t)$-graph of this solution oscillates without any particular pattern. This indeed is what we observe in Figure 5.35.

We can get some idea of why this solution behaves the way it does by looking again at the phase plane of the unforced system (see Figure 5.36). A solution that starts near

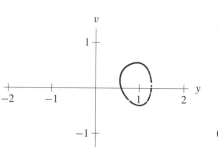

Figure 5.34
Poincaré return map for a solution with initial point near the origin for the forced nonlinear system with $\epsilon = 0.06$.

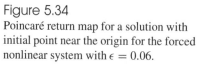

Figure 5.35
Graph of the solution in Figure 5.34 in the ty-plane.

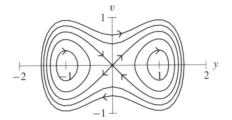

Figure 5.36

Phase plane for the unforced system

$$\frac{dy}{dt} = v$$

$$\frac{dv}{dt} = y - y^3.$$

the origin, say just to the right of $(0, 0)$ near the separatrix leaving the origin, moves away from the origin for a while. Then it loops back and returns close to $(0, 0)$. The forcing term is very small, so it has little effect on the solution when it is far from $(0, 0)$, since the vector field has relatively large magnitude in this region. So we expect that the solution will return to this vicinity near the incoming separatrix. When the solution is close to the origin, the vector field of the unforced system is small, so the forcing becomes more important. When the solution is close to $(0, 0)$, if the forcing term is pushing the solution up (that is, $0.06 \sin t > 0$), then the solution moves toward the region $v > 0$ and hence follows the outgoing separatrix back into the right half-plane. If, on the other hand, the forcing term is negative when the solution is close to the origin, then it is possible that there is enough "push" to move the solution below the incoming separatrix. In this situation the solution proceeds into the left half-plane. Once in the left half-plane, the solution makes a loop around the left equilibrium point.

So, with each loop, the solution of the forced system must "make a choice" when it returns to a neighborhood of the origin. The solution can go into either the right or left half-plane. Which direction the solution goes depends on the position of the solution relative to the origin and the sign of the forcing term when it arrives near the origin. So the choice depends on timing.

As a consequence we cannot predict what happens when the initial condition is changed slightly. The solution moves slowly when it is near $(0, 0)$, so a slight change of initial condition that pushes the solution closer to the origin makes a significant difference in the amount of time the solution spends near $(0, 0)$. This in turn affects the time at which the solution returns to a neighborhood of the origin, and hence it can affect which side of the yv-plane it next enters. Hence, a slight difference in the initial condition can make a radical difference in the long-term behavior of the solution. This is demonstrated in Figure 5.37 where the $y(t)$-graphs of two solutions with initial conditions that are very close together are shown. The solutions stay close for a while, but at some later time they are far enough apart so that they make different decisions

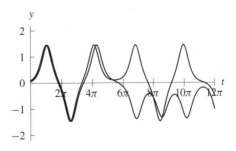

Figure 5.37

The $y(t)$-graphs of two solutions whose initial points are very close together.

about which way to turn when they are near the origin. After this time the solutions are radically different.

Reality check

This sort of behavior is rather unnerving. We know that the system is "deterministic." The behavior of solutions is completely determined by the right-hand side of the system of differential equations. However, when we look at solutions, they behave in a way that does not seem to have any particular pattern. We have even anthropomorphized the solutions, saying things like "the solution decides which way to go." Solutions don't think. They don't have to think because their behavior is determined by the right-hand side of the differential equation. What is going on is that a slight change in the initial conditions can cause a huge change in the long-term behavior of the system.

While this sort of behavior seems unusual compared to the differential equations we have studied in previous sections, it is not at all unusual in nature. Physical systems like the flow of water in a turbulent stream, the weather patterns on the earth, and even the flipping of a coin, all behave this way. These systems are deterministic in that they follow strict laws of physics. This does not mean that they are predictable. A small change in initial conditions can make a radical difference in their behavior.

It is even dangerous to trust numerical simulations of these systems. We know that every numerical method gives only approximations of solutions. There are small errors in each step of the simulation. For a system like the one above, a small error in the numerical approximation gives us an approximate solution that is slightly different from the intended solution. But nearby solutions can have radically different long-term behavior. So a numerical simulation can give results that are very far from the desired solution.

This is one reason that long-range (beyond five days) weather prediction is usually not very accurate. Incomplete knowledge of weather systems and errors in numerical simulations yield predictions that can be far from correct.

The Periodically Forced Pendulum

As a second example of a periodically forced nonlinear system we return to the system modeling the motion of a pendulum. We can think of a pendulum sitting on a table which is being shaken periodically. It turns out that much of the behavior observed above also occurs for this system.

The equations

The equations of the periodically forced pendulum with mass 1 and arm length 1 are

$$\frac{d\theta}{dt} = v$$

$$\frac{dv}{dt} = -g \sin \theta + \epsilon \sin t,$$

where g is the gravitational constant. The forcing term $\epsilon \sin t$ models an external force that periodically pushes the pendulum clockwise and counterclockwise with amplitude ϵ and period 2π. For convenience we assume that units of time and distance have been chosen so that $g = 1$ and our system is

$$\frac{d\theta}{dt} = v$$

$$\frac{dv}{dt} = -\sin\theta + \epsilon \sin t.$$

The return map

We construct the return map for the forced pendulum system exactly as above. The period of the forcing term is again 2π, so we follow solutions in θvt-space starting on the plane $t = 0$ and marking where they cross the plane $t = 2\pi$.

For the examples below, we fix $\epsilon = 0.01$; other values of ϵ are considered in the exercises. In Figure 5.38 we show the Poincaré return map for a solution with initial condition near $(0, 0)$. The resulting thick loop corresponds to a solution that oscillates with varying amplitude. The $y(t)$-graph of the same solution in Figure 5.39 shows this oscillation.

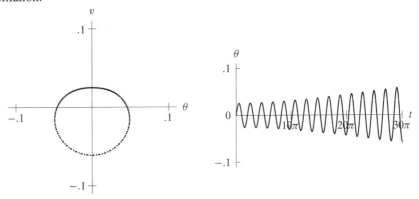

Figure 5.38
Poincaré return map for the periodically forced pendulum system with $\epsilon = 0.01$ for a solution with initial condition near $(0, 0)$.

Figure 5.39
Graph in the $t\theta$-plane of the solution in Figure 5.38. (The amplitude of the oscillation remains bounded for all t.)

Initial conditions near $(0, 0)$ for the forced pendulum system correspond to starting the pendulum at a small angle and with small velocity. In this situation the unforced (undamped) pendulum oscillates forever with small constant amplitude. The addition of the forcing term means that, just as for the forced harmonic oscillator, the forcing sometimes pushes with the direction of motion, making the pendulum swing higher, and sometimes pushes against the direction of motion, making the pendulum swing less. Unlike the harmonic oscillator, the period of the pendulum swing depends on the amplitude. Hence forcing adds and subtracts energy from the system in a way that, over the long term, is quite complicated. (This is not so evident from the picture because the forcing is small.)

Solutions near the saddle equilibrium points

Figure 5.40 shows the Poincaré return map for a solution of the periodically forced pendulum equation ($\epsilon = 0.01$) with initial condition near $(-\pi, 0)$. These points form a

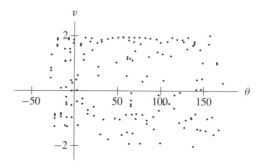

Figure 5.40
Poincaré map for the periodically forced pendulum
system with $\epsilon = 0.01$ for a solution with initial
point near $(-\pi, 0)$.

"cloud" with no particular structure. Also, the θ coordinate becomes quite large. This means that the pendulum arm has completely rotated many times in one direction.

Just as in the forced Duffing system, whenever the solution approaches a saddle equilibrium point it must "decide" which way to turn. If it stays in the upper half-plane then the θ coordinate of the solution increases by a multiple of 2π before returning to the vicinity of another saddle. If it chooses to go into the lower half-plane, then the θ coordinate decreases by 2π before another choice is made. If we graph the θ coordinate of the solution above on the $t\theta$-plane we see that it moves in a very wild way. In particular, it is possible that the θ component of the solution can become very large positive (or negative) if the pendulum repeatedly "decides" to rotate the same direction (see Exercises 5–8).

The decision of which way the solution turns when it is close to a saddle depends on the sign of the forcing term. At that time, the behavior of the solution depends very delicately on the initial condition. If we start two solutions near $(-\pi, 0)$ with almost the same initial condition, then eventually they split apart and become radically different (see Figure 5.41).

From the pictures above, we can deduce the existence of some interesting behavior in the forced pendulum system. A solution of the forced pendulum system with

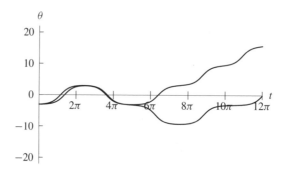

Figure 5.41
The $\theta(t)$-graphs for two solutions of the periodically
forced pendulum equation with almost equal initial
conditions.

initial condition near $(-\pi, 0)$ corresponds to an initial placement of the pendulum that is almost vertical but with very little velocity. Of course the pendulum swings down. During the swing the forcing term has very little effect on the motion of the pendulum.

When the pendulum swings up to the almost vertical position, it slows down again and the effect of the forcing term is more pronounced. When the pendulum is near the top of the swing, the forcing term either "pushes it over the top" so that it makes another turn in the same direction, or "pulls it back" so that it swings back the way it came. Which way the pendulum goes depends on the sign of the forcing term, which in turn depends on the time the pendulum arrives at the top of its swing. Since the pendulum moves very slowly near the top of the swing, a small change in initial conditions can make a big change in the timing and hence cause a change in direction of the pendulum.

We emphasize that the sort of physical argument given above is not meant to replace the Poincaré return map pictures and the analysis of the solutions. Although the physical argument makes sense, it does not say whether a forcing term with $\epsilon = 0.01$ is large enough to cause this sort of behavior.

Moral

The moral of this section is that even systems like the pendulum that we "understand" can become very complicated when additional terms like periodic forcing are added. The periodically forced pendulum system does not appear at first glance to be all that complicated, but we see from the Poincaré return map that its solutions behave quite unpredictably. A small change in the initial position frequently has a radical effect on the behavior of the solution.

If this sort of behavior, which we now call **chaos**, can be observed in a system as simple as the periodically forced pendulum, it should not be surprising that it can also be found in nature. This is not a new discovery. Henri Poincaré first considered the possibility of the existence of "chaos" in nature in the 1880s. He was studying the motion of a small planet (an asteroid) under the gravitational influence of a star (the sun) and a large planet (Jupiter). Poincaré developed the return map to investigate the behavior of this system. What is remarkable is that Poincaré did not have the advantage of looking at numerical simulations of solutions as we do today. Nevertheless, he could see that the return map would behave in a very complicated fashion and he wisely said he would not attempt to draw it.

Exercises for Section 5.5

In Exercises 1–4, Poincaré return map pictures are given for four different orbits with four different values of ϵ and initial conditions for the periodically forced Duffing equation

$$\frac{dy}{dt} = v$$

$$\frac{dv}{dt} = y - y^3 + \epsilon \sin t.$$

described in the text. Also, four $y(t)$-graphs for solutions of this system are given.

(a) Match the Poincaré return map picture with the $y(t)$-graph.

(b) Describe in a brief essay how you made the match and describe the qualitative behavior of the solution.

i)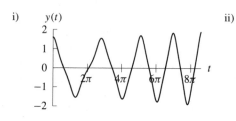

ii)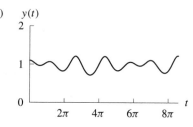

iii)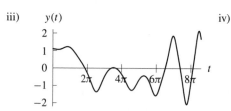

iv)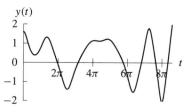

The $y(t)$-graphs for Exercises 1–4.

1. For $\epsilon = 0.1$, $y(0) = 1.1$, $v(0) = 0$.

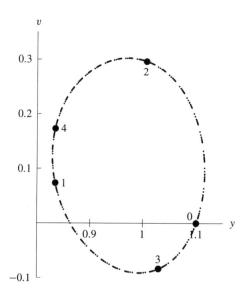

Poincaré return map with 500 iterates. The first
four returns are indicated.

2. For $\epsilon = 0.4$, $y(0) = 1.1$, $v(0) = 0$.

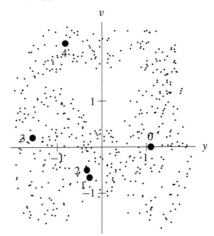

Poincaré return map with 500 iterates. The first four returns are indicated.

3. For $\epsilon = 0.1$, $y(0) = 1.6$, $v(0) = 0$.

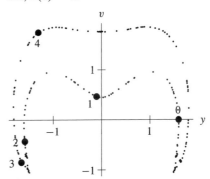

Poincaré return map with 200 iterates. The first four returns are indicated.

4. For $\epsilon = 0.5$, $y(0) = 1.6$, $v(0) = 0$.

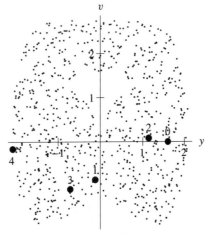

Poincaré return map with 800 iterates. The first four returns are indicated.

In Exercises 5–8, the Poincaré return map pictures are given for four different orbits with four different values of ϵ and initial conditions for the periodically forced pendulum system

$$\frac{d\theta}{dt} = v$$

$$\frac{dv}{dt} = -\sin\theta + \epsilon\sin t$$

described in the text. Also, four $\theta(t)$-graphs for solutions of this system are given.

(a) Match the Poincaré return map picture with the $\theta(t)$-graph.

(b) Describe in a brief essay how you made the match and describe the qualitative behavior of the solution.

(c) Describe the behavior of the pendulum arm when it follows the indicated solution.

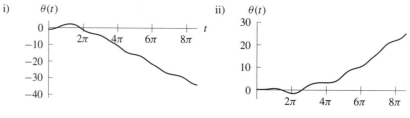
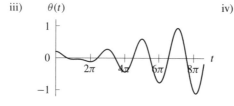
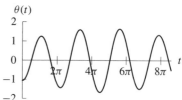

The $\theta(t)$-graphs for Exercises 5–8.

5. For $\epsilon = 0.1$, $\theta(0) = .2$, $v(0) = 0$.

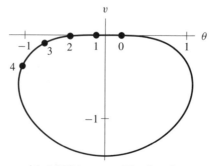

Poincaré return map with 1,000 iterates. The first four returns are indicated.

6. For $\epsilon = 0.5$, $\theta(0) = .2$, $v(0) = 0$.

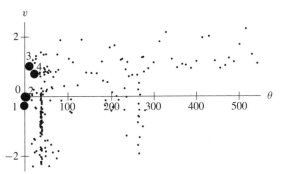

Poincaré return map with 250 iterates. The first four returns are indicated.

7. For $\epsilon = 0.1$, $\theta(0) = -1.06$, $v(0) = 0$.

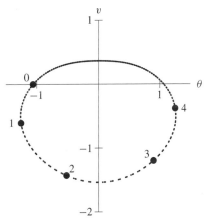

Poincaré return map with 400 iterates. The first four returns are indicated.

8. For $\epsilon = 0.5$, $\theta(0) = -1.06$, $v(0) = 0$.

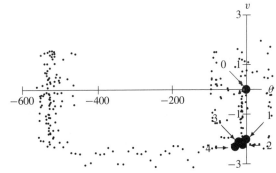

Poincaré return map with 250 iterates. The first four returns are indicated.

9. The Poincaré return map for a solution of the periodically forced harmonic oscillator

$$\frac{dy}{dt} = v$$

$$\frac{dv}{dt} = -3y + 0.2 \sin t$$

is given below.

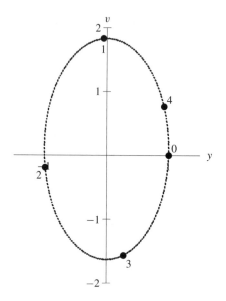

Poincaré return map for the forced harmonic oscillator system with initial conditions $y(0) = 1$, $v(0) = 0$.

(a) Find the solution with the initial condition $y(0) = 1$, $v(0) = 0$.

(b) Discuss why the Poincaré return map for this solution looks the way it does.

(c) What would the Poincaré return map for the solution with $y(0) = 4$, $v(0) = 0$ look like?

(d) Discuss why the Poincaré return map for the forced harmonic oscillator system is different from the Poincaré map of the forced pendulum.

10. Describe and sketch the Poincaré return map for the solution of the forced harmonic oscillator system

$$\frac{dy}{dt} = v$$

$$\frac{dv}{dt} = -4y + 0.1 \sin(2t).$$

[*Hint:* You should be able to do this problem qualitatively, with only a small amount of computation. What is the natural frequency of the system?]

Lab 5.1   A Periodically Forced RLC Circuit

In this lab we continue the study of simple RLC circuits that we began in Lab 3.2. The circuit is shown in Figure 5.42. The parameters are the resistance R, the capacitance C, and the inductance L. The dependent variables we use are v_C, the voltage across the capacitor, and the current i. In this lab we consider the case where the voltage source $v_T = v_T(t)$ is a time-dependent forcing term.

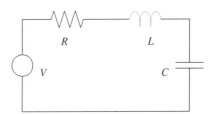

Figure 5.42
An RLC circuit.

From Lab 3.2, we know that the voltage v_C and current i satisfy the system of differential equations

$$\frac{dv_C}{dt} = \frac{i}{C}$$

$$\frac{di}{dt} = -\frac{v_C}{L} - \frac{R}{L}i - \frac{v_T(t)}{L}.$$

This is more commonly written in the form of a second order equation for v_C, that is, in the form

$$LC\frac{d^2v_C}{dt^2} + RC\frac{dv_C}{dt} + v_C = v_T(t).$$

In this lab we consider the possible behavior of solutions to this equation (or the corresponding system) when $v_T(t) = a\sin(\omega t)$.

In your report, consider the following questions:
1. Assuming R, C and L are all nonnegative, what types of long-term behavior are possible for solutions of the equation

$$LC\frac{d^2v_C}{dt^2} + RC\frac{dv_C}{dt} + v_C = a\sin(\omega t)?$$

Describe how the behavior of solutions depends on the parameters a and ω.

2. In a typical circuit, R will be on the order of 1000, C will be on the order of 10^{-6}, and L will be on the order of 1. Does this information help in limiting the possible behaviors of solutions?

3. Describe the solutions for various values of a and ω when $R = 2000$, $C = 2 \times 10^{-7}$ and $L = 1.5$.

Your report: In your report, address each of the items above, justifying all statements and showing all details. Give graphs of solutions as appropriate.

Lab 5.2

A Predator-Prey System with Periodic Immigration

In Chapter 2, we considered the autonomous (unforced) predator-prey system

$$\frac{dR}{dt} = 2R - 1.2RF$$

$$\frac{dF}{dt} = -F + 0.9RF,$$

where R is the number of prey and F is the number of predators, measured in hundreds or thousands. In this lab we consider the same predator-prey system with the additional assumption of periodic emigration and immigation of prey. That is, we add an additional term to the dR/dt equation that is periodic in time. As a first model we use a sine function as the forcing term. Our new model is

$$\frac{dR}{dt} = 2R - 1.2RF + \alpha \sin(\omega t)$$

$$\frac{dF}{dt} = -F + 0.9RF.$$

The amplitude of the emigration and immigration is given by the parameter α; the period is $2\pi/\omega$.

In your report, address the following items:

1. Recall that there is an equilibrium point for the unforced system at $R \approx 1.11$, $F \approx 1.67$ (see Section 2.1). Linearize the system near this point. Using the linearization, can you predict the values of ω that will cause the most dramatic effect on solutions of the forced system

$$\frac{dR}{dt} = 2R - 1.2RF + \alpha \sin(\omega t)$$

$$\frac{dF}{dt} = -F + 0.9RF$$

   with initial point near $(1.11, 1.67)$?

2. Setting $\alpha = 0.2$ and $\omega = 5/(2\pi)$, study the solutions of the forced system

$$\frac{dR}{dt} = 2R - 1.2RF + 0.2 \sin(5t/(2\pi))$$

$$\frac{dF}{dt} = -F + 0.9RF$$

   numerically. What can you say about the long-term behavior of solutions? Interpret your observations in terms of the behavior of the populations. [Hint: Try initial conditions at varying distances from the point $(1.11, 1.67)$. Make sure you follow solutions long enough to be confident you are seeing the "long-term" behavior.]

3. Repeat part 2 with $\alpha = 0.2$ and $\omega = 1$.

Your report: Address each of the items above in your report. In part 1, show all steps and justify all assertions. In parts 2 and 3, you may include phase plane and $R(t)$- and $F(t)$-graphs as appropriate to illustrate your conclusions, but a catalog of graphs of solutions with different initial conditions is definitely insufficient. You should describe the behaviors you observe, giving a few graphs as illustrations of your description.

DISCRETE DYNAMICAL SYSTEMS

6

In this chapter we begin the study of a completely different type of model for processes that evolve in time, namely discrete dynamical systems or difference equations. Unlike differential equations, these models are well suited to situations in which changes occur at specific times, rather than continuously. For example, we often measure population growth at specific times such as the year-end of a reproductive cycle.

The study of discrete dynamical systems most often involves the process of iteration. To *iterate* means to repeat a procedure numerous times. In discrete dynamics, the process that is repeated is the application of a mathematical function. We have encountered several types of iterative processes already. For example, Euler's method for solving a differential equation involves iteration.

As was the case with differential equations, we see that finding exact solutions of discrete systems is rarely possible. As usual we rely on a combination of numerical, analytic, and graphical techniques to understand these systems. Even with these sophisticated techniques, we are unable to describe even one-dimensional, nonlinear iterations completely. The reason is that many discrete nonlinear systems behave in a complex and unpredictable manner. This phenomenon, called chaos by mathematicians, is discussed at length in Sections 6.4 and 6.5.

6.1 THE DISCRETE LOGISTIC EQUATION

The population models we have studied so far all have the property that the rate of change of the population is continuous. This is a reasonable assumption for species that reproduce quickly with respect to the time scale we are considering. However, for some species, all births occur in the spring and most deaths occur during the winter; hence the assumption of continuous population change is not valid. Instead, we should think of time as discrete. If we measure the population once a summer, then we will have an accurate estimate of the population for the entire year.

We begin our study of discrete time systems by recasting several of the population models we discussed in Chapter 1 as discrete dynamical systems.

Exponential Growth Model

The **exponential growth** model is a simple but unrealistic model of population growth. We discuss it first because it illustrates the main ideas behind discrete dynamics. In this model we assume that the population of a certain species in the next generation (or other time period) is directly proportional to the population in the current generation. We let P_n denote the population of the species at the end of the nth time period. The assumption says that, for each time n, the population at the end of the $n + 1^{st}$ time step, P_{n+1}, is proportional to the population at the end of the previous time step, P_n. That is,

$$P_{n+1} = kP_n,$$

where k is the proportionality constant that determines the growth rate. This equation is an example of a **discrete dynamical system** or a **difference equation**. Unlike the differential equation for unlimited population growth,

$$\frac{dP}{dt} = kP,$$

which tells us the rate of change of the population, this difference equation tells us directly the population P_{n+1} in the succeeding generation, provided we know P_n. Thus if we know P_0, the initial population, we can determine the population in each succeeding generation by simply computing the expression kP_n at each stage. There are no integrals to evaluate, no solution curves to approximate. All we have to do is repeatedly calculate the right-hand side of the equation using the output of the previous calculation as the input for the next. This repetition is what we call **iteration**. Thus if we know the initial population P_0, we then compute

$$P_1 = kP_0$$
$$P_2 = kP_1 = k^2 P_0$$
$$P_3 = kP_2 = k^3 P_0$$
$$\vdots$$
$$P_n = kP_{n-1} = k^n P_0.$$

We see that the population at generation n is determined by the formula $P_n = k^n P_0$. Therefore, if $P_0 > 0$, we conclude the following: If $k > 1$, the population explodes, since

$$\lim_{n \to \infty} k^n = \infty.$$

On the other hand, if $k < 1$, the population dies out, since

$$\lim_{n \to \infty} k^n = 0.$$

Finally, if $k = 1$, the population never changes and $P_n = P_0$.

This model has the same drawbacks as the differential equation version. Although it may work well for small populations in large environments, if the population grows at all, then the model predicts unlimited growth. Thus, we modify this model to account for a limited environment.

The Logistic Difference Equation

To make the exponential growth model somewhat more realistic, we add some assumptions that account for overcrowding, just as we did with the logistic differential equation in Chapter 1. The assumptions we make are:

- The population at the end of the next generation is proportional to the population at the end of the current generation when the population is very small.
- If the population is too large, then all resources will be used and the entire population will die out in the next generation and extinction will result.

This last assumption is slightly different from the one we made for the logistic differential equation. Here we assume that there is a maximum population level M that when reached results in extinction of the population in the next generation. Thus M is called the **annihilation parameter**. If the population ever reaches M, the species is doomed.

One model that reflects these assumptions is

$$P_{n+1} = k \left(1 - \frac{P_n}{M} \right) P_n.$$

As before, P_n denotes the population at the end of generation n and P_0 is the initial population. Note that if P_n is small, the term $(1 - P_n/M)$ is approximately 1. So the difference equation becomes $P_{n+1} \approx kP_n$, which is the exponential growth model. On the other hand, if $P_n \geq M$, then $P_{n+1} \leq 0$; that is, the population is nonpositive. We interpret this to mean that the species is extinct.

Rather than deal with the large numbers that often arise in population models, we will assume that P_n represents the percentage or fraction of this maximum population alive at generation n. That is, we assume that $M = 1$ and that P_n lies between 0 and 1, with $P_n = 0$ (or P_n negative) representing extinction and $P_n = 1$ representing the maximum population level. Thus the model becomes

$$P_{n+1} = kP_n(1 - P_n),$$

which we call the **discrete logistic equation** or **logistic difference equation**. As before, k is a parameter that depends on the specific species under investigation.

As always we must insert the caveat that this is an extremely naive model for population growth. We have neglected all kinds of other factors that affect P_n, including the effects of predators, cyclical diseases, and the variable nature of the food supply. Nevertheless this model does provide many more scenarios for changes in population than does the logistic differential equation.

Table 6.1   Successive populations of the logistic model with $P_0 = 0.5$ and $k = 0.5, 1.5, 2, 3.2, 3.5,$ and 3.9.

n	$k = 0.5$	$k = 1.5$	$k = 2$	$k = 3.2$	$k = 3.5$	$k = 3.9$
1	0.1250	0.3750	0.5000	0.8000	0.8750	0.9750
2	0.0546	0.3515	0.5000	0.5120	0.3828	0.0950
3	0.0258	0.3419	0.5000	0.7995	0.8269	0.3355
4	0.0125	0.3375	0.5000	0.5128	0.5008	0.8694
5	0.0062	0.3354	0.5000	0.7995	0.8749	0.4426
6	0.0030	0.3343	0.5000	0.5130	0.3828	0.9621
7	0.0015	0.3338	0.5000	0.7995	0.8269	0.1419
8	7.7×10^{-4}	0.3335	0.5000	0.5130	0.5008	0.4750
9	3.8×10^{-4}	0.3334	0.5000	0.7995	0.8749	0.9725
10	1.9×10^{-4}	0.3333	0.5000	0.5130	0.3828	0.1040
11	9.6×10^{-5}	0.3333	0.5000	0.7995	0.8269	0.3634
12	4.8×10^{-5}	0.3333	0.5000	0.5130	0.5008	0.9022
13	2.4×10^{-5}	0.3333	0.5000	0.7995	0.8749	0.3438
14	1.2×10^{-5}	0.3333	0.5000	0.5130	0.3828	0.8799
15	6.0×10^{-6}	0.3333	0.5000	0.7995	0.8269	0.4120
16	3.0×10^{-6}	0.3333	0.5000	0.5130	0.5008	0.9448
17	1.5×10^{-6}	0.3333	0.5000	0.7995	0.8749	0.2033
18	7.5×10^{-7}	0.3333	0.5000	0.5130	0.3828	0.6316
19	3.7×10^{-7}	0.3333	0.5000	0.7995	0.8269	0.9073
20	1.9×10^{-7}	0.3333	0.5000	0.5130	0.5008	0.3278

Some predictions of the model

As an example of the various possibilities we encounter in the discrete logistic model, we sample the output of this model for a few k values. Suppose we begin with a population that is exactly half the maximum permissible population, that is, $P_0 = 0.5$. Then depending on k, we find very different results when we compute successive values of P_n. In Table 6.1, we list the populations (using only 4 significant digits) when $k = 0.5, 1.5, 2, 3.2, 3.5,$ and 3.9. Note that these different k values yield very different behaviors for the population. When $k = 0.5$, the population tends gradually toward extinction. When $k = 1.5$, the population seems to level out and approach an equilibrium state. When $k = 2$, the population never changes and remains fixed at 0.5. When $k = 3.2$, we see a different result: The population eventually oscillates back and forth between two different values. The population is high one year, approximately 0.7995, low the next, approximately 0.513, and then repeats cyclically. When $k = 3.5$, we see a similar cyclic behavior, but now the populations eventually repeat every four years instead of two. Finally, when $k = 3.9$, there is no apparent pattern to the successive populations.

Iteration

For the discrete logistic model, finding successive populations is the same as iterating a quadratic function of the form

$$L_k(x) = kx(1 - x).$$

This function depends on the parameter k and is often called the **logistic function**. To iterate this function, we begin with an initial population P_0 and then compute in succession

$$P_1 = L_k(P_0)$$
$$P_2 = L_k(P_1)$$
$$P_3 = L_k(P_2)$$
$$\vdots$$
$$P_n = L_k(P_{n-1})$$

and so forth. The list of numbers P_0, P_1, P_2, ... that results from this iteration is called the **orbit** of P_0 under the function L_k. The initial value P_0 is sometimes called the **seed** or initial condition for the orbit.

In discrete dynamics the basic goal is to predict the fate of orbits for a given function. That is, the main question is: What happens to the numbers that constitute the orbit as n tends to infinity?

Sometimes predicting the fate of orbits is easy. For example, if $F(x) = x^2$, then we can determine what happens to all orbits without difficulty. For example, the seeds $x_0 = 0$ and $x_0 = 1$ are **fixed points**, since $F(0) = 0$ and $F(1) = 1$. That is, the orbit of 0 is the constant sequence 0, 0, 0, 0 ... as is the orbit of 1: 1, 1, 1, ... The orbit of -1 under $F(x) = x^2$ is slightly different: this orbit is **eventually fixed** since $F(-1) = 1$, which is a fixed point. The orbit of -1 is $-1, 1, 1, 1, \ldots$.

For any other seed x_0, there are only two possibilities for the orbit of x_0: Either the orbit tends to the fixed point at 0, or the orbit tends to infinity. For example, if $x_0 = 1/2$, the orbit is

$$x_0 = 1/2$$
$$x_1 = 1/4$$
$$x_2 = 1/16$$
$$x_3 = 1/256$$
$$\vdots$$
$$x_n = 1/2^{2^n},$$

which tends to 0 as n tends to infinity. The fate of the orbit is the same for any other seed x_0 with $|x_0| < 1$.

If $|x_0| > 1$, successive applications of the squaring function $F(x) = x^2$ yield larger and larger results. For example, if $x_0 = 2$, we have

$$x_0 = 2$$
$$x_1 = 4$$
$$x_2 = 16$$
$$x_3 = 256$$
$$\vdots$$
$$x_n = 2^{2^n},$$

and we see that the orbit of x_0 tends to infinity.

Cycles

In a typical discrete dynamical system, there are often many different types of orbits. For example, if $G(x) = x^2 - 1$, then the orbit of 0 lies on a **cycle of period** 2 or a **periodic orbit of period** 2 since

$$x_0 = 0$$
$$x_1 = -1$$
$$x_2 = 0$$
$$x_3 = -1$$
$$\vdots$$

This orbit is the repeating sequence $0, -1, 0, -1, \ldots$.

If we choose the seed $x_0 = \sqrt{2}$, then this orbit is **eventually periodic** since we have

$$x_0 = \sqrt{2}$$
$$x_1 = 1$$
$$x_2 = 0$$
$$x_3 = -1$$
$$x_4 = 0$$
$$x_5 = -1$$
$$\vdots,$$

which begins to cycle after the second iteration.

In contrast if we choose the seed $x_0 = 0.5$, the orbit tends to the cycle of period 2 since

$$x_0 = 0.5$$
$$x_1 = (0.5)^2 - 1 = -0.75$$
$$x_2 = -0.4375$$
$$x_3 = -0.8086\ldots$$
$$\vdots$$
$$x_{20} = 0.00000\ldots$$
$$x_{21} = -1.00000\ldots$$
$$x_{22} = 0.00000\ldots$$
$$\vdots,$$

It is important to realize that this orbit never actually reaches the cycle at 0 and -1. Rather, the orbit comes arbitrarily close to these two numbers, but, because of round-off error the calculator or computer eventually displays the numbers 0 and -1 in succession. Thus the orbit is not eventually periodic, it merely tends to the cycle of period 2. This behavior is technically different from that of an orbit that is eventually periodic, although in practice the fate of these orbits is essentially the same.

Discrete dynamical systems may have orbits that cycle with any period. For example, the difference equation

$$x_{n+1} = -\tfrac{3}{2}x_n^2 + \tfrac{5}{2}x_n + 1$$

admits a cycle of period 3, since for the seed $x_0 = 0$, we calculate

$$x_0 = 0$$
$$x_1 = 1$$
$$x_2 = 2$$
$$x_3 = 0$$

Thus the orbit is the repeating sequence $0, 1, 2, 0, 1, 2, \ldots$. In general the orbit of x_0 lies on a cycle of period n if $x_n = x_0$ and n is the smallest positive integer for which this happens. A fixed point would not be regarded as a cycle of period 10, even though $x_{10} = x_0$, since 10 is not the smallest integer for which the orbit repeats. Similarly, if x_0 has period 5, we also have $x_{10} = x_0$, but we would not call this a period 10 orbit either.

Since the orbit of a fixed point is a constant sequence $x_0, x_0, x_0, \ldots$, fixed points for discrete dynamical systems may be regarded as the analogs of equilibrium points for differential equations. Both represent constant solutions in which the given system is at rest. Similarly, cycles for a discrete system are analogous to periodic solutions of differential equations, as both return to their original position after some time.

Time Series and Histograms

One convenient method to describe orbits geometrically is via a **time series**. In these diagrams we plot the iteration count on the horizontal axis and the numerical values of the orbit on the vertical axis. It often helps to draw straight lines connecting successive points in the time series. In Figure 6.1 we plot the time series corresponding to the orbit of -0.5 under $G(x) = x^2 - 1$. Note that this orbit tends to a cycle of period 2. In Figure 6.2 we plot the time series corresponding to the orbit of 0 under $H(x) = x^2 - 1.3$. Here we see that this orbit tends to a cycle of period 4. These time series are the analogs of the $x(t)$- and $y(t)$-graphs we often plot for differential equations.

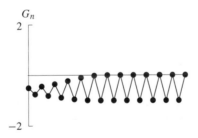

Figure 6.1
Time series for the orbit of -0.5 under $G(x) = x^2 - 1$.

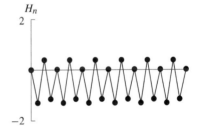

Figure 6.2
Time series for the orbit of 0 under $H(x) = x^2 - 1.3$.

In Figure 6.3 we display the time series for the orbit of 0.5 under iteration of the logistic function $L_{3.9}(x) = 3.9x(1-x)$. This is not fixed or periodic behavior; it is difficult to see any sort of coherent pattern. We study this sort of time series in Section 6.5.

Another visual way of displaying orbits is a **histogram**. In these image we subdivide an interval that contains the points of an orbit into many small subintervals of

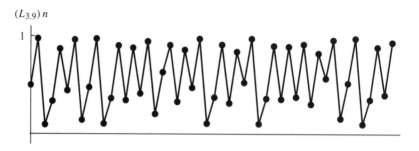

$(L_{3.9})n$

Figure 6.3
Time series for the orbit of 0.5 under $L_{3.9}(x) = 3.9x(1-x)$.

equal length. Each time the orbit enters one of these subintervals, we increment the histogram by one unit over that subinterval. In Figures 6.4 and 6.5 we display the histograms corresponding to the orbit of 0.123 under iteration of the logistic functions $L_{3.83}(x) = 3.83x(1-x)$ and $L_{3.9}(x) = 3.9x(1-x)$. In the first case the orbit quickly settles down on a period 3 cycle. In the second case we see a different view of Figure 6.3. The orbit apparently never settles down. Rather, it visits many of the subintervals, some more than others.

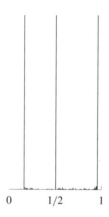

0   1/2   1

Figure 6.4
Histogram for the
orbit of 0.123 under
$L_{3.83}(x) =$
$3.83x(1-x)$.

0   1/2   1

Figure 6.5
Histogram for the
orbit of 0.123 under
$L_{3.9}(x) =$
$3.9x(1-x)$.

Finding Fixed Points

As we see from the examples above, there are many different types of orbits for a discrete dynamical system. Fixed points are the simplest orbits and these are often the most important types of orbits in such a system. As in the case of differential equations, we employ three different methods to find these special orbits: analytic, qualitative, and numerical techniques.

The analytic method involves solving equations. Given a function F, to find the fixed points for F, we need only solve the equation $F(x) = x$. For example, if $F(x) = x^2 - 2$, then we find that the fixed points are the solutions of the equation

$$x^2 - 2 = x,$$

which can be written as

$$x^2 - x - 2 = 0.$$

Factoring turns this into

$$(x + 1)(x - 2) = 0,$$

and the fixed points for F are -1 and 2.

Geometrically we can view fixed points by superimposing the graph of the diagonal line $y = x$ on the graph of F. Note that the diagonal $y = x$ meets the graph of $F(x) = x^2 - 2$ directly over -1 and 2 as shown in Figure 6.6. As another example, $K(x) = x^3$ has 3 fixed points, which occur at the roots of $x^3 - x = 0$, or at 0, -1, and $+1$ (see Figure 6.7).

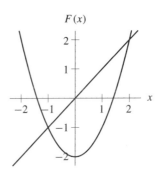

Figure 6.6
Fixed points of $F(x) = x^2 - 2$.

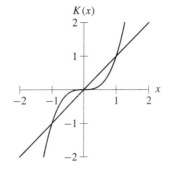

Figure 6.7
Fixed points of $K(x) = x^3$.

For the logistic equation $L_k(x) = kx(1 - x)$, we find the fixed points by solving

$$kx(1 - x) = x,$$

which yields solutions $x = 0$ and $x = (k - 1)/k$. Recall that for our population model we assume that $0 \le x \le 1$, so this second fixed point is positive only if $k > 1$. (When $k < 1$, this fixed point is negative and so not biologically interesting). Also, $(k - 1)/k < 1$ for all $k > 1$. The fixed point at $x = 0$ represents population extinction, and the nonzero fixed point represents a population that never changes from generation to generation. See Figure 6.8, which shows the graph in the case of $k = 4$, with fixed points at 0 and 0.75.

Given a function F, finding the fixed points by solving $F(x) = x$ analytically may be difficult or impossible. If we require a more accurate value for the fixed point than we can obtain from graphical methods, then we can turn to numerical methods. We can use Newton's method to find values of x where $F(x) - x = 0$. For example, from Figure 6.9 we see that $C(x) = \cos x$ has a single fixed point, which we can determine numerically to be given by $0.73908\ldots$ (see the exercises).

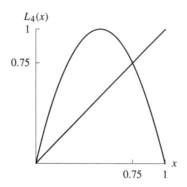

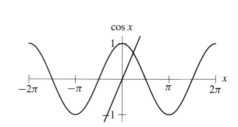

Figure 6.8
Fixed points for the logistic model
$L_4(x) = 4x(1 - x)$.

Figure 6.9
Fixed point for $\cos x$.

Notation for Iteration

To simplify notation, we introduce the expression F^n to denote the nth iterate of F. For example, if $F(x) = x^4$, then

$$F^2(x) = F(F(x)) = F(x^4) = (x^4)^4 = x^{16}$$

and similarly $F^3(x) = x^{64}$. It is important to realize that $F^n(x)$ *does not* mean $F(x)$ raised to the nth power; rather, $F^n(x)$ means to first iterate F exactly n times, then evaluate this new function at x.

Finding Cycles

To find cycles for a discrete dynamical system, we proceed in essentially the same manner as for fixed points. For example, if $H(x) = -x^3$, the only fixed point of H is 0, since the equation for the fixed points is $-x^3 = x$ or $x(1 + x^2) = 0$. For the cycles of period 2, we must compute

$$H(H(x)) = H^2(x) = -(-x^3)^3 = x^9$$

and then solve $x^9 = x$. The solutions here are the fixed point $x = 0$ and two new solutions, $x = 1$ and $x = -1$. These latter two points form a cycle of period 2, since $H(1) = -1$ and $H(-1) = 1$.

Finding cycles is usually more difficult than finding fixed points. For example, to find cycles of period 2 for $F(x) = x^2 - 2$, we must find values of x for which $F(F(x)) = x$. We have

$$F^2(x) = (x^2 - 2)^2 - 2 = x^4 - 4x^2 + 2,$$

so that we must solve

$$x^4 - 4x^2 + 2 = x.$$

That is, we must find the roots of the fourth-degree equation

$$x^4 - 4x^2 - x + 2 = 0.$$

Luckily we know two solutions of this equation already, since we saw above that -1 and 2 are fixed points, and fixed points have orbits that repeat every two iterations as well as every iteration. So we can divide this expression by $(x + 1)(x - 2)$ to find

$$\frac{x^4 - 4x^2 - x + 2}{(x + 1)(x - 2)} = \frac{x^4 - 4x^2 - x + 2}{x^2 - x - 2} = x^2 + x - 1 = 0.$$

This quadratic equation has roots $(-1 \pm \sqrt{5})/2$. These points lie on a cycle of period 2 since we have

$$\left(\frac{-1 + \sqrt{5}}{2}\right)^2 - 2 = \frac{-1 - \sqrt{5}}{2}$$

and

$$\left(\frac{-1 - \sqrt{5}}{2}\right)^2 - 2 = \frac{-1 + \sqrt{5}}{2}.$$

In general, to find cycles of period n, we must first iterate the function F a total of n times. For example, if $L_3(x) = 3x(1 - x)$, then we compute

$$L_3^2(x) = 3[3x(1 - x)][1 - 3x(1 - x)] = -27x^4 + 54x^3 - 36x^2 + 9x$$

and

$$L_3^3(x) = L_3(L_3^2(x))$$
$$= 3\left[-27x^4 + 54x^3 - 36x^2 + 9x\right]\left[1 - (-27x^4 + 54x^3 - 36x^2 + 9x)\right],$$

which we will not bother to simplify. You should note that, when multiplied out, $L_3^3(x)$ has terms that involve x^8, so L_3^3 is an eighth-degree polynomial. Finding the solutions of $L_3^3(x) = x$ is therefore not a pleasant task.

Finding cycles geometrically

In general, finding cycles of period n involves solving the equation $F^n(x) = x$. Even for logistic functions of the form $L_k(x) = kx(1 - x)$, $L_k^n(x)$ is a polynomial of degree 2^n, and so we have little chance of finding explicit solutions. However, geometric information is relatively easy to come by. We can usually discover some information about the number of cycles of period n by sketching the graph of F^n and looking for places where this graph crosses the diagonal line $y = x$.

For example, if $H(x) = -x^3$, the graph of H shows that H has a single fixed point that occurs at 0. But $H^2(x) = x^9$ has a graph that meets the diagonal three times, at the fixed point and at a cycle of period 2, which as we saw above is given by ± 1 (see Figures 6.10 and 6.11).

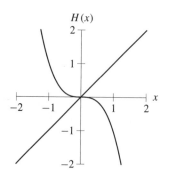

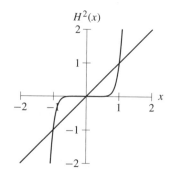

Figure 6.10
The graph of $H(x) = -x^3$.

Figure 6.11
The graph of $H^2(x) = x^9$.

In Figures 6.12–6.14 we sketch the graph of $F(x) = x^2 - 2$ as well as F^2 and F^3. Note that we see two fixed points for F and four fixed points for F^2. Two of these are the fixed points for F, and the other two lie on a cycle of period 2. For F^3 there are eight points where the graph of $y = F^3(x)$ crosses $y = x$. Two of these points are again the fixed points of F, but the other six must be periodic with period 3. The cycle of period 2 does not appear in the graph of F^3 since these orbits do not repeat after 3 iterations.

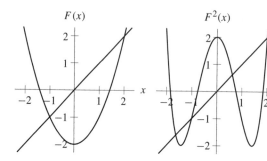

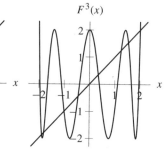

Figure 6.12
The graph of $F(x) = x^2 - 2$.

Figure 6.13
The graph of $F^2(x)$.

Figure 6.14
The graph of $F^3(x)$.

Exercises for Section 6.1

In Exercises 1–8, compute the orbit of $x_0 = 0$ for each of the given difference equations. Determine whether this orbit is fixed, cycles with some period, is eventually periodic, tends to infinity, or is none of the above. (Use calculators as needed.)

1. $x_{n+1} = x_n^2 - 2$

2. $x_{n+1} = \sin(x_n)$

3. $x_{n+1} = e^{x_n}$

4. $x_{n+1} = x_n^2 + 1$

5. $x_{n+1} = -x_n^2 + x_n + 2$

6. $x_{n+1} = \cos(x_n)$

7. $x_{n+1} = 4x_n(1 - x_n) + 1$

8. $x_{n+1} = -x_n^2/2 + 1$

In Exercises 9–21, find all fixed points and periodic points of period 2 for each of the given functions. If you cannot determine these values explicitly, use the graph of F or F^2 to determine how many fixed points and periodic points of period 2 F has.

9. $F(x) = -x + 2$

10. $F(x) = x^4$

11. $F(x) = x^2 + 1$

12. $F(x) = x^2 - 3$

13. $F(x) = \sin x$

14. $F(x) = 1/x$

15. $F(x) = -2x - x^2$

16. $F(x) = e^x$

17. $F(x) = -e^x$

18. $F(x) = x^3$

19. $F(x) = -x$

20. $F(x) = -2x + 1$

21. $F(x) = 2$

22. Describe the fate of the orbit of any seed under iteration of $F(x) = x^3$.

23. Describe the fate of the orbit of any seed under iteration of $F(x) = -x + 4$.

In Exercises 24–35, describe the fate of the orbit of each of the following seeds under iteration of the function

$$T(x) = \begin{cases} 2x, & \text{if } x < \frac{1}{2} \\ 2 - 2x, & \text{if } x \geq \frac{1}{2}. \end{cases}$$

24. $2/3$

25. $1/6$

26. $2/5$

27. $2/7$

28. $3/14$

29. $1/8$

30. $1/9$

31. $6/11$

32. $2/9$

33. 0

34. $1/4$

35. $1/2$

36. Discuss the fate of the orbit of 0 under iteration of $F(x) = x^2 + c$ for each of the following c values: $c = 0.4, 0.3, 0.2, 0.1$. Use the graph and technology as necessary to explain your findings. Is there a change in the fate of the orbit of 0 as c changes? Explain in a sentence or two.

37. Consider the family of functions $F_c(x) = x^2 + c$ that depends on a parameter c. For which values of c does this function have real fixed points? How many?

38. Discuss the fate of the orbit of 0 under iteration of $F(x) = x^2 + c$ for each of the following c values: $c = -0.6, -0.7, -0.8, -0.9$. Use the graph of F and F^2 and technology as necessary to explain your findings. Is there a change in the fate of the orbit of 0 as c changes? Explain in a sentence or two.

39. Consider the family of functions $F_c(x) = x^2 + c$ that depends on a parameter c. For which values of c does this function have periodic points of period 2? How many?

6.2 FIXED AND PERIODIC POINTS

As was the case with equilibrium points for differential equations, fixed points and cycles for difference equations come in several distinct varieties, depending on what happens to nearby orbits. In this section we discuss analytic and qualitative methods for distinguishing different types of fixed points and cycles.

Graphical Iteration

As we saw in the previous section, the graph of a function and the diagonal line $y = x$ give us a geometric method for finding fixed points and cycles. In this section we see that the graph of F together with the diagonal line $y = x$ also provides a convenient way to sketch other orbits geometrically.

Given the function F and a seed x_0, we view the orbit of x_0 as follows: Begin at the point (x_0, x_0) on the diagonal. Draw a vertical line from (x_0, x_0) to the graph of F; we reach the graph at (x_0, x_1), where $x_1 = F(x_0)$. Now draw a horizontal line from this point back to the diagonal. Along this line, the y-values remain constant at $y = x_1$, so we reach the diagonal at (x_1, x_1). We interpret this point as the second point on the orbit.

Now continue in the same fashion. First draw a vertical line from (x_1, x_1) to the graph of F. We reach the graph at (x_1, x_2), where $x_2 = F(x_1)$. Then draw a horizontal line back to the diagonal, reaching this line at (x_2, x_2). This is the second point on the orbit.

Continuing in this manner, the orbit of x_0 is displayed as a sequence of points along the diagonal. (see Figure 6.15) . The resulting picture, which sometimes resembles a staircase and sometimes a cobweb, is often called a **web diagram**. The process of drawing vertical lines to the graph and then horizontal lines to the diagonal is called **graphical analysis** or **graphical iteration**. We sometimes add arrows to the web diagram to indicate the direction of the iteration.

Figure 6.16 depicts the web diagram for $G(x) = x^2 - 0.7$ using the seed $x_0 = 1.4$. We clearly see two fixed points for G, given by the points where the graph of G crosses the diagonal. Note that the orbit of $x_0 = 1.4$ tends away from one of these fixed points and toward the other fixed point.

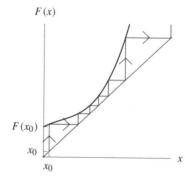

Figure 6.15
Graphical analysis showing the orbit of x_0.

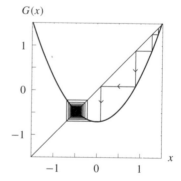

Figure 6.16
Web diagram of $G(x) = x^2 - 0.7$ using the seed $x_0 = 1.4$.

Another important type of orbit for discrete systems is the cycle or periodic orbit. Recall that, for a function F, the seed x_0 lies on a cycle or periodic orbit of period n if $F^n(x_0) = x_0$. (Remember that F^n indicates the nth iterate of F, not the nth power.) Equivalently, x_0 lies on a cycle of period n if the orbit of x_0 is a repeating sequence of length n of the form

$$x_0, x_1, \ldots, x_{n-1}, x_0, x_1, \ldots, x_{n-1}, \ldots .$$

Graphical iteration allows us to visualize cycles without having to compute the graph of F^n. For example, 0 and -1 lie on a cycle of period 2 for $F(x) = x^2 - 1$, since $F(0) = -1$ and $F(-1) = 0$. In Figure 6.17 we see that this cycle is represented geometrically by a square in the web diagram. In Figure 6.17 we also see an orbit that tends to this cycle.

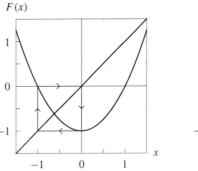
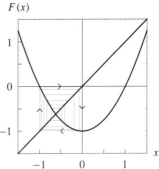

Figure 6.17
A 2-cycle for $F(x) = x^2 - 1$ and an orbit tending to the 2-cycle.

As we saw in the previous section, the logistic difference equation admits cycles of many periods. Figures 6.18 and 6.19 display some of these cycles using graphical iteration.

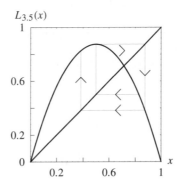
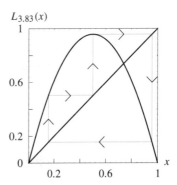

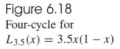

Figure 6.18
Four-cycle for
$L_{3.5}(x) = 3.5x(1 - x)$

Figure 6.19
Three-cycle for
$L_{3.83}(x) = 3.83x(1 - x).$

Attracting and Repelling Fixed Points

Figures 6.20 and 6.21 display the graphical analysis of two additional logistic functions, $L_{2.8}(x) = 2.8x(1 - x)$ and $L_{3.2}(x) = 3.2x(1 - x)$. In the first case observe that the displayed orbit tends to a fixed point, but when the parameter is changed to 3.2, this is no longer the case. The displayed orbit comes close to the fixed point but then tends toward a 2-cycle. This is an example of a bifurcation, a topic we discuss later. For now note that there is a dramatic difference between the behavior of orbits near the nonzero fixed point in these two cases. In the first case the orbit is attracted to the nonzero fixed point, whereas in the second the orbit seems to be repelled away. Just as with equilibrium points for a differential equation, there are different types of fixed points and cycles.

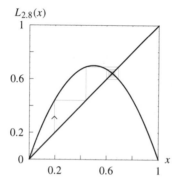

Figure 6.20
An attracting fixed point for $L_{2.8}(x) = 2.8x(1 - x)$.

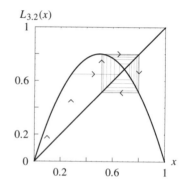

Figure 6.21
A repelling fixed point for $L_{3.2}(x) = 3.2x(1 - x)$.

Definition: A fixed point x_0 of a function F is called **attracting** if there is an interval around x_0 having the property that every seed in this interval has an orbit that remains in the interval and tends to x_0 under iteration of F. The fixed point is called **repelling** if there is an interval around x_0 having the property that every seed in this interval (except x_0) has an orbit that leaves the interval under iteration of F. A fixed point that is neither attracting nor repelling is called **neutral**.

Attracting fixed points are the analogs of sinks for one-dimensional differential equations, whereas repelling fixed points correspond to sources.

Fixed points of linear functions

To see what makes certain fixed points attracting and others repelling, we first consider linear functions. For $F(x) = mx$, $x_0 = 0$ is always a fixed point. Graphical analysis shows that as long as $|m| < 1$, x_0 attracts all orbits (see Figure 6.22). Note that if $-1 < m < 0$, orbits hop back and forth about 0 as they tend to the fixed point.

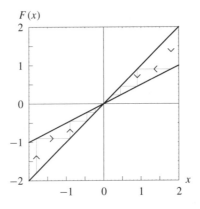
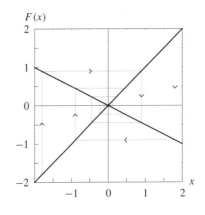

Figure 6.22
The function $F(x) = mx$ has an attracting fixed point at 0 if $|m| < 1$.

When $|m| > 1$, the situation is quite different. Now all orbits tend away from 0; the origin is a repelling fixed point (see Figure 6.23). Again orbits hop from side to side as they leave the origin when $m < -1$.

The intermediate cases $m = \pm 1$ are special. When $m = 1$, all points are fixed, whereas when $m = -1$, all nonzero points are periodic with period 2.

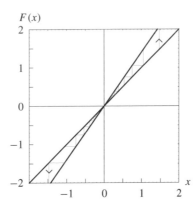
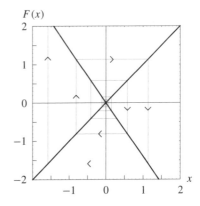

Figure 6.23
The function $F(x) = mx$ has a repelling fixed point at 0 if $|m| > 1$.

Classification of Fixed Points

When F is a nonlinear function, a similar story holds true, at least near fixed points. If F has a fixed point at x_0 and $|F'(x_0)| < 1$, then the graph of F is tangent to a straight line whose slope is less than 1 in absolute value. As in the case of linear functions, graphical analysis shows that nearby orbits must be attracted to x_0. Thus a fixed point x_0 for F for which $|F'(x_0)| < 1$ is attracting (see Figure 6.24). Note that this result is true only close to x_0; far away, orbits may behave very differently.

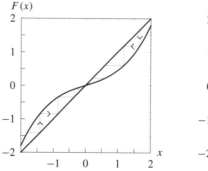

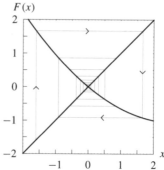

Figure 6.24

If $|F'(x_0)| < 1$ then x_0 is an attracting fixed point.

Similarly, if $|F'(x_0)| > 1$, nearby orbits are repelled away from x_0 just as in the linear case. Therefore if $|F'(x_0)| > 1$ at the fixed point, then x_0 is a repelling fixed point (see Figure 6.25).

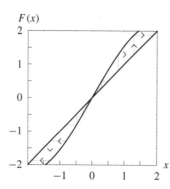

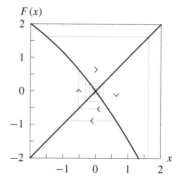

Figure 6.25

If $|F'(x_0)| > 1$ then x_0 is a repelling fixed point.

Fixed points for the logistic function

Consider the logistic function $L_{2.8}(x) = 2.8x(1 - x)$. The fixed points for $L_{2.8}$ are given by solving

$$2.8x(1 - x) = x.$$

Some algebra shows that these fixed points are $x = 0$ and $x = 1.8/2.8 \approx 0.64$. Now

$$L'_{2.8}(x) = 2.8(1 - 2x),$$

so $L'_{2.8}(0) = 2.8$ and $L'_{2.8}(0.64\ldots) = 2.8(-0.28\ldots) \approx -0.78$. Thus 0 is a repelling fixed point and $0.64\ldots$ is an attracting fixed point. We saw this qualitatively in Figure 6.20.

As a second example, suppose we consider instead the logistic function $L_{3.2}(x) = 3.2x(1 - x)$. Solving for the fixed points as above, we find fixed points at $x = 0$ and $x = 2.2/3.2 = 0.6875$. Now we have

$$L'_{3.2}(x) = 3.2(1 - 2x).$$

Thus $L'_{3.2}(0) = 3.2$ and $L'_{3.2}(0.6875) = -1.2$. In this case both fixed points are re-pelling. We saw this also in Figure 6.20.

These two examples are special cases of the general logistic equation $L_k(x) = kx(1-x)$. Recall from the previous section that the fixed points of F occurred at 0 and at $(k-1)/k$, provided $k > 1$. We compute

$$L'_k(x) = k - 2kx.$$

So $L'_k(0) = k$. Therefore 0 is an attracting fixed point when $0 \le k < 1$ and a repelling fixed point when $k > 1$. Also,

$$L'_k\left(\frac{k-1}{k}\right) = k - 2(k-1) = -k + 2.$$

So $(k-1)/k$ is an attracting fixed point when $-1 < -k+2 < 1$, that is, for $1 < k < 3$. When $k > 3$, this fixed point is repelling (see Figures 6.26 and 6.27).

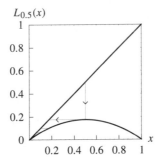

$L_{0.5}(x)$

Figure 6.26
Fixed points for
$L_k(x) = kx(1-x)$ when
$k = 0.5$.

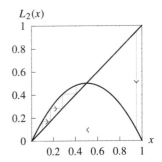

$L_2(x)$

Figure 6.27
Fixed points for
$L_k(x) = kx(1-x)$ when
$k = 2$.

Examples of neutral fixed points

Neutral fixed points can occur only if $F'(x_0) = \pm 1$. Orbits near neutral fixed points may behave in a variety of ways. For example, consider $F(x) = -x + 4$. We have $F(2) = 2$ and $F'(2) = -1$. All other seeds have orbits that lie on cycles of period 2, since $F^2(x) = -(-x+4) + 4 = x$.

As another example, $G(x) = x + x^2$ has a fixed point at $x = 0$. Note that $G'(0) = 1$. The graph of G shows that 0 attracts from the left but repels from the right. Thus 0 is a neutral fixed point (see Figure 6.28). If we consider instead $H(x) = x + x^3$, then again 0 is a fixed point and $H'(0) = 1$. This time, however, $x = 0$ is a repelling fixed point as we see from graphical iteration (see Figure 6.29).

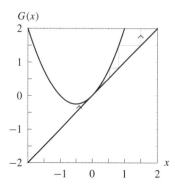

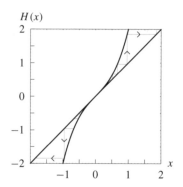

Figure 6.28
$G(x) = x + x^2$ has a neutral fixed
point at $x = 0$.

Figure 6.29
$H(x) = x + x^3$ has a repelling
fixed point at $x = 0$.

Classification of Periodic Points

Periodic points can also be classified as attracting, repelling, and neutral. If x_0 lies on an n-cycle of a given function F, then the graph of F^n meets the diagonal at (x_0, x_0). That is, F^n has a fixed point at x_0. So it is natural to call the cycle attracting, repelling, or neutral depending on whether the fixed point for F^n has this property.

For example, the function $F(x) = x^2 - 1$ has a 2-cycle at 0 and -1. Since $F^2(x) = (x^2 - 1)^2 - 1 = x^4 - 2x^2$, we have

$$(F^2)'(x) = 4x^3 - 4x.$$

Therefore $(F^2)'(0) = 0$, so 0 lies on an attracting 2-cycle, as we have seen in Figure 6.17. Note also that $(F^2)'(-1) = 0$.

The fact that $(F^2)'(x_0)$ is the same at both points on the cycle in this example is no accident. The chain rule tells us why. Suppose x_0 and x_1 lie on a 2-cycle for F. So $F(x_0) = x_1$ and $F(x_1) = x_0$. Then we have

$$(F^2)'(x_0) = F'(F(x_0)) \cdot F'(x_0)$$
$$= F'(x_1) \cdot F'(x_0).$$

That is, the derivative of F^2 at x_0 is just the product of the derivatives along the orbit of x_0. The same is true for x_1. More generally, if $x_0 \ldots, x_{n-1}$ lies on a cycle of period n for F, then

$$(F^n)'(x_0) = F'(F^{n-1}(x_0)) \cdot F'(F^{n-2}(x_0)) \cdot \ldots \cdot F'(x_0)$$
$$= F'(x_{n-1}) \cdot F'(x_{n-2}) \cdot \ldots \cdot F'(x_0),$$

which is again the product of the derivatives along the cycle.

As a check, for $F(x) = x^2 - 1$, we have $F'(0) = 0$ and $F'(-1) = -2$. Consequently,

$$(F^2)'(0) = F'(0) \cdot F'(-1) = 0(-2) = 0$$

as before.
  As a final example, consider the function

$$T(x) = \begin{cases} 2x, & \text{if } x < \frac{1}{2} \\ \\ 2 - 2x, & \text{if } x \geq \frac{1}{2}. \end{cases}$$

The seed $x_0 = 2/7$ leads to a cycle of period 3 for T with orbit $2/7, 4/7, 6/7, 2/7 \ldots$.
We compute

$$T'(2/7) = 2$$
$$T'(4/7) = -2$$
$$T'(6/7) = -2.$$

Therefore we have

$$(T^3)'(2/7) = T'(2/7) \cdot T'(4/7) \cdot T'(6/7) = 8,$$

and so this cycle is repelling. Note that any cycle for this function is repelling, since the
derivative will always be a product of 2s and -2s .

Exercises for Section 6.2

In Exercises 1–11, for each of the given functions, find all fixed points and determine
wether they are attracting, repelling, or neutral.

1. $F(x) = x^2 - 2x$ **2.** $F(x) = x^5$
3. $F(x) = \sin x$ **4.** $F(x) = x^3 - x$
5. $F(x) = \arctan x$ **6.** $F(x) = 3x(1 - x)$
7. $F(x) = (\pi/2) \sin x$ **8.** $F(x) = x^2 - 3$
9. $F(x) = 1/x$ **10.** $F(x) = 1/x^2$
11. $F(x) = e^x$

In Exercises 12–17, for each of the given functions, 0 lies on a periodic orbit. First
determine the period of this orbit. Then decide if this cycle is attracting, repelling, or
neutral.

12. $F(x) = -x^5 + 1$ **13.** $F(x) = 1 - x^2$
14. $F(x) = (\pi/2) \cos x$

15. $F(x) = \begin{cases} x + 1 & \text{if } x < 3.5 \\ 2x - 8 & \text{if } x \geq 3.5 \end{cases}$

16. $F(x) = 1 - x^3$ **17.** $F(x) = |x - 2| - 1$

In Exercises 18–25, each of the given functions has at least one fixed point with derivative ± 1. First find these fixed points. Then, using graphical analysis with an accurate
graph or other means, determine if this fixed point is attracting, repelling, or neutral.

18. $F(x) = x - x^2$ **19.** $F(x) = 1/x$
20. $F(x) = \sin x$ **21.** $F(x) = \tan x$
22. $F(x) = -x + x^3$ **23.** $F(x) = -x - x^3$
24. $F(x) = e^{x-1}$ (fixed point is 1)   **25.** $F(x) = -e \cdot e^x$ (fixed point is -1)

26. Find all fixed points for $F_c(x) = x^2 + c$ for all values of c. Determine for which values of c each fixed point is attracting, repelling, and neutral.

27. How many fixed points does $F(x) = \tan x$ have? Are they attracting, repelling, or neutral? Why?

28. What can you say about fixed points for $F_c(x) = ce^x$ with $c > 0$? What does the graph of F_c tell you about these fixed points? Note that when $c = 1/e$, $F_c(1) = 1$.

29. Consider the function

$$T(x) = \begin{cases} 4x & x < \frac{1}{2} \\ 4 - 4x & x \geq \frac{1}{2} \end{cases}$$

Does T have any attracting cycles? Why or why not?

30. Recall from calculus that Newton's method is an iterative procedure to find the roots of a given function P. Indeed, the Newton iteration is just

$$x_{n+1} = x_n - \frac{P(x_n)}{P'(x_n)}.$$

(a) Show that a root x of P for which $P'(x) \neq 0$ is a fixed point for the Newton iteration.

(b) Suppose $P(x) = x^3 - x$. Determine whether the fixed points for the corresponding Newton iteration are attracting or repelling.

6.3 BIFURCATIONS

Just as with differential equations, discrete dynamical systems may undergo changes in their orbit structure as parameters vary. These changes may include the birth or death of fixed points and cycles or changes in the type of these orbits. As before, such changes are known as bifurcations.

Tangent Bifurcation

One of the simplest types of bifurcation occurs when fixed or periodic points suddenly appear or disappear. As an example, consider the family of quadratic functions $F_c(x) = x^2 + c$. These functions have fixed points when

$$x^2 + c = x.$$

Using the quadratic formula, we see that there are two fixed points for F_c given by

$$p_\pm = \frac{1 \pm \sqrt{1 - 4c}}{2}.$$

These fixed points are real if $1 - 4c \geq 0$, so F_c has fixed points only if $c \leq 1/4$. If $c > 1/4$, F_c has no real fixed points. If $c = 1/4$, the two fixed points are the same since $p_+ = p_- = 1/2$. For $c < 1/4$, this fixed point splits apart into distinct fixed points p_+ and p_-. Thus a bifurcation occurs at $c = 1/4$ since a pair of fixed points is born as c decreases through $1/4$.

Qualitatively, the graph of F_c lies above the diagonal $y = x$ when $c > 1/4$, so F_c has no fixed points. When $c = 1/4$, the graph is tangent to the diagonal at one point. As we see in Figure 6.30, this is a neutral fixed point. When $c < 1/4$, the graph meets the diagonal at two distinct fixed points $p_\pm$.

To determine the character of these fixed points, we compute $F_c'(x) = 2x$, so that

$$F_c'(p_+) = 1 + \sqrt{1 - 4c}.$$

Now $\sqrt{1 - 4c} > 0$ when $c < 1/4$, so it follows that $F_c'(p_+) > 1$ for these c values. Therefore p_+ is a repelling fixed point for all $c < 1/4$. For p_-, the situation is more complicated. We have

$$F_c'(p_-) = 1 - \sqrt{1 - 4c},$$

so $F_c'(p_-) < 1$ for all values of $c < 1/4$. To guarantee that p_- is attracting, we must also have $F_c'(p_-) > -1$. This happens provided

$$-1 < 1 - \sqrt{1 - 4c} = F_c'(p_-)$$
$$2 > \sqrt{1 - 4c}$$
$$-3/4 < c.$$

Therefore the fixed point p_- is attracting for $-3/4 < c < 1/4$ and repelling for $c < -3/4$.

Graphical analysis provides an alternative view of this bifurcation. In Figure 6.30 we sketched some orbits before, at, and after the bifurcation when $c = 0.5, 0.25$, and 0.1. Of course, there really is no "before" and "after" the bifurcation. We could just as easily view this change as c increases, in which case this bifurcation consists of two fixed points that coalesce when $c = 1/4$ and then disappear. Note that this bifurcation is quite similar to some of the bifurcations of equilibrium points that we encountered for differential equations.

As with differential equations, a picture is a useful way to display the changes that occur in a bifurcation.

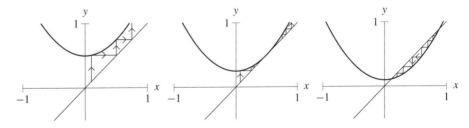

Figure 6.30
Graphical analysis of $F_c(x) = x^2 + c$ for $c = 0.5, 0.25$, and 0.1.

In Figure 6.31 we sketch the bifurcation diagram corresponding to the bifurcation above. In this picture we plot the fixed points $p_\pm$ as functions of c. The repelling fixed points are indicated by gray lines and the attracting fixed points are black. As c decreases through $1/4$ we see the bifurcation: A neutral fixed point is born, which then immediately splits apart into a pair of fixed points, one attracting and one repelling. Any bifurcation of fixed points or cycles that proceeds in this manner is called a **tangent** or **saddle node** bifurcation.

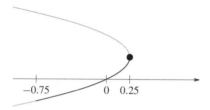

Figure 6.31
The tangent bifurcation diagram.

Pitchfork Bifurcation

The fixed-point structure of a discrete dynamical system may change in other ways as well. For example, consider the family of functions $G_\alpha(x) = x^3 - \alpha x$ depending on the parameter α. The fixed points of G_α are the solutions of $x^3 - \alpha x = x$, which are $x = 0, \pm\sqrt{1+\alpha}$. For $\alpha \le -1$, G_α has only one fixed point, at $x = 0$, but when $\alpha > -1$, there are three fixed points. Hence a bifurcation occurs at $\alpha = -1$. Note that $G'_\alpha(0) = -\alpha$, so the fixed point at 0 is attracting if $|\alpha| < 1$ and repelling if $|\alpha| > 1$. Also, $G'_\alpha(\pm\sqrt{1+\alpha}) = 3 + 2\alpha$. Since $\alpha > -1$, it follows that both of these fixed points are repelling. Thus the bifurcation is as shown in Figure 6.32. Because of the shape of this diagram, this bifurcation is called a **pitchfork bifurcation**.

Note that two things happen at the bifurcation point. As α increases through -1, the fixed point at $x = 0$ suddenly changes from repelling to attracting. Meanwhile, a pair of new repelling fixed points is born. It is also possible for a fixed point to change from attracting to repelling and simultaneously give birth to a pair of new attracting fixed points. This type of bifurcation is also called a pitchfork bifurcation.

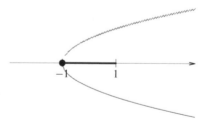

Figure 6.32
The pitchfork bifurcation diagram.

Period-Doubling Bifurcation

One of the most important types of bifurcations for discrete dynamical system is the **period doubling bifurcation**. This bifurcation produces a new cycle having twice the period of the original cycle. For example, consider again the family $F_c(x) = x^2 + c$. Recall that this function has two fixed points given by

$$p_\pm = \frac{1 \pm \sqrt{1 - 4c}}{2}.$$

The fixed point p_- is attracting if $-3/4 < c < 1/4$ and repelling if $c < -3/4$. The fact that $F_c'(p_-) = -1$ at $c = -3/4$ is our signal that a bifurcation may occur. Graphical analysis gives us a hint about what happens. In Figure 6.33, we display the web diagrams for F_c for the nearby c-values $c = -0.6$ and $c = -0.9$. Note that orbits tend to the attracting fixed point when $c = -0.6$, but they tend to an attracting cycle of period 2 when $c = -0.9$.

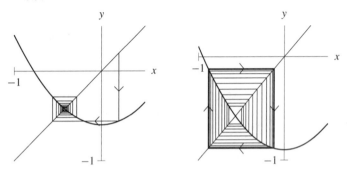

Figure 6.33
Graphical analysis of $F_c(x) = x^2 + c$ when $c = -0.6$ and $c = -0.9$.

To find cycles of period 2 for F_c, we must solve the equation $F_c^2(x) = x$. Writing this equation out, we find

$$(x^2 + c)^2 + c = x,$$

or in simplified form,

$$x^4 + 2cx^2 - x + c^2 + c = 0.$$

This equation looks tough to solve. However, recall that any fixed point for F_c must also solve this equation, for if $F_c(x_0) = x_0$, then certainly $F_c^2(x_0) = x_0$ as well. But we know the fixed points already; they are $p_\pm$. Hence $(x - p_+)(x - p_-) = x^2 - x + c$ is a factor of $F^2(x) - x$.

After some long division, we find

$$\frac{x^4 + 2cx^2 - x + c^2 + c}{x^2 - x + c} = x^2 + x + c + 1.$$

Therefore to find the cycles of period 2, we need only solve the quadratic equation

$$x^2 + x + c + 1 = 0.$$

The roots of this equation are

$$q_\pm = \frac{-1 \pm \sqrt{-3 - 4c}}{2},$$

as given by the quadratic formula.

Note that $q_\pm$ are real numbers only if $c < -3/4$. Thus a cycle of period 2 appears precisely when c decreases through $c = -3/4$. We leave it to the reader to check that this cycle is attracting for c in the interval $-5/4 < c < -3/4$ and repelling for $c < -5/4$ (see the exercises) Thus we may augment the bifurcation diagram in Figure 6.32 as in Figure 6.34.

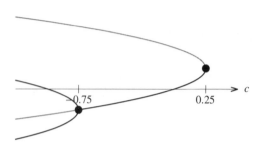

Figure 6.34
The bifurcation diagram for $F_c(x) = x^2 + c$ showing fixed points and 2-cycles

Any bifurcation in which a given cycle changes from attracting to repelling (or vice versa) with derivative passing through -1 and that is accompanied by the birth of a new cycle having twice the original period is called a **period-doubling bifurcation**.

A qualitative view of this period-doubling bifurcation is provided by the graphs of the second iterate of F_c, namely $F_c^2(x) = x^4 + 2cx^2 + c^2 + c$. In Figure 6.35 we sketch these graphs for $c = -0.6$, $c = -0.75$ and, $c = -0.9$. Note how the graph of F_c^2 twists through the diagonal at $c = -0.75$ to produce a pair of new fixed points for F_c^2 when $c < -0.75$.

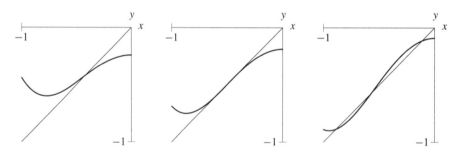

Figure 6.35
The graphs of $F_c^2(x)$ for $c = -0.6, c = -0.75$, and $c = -0.9$.

The Logistic Equation

The types of bifurcations discussed in this section are just the beginning of a long story of how discrete dynamical systems evolve as parameters are varied. To illustrate this in a concrete setting, we return to the logistic family of functions given by $L_k(x) = kx(1 - x)$. Recall that x represents the fraction of the maximum possible population of a species, so we only consider x values in the interval $0 \le x \le 1$.

We will use a combination of qualitative, analytic, and numerical methods to analyze this family of functions. However, the fate of orbits for many parameter values in this system is still an unsolved problem, so we are not be able to describe completely what happens for every parameter value and seed.

The first step in analyzing this family of functions is to find the fixed points. As we have seen, L_k has fixed points at 0 and and at the point $p_k = (k - 1)/k$. Note that p_k lies in the interval $0 \le x \le 1$ only if $k \ge 1$; otherwise, p_k is negative and so is not of interest in terms of a population model.

Differentiating L_k, we find $L_k'(x) = k(1 - 2x)$. Hence $L_k'(0) = k$. We conclude that 0 is an attracting fixed point for $0 \le k < 1$ and a repelling fixed point if $k > 1$. Graphical iteration gives us much more information. We see from Figure 6.36 that, as long as $k < 1$, the graph of L_k meets the diagonal at the fixed point 0 and lies below the diagonal for all other x values. This means that all orbits tend to the fixed point. In terms of our population, we conclude that if $k < 1$, the population eventually dies out.

At the other fixed point, we compute

$$L_k'(p_k) = k\left(1 - 2\left(\frac{k-1}{k}\right)\right) = -k + 2.$$

Therefore p_k is attracting for $-1 < -k + 2 < 1$ or $1 < k < 3$. Again using graphical iteration, we see from Figure 6.36 that all seeds in the interval $0 < x < 1$ have orbits that tend to this attracting fixed point. In terms of populations, we conclude that when $1 < k < 3$, the population of our species eventually levels off at the constant value $(k - 1)/k$.

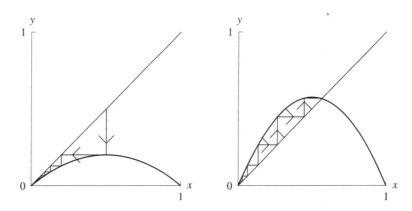

Figure 6.36
Populations die out when $k = 0.8$ but level off when $k = 2.3$.

If $k = 3$, we have $L_k'(p_k) = -1$, and if $k > 3$, we find $L_k'(p_k) < -1$. Thus p_k changes from an attracting to a repelling fixed point as k increases through 3. Thus we expect a period doubling bifurcation at this point. In Exercise 10 we ask you to verify this fact analytically. As usual, we can use graphical iteration to see this qualitatively.

In Figure 6.37 we display orbits of L_k when $k = 2.8$ and $k = 3.2$. Note that these orbits tend to the attracting fixed point when $k = 2.8$, but to a 2-cycle when $k = 3.2$. In terms of populations, when k is slightly larger than 3, we expect that the populations will eventually cycle in a biennial cycle. One year the population will be high; the next year it will be low.

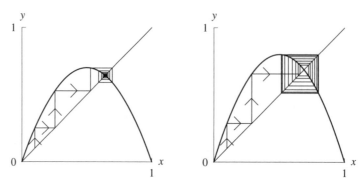

Figure 6.37
The period-doubling bifurcation in the logistic family when $k = 2.8$ and $k = 3.2$.

The period-doubling route to chaos

It is impossible to use analytic methods to describe much of what occurs for the logistic family as k continues to increase. So we resort to qualitative and numerical methods. In Figure 6.38 we provide a "movie" of the graph of L_k for 6 k values in the range $1 \leq k \leq 4$. As k increases, note how the slope of the tangent line to the graph of L_k at p_k decreases, eventually passing through the period-doubling point at which the slope is -1. Dynamically we see p_k change from attracting to repelling at this bifurcation point.

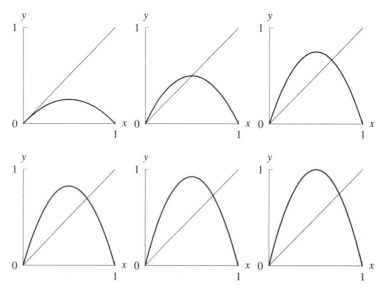

Figure 6.38
Graphs of $L_k(x)$ for $k = 1, 2, 3, 3.3, 3.7$, and 4

In Figure 6.39 we provide another movie, this time of the graphs of L_k^2 for 6 other k values. We have superimposed a small box on each of these graphs. These are not web diagrams; rather, we wish to draw your attention to the piece of the graph of L_k^2 inside this box. Note that these graphs look exactly like the graphs of L_k in Figure 6.38, only smaller and upside down. As we watch k change, we see these small pieces of the graph of L_k^2 behave just as the graphs of L_k do. Hence we expect L_k^2 to undergo the exact same bifurcations that L_k did in the previous pictures.

To make this precise, note that there is a fixed point for L_k^2 at the right-hand edge of the small boxes. This is actually a fixed point for L_k, not a periodic point of period 2. As k increases in these pictures, we first see the birth of a new fixed point for L_k^2 (this point lies on a cycle of period 2 for L_k). Call this point q_k. The point q_k is at first attracting, then repelling, as the derivative of L_k^2 at q_k passes through -1. Thus we expect that L_k^2 will undergo a period-doubling bifurcation as k increases. That is, for some intermediate k value, the period 2 cycle containing q_k will cease to be attracting and become repelling. Meanwhile a new attracting cycle of twice the period, a 4-cycle, will be born.

Another way to view this would be to draw the graphs of L_k^4 over the intervals of x values contained in each of the small boxes in Figure 6.39. If we do this, we see that the graphs of L_k^4 resemble the corresponding graphs of L_k^2 in Figure 6.39, only much smaller and turned upside down. So if we magnify these pictures and turn them around, we see that the pieces of the graphs of L_k^2 and L_k^4 look exactly like those of L_k and L_k^2 in Figures 6.38 and 6.39 respectively.

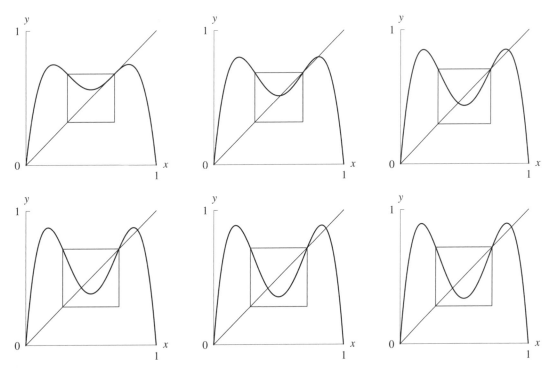

Figure 6.39
Graphs of $L_k^2(x)$ for $k = 3, 3.2, 3.4, 3.5, 3.55,$ and 3.58.

This process of iterating and zooming in to small portions of a graph is called **renormalization**. Borrowed from physics, this process allows us to recognize certain patterns that occur over and over. For example, we could now renormalize the graph of L_k^4. We would see that, over a smaller interval of k values, the 4-cycle discussed above would itself undergo a period-doubling bifurcation. That is, the attracting 4-cycle would become repelling and a new attracting 8-cycle would be born. And this scenario continues, with the successive births of attracting cycles of periods 16, 32, 64, and so forth.

Bifurcation diagram for the logistic model

Thus we see that the dynamical behavior of the discrete logistic model is significantly different from that of the logistic differential equation. We can find parameter values where a typical seed has an orbit that tends to a cycle of period 2^n for any n. However, there is much more to the story. In Figure 6.40 we plot a version of the bifurcation diagram discussed earlier. As usual, we plot the parameter values $0 \le k \le 4$ on the horizontal axis. Above each k value, we plot the **asymptotic orbit** of a particular seed, namely $x_0 = 1/2$. By this we mean given a k value, we compute 200 points on the orbit of the seed $1/2$ for L_k but display only the last 100 such points. If the orbit of $1/2$ tends to an attracting cycle, then that is what we see. The **transient behavior** of the orbit has been eliminated. Note that we do not see any of the repelling periodic points that we discovered above in this diagram.

Note that we see many **windows** in this picture. In Figure 6.41 we magnify one of these windows, specifically for the parameter values $3.82 \le k \le 3.865$. This is the **period 3 window**, for inside we see an attracting cycle of period 3, which then undergoes a similar pattern of period doubling as k increases.

We also see other regions in the bifurcation diagram where the orbit of $x = 1/2$ does not seem to settle down on an attracting cycle of some period. These include parameter values for which the logistic function behaves chaotically. In terms of the model, the population of the species changes dramatically and without pattern for these k values.

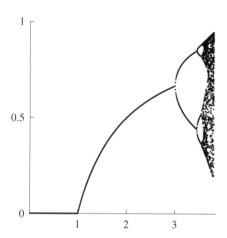

Figure 6.40
The bifurcation diagram for the logistic family

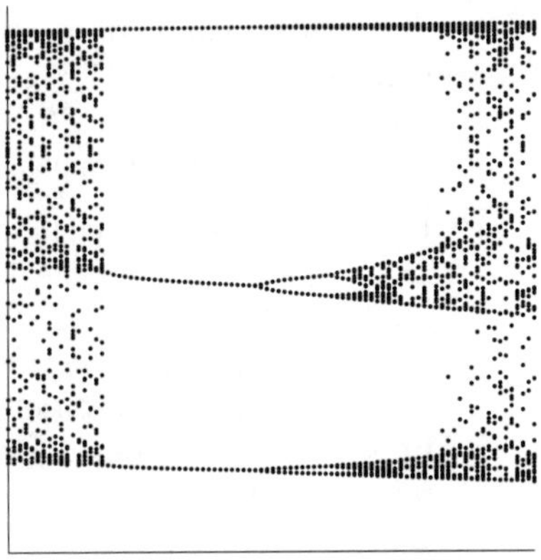

Figure 6.41
A magnification of the bifurcation
diagram for the logistic family.

Exercises for Section 6.3

In Exercises 1–7, each of the given functions undergoes a bifurcation of fixed points at the given parameter value. In each case, use analytic or qualitative methods to identify this bifurcation as a tangent, pitchfork, or period-doubling bifurcation or as none of these. Discuss the behavior of orbits near the fixed point in question at, before, and after the bifurcation.

1. $F_\alpha(x) = x + x^2 + \alpha, \quad \alpha = 0$

2. $F_\alpha(x) = \alpha x, \quad \alpha = -1$

3. $F_\alpha(x) = \alpha \sin x, \quad \alpha = 1$

4. $F_\alpha(x) = \alpha \sin x, \quad \alpha = -1$

5. $F_\alpha(x) = \alpha - x^2, \quad \alpha = -1/4$

6. $F_\alpha(x) = \alpha \arctan x, \quad \alpha = 1$

7. $F_\alpha(x) = \alpha(x + x^2), \quad \alpha = 0$

8. Consider the family of functions $F_\beta(x) = x^3 + \beta$. Find all β-values for which this family undergoes a bifurcation of fixed points. Determine the type of bifurcation that occurs at each bifurcation point. Then sketch the bifurcation diagram for this family.

9. Consider the family of functions given by

$$T_\mu(x) = \begin{cases} \mu x, & \text{if } x < \frac{1}{2} \\ \mu(1 - x), & \text{if } x \geq \frac{1}{2}. \end{cases}$$

Discuss in detail the bifurcation of fixed points that occurs at $\mu = 1$ for this family. Sketch the fixed-point bifurcation diagram.

10. For the logistic family $F_k(x) = kx(1 - x)$, show explicitly that there exists a cycle of period 2 for $k > 3$.

11. Consider the family of functions $F_c(x) = x^2 + c$. Sketch the graphs of F_c for $c = 0.25, 0, -0.75, -1, -1.5$, and -2 for $-2 \leq x \leq 2$. Then sketch the graphs of F_c^2 for $c = -0.75, -1, -1.25, -1.3, -1.35$, and -1.4. Is there any similarity between these two families of graphs? Discuss in a brief essay complete with pictures.

12. Use a computer to sketch the bifurcation diagram for the family of functions $F_c(x) = x^2 + c$. To do this, compute the fate of the orbit of 0 for 100 (or more) equally spaced c values in the interval $-2 \leq c \leq 0.25$. That is, compute 200 points on the orbit of 0 for each c, but display only the last 150 points on the orbit. You should plot c values horizontally and the orbit of 0 vertically, as in this section. Is there any similarity between the bifurcation diagram for F_c and that of the logistic family? Discuss in a brief essay with pictures.

13. What is the fate of the orbit of 0 under iteration of $F_c(x) = x^2 + c$ for $c < -2$?

14. Use a computer to sketch the bifurcation diagram for the family of functions $S_\alpha(x) = \alpha \sin x$. To do this, compute the fate of the orbit of the seed $x_0 = \pi/2$ for 100 (or more) equally spaced α values in the interval $0 \leq \alpha \leq \pi$. That is, compute 200 points on the orbit of $\pi/2$ for each α, but display only the last 150 points on the orbit. You should plot α values horizontally and the orbit of $\pi/2$ vertically as in this section. Is there any similarity between the bifurcation diagram for S_α and that of the logistic family? Discuss in a brief essay with pictures.

15. For the logistic family $L_k(x) = kx(1 - x)$, we have seen in this section that in the bifurcation diagram a period 3 window abruptly opens as k increases. This occurs when $k \approx 3.828$. Using the graphs of L_k and L_k^3 for nearby k values, discuss the bifurcation that leads to the sudden appearance of this window.

6.4 CHAOS

In this section we begin the study of chaotic behavior. Until a few years ago most scientists and mathematicians thought that typical differential and difference equations did not exhibit the kind of behavior we now call chaos – that this kind of behavior was extremely unlikely in simple mathematical models. Now we know that this is not the case. There are many models whose behavior is quite erratic, and this behavior persists when parameters are varied. Moreover, mathematicians have begun to develop techniques to predict the onset of chaos and to determine the scope or limits of this behavior. In this chapter we use a number of qualitative and numerical techniques to explain and illustrate the concept of chaos.

An Example without Chaos

Roughly speaking, chaos means unpredictability. Thus far most of the differential and difference equations that we have encountered have been completely predictable. For example, consider the simple discrete logistic model

$$P_{n+1} = 2P_n(1 - P_n).$$

As we have seen, this system has two fixed points, at 0 and at $1/2$. Graphical analysis shows that the orbit of any seed x_0 in the interval $0 < x_0 < 1$ tends to $1/2$ (see Figure 6.42). The only other point, $x_0 = 1$, has an orbit that is eventually fixed. So we know the fate of all orbits.

Now in real life we rarely know the seed for our orbit with complete accuracy. For example, we most likely do not know the precise value of the population at time 0; we probably miss a few individuals in our initial count. If our model is the simple logistic model above, however, then this inaccuracy does not really matter. A small fluctuation in our initial seed in the interval $0 < x_0 < 1$ does not alter our predictions at all.

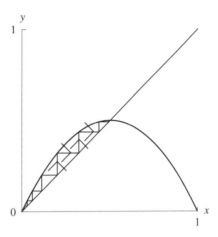

Figure 6.42
Graphical analysis of $F(x) = 2x(1 - x)$.

A Chaotic Logistic Function

Now let's contrast the orbits of the logistic model above with those of

$$P_{n+1} = 4P_n(1 - P_n),$$

where we have simply changed the growth parameter from 2 to 4. As always, studying this system is the same as iterating the function $L_4(x) = 4x(1 - x)$. This function has fixed points at 0 and $3/4$ and eventually fixed points at $1/2$ and 1. But consider the fate of virtually any other orbit. In Figure 6.43 we compute the web diagram for the first 200 points on the orbit of 0.123 for this function. Note that it is hard to distinguish any pattern in this picture. A histogram and a time series for this orbit further confirm this erratic behavior (see Figures 6.44 and 6.45). In the histogram we see that the orbit seems to visit every small subinterval in $0 \le x \le 1$ many times.

If you repeat this experiment for a variety of seeds in the interval $0 < x < 1$, you will see essentially the same results in each case. Of course, the fixed points and eventually fixed points yield quite different diagrams. However, virtually any other seed yields figures that resemble those in Figures 6.44 and 6.45 in their complexity.

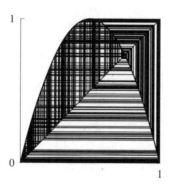

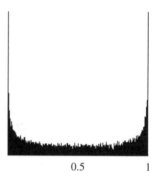

Figure 6.43
Web diagram of the orbit of
0.123 under $L_4(x) = 4x(1 - x)$.

Figure 6.44
Histogram of the same orbit.

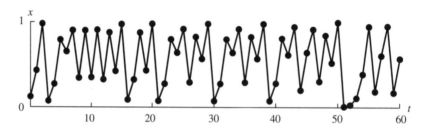

Figure 6.45
Time series for the orbit of 0.123 under $L_4(x) = 4x(1 - x)$.

These computer experiments suggest that most orbits of $L_4(x) = 4x(1-x)$ behave erratically. However, there are many cycles for L_4 as well. In Figure 6.46 we plot the graphs of L_4^2, L_4^3, and L_4^4. Note that the graph of L_4^n appears to cross the diagonal $y = x$ at 2^n points, yielding infinitely many cycles in the interval $0 \leq x \leq 1$ of arbitrarily high period.

Another important element in our discussion of chaotic behavior is **sensitive dependence on initial conditions**. This is illustrated in Figure 6.47, where we plot the time series for the orbits of 0.123 and 0.124. Note that these orbits initially are very

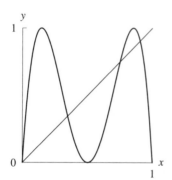

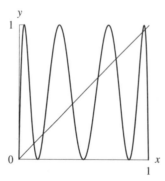

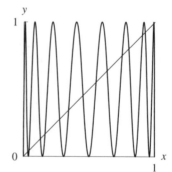

Figure 6.46
Graphs of L_4^2, L_4^3, and L_4^4

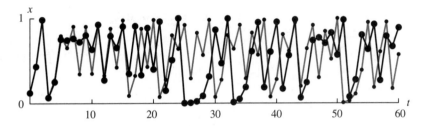

Figure 6.47
Time series for the orbit of 0.123 in grey; for 0.124 in black.

close, but after very few iterations, the orbits separate and thereafter bear very little resemblance to one another. Just a small change in the third decimal place of our seed has resulted in a large change in the ensuing orbit.

A Model Chaotic System

To characterize the chaotic behavior we observed for the logistic function $L_4(x) = 4x(1 - x)$, we turn to a discrete dynamical system that is much easier to understand. Consider the function T defined for $0 \leq x < 1$ by

$$T(x) = 10x \mod 1.$$

This notation means that to compute $T(x)$ we first multiply x by 10 and then drop the integer part, keeping only the fractional part of the result. For example

$$T(1/3) = 10/3 - 3 = 1/3,$$

$$T(0.232323\ldots) = 2.32323\ldots - 2 = 0.32323\ldots,$$

and

$$T(0.1111\ldots) = 1.111\ldots - 1 = 0.111\ldots.$$

More generally if $0 \leq x < 1$, we can write the decimal expansion of x in the form

$$x = .a_1 a_2 a_3 \ldots,$$

where the a_i are digits between 0 and 9. Then

$$10x = a_1.a_2 a_3 a_4 \ldots$$

so that

$$T(x) = 10x - a_1 = 0.a_2 a_3 a_4 \ldots$$

That is, T simply drops or chops off the first digit in the decimal expansion of x. For this reason, T is called the **chopping function**. For example,

$$T(0.\overline{123}) = 0.\overline{231}$$

and

$$T^2(.25) = T(.5) = 0.$$

The notation $0.\overline{123}$ denotes the repeating sequence $0.123123123\ldots$.

Density of periodic points

Because of the special definition of the chopping function, periodic points of T are easy to find. Any x whose decimal expansion is repeating lies on a cycle for T. For example, 2/9 is a fixed point, since

$$\frac{2}{9} = 0.222\ldots$$

and $T(0.222\ldots) = 0.222\ldots$. Similarly, $0.\overline{37}$ lies on a 2-cycle and $0.\overline{1234}$ lies on a 4-cycle. Clearly T has infinitely many periodic points in $[0, 1)$, and the periods of these points may be arbitrarily large. Moreover, unlike the logistic functions, we can write down all of the periodic points for T explicitly.

There is more to this story however. Given any x in the interval $[0, 1)$, we can find a periodic point of T arbitrarily close to x. This follows since, if x has the decimal expansion

$$x = 0.a_1 a_2 a_3 \ldots,$$

then the point

$$y = 0.\overline{a_1 a_2 \ldots a_n}$$

is close to x (within $1/10^n$ units) and moreover is periodic with period $n + 1$.

For example, suppose we are given the number

$$x = \frac{1}{\sqrt{2}} = 0.707106781\ldots$$

and are asked to find a periodic point within 0.001 units of x. We could choose $x_1 = 0.\overline{707}$, since

$$|0.707106781\ldots - 0.\overline{707}| = 0.00060092\ldots.$$

and x_1 has period 3. But there are many other choices that work as well, including $x_2 = 0.\overline{707106}$, which is much closer to x than x_1 and has period 6. Alternatively, we could choose $0.\overline{7071999222555}$, which is also within 0.001 units of x and is periodic with period 13.

Extrapolating from this example we see that no matter how small a subinterval we choose in $[0, 1)$, we can always find periodic points inside this subinterval. In fact we can always find infinitely many distinct periodic points inside the subinterval. A subset of $[0, 1)$, which contains points in every subinterval of $[0, 1)$, no matter how small, is said to be a **dense subset** of the interval. For example, the set of all rational numbers between 0 and 1 is a dense subset of $0 \leq x < 1$. So too are the irrationals in this interval. The above argument shows that the set of periodic points for T is also a dense subset of $[0, 1)$.

Periodic orbits are by no means the only types of orbits for the chopping function. There are many eventually periodic points as well. Any point whose decimal expansion eventually begins to repeat is eventually periodic under T. For example,

$$x = 0.123\overline{4}\ldots$$

is a point whose orbit becomes fixed after 3 iterations of T. Similarly, the orbit of

$0.123\overline{4567}$ is eventually periodic with period 3. Now recall that any rational number has decimal expansion that either repeats, eventually repeats, or terminates (which means it ends with a sequence of all zeroes). As a consequence, we see that any rational number in $[0, 1)$ has orbit under T which is either periodic or eventually periodic. On the other hand, any irrational number in $[0, 1)$ has an orbit that never cycles, since the decimal expansions of irrational numbers never repeats. For example, the orbit of $1/\sqrt{2}$ is not periodic under T since $\sqrt{2}$ is an irrational number.

A dense orbit

Some of these irrational numbers have very interesting orbits. For example, consider the number x_0 whose decimal expansion begins

$$x_0 = .0123456789.$$

Suppose the next 200 terms of this expansion consist of all possible 2-blocks of digits, that is,

$$x_0 = 0. \underbrace{0123\ldots9}_{\text{all 1-blocks}}\ \underbrace{00\ 01\ 02\ldots10\ 11\ 12\ldots20\ 21\ldots99}_{\text{all 2-blocks}}$$

followed by all possible 3-blocks, 4-blocks, and so on, so we have

$$x_0 = 0.\ \underbrace{0\ldots9}_{\text{all 1-blocks}}\ \underbrace{00\ldots99}_{\text{all 2-blocks}}\ \underbrace{000\ldots999}_{\text{all 3-blocks}}\ldots$$

What can we say about the orbit of x_0? Recall that each iteration of T simply drops the leading digit from this sequence. Thus given any x in $[0, 1)$, we see that there is a point on the orbit of x_0 that comes arbitrarily close to x. Indeed, we need only iterate T enough times so that the first n digits of the decimal expansion of x appear as the leading terms of this point on the orbit of x_0. Then these two points are within $1/10^n$ units of each other. For example, suppose we are given the number $1/3$ and are asked to find a point on the orbit of x_0 that lies within 0.001 of $1/3$. To do this, we simply iterate T until the 3-block that begins with 333 appears as the first entry of the sequence. Now it takes 10 iterations to remove the entries $0.012\ldots9$, and then another 200 iterations to remove $00\ 01\ 02\ldots$
98 99, and then finally $3 \cdot 333$ more iterations to bring 333 to the head of the list. Thus, $T^{1209}(x_0)$ lies within the required distance of $1/3$.

Of course, there are many other points on the orbit of x_0 that lie within 0.001 units of $1/3$. All we need do is iterate T enough times until any k-block of the form $33\ldots3$ is first in line and this point is close enough to $1/3$. As a consequence, the set of points that comprises the orbit of x_0 is also a dense subset of $[0, 1)$.

Sensitive dependence on initial conditions

The final observation we wish to make about T is that this dynamical system exhibits **sensitive dependence on initial conditions**. By this we mean the following. Consider any x_0 in $[0, 1)$. Then there are points y_0 arbitrarily close to x_0 so that the orbit of y_0 is eventually "far" from that of x. For example, if we consider the fixed point $x_0 = 1/9 = 0.\overline{1}$ of T, then there are nearby points whose orbits behave very differently. For example, the seed $y_0 = 0.11\ldots 1\overline{8}$ eventually lands on the fixed point $0.\overline{8} = 8/9$. How close y_0 is to $1/9$ depends upon how many initial 1s are present in the sequence. Although these two seeds are close together initially, their orbits eventually are far apart in the interval $[0, 1)$ (see Figure 6.48). As another example, we can consider a seed of the form $0.11\ldots 1\overline{35}$, which is eventually periodic with period 2, or $0.11\ldots 1\overline{23456}$, which eventually has period 5. Both of these seeds are also close to x_0, but their orbits behave quite differently. Indeed, arbitrarily close to the given seed $x_0 = 1/9$, we can find another seed whose orbit eventually has any behavior whatsoever. All we need to do is precede a given orbit by a sufficiently long string of 1s. Again, this orbit is initially close to $1/9$, but it eventually diverges from this fixed point.

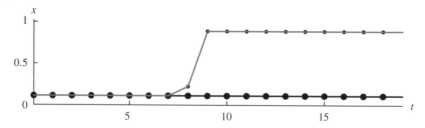

Figure 6.48
Time series for the orbit of $0.\overline{1}$ (in blue) and for $0.11\ldots 1\overline{8}$ (in black).

The general situation is similar. Given any seed x_0 whose decimal expansion is

$$x_0 = 0.a_1 a_2 a_3 \ldots,$$

we consider the nearby seed

$$y_0 = 0.a_1 a_2 \ldots a_n b_1 b_2 b_3 \ldots,$$

where $b_1 b_2 b_3 \ldots$ are arbitrary digits. Then the seed y_0 is close to x_0 (how close depends on n, the number of decimal places to which x_0 and y_0 agree), but the fate of the two orbits can be radically different. Thus T exhibits sensitive dependence throughout the interval $[0, 1)$.

The existence of sensitive dependence in dynamical systems has profound implications for scientists and mathematicians who use difference or differential equations as mathematical models. If a given system exhibits sensitive dependence on initial conditions, then numerical predictions about the fate of orbits or solutions are to be totally distrusted.

For we can never know the exact seed or initial condition for our orbit or solution because we cannot make physical measurements with infinite precision. Even if we had exact measurements, we could never carry out the necessary computations. The small numerical errors that are always introduced in such numerical procedures throw us off our original orbit and onto another whose ultimate behavior may be radically different.

Unlike T, for which we have sensitive dependence at all points in its domain, many functions exhibit sensitivity only at isolated points. Consider for example the logistic function $L_{3.2}(x) = 3.2x(1 - x)$. As shown in Figure 6.49, most orbits of this function tend to an attracting cycle of period 2. However, not all orbits share this fate. As we see in Figure 6.49, there are 2 repelling fixed points for $L_{3.2}$, and there are infinitely many points whose orbits eventually land on the nonzero fixed point for $L_{3.2}$. The function $L_{3.2}$ exhibits sensitive dependence at each of these seeds, since we can change the seed only slightly and find an orbit that is neither fixed nor eventually fixed but rather tends to the attracting cycle of period 2. This behavior is much different from what we see for T; it is commonly called *unstable behavior*. Chaos occurs when we have unstable behavior everywhere, or at least on a large set.

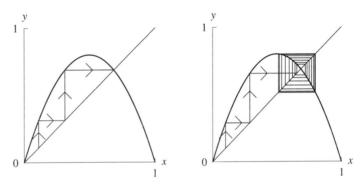

Figure 6.49
Two nearby orbits for $F(x) = 3.2x(1 - x)$; one is eventually fixed, the other tends to the 2-cycle.

Summary

We have shown that our model system T possesses an amazing amount of complexity in the interval $[0, 1)$. The periodic points of T form a dense subset of this interval, and there is also a single orbit that forms another dense subset of $[0, 1)$. Moreover, this function exhibits sensitive dependence on initial conditions, which renders precise calculations of orbits impossible. The combination of these three properties makes this system quite unpredictable, or **chaotic**.

It may seem to be the case that this type of complexity is special to our model function T because T has constant slope 10. However, over the past few years mathematicians have verified that a number of dynamical systems — both discrete systems and systems of differential equations — exhibit this type of behavior. For example, there is a large set of k values for which the logistic function $L_k(x) = kx(1 - x)$ is chaotic. The histograms and time series we displayed earlier give qualitative evidence for this behavior.

Exercises for Section 6.4

1. For the logistic function $L_4(x) = 4x(1 - x)$ compute a histogram of the first 10,000 points on the orbit of $x_0 = 0.3$ by dividing the interval $[0, 1]$ into 100 subintervals, each of length 0.01, and then calculating the successive subintervals into which the orbit of x_0 falls. Then repeat this calculation using the seed $y_0 = 0.3001$. Do you see any difference? Now compute the time series for the first 100 points on the orbits of x_0 and y_0 and compare the results. Do you see any differences? Discuss your findings in a brief essay. **Do not attempt this exercise without access to computing and graphing technology.**

Exercises 2–11 deal with the doubling function given by

$$
T(x) = \begin{cases} 2x, & \text{if } 0 \le x < \frac{1}{2} \\[2mm] 2x - 1, & \text{if } \frac{1}{2} \le x < 1. \end{cases}
$$

2. Compute the orbit of the following seeds under T. Which seeds are periodic and which are eventually periodic?

   (a) $x_0 = 1/5$
   (b) $x_0 = 2/7$
   (c) $x_0 = 3/11$
   (d) $x_0 = 1/10$
   (e) $x_0 = 1/6$
   (f) $x_0 = 4/14$
   (g) $x_0 = 4/15$

3. What can you say about the orbit of the seed $x_0 = p/2^n$ under iteration of T?

4. Sketch the graphs of T^2, T^3, and T^4. What do you expect the graph of T^n to look like? How many fixed points should T^n have?

5. Using the graphs drawn in Exercise 4, discuss the question of density of the subset of periodic points of T in the interval $0 \le x < 1$.

6. Suppose x_0 has binary representation $.a_1 a_2 a_3 \ldots$. That is,

$$
x_0 = \frac{a_1}{2} + \frac{a_2}{2^2} + \frac{a_3}{2^3} + \cdots.
$$

   What is the binary expansion of $T(x_0)$? Of $T^n(x_0)$?

7. Using the results of Exercise 6, find all points in the interval $[0, 1)$ that are eventually fixed at 0 by T^n.

8. Using the results of Exercise 6, find all points in the interval $[0, 1)$ that are periodic with period 2, 3, or 4 under T.

9. How many points in $[0, 1)$ are fixed by T^n for each n?

10. Does T exhibit sensitive dependence on initial conditions?

11. Write down an expression for a seed x_0 in $[0, 1)$ whose orbit under T forms a dense subset of this interval.

Exercises 12–14 deal with the doubling function given by

$$T(x) = \begin{cases} 2x, & \text{if } 0 \le x < \frac{1}{2} \\[2mm] 2 - 2x, & \text{if } \frac{1}{2} \le x \le 1. \end{cases}$$

12. Sketch the graphs of T, T^2, and T^3 over the interval $[0, 1]$. How many fixed points does T^n have for each n?

13. Discuss the question of density of the set of periodic points of T.

14. Does T exhibit sensitive dependence on initial conditions?

6.5 CHAOS IN THE LORENZ SYSTEM

In this section we complete the discussion of the Lorenz system of differential equations started in Section 2.7 and continued in Section 4.4. Recall that this system is a nonlinear system of three differential equations given by

$$\frac{dx}{dt} = 10(y - x)$$
$$\frac{dy}{dt} = 28x - y - xz$$
$$\frac{dz}{dt} = -\frac{8}{3}z + xy.$$

We will first show how to reduce this problem to a two-dimensional model, and then we will further simplify the system to a one-dimensional iteration. Using techniques that we developed in the last section, we will then show how this simplified Lorenz system exhibits chaotic behavior.

The Lorenz Attractor

As we saw in Chapter 2, virtually every solution curve of the Lorenz system eventually has similar behavior: After initially meandering around three-dimensional space, the solution curve eventually winds in complicated fashion around a certain region in three dimensional space. Our goal is to create a model of this behavior that will enable us to see more clearly how the solutions behave.

For example, in Figure 6.50 we plot the solution curve starting at the initial condition $(0, 1, 0)$. After a short period of time, the solution settles down on a region where it winds around two "holes" in seemingly random manner.

The object toward which solutions of the Lorenz equation tend is called a **strange attractor**. It is an "attractor" because all solutions tend to it. It is "strange" since it is very different from the simpler attractors we have encountered thus far: equilibrium points (sinks) and certain periodic orbits.

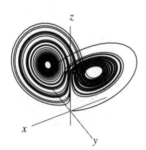

Figure 6.50
The solution of the Lorenz equation with initial condition $(0, 1, 0)$.

Projections

To understand this attractor, we will try to view it from many different vantage points. The easiest way to do this is to project the solution curve onto a coordinate plane. Geometrically this means that we simply forget one of the coordinates of the solution curve. For example, to project the solution to the xy-plane, we merely plot the curve $(x(t), y(t))$ in the plane, forgetting about the third coordinate of the solution, $z(t)$. In Figure 6.51, we sketch the xy-, yz-, and xz-projections of the solution tending to the Lorenz attractor. Note that in each of these projections the solution curve apparently crosses itself. By the Uniqueness Theorem, this does not happen in the full xyz-space. When the solution crosses itself in the projection, the missing coordinate of the solution assumes different values, so the real solution curve passes "above" or "below" itself in three dimensions.

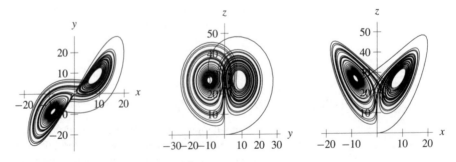

Figure 6.51
The xy, yz, and xz-projections of the solution in Figure 6.50

The advantage of projections is that they can easily be viewed on computer screens, whereas three-dimensional images require some geometric intuition and perspective when viewed on a screen. The trick is to use various projections to piece together the behavior of solutions.

The Lorenz Template

The projections of the Lorenz attractor indicate that solution curves tend toward a region in the three-dimensional space that resembles the surface drawn in Figure 6.52. This image is called the **Lorenz template**, although it is sometimes called the Lorenz mask for obvious reasons. Note that the two lobes of the template are joined along a straight line, with one lobe bending backward and the other forward of the line.

The attractor for the Lorenz system is actually a much more complicated object. It consists of infinitely many sheets packed tightly together in a complicated fashion. However, this template allows us to get a good idea of the mechanism that produces the chaotic behavior in the Lorenz system. So what follows is not an exact description of the solutions of the equation. Rather, it should be viewed as a simplification or model of the full system.

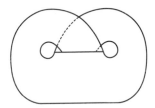

Figure 6.52
The Lorenz template.

To understand the full phase portrait of the Lorenz system, recall from Section 4.4 that there are three equilibrium points for the Lorenz system. There is a saddle at the origin with one positive and two negative eigenvalues. There is also a pair of equilibrium points that are spiraling saddles. Each of these points have two complex eigenvalues with positive real parts and a single negative eigenvalue. Figure 6.53 indicates the positions of these equilibrium points relative to the template. Note that at each of the equilibrium points there is a stable eigenvector pointing toward the template. This forces nearby solutions to tend toward the attractor.

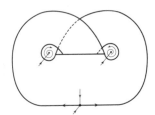

Figure 6.53
The equilibrium points for the
Lorenz system and the template.

The Poincare return function

Now we add a typical solution curve to the template. As we have seen, solutions wind about the attractor, sometimes circling about the left lobe, at other times about the right lobe. We think of this solution as actually lying on the template, wandering back and forth between lobes (see Figure 6.54.) The solution depicted in this figure is really fictitious. It is intended as a representation of the behavior of a real solution of the equations.

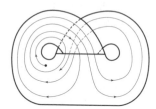

Figure 6.54
A fictitious solution lying on the
Lorenz template

Note that the typical solution on the template repeatedly crosses the line of intersection of the two lobes. We call this line L. To understand the qualitative behavior of solutions near the template, we will try to understand how this solution returns to L. Note that there is a point p plate on L through which the solution curve never returns to L. This is a solution that lies in both the template and the stable separatrix for the equilibrium point at $(0, 0, 0)$ (see Figure 6.55). Using the linearization near this equilibrium point we see that solutions on either side of this solution through p evolve in different directions about the lobes (see Figure 6.56). Thus, this "disappearing" solution through p forms the boundary between those curves on the template that circle around the left and those that circle around right lobes.

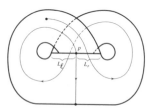

Figure 6.55
A disappearing solution on the
Lorenz template.

Let L_ℓ denote the portion of L to the left of p (in Figure 6.56), and L_r the portion to the right of p. We wish to understand how solution curves that begin in L_ℓ or L_r return to L. To that end we will construct a function that assigns to any point in L_ℓ or L_r its next point of intersection with L. This function is called the Poincaré function or the (first) return function on L (see Section 6.4). We denote it by ϕ. To understand how solution curves repeatedly intersect L, we need only iterate the function ϕ. It is impossible to give an analytic formula for ϕ. After all, the template itself is really fictitious. However, by observing the behavior of solutions on the $xz-$ or yz-projections we can give a good qualitative description of ϕ.

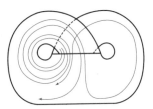

Figure 6.56
Several solutions on the Lorenz template.

In Figure 6.57 we juxtapose the yz-projection and a model for ϕ. Note that ϕ is undefined at p. As indicated in the projection, ϕ takes the interval L_ℓ to an interval that covers L_r and a portion of L_ℓ. Also, ϕ takes L_r to an interval that covers L_ℓ and a portion of L_r. Therefore we can construct an approximation of the graph of ϕ as shown in Figure 6.57.

Note that the images of both L_ℓ and L_r when they return to L are longer than L_ℓ or L_r. That is, ϕ expands these intervals. This is the reason that we have drawn the graph of ϕ with slope larger than 1.

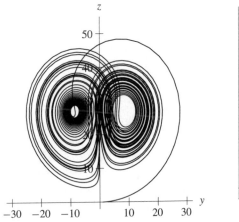
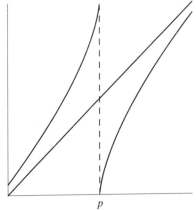

Figure 6.57
The yz-projection and a model for ϕ.

Iteration of the Poincare Return Function

To understand the evolution of solutions as they continually reintersect L, we must iterate ϕ. As in the previous section, we see that ϕ exhibits sensitive dependence on initial conditions, since $\phi' > 1$ on L. Graphical iteration shows that there are a number of periodic points for ϕ as well (see Figure 6.58).

We need to be careful when we interpret the meaning of the observed behavior of ϕ. For example, consider the periodic orbit and the corresponding cycle for ϕ displayed in Figure 6.53. Presumably there is a real orbit for the Lorenz system that exhibits this behavior. However, such an orbit is impossible to find using numerical methods such as Euler's method, mainly because of the expansion in the direction along L. Moreover, since the Lorenz attractor is really much thicker than the template, there will be other orbits in the attractor that tend to this periodic solution. For example, note how far away solutions may eventually match up with the given periodic solution in the template. These correspond to eventually periodic points for ϕ.

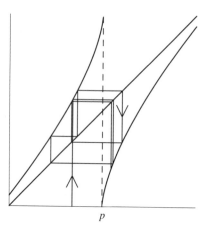

Figure 6.58
Graphical iteration of ϕ.

Summary

In this section we have described an important technique in the study of differential equations. When we observed complicated behavior of solutions in a higher dimensional setting, we tried to describe this behavior by reducing the dimension of the system and viewing it as a discrete dynamical system. Although it is not always possible to carry out a procedure, this type of process does work in a number of important systems of differential equations.

Lab 6.1   Newton's Method as a Difference Equation

The **roots** of a function $f(x)$ are the values of x for which $f(x) = 0$. Given a function f we can find approximate values for its roots using Newton's method. To use Newton's method, we first make an initial guess for the value of the root, say x_0. If this is not the root (that is, $f(x_0)$ is not 0), we can (hopefully) improve the guess by first computing the tangent line to the graph of f at x_0, then compute the point x_1, where this tangent line intersects the x-axis. The relationship between x_0 and x_1 (a good calculus review) is given by

$$x_1 = x_0 - \frac{f(x_0)}{f'(x_0)}.$$

If $f(x_1)$ is not sufficiently close to zero, then we improve the guess again by computing

$$x_2 = x_1 - \frac{f(x_1)}{f'(x_1)},$$

and so forth. What we are doing is creating the orbit of the seed x_0 under the Newton's method function

$$N(x) = x - \frac{f(x)}{f'(x)}.$$

The initial guess is the seed x_0, the first improvement is $x_1 = N(x_0)$, the second improvement is $x_2 = N(x_1)$, and so on. The hope is that the improved guesses really are improvements and that the sequence $x_0, x_1, \ldots$ approaches a root of the function f. In this lab we consider how the choice of seed affects the long-term behavior of the orbits of the Newton's method function by studying a particular example: Let

$$f(x) = x^3 + 3x^2 + 2x = x(x-1)(x-2).$$

This function has roots at $x = 0$, $x = 1$, and $x = 2$.

In your report, consider the following items:

1. For the function $f(x)$ chosen above, compute the corresponding Newton's method function $N(x) = x - f(x)/f'(x)$. Verify that this function has fixed points at the roots $x = 0$, $x = 1$ and $x = 2$. Verify that these are attracting fixed points. What does this imply for orbits of $N(x)$ for seeds x_0 near 0, 1, or 2?

2. Compute the first 20 points of the orbits under $N(x)$ for the initial seeds

$$x_0 = 0, 0.05, 0.10, 0.15, 0.20, \ldots, 0.90, 0.95, 1.$$

   Sketch the segment $0 \leq x_0 \leq 1$. For each choice of x_0, if the 20th point x_{20} in its orbit under N is within 0.01 of 0, color the point red; if x_{20} is within 0.01 of 1, color it blue; if x_{20} is within 0.01 of 2, color it green; and if none of these conditions holds, color it black. Is this figure consistent with your conclusions from part 1? What implications does this picture have concerning the choice of the seed x_0 for Newton's method?

3. Repeat part 2 for the points

$$x_0 = 0.30, 0.32, 0.34, 0.36, 0.38, \ldots, 0.46, 0.48, 0.50.$$

Your report: In your report, address each of the items above. Parts 2 and 3 require a computer or calculator. Do **not** hand in lists of numbers. What is important is the color-coded sketches of the x-axis and your interpretation of these sketches.

Lab 6.2

The Delayed Logistic and Iteration in Two Dimensions

In Section 6.1, the logistic equation

$$L_k(y) = ky(1 - y)$$

was used to model populations which reproduce on a discrete time scale. In particular, if y_n represents the population (or population density) of some species in the nth generation, then the population in the $(n + 1)^{st}$ generation is given by

$$y_{n+1} = L_k(y_n) = ky_n(1 - y_n).$$

We have seen that the behavior of orbits can depend in a dramatic way on the value of the value of the growth rate parameter k and on the seed y_0.

One hidden assumption in the discrete logistic model is that the population in the $(n + 1)^{st}$ generation depends only on the population in the nth. This may not always be the case for species for which there is a long maturation time or for species which migrate to nesting or breading areas. For example, suppose a species returns to the same breeding area each year. The amount of food available to the nth generation in that area might depend on how much was eaten in the previous year. In this case, the population of the $(n + 1)^{st}$ generation depends on the population in the nth and the $(n - 1)^{st}$ generations. This motivates the following model of population growth called the **delayed logistic** population model,

$$y_{n+1} = ky_n(1 - y_{n-1}).$$

As with the logistic model, y_n represents the population density in the nth generation. We restrict attention to the case where $0 \le y_n \le 1$. For the delayed logistic model, the population in the $(n + 1)^{st}$ generation is proportional to the population in the nth generation as well as to how close the population in the $(n - 1)^{st}$ generation was to the maximum possible population density.

In order to study the delayed logistic model, we perform an operation very similar to the conversion of second order differential equations into first order systems. We introduce a new variable x_n, letting

$$x_n = y_{n-1}.$$

That is, x_n is the population density in the $(n - 1)^{st}$ generation. We then rewrite the delayed logistic equation as

$$x_{n+1} = y_n$$

$$y_{n+1} = ky_n(1 - x_n).$$

As with the conversion of second order equations to first order systems, the system is simpler in one way because the values x_{n+1} and y_{n+1} depend only on the values of x_n and y_n. However, it is more complicated because we have had to introduce a new variable.

We can interpret this model as an iteration by letting

$$F_k(x, y) = (y, ky(1 - x)).$$

514

Then
$$(x_{n+1}, y_{n+1}) = F_k(x_n, y_n),$$

that is, orbits of an initial seed (x_0, y_0) are made up of a sequence of points in the plane. Hence
$$(x_1, y_1) = F_k(x_0, y_0) = (y_0, ky_0(1 - x_0)),$$
$$(x_2, y_2) = F_k(x_1, y_1) = (y_1, ky_1(1 - x_1)),$$

and so on. In this lab we study the behavior of orbits for this two dimensional interation. We can graph orbits in two different ways. We can form "time-series" graphs with n on the horizontal axis and x_n or y_n on the vertical axis. Alternately, we can make "phase plane" graphs, plotting the sequence of points (x_n, y_n) in the xy-plane. You should make both types of graphs for each orbit you compute.

In your report, consider the following items:

1. For several different values of k between 1.5 and 2.5 compute the first 200 iterates of the orbit with seed $(x_0, y_0) = (0.1, 0.2)$. For each of the values of k that you choose, describe the long-term behavior of the orbit. What are the biological implications of your conclusions? (For some values of k you may need more than 200 iterates.)

2. How do your results from part 1 change if you change the initial seed? What are the biological implications of your conclusions? (Remember to keep the initial seed in the "physically meaningful" range $0 \le x_0 \le 1, 0 \le y_0 \le 1$.)

Your report: In your report, describe your discoveries based on these numerical experiments. You may include a **limited** number of graphs and/or phase plane pictures of orbits to illustrate your description. You should **not** include a catalog of orbits for different initial conditions and values of k. The goal of the lab is to interpret the results of your experiments.

NUMERICAL METHODS

7

Errors in numerical approximations of solutions to initial-value problems are inevitable. Usually, these errors can be controlled by reducing the step size, but reducing the step size increases the arithmetic. Consequently, there is a trade-off between accuracy and computation time.

We begin this chapter with an analysis of the errors involved with Euler's method to obtain an understanding of the relationship between error and step size. We see what is meant by the statement that Euler's method is a first-order approximation scheme.

In the second and third sections, we give methods that achieve a given accuracy with fewer steps and less arithmetic. The methods we consider are improved Euler's method and Runge-Kutta. These methods generally produce much more accurate results for a given step size.

Computers and calculators can use only a finite amount of memory to store each number in a computation. However, most numbers have infinite decimal expansions. Hence, computers and calculators make small errors in each calculation due to the representation of numbers by the machines involved. In the final section we consider the effect of these errors on our approximations.

7.1 NUMERICAL ERROR IN EULER'S METHOD

Of the three approaches to the study of differential equations — the analytic, the qualitative, and the numeric — the most commonly used technique in science and engineering is numerical approximation. This comes as no surprise since there are many efficient software packages available that automate a great deal of the process of obtaining numerical solutions. Even when relatively simple analytic or qualitative techniques exist, there is a temptation to "just let the computer do it."

There is nothing wrong with this mode of operation, as long as two important points are kept in mind. First, the computer gives numbers and graphs, not interpretation and insight. Understanding the qualitative behavior of an approximate solution and its implications for the physical system under consideration are still the job of the human.

Second, numerical methods give *approximate* solutions. There is always a certain degree of error, and if the time interval over which the solution is appropriated is lengthened, the error will usually increase. While the error may remain insignificant for some differential equations, we have seen "chaotic" examples where small errors amplify very quickly (see Sections 2.7, 5.5, 6.4, and 6.5).

So far we have avoided any discussion of the errors that arise with numerical procedures such as Euler's method. However, if we are going to rely on computers, we should be aware of these errors, and we should know how to determine if our appropriate solutions have errors that are within acceptable ranges.

When we use a computer to approximate the solution of a differential equation, there are two very different ways in which we introduce error into the computation. One is the error that is inherent in the approximation scheme that we use, and the other is the error that results from the use of finite arithmetic during the computations. In this chapter, we primarily discuss the errors that arise from the approximation procedure. However, in Section 7.4, we discuss some of the interesting and practical issues that stem from the arithmetic of the computer.

Step Size and Euler's Method

From Section 1.4, recall that Euler's method is an iterative scheme that approximates the solution to an initial-value problem of the form

$$\frac{dy}{dt} = f(t, y), \quad y(t_0) = y_0.$$

The approximate solution is given by the iteration

$$y_{k+1} = y_k + f(t_k, y_k)\,\Delta t$$

where Δt is the step size and y_0 is the initial value $y(0)$. There is a simple relationship between the step size Δt, the length of the interval over which we would like to approximate the solution, and the number of steps involved in the computations. In particular, if we want to approximate the solution over the interval from t_0 to t_n (using n steps), we divide the total length of the interval into n equal-length subintervals. In other words,

$$\Delta t = \frac{t_n - t_0}{n}.$$

The rule-of-thumb discussed in Section 1.4 is that the smaller the step size the more accurate the approximate solution. However, the smaller the step size, the larger the number of steps, so the longer it takes the computer to approximate the solution. It is useful to be able to predict roughly how large a step size to take in order to achieve a particular accuracy.

We begin by studying how the accuracy of Euler's method changes with step size for a particular example. When we introduced Euler's method for first-order equations in Section 1.4, we considered the initial-value problem

$$\frac{dy}{dt} = -2ty^2, \quad y(0) = 1.$$

This equation is separable and the solution is

$$y(t) = \frac{1}{1 + t^2}.$$

Since we have a formula for the solution, we can compare the results of Euler's method for different step sizes to the exact values of the solution.

In Section 1.4, we applied Euler's method to this initial-value problem over the interval $0 \leq t \leq 2$ using $\Delta t = 0.1$ (20 steps). For this differential equation, the Euler's method iteration scheme is

$$y_{k+1} = y_k - 2t_k y_k^2 \Delta t,$$

and we obtained the approximation $y_{20} = 0.1933\ldots$, which is a good approximation of the exact value $y(2) = 1/(1 + 2^2) = 0.2$. Increasing the number of steps to 2000 ($\Delta t = 0.001$) yields the more accurate approximation $y_{2000} = 0.199937\ldots$, which is an even better approximation of $y(2) = 0.2$. In Section 1.4, we simply observed that increasing the number of steps in the overall computation seems to improve the accuracy. Now we would like to be more precise about exactly how the accuracy is improved as the number of steps is increased.

First, we take a closer look at the result of the computation that used 2000 steps over the interval $0 \leq t \leq 2$. Denote the error in the computation by e_{2000}. We have

$$e_{2000} = |0.2 - y_{2000}| = |0.2 - 0.199937| = 0.000063.$$

Now suppose we double the number of steps. After starting at the same initial condition and taking 4000 steps with $\Delta t = 0.0005$, we obtain the approximation $y_{4000} = 0.199969$. In this case we have the error

$$e_{4000} = |0.2 - y_{4000}| = |0.2 - 0.199969| = 0.000031.$$

In this example, we see that the ratio of the errors is

$$\frac{e_{4000}}{e_{2000}} = \frac{0.000031}{0.000063} \approx 0.492.$$

When we doubled the number of steps in this example, we halved the error.

This straightforward computation leads us to wonder how the accuracy of the approximation depends on the step size and the number of steps. Since Euler's method is so easy to implement, we can play around with various values to see what what happens. Using the computer, we first compute the value of y_n various choices of the number n of steps. Then we calculate the error e_n for these values. In Figure 7.1 we plot the results for $n = 100, 200, 300, \ldots, 6000$. This graph is a plot of the accuracy for Euler's method applied to the initial-value problem

$$\frac{dy}{dt} = -2ty^2, \quad y(0) = 1,$$

as a function of step size. If we ignore the scales on the axes for a moment, we note that this graph looks especially familiar. In fact, if we let n denote the total number of steps, this graph looks remarkably like the graph of $1/n$. Indeed, in Figure 7.2, we plot a graph of a function of the form K/n. (Actually $K = 0.126731$, but it is not important how we came up with this number.) Note the agreement between the data points and the graph of this function. The remainder of this section is devoted to an explanation of why these results are typical.

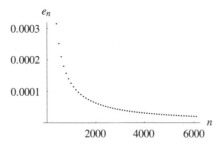

Figure 7.1
The error in Euler's method as a function of the number n of steps used in the approximation. Euler's method is applied to $dy/dt = -2ty^2$, $y(0) = 1$, over the interval $0 \leq t \leq 2$.

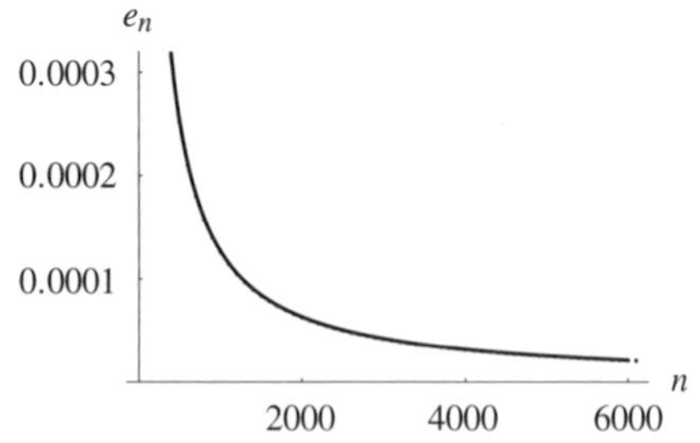

Figure 7.2
The error in Euler's method from Figure 7.1 fit to the function $0.126731/n$. Note that it is almost impossible to distinguish the graph of $0.126731/n$ from the data points e_n.

The Error in the First Step of the Method

Given an initial-value problem of the form

$$\frac{dy}{dt} = f(t, y), \quad y(t_0) = y_0,$$

and a step size Δt, we know that the value of y_1 in the first Euler step is determined by assuming that the graph of the solution is a line segment that starts at (t_0, y_0) and has slope $f(t_0, y_0)$. First, we need to determine how to estimate the error made in this initial step. The tool that lets us accomplish this estimate is Taylor's Theorem with remainder.

Taylor's Theorem says that

$$y(t_1) = y(t_0) + y'(t_0) \, \Delta t + y''(\xi_1) \frac{(\Delta t)^2}{2}$$

for some number ξ_1 between t_0 and t_1. Since we know that $y(t)$ is a solution to the initial-value problem, we have $y'(t_0) = f(t_0, y_0)$ and $y(t_0) = y_0$. In other words,

$$y(t_1) = y_0 + f(t_0, y_0)\,\Delta t + y''(\xi_1)\frac{(\Delta t)^2}{2}.$$

Euler's method corresponds to dropping the $(\Delta t)^2$ term, that is, taking the linear approximation of the solution from t_0 to t_1. Thus, the difference $|y(t_1) - y_1|$ between the first Euler approximation y_1 and the actual value $y(t_1)$ of the solution is

$$\left| y''(\xi_1)\frac{(\Delta t)^2}{2} \right|.$$

It is convenient to have some notation for this term, so we denote it τ_1 to represent the **truncation** error. It is called the truncation error because it results from a "truncation" of the Taylor series for $y(t_1)$. Euler's method is based on the Taylor approximation of $y(t_1)$, but it throws out all nonlinear terms (terms of degree 2 or higher).

For the first step of Euler's method, the truncation error is the error, but that won't be true for subsequent steps. So we introduce notation for the **total error** after k steps. By definition, the total error after k steps is the difference between the kth Euler approximation and the actual value $y(t_k)$ at time t_k, and we denote it by e_k. More precisely,

$$e_k = |y(t_k) - y_k|.$$

Note that for the first step ($k = 1$), the truncation error and the total error are identical. That is,

$$e_1 = \tau_1.$$

Usually, the truncation error and the total error agree only at the first step.

At this point, it is natural to wonder if the formula

$$\tau_1 = \left| y''(\xi_1)\frac{(\Delta t)^2}{2} \right|$$

is something that we can really compute if we don't already know the solution $y(t)$. The key point to keep in mind is the fact that $y(t)$ is a solution to the differential equation. Since we know that $dy/dt = f(t, y)$, we can compute the second derivative of $y(t)$ using the multivariable chain rule. Since

$$\frac{dy}{dt} = f(t, y(t)),$$

we have

$$\frac{d^2 y}{dt^2} = \frac{\partial f}{\partial t}\frac{dt}{dt} + \frac{\partial f}{\partial y}\frac{dy}{dt}$$

$$= \frac{\partial f}{\partial t} + \frac{\partial f}{\partial y}\,f(t, y).$$

Therefore, we can bound the error e_1 using an upper bound M_1 for the quantity

$$\frac{\partial f}{\partial t} + \frac{\partial f}{\partial y}\,f(t, y)$$

along the graph of $y(t)$ between $y(t_0)$ and $y(t_1)$. Since we are given $f(t, y)$, we can actually compute the quantities in this expression.

Suppose M_1 is a constant that is large enough so that

$$\left| \frac{\partial f}{\partial t} + \frac{\partial f}{\partial y} f(t, y) \right| < M_1$$

along the graph of $y(t)$ (or at least the part of the graph we are considering). If we are able to find such a bound M_1, then the error in the first step is bounded according to the inequality

$$e_1 = \tau_1 \le M_1 \frac{(\Delta t)^2}{2}$$

(see Figure 7.3).

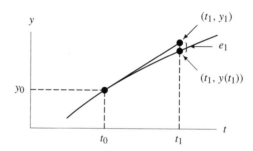

Figure 7.3
The error e_1 in the first step of Euler's method.

The Error in the Second Step

Unfortunately, once we have taken the first step, we are no longer on the graph of $y(t)$, and consequently our error analysis is more complicated. Basically there are two sources of error in the second step. If we knew the value of $y(t_1)$ exactly, then the error e_2 in the second step would again be the truncation error τ_2. In other words, from Taylor's Theorem again we have

$$y(t_2) = y(t_1) + y'(t_1)\, \Delta t + y''(\xi_2) \frac{\Delta t}{2}$$

for some number ξ_2 between t_1 and t_2, and using the differential equation to replace $y'(t_1)$ with $f(t_1, y(t_1))$, we would have

$$y(t_2) = y(t_1) + f(t_1, y(t_1))\, \Delta t + y''(\xi_2) \frac{(\Delta t)^2}{2}.$$

The truncation error at this step is therefore

$$\tau_2 = \left| y''(\xi_2) \frac{(\Delta t)^2}{2} \right|.$$

Comparing the formula

$$y(t_2) = y(t_1) + f(t_1, y(t_1)) \, \Delta t + y''(\xi_2)\frac{(\Delta t)^2}{2}$$

from Taylor's Theorem with the formula

$$y_2 = y_1 + f(t_1, y_1) \, \Delta t$$

for the second approximation y_2 from Euler's method, we see that, not only are we missing the quadratic term, but Euler's method uses the slope $f(t_1, y_1)$ at (t_1, y_1) rather than the slope $f(t_1, y(t_1))$ at $(t_1, y(t_1))$ (see Figure 7.4). Therefore, when we estimate the error e_2, we have

$$e_2 = |y(t_2) - y_2|$$

$$\leq |y(t_1) - y_1| + |f(t_1, y(t_1)) - f(t_1, y_1)| \, \Delta t + |y''(\xi_2)|\frac{(\Delta t)^2}{2}$$

$$\leq e_1 + |f(t_1, y(t_1)) - f(t_1, y_1)| \, \Delta t + M_1\frac{(\Delta t)^2}{2}$$

if we can find a bound M_1 on the second derivative as we did when we estimated τ_1. In other words, the total error after the second step is bounded by the total error after the first step, the truncation error τ_2 associated to the second step, and a third term, $|f(t_1, y(t_1)) - f(t_1, y_1)| \, \Delta t$, which accounts for the fact that we are using the slope $f(t_1, y_1)$ at the point (t_1, y_1) rather than the slope $f(t_1, y(t_1))$ at the point $(t_1, y(t_1))$ on the actual solution curve. Figure 7.4 illustrates the contribution that each of these three terms makes to the overall error.

To obtain an estimate for the error e_2 in a relatively convenient form, we need to bound the difference between $f(t_1, y(t_1))$ and $f(t_1, y_1)$. Since these two values of f come from points along the line $t = t_1$, we have the estimate

$$|f(t_1, y(t_1)) - f(t_1, y_1)| \leq M_2 \cdot e_1,$$

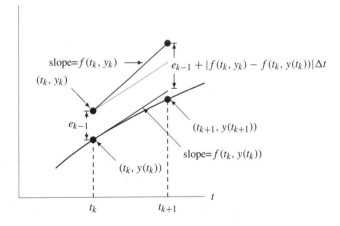

Figure 7.4
The relationship between the error e_{k-1} and the error e_k.

where M_2 is the maximum of the partial derivative $\partial f/\partial y$ along the line segment between $(t_1, y(t_1))$ and (t_1, y_1). We therefore obtain the estimate

$$e_2 \leq e_1 + M_2 \, e_1 \, \Delta t + M_1 \frac{(\Delta t)^2}{2} = (1 + M_2 \, \Delta t) \, e_1 + M_1 \frac{(\Delta t)^2}{2}.$$

The analysis of the error e_k at the kth is essentially the same as the analysis that we just completed for the second step. If M_1 is a bound for the second derivative along the graph of the solution and if M_2 is a bound for the partial derivative $\partial f/\partial y$, then the error e_k is bounded by

$$e_k \leq (1 + M_2 \, \Delta t)e_{k-1} + M_1 \frac{(\Delta t)^2}{2}.$$

Note that this formula is recursive in that it expresses e_k in terms of e_{k-1} (it is really a discrete dynamical system). The term

$$M_1 \frac{(\Delta t)^2}{2}$$

accounts for the truncation error τ_k at the kth step, and the term

$$(1 + M_2 \, \Delta t)e_{k-1}$$

represents the magnification of the error at the $k - 1^{\text{st}}$ step by a factor due to the fact that we are computing the right-hand side of the differential equation at the Euler approximation (t_k, y_k) rather than at the actual point $(t_k, y(t_k))$ on the solution.

The Error After *n* Steps

If we are going approximate the solution $y(t)$ using n steps, then we would definitely like a compact formula for the error e_n in terms of quantities that are easy to compute directly from the differential equation, that is, directly from the right-hand side $f(t, y)$. The recursive formula that we just derived does not really fit these criteria. However, in Exercise 11 we show that, if we find maximums for the expressions

$$M_1 = \left| \frac{\partial f}{\partial t} + \frac{\partial f}{\partial y} f(t, y) \right| \quad \text{and} \quad M_2 = \left| \frac{\partial f}{\partial y} \right|$$

that come up in the estimates, then we can use this recursive formula to bound the total error in Euler's method. Over the interval $t_0 \leq t \leq t_n$, we can derive a bound of the form

$$e_n \leq C \cdot \Delta t,$$

where C is a constant that is determined by the values of M_1, M_2, and the total length of the t-interval over which the solution is approximated. Note that the value of C *does not depend* on the number of steps used in Euler's method. Moreover, since $\Delta t = (t_n - t_0)/n$, we can rewrite this estimate as

$$e_n \leq \frac{K}{n},$$

where K is the product of the constant C and the length of the total interval $t_0 \leq t \leq t_n$. From our point of view, K is just another constant that does not depend on the number of steps used in Euler's method.

What's most important is that we have theoretical bounds that can be expressed either as

$$e_n \leq C \cdot \Delta t \quad \text{or as} \quad e_n \leq \frac{K}{n}.$$

Roughly speaking, if we halve Δt, then we halve the error in Euler's method. Expressed in terms of the number of steps, this estimate says that if we double the number of steps, then we double the accuracy of the approximation.

The Initial-Value Problem
$dy/dt = -2ty^2, \, y(0) = 1$

At the beginning of this section, we compared the results of Euler's Method with the analytic solution to the initial-value problem

$$\frac{dy}{dt} = -2ty^2, \quad y(0) = 1.$$

We saw that the error was roughly proportional to the step size Δt. This is consistent with the theoretical estimates that we just discussed. However, using the theoretical estimates mentioned above (and developed in more detail in Exercise 11) to choose the appropriate step size to guarantee a desired error is usually overkill.

To see why, let's consider Euler's method applied to the initial-value problem

$$\frac{dy}{dt} = -2ty^2, \quad y(0) = 1,$$

more carefully. We have been estimating the solution over the interval $0 \leq t \leq 2$. Without even trying to derive an analytic solution to the problem, we know that the solution is decreasing (why?) and that it is never zero, because $y(t) = 0$ for all t is an equilibrium solution. Therefore we know that the graph of the actual solution stays in the rectangle

$$\{(t, y) \,|\, 0 \leq t \leq 2, \, 0 \leq y \leq 1\}$$

in the ty-plane. Using ideas similar to those discussed above, we can determine that the approximate values $y_1, y_2, y_3, \ldots, y_n$ produced by Euler's method must also lie in the same rectangle. In theory we can now estimate the expressions M_1 and M_2 involving the relevant partial derivatives over this rectangle and come up with a bound on the total error in the approximation. In fact, we do exactly this in Exercise 12. In the end we obtain the bound

$$e_n \leq (16, 661, 456) \, \Delta t.$$

In other words, we need to take roughly 34 million steps to guarantee an accuracy of 1 over this interval of length 2. Looking at the data from our experiments at the beginning of the section, we see that this is an awful estimate. With ten steps we get an error that is less than 0.015.

Why is the theoretical estimate so bad? The problem comes from trying to estimate the total error e_n using upper bounds on M_1 and M_2 that hold throughout the rectangle even though it turns out that the actual solution and the Euler approximates remain close, and their graphs avoid the places where the values of M_1 and M_2 are large. In general, the problem is that we don't have a very good idea of where the graph of the solution is going to be until we make an accurate approximation.

In practice, the theoretical bounds for the total error are replaced by a running total of estimates of the error that may be introduced at each step. Although this analysis is not 100% trustworthy, it tends to give a better indication of the accuracy of our approximation.

To illustrate the point, we return to the initial-value problem

$$\frac{dy}{dt} = -2ty^2, \quad y(0) = 1,$$

and we estimate the solution over the interval $0 \leq t \leq 2$ with 100 steps. Since we have the analytic solution in this case, we can compute the error $e_k = |y(t_k) - y_k|$ at each step. The results of this calculation are shown in Figure 7.5. In addition, if we attempt to keep a running estimate of the total error by appealing to the recursive formula

$$e_k \leq (1 + M_2 \Delta t)\, e_{k-1} + M_1 \frac{(\Delta t)^2}{2}$$

derived above. However, at each step we estimate

$$M_1 = \left| \frac{\partial f}{\partial t} + \frac{\partial f}{\partial y} f(t, y) \right| \quad \text{and} \quad M_2 = \left| \frac{\partial f}{\partial y} \right|$$

based on the current value of y_k. That is, after we compute y_k from (t_{k-1}, y_{k-1}), we then calculate M_1 at the point $(t_{k-1} + \Delta t/2, (y_{k-1} + y_k)/2)$ and M_2 at the point $(t_{k-1}, (y_{k-1} + y_k)/2)$. (These points are "halfway" points among the set of all possibilities.) Then given the resulting values for M_1 and M_2, we use the approximation

$$e_k \approx (1 + M_2 \Delta t)\, e_{k-1} + M_1 \frac{(\Delta t)^2}{2}.$$

This gives us a running estimate for the total error e_k, and the estimate is based on the same principles that justify our theoretical results. A plot of these estimates is shown in Figure 7.6. Note that this graph is almost identical to the graph of the actual error as shown in Figure 7.5.

In general this type of local estimation of the errors involved in the calculation is much easier to achieve. In fact, the calculation can be left to the computer, and the estimates of the errors e_k can be one of the results of the computation.

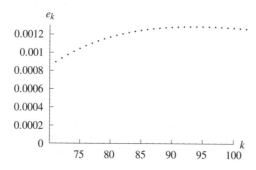

Figure 7.5
The cumulative error e_k after the kth step in the Euler approximation for the initial-value problem

$$\frac{dy}{dt} = -2ty^2, \quad y(0) = 1,$$

over the interval $0 \leq t \leq 2$ with $n = 100$ steps.

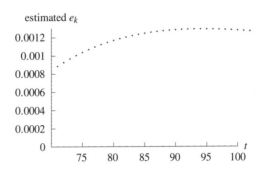

estimated e_k

Figure 7.6
Estimates of the errors e_k as shown in Figure 7.5 based on the evaluation of relevant partial derivatives.

The Order of Euler's Method

The most important lesson of this section is that the size of the error in Euler's method for an initial-value problem over a fixed interval of time is proportional to the step size, or in other words, it is inversely proportional to the number of steps that we use over a fixed interval. We say that Euler's method is a **first-order** method because the error is, at worst, proportional to the first power of the step size — note the 1 in the exponent of the bound

$$e_n \leq C \cdot (\Delta t)^1.$$

In the next two sections we introduce methods that have higher orders. Higher-order methods are more typical of professionally written numerical solvers.

Exercises for Section 7.1

1. Reproduce Figure 7.1 by performing the following calculations:
   (a) Solve the initial-value problem $dy/dt = -2ty^2$, $y(0) = 1$.
   (b) Given the interval $0 \leq t \leq 2$ and $n = 20$, calculate the Euler approximation y_{20} to $y(2)$. (Double-check your calculations by comparing your results to those given at the beginning of the section.)
   (c) Calculate the total error e_{20} involved in your approximation.
   (d) Repeat the calculations in parts (b) and (c) using $n = 1000, 2000, \ldots, 6000$.
   (e) If your computer is up to the task, compute e_n for additional intermediate values of n. (Figure 7.1 was done with n ranging from 100 to 6000 by steps of 100.)
   (f) Some computer math systems can determine the best fit for this data to a function of the form k/n. If your computer can do this, determine the best value of k for this initial-value problem. Plot the data and the graph of k/n on the same axes.

In Exercises 2–5, repeat the steps in Exercise 1 to arrive at plots like Figures 7.1 and 7.2 for the initial-value problem and interval specified.

2. $\dfrac{dy}{dt} = 1 - y$, $\quad y(0) = 0$, $\quad 0 \leq t \leq 1$

3. $\dfrac{dy}{dt} = ty$, $\quad y(0) = 1$, $\quad 0 \leq t \leq \sqrt{2}$

4. $\dfrac{dy}{dt} = -y^2, \quad y(0) = 1/2, \quad 0 \le t \le 2$

5. $\dfrac{dy}{dt} = 1 - t + 3y, \quad y(0) = 1, \quad 0 \le t \le 1$

6. In this section, we state that the error e_{2000} is 0.000063 for the Euler approximation of the initial-value problem $dy/dt = -2ty^2$, $y(0) = 1$ over the interval $0 \le t \le 2$. Estimate how many steps n are needed to obtain an approximation y_n that has an error e_n no larger than 0.000001.

7. Reproduce Figure 7.6 by performing the following calculations:

   (a) Calculate the expressions

$$M_1 = \frac{\partial f}{\partial t} + \frac{\partial f}{\partial y} f(t, y) \quad \text{and} \quad M_2 = \frac{\partial f}{\partial y}$$

   for $f(t, y) = -2ty^2$.

   (b) Given the initial-value problem $dy/dt = -2ty^2$, $y(0) = 1$, the interval $0 \le t \le 2$, and $n = 100$ steps, calculate the point (t_1, y_1) that corresponds to the first step of Euler's method. Evaluate M_1 at the point $(t_0 + \Delta t/2, (y_0 + y_1)/2)$, and use this value to estimate the error e_1. How does your estimate compare with the true error e_1?

   (c) Calculate the point (t_2, y_2) that corresponds to the second step of Euler's method. Evaluate M_1 at the point $(t_1 + \Delta t/2, (y_1 + y_2)/2)$ and M_2 at the point $(t_1, (y_1 + y_2)/2)$. Use these values to estimate the error e_2. How does your estimate compare with the true error e_2?

   (d) Repeat part (c) for the remaining 98 steps.

   (e) Make two plots (similar to Figures 7.5 and 7.6) that compare the true errors e_k with the estimates.

In Exercises 8–9, repeat parts (b)–(d) of Exercise 7 to estimate the errors e_k for the initial-value problem and interval specified. Why don't we ask for a comparison of this estimate of e_k with the true value of e_k?

8. $\dfrac{dy}{dt} = t - y^3, \quad y(0) = 1, \quad 0 \le t \le 1$

9. $\dfrac{dy}{dt} = \sin ty, \quad y(0) = 3, \quad 0 \le t \le 3$

10. Using what you learned in calculus, show that $1 + \alpha < e^{\alpha}$ if $\alpha > 0$.

11. In this exercise, we derive the theoretical bounds mentioned in the section. Let R be a rectangle $\{(t, y) \mid a \le t \le b, c \le y \le d\}$ in the ty-plane and suppose that $f(t, y)$ is continuously differentiable on R. Given an initial-value problem

$$\frac{dy}{dt} = f(t, y), \quad y(t_0) = y_0,$$

and an interval $t_0 \le t \le t_n$ such that the point (t_0, y_0) is in R and $t_n \le b$, then we can bound the error e_n involved in the Euler approximation assuming the Euler approximate values $y_1, y_2, \ldots, y_n$ all lie between c and d. To do so, let

$$M_1 = \max \left| \frac{\partial f}{\partial t} + \frac{\partial f}{\partial y} f \right| \text{ on } R$$

and let

$$M_2 = \max \left| \frac{\partial f}{\partial y} \right| \text{ on } R.$$

(a) Show that

$$e_1 \leq M_1 \frac{(\Delta t)^2}{2}.$$

(b) Show that

$$e_2 \leq e_1 + M_2 e_1 \Delta t + M_1 \frac{(\Delta t)^2}{2}$$

and explain the significance of each of these three terms.

(c) Explain why

$$e_{k+1} \leq (1 + M_2 \Delta t) e_k + M_1 \frac{(\Delta t)^2}{2}.$$

(d) Let $K_1 = 1 + M_2 \Delta t$ and $K_2 = M_1 (\Delta t)^2 / 2$. Show that

$$e_3 \leq (K_1^2 + K_1 + 1) K_2.$$

(e) Show that

$$e_n \leq \left(K_1^{n-1} + K_1^{n-2} + \cdots + K_1 + 1 \right) K_2.$$

(f) Explain why

$$K_1^{n-1} + K_1^{n-2} + \cdots + K_1 + 1 = \frac{K_1^n - 1}{K_1 - 1},$$

and thus

$$e_n \leq \left(\frac{K_1^n - 1}{K_1 - 1} \right) K_2.$$

(g) Using the definitions of K_1 and K_2, show that

$$e_n \leq \frac{M_1}{2M_2} \left((1 + M_2 \Delta t)^n - 1 \right) \Delta t.$$

(h) Use the result of Exercise 10 to conclude

$$e_n \leq \frac{M_1}{2M_2} \left(e^{(M_2 \Delta t) n} - 1 \right) \Delta t.$$

(i) Finally, conclude that

$$e_n \leq \frac{M_1}{2M_2} \left(e^{M_2 (t_n - t_0)} - 1 \right) \Delta t.$$

(j) Explain why this justifies the inequality

$$e_n \leq C \cdot \Delta t$$

given in this section.

(k) In what way is this bound for the total error different from the "estimates" shown in Figure 7.6 and computed in Exercises 7–9?

12. Given the initial-value problem $dy/dt = -2ty^2$, $y(0) = 1$, and the interval $0 \leq t \leq 2$, carry out the theoretical bound on e_n given by the results of Exercise 11 as follows:

(a) Let R be the rectangle $\{(t, y) \mid 0 \leq t \leq 2, 0 \leq y \leq 1\}$. Determine the maximum values for M_1 and M_2 on R.

(b) Using the results of Exercise 11, derive the constant C for which we are certain that the inequality

$$e_n \leq C \cdot \Delta t$$

holds.

(c) Using the value of C determined in part (b), find K such that

$$e_n \leq \frac{K}{n}.$$

(d) Explain why you think these two estimates are so conservative compared to what we know from computations like those given at the beginning of the section.

7.2 IMPROVING EULER'S METHOD

Euler's method is a convenient numerical algorithm in many ways. It is easy to understand and to implement. However, for numerical work where a high degree of accuracy is essential, Euler's method is not the algorithm of choice. There are algorithms that usually are more accurate and that require fewer calculations to attain that accuracy. In this section and the subsequent one, we present two of these algorithms in order to illustrate how we can derive and implement more accurate algorithms.

Higher-Order Algorithms

As we mentioned last section, we can interpret Euler's method in terms of Taylor approximation. Given a solution to the initial-value problem

$$\frac{dy}{dt} = f(t, y), \quad y(t_0) = y_0,$$

then its Taylor series

$$y(t_1) = y_0 + y'(t_0)\Delta t + \frac{y''(t_0)}{2}(\Delta t)^2 + \ldots$$

can be rewritten as

$$y(t_1) = y_0 + f(t_0, y_0)\Delta t + \frac{y''(t_0)}{2}(\Delta t)^2 + \ldots.$$

Euler's method can be interpreted as the approximation that results from truncating this series at the linear term. That is,

$$y(t_1) \approx y_1 = y_0 + f(t_0, y_0)\Delta t.$$

One way to improve the accuracy of Euler's method is to use more of the Taylor series. Rather than truncating at the linear term, we include the quadratic term and obtain a more accurate approximation. In other words, we can approximate $y(t_1)$ by a new y_1 where

$$y(t_1) \approx y_1 = y_0 + f(t_0, y_0)\Delta t + \frac{y''(t_0)}{2}(\Delta t)^2.$$

To do this, we need to know $y''(t_0)$, the second derivative of the solution. The only information we have about the solution y is that it satisfies the differential equation $dy/dt = f(t, y)$. We can differentiate both sides to obtain an expression for $y''(t_0)$, that is

$$\frac{d^2 y}{dt^2} = \frac{\partial f}{\partial t}\frac{dt}{dt} + \frac{\partial f}{\partial y}\frac{dy}{dt}$$

$$= \frac{\partial f}{\partial t} + \frac{\partial f}{\partial y}\, f(t, y).$$

In principle there is no problem performing this calculation and using the results to implement a numerical algorithm that is more accurate than Euler's method. In practice this is seldom done because the calculation requires knowledge of the partial derivatives of f, and if we want to program a "black box" in a traditional computer language such as Fortran or C, we would have to provide these partial derivatives as well as the original function f from the differential equation. These days, there exist computer languages that can calculate these derivatives, so this restriction is no longer a problem. Nevertheless, some very clever algorithms have been developed that provide approximation schemes with equivalent accuracy without using the partial derivatives of f, and these algorithms are commonly used. They will be the ones that we focus on in this chapter.

Numerical Approximation and Numerical Integration

To understand how these methods were developed, it is useful to think about Euler's method in terms of the numerical integration techniques that are used to approximate the definite integral. By the Fundamental Theorem of Calculus, we know that the solution of the initial-value problem

$$\frac{dy}{dt} = f(t, y), \quad y(t_0) = y_0$$

satisfies the equation

$$y(t_1) = y(0) + \int_{t_0}^{t_1} f(\tau, y(\tau))\, d\tau$$

because $dy/dt = f(t, y)$. In other words,

$$y(t_1) - y(t_0) = \int_{t_0}^{t_1} f(\tau, y(\tau))\, d\tau.$$

This is the "integral equation" equivalent of the original differential equation. In Euler's method we approximate this difference by the step

$$y_1 - y_0 = f(t_0, y_0)\, \Delta t.$$

Geometrically this approximation is really the approximation of the integral — the "area" under the graph of $f(t, y(t))$ — by the area of a rectangle with height $f(t_0, y_0)$ and width Δt (see Figure 7.7). We use the value of the derivative $y'(t_0) = f(t_0, y_0)$ at the left-hand endpoint of the interval $t_0 \le t \le t_1$ to approximate the value of $y(t_1)$. Thus Euler's method is analogous to approximating a Riemann integral by left-hand Riemann sums.

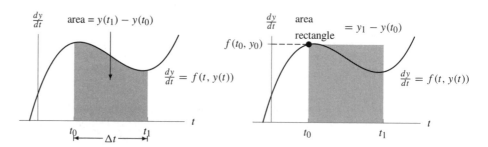

Figure 7.7
The first step of Euler's method interpreted as a Riemann approximation to an area.

From your younger days when you took calculus, you may recall that the left-hand Riemann sums converge to the value of the integral as the number of subdivisions increases, but that there are better ways to approximate this integral. A simple variation on left-hand approximation is trapezoidal approximation. In this case we try to approximate the area under the graph of $f(t, y(t))$ by trapezoids. The width of each trapezoid is Δt, and the heights are given by the values $f(t_0, y(t_0))$, $f(t_1, y(t_1))$, $f(t_2, y(t_2))$, ... (see Figure 7.8). The area of the kth trapezoid is

$$\left(\frac{f(t_{k-1}, y(t_{k-1})) + f(t_k, y(t_k))}{2} \right) \Delta t.$$

It follows that

$$y(t_k) \approx y(t_{k-1}) + \left(\frac{f(t_{k-1}, y(t_{k-1})) + f(t_k, y(t_k))}{2} \right) \Delta t.$$

This formula for $y(t_k)$ suggests the approximation scheme

$$y_k = y_{k-1} + \left(\frac{f(t_{k-1}, y_{k-1}) + f(t_k, y_k)}{2} \right) \Delta t.$$

There is only one thing wrong with this scheme: The number y_k appears on both sides of the equation. In other words, you need to know y_k to compute y_k. (But if you already knew it, you wouldn't need the formula.) We have to get rid of the y_k on the right-hand side of the approximation scheme.

One way to turn this equation into a useful approximation scheme is to replace the y_k on the right-hand side with some other reasonable value. At this point, the only way

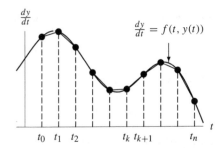

Figure 7.8
Approximating the area under the graph of $dy/dt = f(t, y(t))$ using trapezoidal approximation.

we know to approximate y_k is to use Euler's method, so that is what we do. In other words, we replace y_k on the right-hand side by the Euler approximation to y_k. This yields the approximation scheme

$$y_k = y_{k-1} + \left(\frac{f(t_{k-1}, y_{k-1}) + f(t_k, y_{k-1} + f(t_{k-1}, y_{k-1})\Delta t)}{2} \right) \Delta t.$$

This complicated formula can be more clearly thought of as a sequence of steps.

Improved Euler's Method

Given the initial condition $y(t_0) = y_0$ and the step size Δt, compute the point (t_{k+1}, y_{k+1}) from (t_k, y_k) as follows:

1. *Use the differential equation to compute the slope*

$$m_k = f(t_k, y_k).$$

2. *Calculate the value $\tilde{y}_{k+1}$ that results from one application of Euler's method. That is,*

$$\tilde{y}_{k+1} = y_k + f(t_k, y_k)\Delta t.$$

3. *Calculate $t_{k+1} = t_k + \Delta t$ and use the differential equation to compute the slope*

$$n_k = f(t_{k+1}, \tilde{y}_{k+1})$$

at the point $(t_{k+1}, \tilde{y}_{k+1})$.

4. *Compute y_{k+1} by*

$$y_{k+1} = y_k + \left(\frac{m_k + n_k}{2} \right) \Delta t.$$

Improved Euler's Method for
$f(t, y) = -2ty^2, \ y(0) = 1$

In Section 1.4, we applied Euler's method to the initial-value problem

$$\frac{dy}{dt} = -2ty^2, \quad y(0) = 1,$$

and in the last section we used this example to illustrate the typical errors involved in Euler's method. Now let's compare our previous results with the results from improved Euler's method.

We begin with $\Delta t = 0.1$. In this example the function $f(t, y) = -2ty^2$, and its value at the initial condition is $f(0, 1) = 0$. Therefore the initial slope is $m_0 = 0$. One step of Euler's method yields the point $(t_1, \tilde{y}_1) = (0.1, 1.0)$. Improved Euler's method uses the value of $f(t, y)$ at this new point to help determine the value of y_1. In other words, we compute $n_0 = f(0.1, 1.0) = -0.2$, and average this slope with m_0 to obtain the slope that we use to compute y_1. In this case, the average is

$$\frac{m_0 + n_0}{2} = \frac{0.0 - 0.2}{2} = -0.1.$$

Hence we use this slope to calculate y_1. We obtain

$$\begin{aligned} y_1 &= y_0 + (-0.1)\Delta t \\ &= 1.0 + (-0.1)(0.1) = 0.99. \end{aligned}$$

Before computing another step by hand, we compare the result of this calculation to the result that Euler's method provides. At the initial condition, the differential equation yields the value 0 for the slope. Therefore Euler's method assumes that the slope over the subinterval $0 \le t \le 0.1$ vanishes, and therefore $y_1 = y_0 = 1.0$. On the other hand, improved Euler's method considers the slope at the point $(0.1, 1.0)$ as well as the slope at $(0.0, 1.0)$. When these two slopes are averaged, we obtain a (small) negative value for the slope to use over the subinterval $0 \le t \le 0.1$. Using this average slope makes sense because we can see from the differential equation that $f(t, y) = -2ty^2$ is negative for all nonzero t throughout the subinterval. Of course, there is no reason to believe that this average of the two slopes is going to be a better choice than the slope from Euler's method, but it is a good idea to try this average.

The process is basically the same for the next step. The first step yielded the point $(t_1, y_1) = (0.1, 0.99)$. Consequently, we use the differential equation to produce the slope that Euler's methods uses. We have

$$m_1 = f(t_1, y_1) = f(0.1, 0.99) = -0.19602.$$

So starting at $(0.1, 0.99)$, we take one step of Euler's method via the computation

$$\begin{aligned} \tilde{y}_2 &= y_1 + m_1 \Delta t \\ &= 0.99 + (-0.19602)(0.1) \\ &= 0.970398. \end{aligned}$$

Thus, for our other slope n_1, we compute $n_1 = f(t_2, \tilde{y}_2) = -0.37669$. We average these two slopes and obtain

$$\frac{m_1 + n_1}{2} = \frac{(-0.19602) + (-0.37669)}{2} = -0.286344.$$

Using this average, we can now compute the second step

$$\begin{aligned} y_2 &= y_1 + (-0.286344)\Delta t \\ &= 0.99 + (-0.286344)(0.1) \\ &= 0.961366. \end{aligned}$$

Table 7.1   Improved Euler's method for
$dy/dt = -2ty^2$ with $\Delta t = 0.1$

k	t_k	y_k
0	0.	1.
1	0.1	0.99
2	0.2	0.961366
3	0.3	0.917246
⋮	⋮	⋮
19	1.9	0.21767
20	2.	0.200695

That's enough by hand. Turning to the computer we obtain the results shown in Table 7.1.

Recall that we know the exact solution $y(t) = 1/(1 + t^2)$ for this initial-value problem, and its value at $t = 2$ is $y(2) = 0.2$. Thus, the error in this computation is

$$|0.2 - 0.200695| = 0.000695.$$

When we used Euler's method with the same step size, we obtained the approximation $y(2) \approx 0.193342$ (see page 519). The error in that approximation was

$$|0.2 - 0.193342| = 0.00658.$$

Note that improved Euler's method is roughly 10 times more accurate than Euler's method in this example.

Cost

For a given step size, using improved Euler's method instead of Euler's method yields a good improvement in accuracy. The down side is that each step of improved Euler's method involves two steps of Euler's method. The first step is used to get the value of $\tilde{y}_{k+1}$, and the second computes y_{k+1}. Each step of improved Euler's method takes roughly twice the amount of arithmetic as one step of Euler's method.

In the example above, we gained ten times the accuracy by using the improved Euler's method over the Euler's method at a cost of doing twice the arithmetic. This is a good return (particularly if a computer is doing the arithmetic).

The Order of Improved Euler

In the previous section we saw that Euler's method has order 1. In other words, the error is roughly proportional to the reciprocal of the number of subdivisions employed in the approximation. In order to compare the two methods, we need to know how the accuracy of improved Euler's method behaves as a function of the step size. As in the previous section, we consider the initial-value problem

$$\frac{dy}{dt} = -2ty^2, \quad y(0) = 1,$$

which has the exact solution $y(t) = 1/(1 + t^2)$. Using the computer, we approximate the value $y(2) = 0.2$ twice. First we compute the approximation with 1000 steps. The error e_{1000} in this approximation is

$$e_{1000} = 2.59 \times 10^{-7}.$$

Then we repeat the computation using twice as many steps. In this case we get a more accurate answer with an error of

$$e_{2000} = 6.48 \times 10^{-8}.$$

These errors are definitely small, but to get a sense of how this method depends on the step size, we compute the ratio

$$\frac{e_{1000}}{e_{2000}} \approx 4.003.$$

We see that, if we double the number of steps (that is, halve the step size), we decrease the error by a factor of 4 ($= 2^2$). Consequently, we say that improved Euler's method is a numerical method with order 2.

Using estimates like the ones that we discussed in the last section, it can be shown that, given appropriate bounds on f and its derivatives, the error for improved Euler's method behaves like a quadratic function in its step size. That is, the error in a given approximation using improved Euler's method is proportional to $(\Delta t)^2$. Since Δt is simply the length of the t-interval in question divided by the number of steps n, we can also express the error as

$$e_n = \frac{K}{n^2}.$$

for some constant K. In Figure 7.9, we graph the error as a function of the number of steps for Euler's method and improved Euler's method. This plot illustrates the advantage of a second-order method over a first-order method.

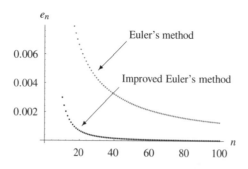

Figure 7.9
The errors related to Euler's method and improved Euler's method for the initial-value problem $dy/dt = -2ty^2$, $y(0) = 1$, over the interval $0 \le t \le 2$.

Exercises for Section 7.2

In Exercises 1–8, use improved Euler's method with the given number n of steps to approximate the solution to the initial-value problem specified. Your answer should include a table of approximate values of the dependent variable. It should also include a sketch of the graph of the approximate solution. These initial-value problems are the same as those in Exercises 1–8 of Section 1.4. Compare the graphs that you get from improved Euler's method with those that you get from Euler's method. If your computer system has a built-in routine for the numerical solution of differential equations, compare these graphs with those that come from the built-in numerical solver.

1. $\dfrac{dy}{dt} = 2y + 1,$ $y(0) = 3,$ $0 \le t \le 2,$ $n = 4$

2. $\dfrac{dy}{dt} = t - y^2,$ $y(0) = 1,$ $0 \le t \le 1,$ $n = 4$

3. $\dfrac{dy}{dt} = y^2 - 2y + 1,$ $y(0) = 1.4,$ $0 \le t \le 2,$ $n = 4$

4. $\dfrac{dy}{dt} = \sin y,$ $y(0) = 1,$ $0 \le t \le 3,$ $n = 6$

5. $\dfrac{dw}{dt} = (3 - w)(w + 1),$ $w(0) = 0,$ $0 \le t \le 5,$ $n = 5$

6. $\dfrac{dw}{dt} = (3 - w)(w + 1),$ $w(0) = 0,$ $0 \le t \le 5,$ $n = 10$

7. $\dfrac{dy}{dt} = e^{(2/y)},$ $y(0) = 2,$ $0 \le t \le 2,$ $n = 4$

8. $\dfrac{dy}{dt} = e^{(2/y)},$ $y(1) = 2,$ $1 \le t \le 3,$ $n = 4$

In Exercises 9–12, use improved Euler's method to approximate the solution to the given initial-value problem over the interval specified. For each initial-value problem,

(a) determine an analytic solution (see Exercises 2–5 in Section 7.1); and

(b) using improved Euler's method with $n = 4$, determine an approximate value y_4 and the total value e_4.

(c) How many steps of improved Euler's method would you expect to use to obtain an approximation for which $e_n \le 0.0001$?

9. $\dfrac{dy}{dt} = 1 - y,$ $y(0) = 0,$ $0 \le t \le 1$

10. $\dfrac{dy}{dt} = ty,$ $y(0) = 1,$ $0 \le t \le \sqrt{2}$

11. $\dfrac{dy}{dt} = -y^2,$ $y(0) = 1/2,$ $0 \le t \le 2$

12. $\dfrac{dy}{dt} = 1 - t + 3y,$ $y(0) = 1,$ $0 \le t \le 1$

13. In this section, we showed that the total error e_{20} in the approximation of the initial-value problem $dy/dt = -2ty^2$, $y(0) = 1$, over the interval $0 \le t \le 2$ is

$$e_{20} = 0.000695.$$

(a) How many steps should you use to get an approximate value y_n that has an associated error e_n no greater than 0.0001?

(b) Calculate that approximate value y_n.

(c) Calculate the total error in that case.

In Exercises 14–17, produce a plot like Figure 7.9 for the given initial-value problem over the interval specified. See the directions to Exercise 1 in Section 7.1 for more details. However, restrict the values of n to numbers in the interval $10 \le n \le 100$.

14. $\dfrac{dy}{dt} = 1 - y, \quad y(0) = 0, \quad 0 \le t \le 1$

15. $\dfrac{dy}{dt} = ty, \quad y(0) = 1, \quad 0 \le t \le \sqrt{2}$

16. $\dfrac{dy}{dt} = -y^2, \quad y(0) = 1/2, \quad 0 \le t \le 2$

17. $\dfrac{dy}{dt} = 1 - t + 3y, \quad y(0) = 1, \quad 0 \le t \le 1$

7.3 THE RUNGE-KUTTA METHOD

Improved Euler's method shows us that if we intelligently modify Euler's method and do a little bit more work in each step, we can get a significantly more accurate approximation to our solution. This observation leads us to wonder if there is a "standard" method for the numerical approximation of solutions of differential equations. We want an algorithm that minimizes the number of computations that have to be performed while it maximizes the accuracy.

In fact for most practical purposes, there is such an algorithm. It is called the (fourth-order) Runge-Kutta method. It is named after the two German mathematicians that developed the method approximately 100 years ago. There are more sophisticated algorithms that are required for specialized problems, but the Runge-Kutta method is often the numerical method used for the numerical approximation of solutions.

The Runge-Kutta Method

In improved Euler's method applied to the differential equation $dy/dt = f(t, y)$, we used an average of two slopes to compute each value. In other words, to compute y_k from y_{k-1}, we used the average of values of the right-hand side $f(t, y)$ at $t = t_{k-1}$ and $t = t_k$. By analogy with numerical integration, improved Euler's method is similar to the trapezoidal rule.

For numerical integration, there are algorithms that are usually more efficient than the trapezoidal rule. Simpson's rule approximates the area under the graph using parabolic interpolation and generally gives a better estimate for the integral. In fact Simpson's rule can be interpreted as a weighted average of values where twice as much weight is given to the values of the function at the midpoints of the subintervals than is given to the endpoints of the subintervals. The Runge-Kutta method is similar to Simpson's rule in the way that it considers a weighted average.

We begin with a description of the algorithm and then illustrate its implementation and accuracy using the example from previous sections. Refer to Figure 7.10 as we describe how the various slopes used in the method are derived.

To calculate the value y_{k+1} from y_k, we use four slopes given by the function $f(t, y)$ that defines the differential equation. These slopes are denoted m_k, n_k, q_k,

Given $\tilde{y}_0$ and $\tilde{t} = t_0 + \Delta t/2 = 0 + 0.05 = 0.05$, we obtain the second slope we need by

$$n_0 = f(\tilde{t}, \tilde{y}_0) = f(0.05, 1.0) = -0.1.$$

In turn this slope determines the y-value $\hat{y}_0$ via

$$\begin{aligned}
\hat{y}_0 &= y_0 + n_0 \frac{\Delta t}{2} \\
&= 1.0 + (-0.1)(0.05) \\
&= 0.995.
\end{aligned}$$

Applying $f(t, y)$ to the point $(\tilde{t}, \hat{y}_0)$ determines a third slope

$$q_0 = f(\tilde{t}, \hat{y}_0) = -0.0990025,$$

and using this slope, we obtain the fourth point with y-value $\overline{y}_0$ by

$$\overline{y}_0 = y_0 + q_0 \Delta t = 1.0 + (-0.0990025)(0.1) = 0.9901.$$

Given this fourth point we compute the slope at that point by

$$p_0 = f(t_{k+1}, \overline{y}_0) = -0.19606.$$

Now that we have these four slopes we are able to compute the weighted average that we use to compute the slope that determines the first step. We have

$$\begin{aligned}
\frac{m_0 + 2n_0 + 2q_0 + p_0}{6} &= \left(\frac{0 + 2(-0.1) + 2(-0.0990025) + -0.19606}{6} \right) \\
&= -0.0990108.
\end{aligned}$$

This weighted average yields the first step

$$y_1 = y_0 - 0.0990108\,\Delta t = 1.0 + (-0.0990108)(0.1) = 0.990099.$$

(If you try to reproduce these calculations on your calculator and you get a slightly different answer, don't be concerned. Your calculator may carry a different number of significant digits in the calculation, and this difference may affect your computation slightly.)

Doing these calculations by hand is definitely not the way to go now that we have easy access to programmable calculators, spreadsheets, and general systems for doing mathematics by computers such as Maple, Mathematica, and Matlab. Later in this section we provide codes that can be used in various computing environments to implement Runge-Kutta. They are not too complicated, and using them definitely beats doing these computations by hand.

Computing 20 steps via the Runge-Kutta method yields the results shown in Table 7.2. Note that the approximation of $y(2) = 0.2$ is very good, especially considering the relatively large step size that we used. Compare these results to those found in Table 7.1. The error obtained in improved Euler's method was 0.000658, whereas the error here is 0.000001. Switching to the Runge-Kutta method in this example improves the accuracy by a factor of 658. (Actually the improvement is more like a factor of 1000, but that is not clear from Table 7.2 because we show only the results to six significant digits.)

Table 7.2 Runge-Kutta method for
$dy/dt = -2ty^2$ with $\Delta t = 0.1$

k	t_k	y_k
0	0.	1.
1	0.1	0.990099
2	0.2	0.961538
3	0.3	0.917431
⋮	⋮	⋮
19	1.9	0.21692
20	2.	0.200001

The Order of the Runge-Kutta Method

Euler's Method has order 1 and improved Euler's method has order 2. Now let's see if we are right about Runge-Kutta's order. To find the order of the Runge Kutta method, we recall that the solution of the initial-value problem

$$\frac{dy}{dt} = -2ty^2, \quad y(0) = 1$$

is $y(t) = 1/(1 + t^2)$. If we approximate this solution over the interval $0 \leq t \leq 2$ using the Runge-Kutta method with ten steps, we find that the error in this approximation is

$$e_{10} = 1.095 \times 10^{-5},$$

and if we approximate the solution using 20 steps, we get an error of

$$e_{20} = 6.541 \times 10^{-7}.$$

The ratio

$$\frac{e_{10}}{e_{20}} = 16.75$$

indicates that if we double the number of steps, the error improves by roughly a factor of 16 ($= 2^4$). This is what we expect from a numerical procedure that has order 4.

Programming Runge-Kutta

Although the Runge-Kutta method seems somewhat complicated at first, it really isn't especially difficult to program. Here we give three implementations — one for a graphing calculator, one for *Mathematica*, and one in the C programming language. Some of the exercises at the end of this section require that you implement the method using whatever technology is available to you. These three implementations below should give you some guidance as to how to proceed.

The TI graphing calculator

This code assumes that the initial t-value is stored in A, the end value is stored in B, the number of steps is stored in N, the initial y-value is stored in Y, and $Y_1 = f(X,Y)$. The code uses X to represent t and Y to represent y in our notation.

```
:(B - A) -> H
:A -> X
:0 -> I
:Lbl 1
:Disp Y
:Pause
:Y -> Z
:Y1 -> C
:X + H/2 -> X
:Z + C*H/2 -> Y
:Y1 -> D
:Z + D*H/2 -> Y
:Y1 -> E
:X + H/2 -> X
:Z + E*H -> Y
:Y1 -> F
:Z + (C + 2*D + 2*E + F)*H/6 -> Y
:IS>(I,N)
:Goto 1
```

Mathematica Code

The *Mathematica* code is more compact, but using it effectively requires some knowledge of commands such as `ListPlot` and `NestList`. We begin by defining a function that takes one step of the Runge-Kutta method.

```
RungeKuttaStep[f_, vars_, {tk_, yk_}, deltat_]:=
Module[{k1, k2, k3, k4, tmiddle, tend},
    k1 = f /. {vars[[1]] -> tk, vars[[2]] -> yk};
    tmiddle = tk + deltat/2;
    k2 = f /. {vars[[1]] -> tmiddle,
               vars[[2]] -> yk + k1 deltat/2};
    k3 = f /. {vars[[1]] -> tmiddle,
               vars[[2]] -> yk + k2 deltat/2};
    tend = tk + deltat;
    k4 = f /. {vars[[1]] -> tend,
               vars[[2]] -> yk + k3 deltat};
    {tend, yk + (k1 + 2 k2 + 2 k3 + k4) deltat/6}
    ]
```

For example, the command

```
RungeKuttaStep[-2 t y^2, {t, y}, {0,1}, 0.1]
```

returns

```
{0.1, 0.990099}
```

We then approximate the solution to the initial-value problem

$$\frac{dy}{dt} = -2ty^2, \quad y(0) = 1$$

over the interval $0 \leq t \leq 2$ using 20 steps by the command

```
RKresults =
    NestList[RungeKuttaStep[-2 t y^2, {t, y}, #, 0.1]&,
        {0, 1}, 20]
```

This command stores the resulting points in the variable `RKresults`, and hence we are able to plot the approximate solution using `ListPlot`. For example,

```
ListPlot[RKresults,
        PlotJoined -> True,
        AxesOrigin -> {0,0}]
```

C code for Runge-Kutta

The following is a C program that applies the Runge-Kutta method to the initial-value problem

$$\frac{dy}{dt} = -2ty^2, \quad y(0) = 1.$$

It does the approximation over the interval $0 \leq t \leq 2$ with 20 steps. You should change the values of `T0`, `Tn`, `NUMSTEPS`, and `Y0` at the top and the definition of $f(t, y)$ in the return statement to apply the code to a different initial-value problem.

```
/*
                    Runge-Kutta approximation of
                        dy/dt = f(t,y)
*/

/* need to set these values and the function f(t,y) below
   before compilation */
#define T0 0.0
#define Tn 2.0
#define NUMSTEPS 20
#define Y0 1.0

#include <stdio.h>
#include <math.h>

/* this subroutine contains the formula for the first-order
   differential equation (i.e., f(t,y))
*/
double f(t,y)
     double t,y;
{
  return(-2.0 * t * y * y);
}
```

```
main()
{
  double tk = T0, tmid, yk = Y0,
         k1, k2, k3, k4,
         deltat = (double)(Tn - T0)/NUMSTEPS,
         f();

  printf("%f %f\n", tk, yk);
  while (tk < Tn){
    k1 = f(tk, yk);
    tmid = tk + deltat/2.0;
    k2 = f(tmid, yk + k1 * deltat/2.0);
    k3 = f(tmid, yk + k2 * deltat/2.0);
    tk = tk + deltat;
    k4 = f(tk, yk + k3 * deltat);
    yk = yk + (k1 + 2.0 * k2 + 2.0 * k3 + k4) * deltat/6.0;
    printf("%f %f\n", tk, yk);
  }
}
```

The Runge-Kutta Method for Systems

The generalization of the Runge-Kutta method to first-order systems is straightforward if we take advantage of our vector notation for systems. Moreover, once we implement Runge-Kutta for systems, we also have an efficient method for the numerical study of higher-order differential equations, since higher-order equations can be rewritten as first-order systems. In fact, this is the only way to study second- and higher-order differential equations numerically

Consider the first-order system

$$\frac{d\mathbf{Y}}{dt} = \mathbf{F}(t, \mathbf{Y}), \quad \mathbf{Y}(t_0) = \mathbf{Y}_0.$$

To apply the Runge-Kutta method to this system, we need to compute four vectors $\mathbf{K}_1$, $\mathbf{K}_2$, $\mathbf{K}_3$, and $\mathbf{K}_4$ at each step of the algorithm. These vectors are analogous to the four slopes that we computed for first-order equations, and they are defined in essentially the same way. Each of these vectors also depends on the index k, but we will not indicate this dependence to keep the formulas more readable. That is, suppose we have the point $(t_k, \mathbf{Y}_k)$. Then the vectors are

$$\mathbf{K}_1 = \mathbf{F}(t_k, \mathbf{Y}_k)$$
$$\mathbf{K}_2 = \mathbf{F}(t_k + \Delta t/2, \mathbf{Y}_k + (\Delta t/2)\mathbf{K}_1)$$
$$\mathbf{K}_3 = \mathbf{F}(t_k + \Delta t/2, \mathbf{Y}_k + (\Delta t/2)\mathbf{K}_2)$$
$$\mathbf{K}_4 = \mathbf{F}(t_k + \Delta t, \mathbf{Y}_k + \Delta t\,\mathbf{K}_3).$$

Once we have these vectors, then we can form a weighted average just as in the case of first-order equations. We get the vector

$$\mathbf{K} = \left(\frac{\mathbf{K}_1 + 2\mathbf{K}_2 + 2\mathbf{K}_3 + \mathbf{K}_4}{6}\right),$$

and we use this vector **K** to make the step from $(t_k, \mathbf{Y}_k)$ to

$$(t_{k+1}, \mathbf{Y}_{k+1}) = (t_k + \Delta t, \mathbf{Y}_k + \Delta t \, \mathbf{K}).$$

If we are plotting solution curves in the phase plane, then we simply plot the points $\mathbf{Y}_k$.

Figures 7.11 and 7.12 illustrate the results of Euler's method versus the Runge-Kutta method on the predator-prey system we discussed at length in Section 2.1. Note that although Euler's method is not accurate enough to catch the closed orbits in the phase plane, the Runge-Kutta method does a nice job of capturing the periodic nature of this system.

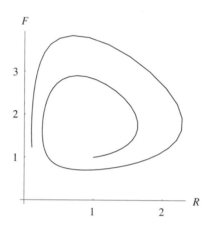

Figure 7.11
Euler's method for the solution with initial condition $\mathbf{Y}(0) = (1, 1)$ with $\Delta t = 0.1$ and 80 steps for the system $dR/dt = 2R - 1.2RF,\quad dF/dt = -F + 0.9RF.$

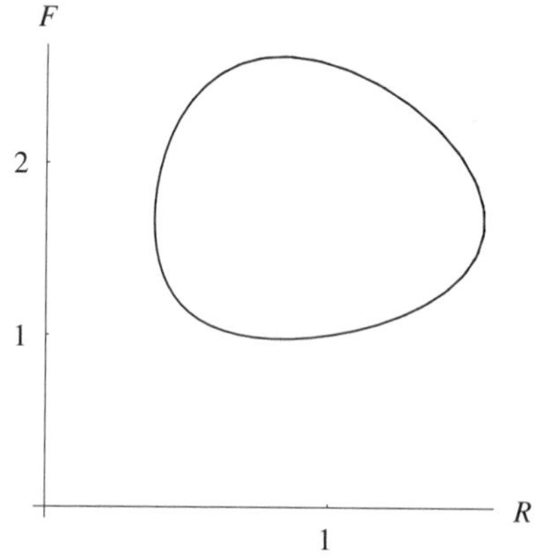

Figure 7.12
Runge-Kutta applied to the solution with initial condition $\mathbf{Y}(0) = (1, 1)$ with $\Delta t = 0.1$ and 80 steps for the same system as Figure 7.11.

Exercises for Section 7.3

In Exercises 1–5, use the Runge-Kutta method with the given number n of steps to approximate the solution to the initial-value problem specified. Your answer should include a table of approximate values of the dependent variable. It should also include a sketch of the graph of the approximate solution. These initial-value problems also appear in Section 7.2. Compare the graphs that you get from the Runge-Kutta method to those that come from Euler's method and improved Euler's method. If your computer has a built-in routine for the numerical solution of differential equations, compare these graphs with those that come from the built-in numerical solver.

1. $\dfrac{dy}{dt} = 2y + 1$, $y(0) = 3$, $0 \le t \le 2$, $n = 4$

2. $\dfrac{dy}{dt} = t - y^2$, $y(0) = 1$, $0 \le t \le 1$, $n = 4$

3. $\dfrac{dy}{dt} = (3 - y)(y + 1)$, $y(0) = 0$, $0 \le t \le 5$, $n = 10$

4. $\dfrac{dy}{dt} = e^{2/y}$, $y(0) = 2$, $0 \le t \le 2$, $n = 4$

5. $\dfrac{dy}{dt} = e^{2/y}$, $y(1) = 2$, $1 \le t \le 3$, $n = 4$

6. Consider the initial-value problem

$$\frac{dy}{dt} = -2ty^2, \quad y(0) = 1$$

over the interval $0 \le t \le 2$.

(a) Calculate the Runge-Kutta approximation to the solution with $n = 4$ steps.

(b) Calculate the total error e_4 associated with this approximation.

(c) How many steps are necessary to approximate the solution with an error of less than 0.0001?

7. Consider the predator-prey system

$$\frac{dR}{dt} = 2R - 1.2RF$$

$$\frac{dF}{dt} = -F + 1.2RF$$

and the initial condition $(R_0, F_0) = (1.0, 1.0)$.

(a) Calculate the Runge-Kutta approximation to this system for five steps with $\Delta t = 1.0$. Plot the results in the phase plane.

(b) With the aid of a calculator or a computer, repeat the calculation in part (a) with $\Delta t = 0.1$ and $n = 80$ steps. Plot the result in the phase plane.

(c) Using the results of part (b), plot the $R(t)$- and $F(t)$-graphs for this solution.

8. Use Runge-Kutta to approximate the solution to the second-order equation

$$\frac{d^2y}{dt^2} + (y^2 - 3)\frac{dy}{dt} + 5y = 0$$

that satisfies the initial condition $(y(0), y'(0)) = (1, 0)$. Use $\Delta t = 0.1$ and $n = 100$. Plot your results both in the yv-phase plane (where $v = y'$) and as the graph of $y(t)$.

7.4 THE EFFECTS OF FINITE ARITHMETIC

In theory, the errors involved in the use of numerical approximation procedures decrease as the step sizes decrease, and our preference for the Runge-Kutta method stems from the fact that this method is a fourth-order method that isn't too complicated to implement. It is tempting to assume that the computer or calculator is doing exactly what it was told to do. Unfortunately, this is not the case. Floating-point arithmetic has only a finite amount of accuracy, and this limitation introduces another source of error in each step of the numerical method.

In this section we briefly take a look at a couple of informative examples so that you are aware of some of the pitfalls that are associated with computer arithmetic. As always, we start with our favorite example so that we know the exact solution, and then we discuss what should be done when we do not know the solution. Of course, the latter case is the situation in which we typically employ these algorithms.

The Initial Value Problem $dy/dt = -2ty^2$, $y(0) = 1$ One More Time

In the previous section, we found that the Runge-Kutta method gives us a very accurate approximation to the solution $y(t) = 1/(1 + t^2)$ over the interval $0 \leq t \leq 2$ using only the step size $\Delta t = 0.1$. This is a fairly large step size. In theory, we can make the error as small as we like by decreasing the step size. To see if this works in practice, we try an experiment. (Such experiments are an excellent use of home computers, a much better use of electricity than the drawing of flying toasters.)

We start by subdividing the interval $0 \leq t \leq 2$ into two subintervals (so there are two steps and $\Delta t = 1.0$), and we calculate the error e_2. The Runge-Kutta method with $\Delta t = 1.0$ yields the approximation $y_2 = 0.182588$ of the exact value $y(2) = 0.2$, and therefore the error is

$$e_2 = |0.2 - 0.182588| = 0.0174118.$$

It's remarkable that Runge-Kutta gives such an accurate approximation with such a large Δt.

The theory says that, if we double the number of steps in the Runge-Kutta method, the error should decrease by a factor of 16. Therefore, we double the number of steps (and halve the step size) and recompute the approximate solution. Repeating this procedure, doubling the number of steps each time, we obtain the data in Table 7.3.

Note that the error improves as expected for the first 12 doublings. Note that, for $k = 13$, we have an approximation that agrees with the exact value to 15 decimal places. Then, at $k = 14$, the error starts to increase. Calculating more steps does not seem to help (see Table 7.4).

Table 7.3 The error in the value of $y(2)$ associated with the Runge-Kutta method for $dy/dt = -2ty^2$, $y(0) = 1$ for 2^k subdivisions of the interval $0 \leq t \leq 2$.

k	2^k	e_{2^k}	k	2^k	e_{2^k}
1	2	0.01741182739635213639	10	1024	0.00000000000008953949
2	4	0.00040567218499903968	11	2048	0.00000000000000552336
3	8	0.00002714430679354174	12	4096	0.00000000000000080491
4	16	0.00000161801719431032	13	8192	0.00000000000000024980
5	32	0.00000009769105269175	14	16384	0.00000000000000094369
6	64	0.00000000598843688526	15	32768	0.00000000000000072164
7	128	0.00000000037049122104	16	65536	0.00000000000000077716
8	256	0.00000000002303576774	17	131072	0.00000000000000144329
9	512	0.00000000000143582368	18	262144	0.00000000000000355271

Table 7.4 The error associated with the Runge-Kutta method for $dy/dt = -2ty^2$ for 2^k subdivisions

k	2^k	e_{2^k}	k	2^k	e_{2^k}
12	4096	0.0000000000000080491	20	1048576	0.0000000000000224820
13	8192	0.0000000000000024980	21	2097152	0.0000000000000621725
14	16384	0.0000000000000094369	22	4194304	0.0000000000001271205
15	32768	0.0000000000000072164	23	8388608	0.0000000000001187939
16	65536	0.0000000000000077716	24	16777216	0.0000000000001057487
17	131072	0.0000000000000144329	25	33554432	0.0000000000002037259
18	262144	0.0000000000000355271	26	67108864	0.0000000000000949241
19	524288	0.0000000000000130451			

The Effects of Finite Arithmetic

At a certain point, the efficiency of the numerical method is defeated by the effects of round-off errors in the arithmetic of the computer. Unless we are lucky for some reason, we cannot expect to get an answer that is any more accurate than the arithmetic allows. This is like the concept of significant digits in the sciences.

There are various ways to go about modeling the accuracy of the computer arithmetic, and like our models, they lead to somewhat different predictions about the effects of finite arithmetic. For our purposes, we'll adopt one of the standard models that says the total error, denoted by $e(\Delta t)$, as a function of step size Δt for the Runge-Kutta method is roughly

$$e(\Delta t) = c_1 (\Delta t)^4 + \frac{c_2}{\Delta t},$$

where c_1 and c_2 are constants determined by the differential equation, its initial condition, and interval of which the solution is being approximated. The first term, $c_1 (\Delta t)^4$, is the term that says that Runge-Kutta is a fourth-order method. The second term, $c_2/(\Delta t)$, is the term that measures the round-off errors for small values of Δt. Presumably, the constant c_2 is very small, and thus Δt must be very small before the effects of this term are noticeable.

We can observe the effects of round-off if we repeat the calculations done above with single-precision arithmetic (see Table 7.5). Note that, in this case, we are better off with 128 subdivisions than we are with millions of subdivisions. In fact, by the time we get to 16 million subdivisions, our approximation is about as accurate as it was with 2 subdivisions.

What to Do in Practice

In practice we want to use these numerical procedures when we do not know analytic representations for the solutions of a differential equation. Since it is usually impossible to carry out meaningful estimates from the theory, how do we decide what is a reasonable strategy to follow? If we accept the assertion that the error is roughly determined by an expression of the form

$$e(\Delta t) = c_1 (\Delta t)^4 + \frac{c_2}{\Delta t},$$

Table 7.5 The error associated to the Runge-Kutta method implemented in single precision for $dy/dt = -2ty^2$ for 2^k subdivisions

k	2^k	e_{2^k}	k	2^k	e_{2^k}
1	2	0.01741183	13	8192	0.00000031
2	4	0.00040567	14	16384	0.00000002
3	8	0.00002714	15	32768	0.00000001
4	16	0.00000161	16	65536	0.00000035
5	32	0.00000012	17	131072	0.00000324
6	64	0.00000002	18	262144	0.00000173
7	128	0.00000001	19	524288	0.00002292
8	256	0.00000007	20	1048576	0.00000132
9	512	0.00000001	21	2097152	0.00062698
10	1024	0.00000005	22	4194304	0.00052327
11	2048	0.00000012	23	8388608	0.00506006
12	4096	0.00000004	24	16777216	0.01590731

then we want to use a value of Δt that minimizes the first term as much as possible before the second term starts to have an effect. However, since we don't know the actual value of the solution, how can we predict where this minimum is going to occur?

We compute multiple approximations where we double the number of steps, and we look for our approximations to "converge" to a certain value. For example, if we apply Runge-Kutta to the initial value problem

$$\frac{dy}{dt} = y^2 - t, \quad y(1) = 1,$$

over the interval $1 \leq t \leq 3$ with various numbers of subdivisions, we get the results given in Table 7.6.

Note that we have the approximation $y(3) \approx -1.49389744$ with both 64 and 128 subdivisions. Given the repetition, we compute one more approximation with 256 subdivisions, and we are fairly confident that the actual value of the solution to five decimal

Table 7.6 Runge-Kutta applied to the initial-value problem $dy/dt = y^2 - t, y(1) = 1.$

k	$y_k \approx y(3)$
2	−1.39687991
4	−1.47455096
8	−1.49300539
16	−1.49384940
32	−1.49389482
64	−1.49389744
128	−1.49389744
256	−1.49389791
512	−1.49389756

places is -1.49389. It's also a relatively safe bet that the next digit in the decimal expansion is 7, but if we need a more precise answer, we repeat the calculations using more precision.

Exercises for Section 7.4

In Exercises 1–4, use Runge-Kutta to produce a table similar to Table 7.6 to approximate the value of the solution at the right-hand endpoint of the interval specified. What number of steps corresponds to the optimal approximation?

1. $\dfrac{dy}{dt} = t - y^3$, $y(0) = 2$, $0 \le t \le 1$

2. $\dfrac{dy}{dt} = \sin ty$, $y(0) = 3$, $0 \le t \le 3$

3. $\dfrac{dy}{dt} = (\cos y)(t^2 - y)$, $y(0) = 0$, $0 \le t \le 2$

4. $\dfrac{dy}{dt} = (t - y^2)\sqrt{y}$, $y(0) = 0$, $0 \le t \le 3$

Lab 7.1   Errors of Numerical Approximations

In this chapter, we studied three different methods for finding numerical approximations of solutions of differential equations: Euler, Improved Euler, and Runge-Kutta. The accuracy of these methods depends on the step size Δt. In this lab, we obtain additional insight into the relationship between the error in numerical approximations and the step size by studying some special equations and solutions.

In your report, address the following items:

1. Use Euler's method to approximate the solution of the initial value problem

$$\frac{dy}{dt} = y^2 - 4ty + 4t^2y - 4y + 8t - 3, \quad y(0) = -1.$$

How does the approximation change when the step size is changed? Interpret your results. [*Hint*: Look at the slope field for this equation. Also, look at Section 1.9.]

2. Use Euler's method, Improved Euler's method, and Runge-Kutta to approximate the solution of the initial value problem

$$\frac{dy}{dt} = 2\sqrt{y}, \quad y(0) = 1.$$

Find the formula for the solution and evaluate the error as a function of the step size for each of the methods. Are any of the methods better or worse than you would have expected? [*Hint*: The geometry of the slope field will not be as helpful here, but the special form of the formula of the solution is helpful.]

3. Repeat part 2 for the initial value problem

$$\frac{dy}{dt} = y + t^3 - 6, \quad y(0) = 0.$$

Your report: In your report, you should address each of the items above. You should **not** include any tables of numerical solutions. You may include a limited number of tables and/or graphs of how the error in your approximate solutions depends on the step size. The goal of this lab is to interpret why the error depends on the step size the way it does.

Lab 7.2   Lost in Space

The motion of planets and satellites in space is governed by the Newton's laws of motion and gravitational attraction. Newton's law of motion states that the total force is the product of the mass and the acceleration, and Newton's law of gravitation asserts that the force due to gravity between two bodies is inversely proportional to the distance between the bodies. This force is an attracting force, pulling each body directly toward the other. As in the case of the harmonic oscillator in Chapter 2 and the pendulum in Chapter 4, these laws give rise to second-order differential equations for the positions of the bodies as a function of time.

It is frequently said that the differential equations governing the motion of the moon and the planets were the first differential equations. The ability to compute the position of heavenly bodies (like the moon) was of great financial importance during the time of Newton. Knowing the precise location of the moon and planets at any given time was essential in navigation of the open sea. These equations are still of great theoretical importance in the development of techniques for the study of differential equations. They are also of great practical importance for space navigation. The type of numerical approximation techniques we have discussed are used for computing orbits of satellites to the moon and planets (and back — see the movie Apollo 13).

In this lab, suppose you are on a mission in deep space, and after a misunderstanding, your roommate leaves you adrift in your space suit with no fuel. Not being completely without a heart, your roommate also leaves behind an escape capsule. However, you and the escape capsule are 100 meters apart and you are both stationary. "Not to worry," you think, "two objects attract each other via the force of gravity, so the escape capsule and I will come together eventually." Suppose your mass (in the space suit) is 150 kg and the mass of the escape capsule is 10^6 kg. Suppose that you must be within 1 meter of the capsule in order to grab it. How long will it take you to reach the capsule?

First we need the differential equations. Since both you and the capsule start at rest and the only force is gravity which acts along the line connecting you with the capsule, both you and the capsule will remain on the straight line connecting you. Consequently, we only need one number to denote the relative positions of you and the capsule. Let $x_1(t)$ be the position of the capsule at time t and $x_2(t)$ denote your position at time t. We assume that coordinates have been chosen so that $x_2 > x_1$. We are most interested in the distance between you and the capsule, so let $x(t) = x_2(t) - x_1(t)$. By Newton's laws we know that the product of your mass and your acceleration is equal to the force of gravity between you and the capsule and that this force is proportional to the product of your masses divided by the square of the distance between you. That is,

$$150 \frac{d^2 x_2}{dt^2} = -G \frac{150 \times 10^6}{|x_2 - x_1|^2}$$

where the minus sign indicates that the force is pulling in the direction of decreasing x_2 and G is the "gravitation constant." This is the same as

$$\frac{d^2 x_2}{dt^2} = -G \frac{10^6}{|x_2 - x_1|^2}.$$

Similarly, the acceleration of the capsule is given by

$$\frac{d^2 x_1}{dt^2} = G \frac{150}{|x_2 - x_1|^2}.$$

Subtracting these equations and using $x = x_2 - x_1$, we obtain the differential equation

$$\frac{d^2 x}{dt^2} = -G \frac{10^6 + 150}{x^2}$$

for $x(t)$. This is a nonlinear, second-order differential equation. The gravitation constant G has been carefully measured by experiment and is given by $G \approx 6.67 \times 10^{-17}$. (The units are kg m^3/sec^2.)

In your report, address the following items:

1. Convert the second-order differential equation

$$\frac{d^2 x}{dt^2} = -G \frac{10^6 + 150}{x^2}$$

into a first-order system.

2. Using numerical techniques, approximate the time necessary for you to come to within 1 meter of the escape capsule given that you started 100 meters away and both you and the capsule have initial velocity 0. At what velocity are you and the capsule coming together when you are 1 meter apart? (You may use whatever numerical method you like.)

3. Verify that your computation of the time until you are within 1 meter of the escape capsule is accurate to within 1 second.

Your report: In your report you should address the items above. Be sure to explain your methods, particularly for part 3. You may use graphs and/or charts to illustrate your methods, but do not submit long lists of numbers.

LAPLACE TRANSFORMS

8

In this chapter we study an analytic technique for finding formulas for solutions of certain differential equations using an operation called the Laplace transform. This technique is particularly effective on linear, constant-coefficient, nonhomogeneous equations. It also allows us to deal very easily with equations that feature discontinuities, a topic that we have not encountered very often earlier.

The Laplace transform is important for two reasons. First, it "turns calculus into algebra" in a way that is made precise in Section 8.1. Second, it gives a way of extending the qualitative techniques associated with eigenvalues for homogeneous linear equations to nonhomogeneous linear equations. We discuss the use of Laplace transforms in the qualitative analysis of differential equations in Section 8.4.

8.1 LAPLACE TRANSFORMS

Integral Transforms

In this chapter we begin the study of a tool for attacking differential equations known as the **Laplace transform**. The Laplace transform is one of many different types of *integral transforms*. The goal of integral transforms is to address the question: How much is a given function $y(t)$ "like" one of the standard functions?

For example, if $y(t)$ represents a radio signal, we might want to compare it to the function $\sin(\omega t)$, a sine wave with freqency $\omega/2\pi$. By adjusting the parameter ω, we can test how well $y(t)$ fits sine waves with different frequencies. Ideally, for each value of ω we would like a number that indicates how much $y(t)$ looks like $\sin(\omega t)$. One way to accomplish this comparison is to compute the integral

$$\int_{-N}^{N} y(t) \sin(\omega t)\, dt$$

for large N. If $y(t)$ is oscillating with freqency $\omega/2\pi$ and is positive when $\sin(\omega t)$ is positive, then this integral is very large. If $y(t)$ has some other frequency, then the signs of $y(t)$ and $\sin(\omega t)$ differ at some times t, so there is cancellation in the integral and its value is smaller.

To use this idea in differential equations, it is natural to use as our standard the function that comes up most often, the exponential function. That is, we compute

$$\int_{-\infty}^{\infty} y(t) e^{-zt}\, dt,$$

where $z = s + i\omega$ is a complex parameter. It turns out to be easier to consider the real and imaginary cases of the parameter z separately. The integral

$$\int_{-\infty}^{\infty} y(t) e^{-i\omega t}\, dt$$

is called the *Fourier transform* of the function $y(t)$. The value of the Fourier transform of y at a particular value of ω is a measure of the extent to which $y(t)$ is oscillating with frequency $\omega/2\pi$. The imaginary part of this quantity is the comparison of $y(t)$ to $\sin(\omega t)$ discussed above. The real part is the comparison of $y(t)$ to $\cos(\omega t)$. Although Fourier transforms have many applications in differential equations, we will not discuss it further in this book. Rather, we focus on the case where $z = s$ (that is, $\omega = 0$).

Laplace Transforms

The Laplace transform of a function $y(t)$ gives a comparison of $y(t)$ to e^{st} where s is real. That is,

Definition: The **Laplace transform** $\mathcal{L}[y] = Y$ of y is the new function defined by

$$\mathcal{L}[y] = Y(s) = \int_{0}^{\infty} y(t) e^{-st}\, dt$$

for all numbers s for which this improper integral converges.

Before we worry about convergence of this improper integral, note that in this expression we integrate with respect to t. The result of this integration is therefore a function that depends on the new variable s. That is, the Laplace transform takes a function $y(t)$ and converts it into a new function $\mathcal{L}[y] = Y(s)$. This transformation is not unlike the translation of a sentence in English into a sentence in Chinese. Both sentences have the same meaning, but the words and grammar are very different. We can express this by saying that $y(t)$ is a representation of a function in the time or t-domain while its Laplace tranform, $Y(s) = \mathcal{L}[y]$, is the same function in the s-domain. We denote the Laplace transform of a function by either $\mathcal{L}$ or by the corresponding capital letter. Thus the Laplace transform of $f(t)$ is denoted by $\mathcal{L}[f]$ or F.

In order to be certain that the Laplace transform exists (that is, the improper integral converges) for at least some values of s, we must restrict attention to functions $y(t)$ for which there is a constant $M > 0$ (which depends on y) such that $y(t) < e^{Mt}$. Such functions are said to have no more than "exponential growth." As a practical matter, most functions encountered in applications have this property.

Computation of Laplace Transforms

For a function $y(t)$ and a given number s, the value of the Laplace transform at s is supposed to be a measure of how much the function $y(t)$ resembles the function e^{st} on the interval $[0, \infty)$. To check this we should compute the Laplace transform of an exponential function. If $y(t) = e^{at}$, then

$$\mathcal{L}[e^{at}] = \int_0^\infty e^{at} e^{-st} \, dt$$

$$= \int_0^\infty e^{(a-s)t} \, dt.$$

Recall that an improper integral is actually a limit of integrals as the upper limit of integration goes to infinity. Hence

$$\mathcal{L}[e^{at}] = \lim_{b \to \infty} \int_0^b e^{(a-s)t} \, dt$$

$$= \lim_{b \to \infty} \left[\frac{1}{a-s} e^{(a-s)t} \Big|_0^b \right]$$

$$= \lim_{b \to \infty} \frac{1}{a-s} \left[e^{(a-s)b} - e^0 \right]$$

$$= \frac{1}{s-a} \quad \text{if } s > a.$$

If $s \leq a$, then the improper integral diverges. So the Laplace transform of, for example, the exponential function $y(t) = e^{2t}$ is the rational function

$$\mathcal{L}[e^{2t}] = \begin{cases} \dfrac{1}{s-2}, & \text{if } s > 2, \\ \text{undefined}, & \text{otherwise.} \end{cases}$$

The Laplace transform does a good job of comparing e^{2t} with e^{st}, since as s approaches 2, $1/(s-2)$ goes to infinity.

We refer back to these computations and to the properties of the Laplace transform often. A table of Laplace transforms of frequently encountered functions and some properties of the Laplace transform operation are collected in a table on page 594.

Laplace transform of a Heaviside function

One of the most useful aspects of the Laplace transform is that it can be applied to discontinuous functions. For example, let

$$u_a(t) = \begin{cases} 0 & \text{if } t < a \\ \\ 1 & \text{if } t \geq a \end{cases}$$

The function $u_a(t)$ jumps from zero to one at time $t = a$. For example, the function $u_2(t)$ has a discontinuity at $t = 2$, where it jumps from zero to one (see Figure 8.1). A function of this form is called a step function "Heaviside function" (named for the engineer Oliver Heaviside). It is useful when modeling discontinuous processes such as turning on a light switch.

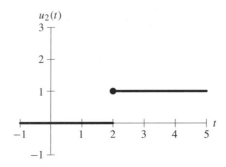

Figure 8.1
Graph of the Heaviside function $u_2(t)$.

The Laplace transform of $u_a(t)$ is

$$\mathcal{L}[u_a] = \int_0^\infty u_a(t)e^{-st}dt.$$

To compute this integral, we use the definition of u_a and break the computation into two pieces,

$$\mathcal{L}[u_a] = \int_0^a u_a(t)e^{-st}dt + \int_a^\infty u_a(t)e^{-st}dt.$$

The first integral is zero because $u_a(t)$ is zero for $t < a$. We can simplify the second integral because $u_a(t) = 1$ for $t \geq a$, hence

$$\mathcal{L}[u_a] = \int_a^\infty e^{-st}dt$$

$$= \lim_{b \to \infty} \int_a^b e^{-st} dt.$$

$$= \lim_{b \to \infty} \left. \frac{e^{-st}}{-s} \right|_a^b$$

$$= \lim_{b \to \infty} \frac{e^{-sb}}{-s} - \frac{e^{-as}}{-s}$$

$$= 0 + \frac{e^{-as}}{s}.$$

We have established the formula

$$\mathcal{L}[u_a] = \frac{e^{-as}}{s}.$$

Even though the original function u_a has a discontinuity at $t = a$, the Laplace transform is continuous for all $s > 0$. This *smoothing property* is a very useful aspect of Laplace transforms.

Properties of the Laplace Transforms

The reason the function e^{at} arises so frequently in differential equations is because its derivative, $d(e^{at})/dt = ae^{at}$, is related to the original function in a simple way. This relationship between exponentials and their derivatives carries over to the Laplace transform.

Laplace Transform of Derivatives

Given a function $y(t)$ with Laplace transform $Y(s)$, then the Laplace transform of dy/dt is $sY(s) - y(0)$. In other words,

$$\mathcal{L}\left[\frac{dy}{dt}\right] = s\mathcal{L}[y] - y(0)$$

To verify this theorem we must compute

$$\mathcal{L}\left[\frac{dy}{dt}\right] = \int_0^\infty \frac{dy}{dt} e^{-st} dt.$$

Using integration by parts with $u = e^{-st}$ and $dv = (dy/dt)\, dt$, we find

$$\mathcal{L}\left[\frac{dy}{dt}\right] = \left. y(t)e^{-st}\right|_0^\infty + \int_0^\infty y(t)se^{-st} dt.$$

Here is where the assumption that $y(t)$ has at most exponential growth becomes important. Since $y(t) < e^{Mt}$ for some constant M, it follows that $\lim_{t \to \infty} y(t)e^{-st} = 0$ for sufficiently large s. Hence,

$$\mathcal{L}\left[\frac{dy}{dt}\right] = -y(0) + \int_0^\infty y(t)se^{-st}\, dt$$

$$= -y(0) + s\mathcal{L}[y],$$

so

$$\mathcal{L}\left[\frac{dy}{dt}\right] = s\mathcal{L}[y] - y(0).$$

Laplace transforms replace the operation of differentiation in the time domain with the algebraic operation of multiplication by s in the s domain. Informally we say that Laplace transforms replaces derivatives with algebra, turning a problem in calculus into a problem in algebra.

Turning a differential equation problem into a problem in algebra would not be very useful if the algebra were impossible to deal with, so the following facts are very important.

Linearity of the Laplace Transform

Given the functions f and g and the constant c,

$$\mathcal{L}[f + g] = \mathcal{L}[f] + \mathcal{L}[g]$$

$$\mathcal{L}[cf] = c\mathcal{L}[f].$$

In other words, the "transform operation" $\mathcal{L}$ is "linear."

Frequently the Laplace transform $\mathcal{L}$ is referred to as a *linear operator*.

To verify these properties, we use the linearity properties of integration, that is,

$$\mathcal{L}[f+g] = \int_0^\infty (f(t)+g(t))e^{-st}\,dt = \int_0^\infty f(t)e^{-st}\,dt + \int_0^\infty g(t)e^{-st}\,dt = \mathcal{L}[f]+\mathcal{L}[g]$$

and

$$\mathcal{L}[cf] = \int_0^\infty cf(t)e^{-st}\,dt = c\int_0^\infty f(t)e^{-st}\,dt = c\mathcal{L}[f].$$

With the derivative formula and the linearity of $\mathcal{L}$ established, we are in a position to use the Laplace transform to solve differential equations.

Solving Differential Equations Using the Laplace Transform

Consider the initial-value problem

$$\frac{dy}{dt} = 2y + 3e^{-t}, \quad y(0) = 4.$$

This is a linear equation, so we could use integrating factors (see Section 1.8). However, the Laplace transform provides an alternate method of solution. The first step is to take the Laplace transform of both sides of the differential equation, which yields

$$\mathcal{L}\left[\frac{dy}{dt}\right] = \mathcal{L}[2y + 3e^{-t}].$$

Using the previous formula for the Laplace transform of a derivative to simplify the left-hand side and the linearity of Laplace transform on the right-hand side, we obtain

$$s\mathcal{L}[y] - y(0) = 2\mathcal{L}[y] + 3\mathcal{L}[e^{-t}].$$

At this early stage we can substitute the initial condition $y(0) = 4$ to get

$$s\mathcal{L}[y] - 4 = 2\mathcal{L}[y] + 3\mathcal{L}[e^{-t}].$$

In the example above we computed that

$$\mathcal{L}[e^{at}] = \begin{cases} \dfrac{1}{s-a}, & \text{if } s > a, \\ \text{undefined}, & \text{otherwise}, \end{cases}$$

so we can apply this formula with $a = -1$ to obtain

$$s\mathcal{L}[y] - 4 = 2\mathcal{L}[y] + \frac{3}{s+1}.$$

The unknown of the original differential equation is the function y, so we solve this equation for the Laplace transform of y, obtaining

$$\mathcal{L}[y] = \frac{4}{s-2} + \frac{3}{(s-2)(s+1)}.$$

This gives us the Laplace transform of the solution of the initial-value problem. We used only arithmetic (and no calculus) to obtain $\mathcal{L}[y]$. In a sense we have solved the initial-value problem.

Unfortunately we are not looking for the Laplace transform $\mathcal{L}[y]$ of the solution, we are looking for the solution $y(t)$. Somehow we must "undo" or take an "inverse" Laplace transform. That is, we must figure out what function $y(t)$ has the function

$$\frac{4}{s-2} + \frac{3}{(s-2)(s+1)}$$

as its Laplace transform.

Recalling the formula

$$\mathcal{L}[e^{at}] = \begin{cases} \dfrac{1}{s-a}, & \text{if } s > a; \\ \text{undefined}, & \text{otherwise} \end{cases}$$

and that Laplace transform is linear, we see that

$$\mathcal{L}[4e^{2t}] = \frac{4}{s-2}.$$

So a function whose Laplace transform is $4/(s-2)$ is $4e^{2t}$.

A function whose Laplace transform is $3/((s-2)(s+1))$ is not as easy to guess. The only general formula we have for Laplace transforms, the Laplace transform of e^{at}, has only linear terms in s in the denominator. A reasonable step is to try to rewrite $3/((s-2)(s+1))$ as a sum where only linear terms in s appear in the denominator. This can be accomplished by partial fractions, yielding

$$\frac{3}{(s-2)(s+1)} = \frac{1}{s-2} + \frac{-1}{s+1}.$$

Now we know the functions whose Laplace transforms appear here:

$$\mathcal{L}[e^{2t}] = \frac{1}{s-2} \quad \text{and} \quad \mathcal{L}[-e^{-t}] = \frac{-1}{s+1}.$$

Putting these together using linearity we have

$$\mathcal{L}[4e^{2t} + e^{2t} - e^{-t}] = \frac{4}{s-2} + \frac{3}{(s-2)(s+1)}.$$

We have discovered a function which has the desired transform. It follows that

$$y(t) = 4e^{2t} + e^{2t} - e^{-t} = 5e^{2t} - e^{-t}$$

is the solution of the initial-value problem. This can be checked as usual by substituting $y(t)$ back into the differential equation.

An RC Circuit Example

Consider the RC circuit in Figure 8.2. We let $v_c(t)$ denote the voltage across the capacitor; R and C denote the resistance and capacitance, respectively; and $V(t)$ the voltage supplied by the voltage source. From electric circuit theory, we know that $v_c(t)$ is a solution of the differential equation

$$RC\frac{dv_c}{dt} + v_c = V(t).$$

Suppose the voltage source $V(t)$ is not constant. Instead, suppose $V(t)$ jumps at time $t = 2$ from 0 to 3, that is

$$V(t) = \begin{cases} 0 & \text{if} \quad t < 2 \\ \\ 3 & \text{if} \quad t \ge 2 \end{cases}$$

(the voltage source is "turned on" at time $t = 2$). We can write $V(t) = 3u_2(t)$, where $u_2(t)$ is the Heaviside function defined above. See Figure 8.3. With the quantities R and C as in Figure 8.2 (rather unrealistic quantities from the point of view of circuits), the equation is

$$2\frac{dv_c}{dt} + v_c = 3u_2(t).$$

Suppose that $v_c(0) = 4$. We could attack this as two equations, one for $t < 2$ and one for $t \ge 2$, but Laplace transforms allow us to work with both t intervals at once.

As before, we begin by applying the Laplace transform to both sides of the differential equation

$$2\mathcal{L}\left[\frac{dv_c}{dt}\right] + \mathcal{L}[v_c] = 3\mathcal{L}[u_2].$$

Using the formula for the Laplace transform of a derivative and substituting the expression computed for $3\mathcal{L}[u_2]$, we obtain

$$2(s\mathcal{L}[v_c] - v_c(0)) + \mathcal{L}[v_c] = \frac{3e^{-2s}}{s}.$$

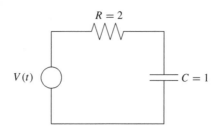

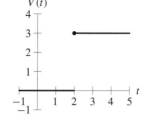

Figure 8.2
Circuit diagram where voltage source is not constant.

Figure 8.3
Graph of the voltage source $V(t) = 3u_2(t)$.

Using the initial condition $v_c(0) = 4$ and solving for $\mathcal{L}[v_c]$ gives

$$\mathcal{L}[v_c] = \frac{8}{2s + 1} + \frac{3e^{-2s}}{s(2s + 1)}.$$

Thus, we have found the Laplace transform of the solution of the given initial-value problem. All that remains is to find a function whose Laplace transform is the right-hand side of this equation.

Unfortunately this is not so easy. The first term can be written

$$\frac{8}{2s + 1} = \frac{4}{s + 1/2},$$

which is the Laplace transform of $4e^{-t/2}$. The second term can be simplified using partial fractions to

$$\frac{3e^{-2s}}{s(2s + 1)} = \frac{3e^{-2s}}{s} - \frac{3e^{-2s}}{s + 1/2},$$

where the first term is the Laplace transform of $3u_2(t)$. However, the term

$$\frac{3e^{-2s}}{s + (1/2)}$$

has not arisen as a Laplace transform yet. To complete this example, we need another property of Laplace transforms.

Shifting the Origin on the t-Axis

Given a function $f(t)$, suppose that we wish to consider a function $g(t)$, which is the same as the function $f(t)$ but shifted so that time $t = 0$ for f corresponds to some later time, say $t = a$, for g. For time $t \le a$ we let $g(t) = 0$. (So g is the same as f except that g is turned on at time $t = a$.) An efficient way to write $g(t)$ is

$$g(t) = u_a(t)f(t - a).$$

Note that $g(a) = u_a(a)f(a - a) = f(0)$ and, if $b > 0$, $g(a + b) = u_a(a + b)f(b) = f(b)$ as desired. For example, if $f(t) = e^{-t/5}$ and $a = 4$, then $g(t) = u_4(t)e^{-(t-4)/5}$ (see Figures 8.4 and 8.5).

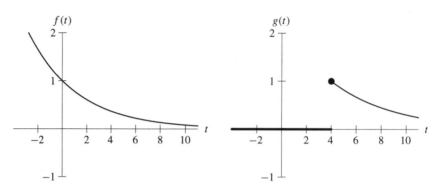

Figure 8.4
Graph of $f(t) = e^{-t/5}$.

Figure 8.5
Graph of $g(t) = u_4(t)e^{-(t-4)/5}$.

To compute the Laplace transform of $g(t)$, we return to the definition

$$\mathcal{L}[g] = \int_0^\infty g(t)e^{-st}dt.$$

Using the fact that $g(t) = 0$ for $t < a$ and that $g(t) = f(t-a)$ for $t \geq a$, we obtain

$$\mathcal{L}[g] = \int_a^\infty f(t-a)e^{-st}dt.$$

The u-substitution $u = t - a$ in the integral gives

$$\mathcal{L}[g] = \int_0^\infty f(u)e^{-s(u+a)}du$$

$$= e^{-sa}\int_0^\infty f(u)e^{-su}du$$

$$= e^{-sa}\mathcal{L}[f] = e^{-sa}F(s).$$

Hence we can express the Laplace transform of $g(t) = u_a(t)f(t-a)$ in terms of the Laplace transform of $f(t)$ by the following rule:

If $\mathcal{L}[f] = F(s)$, then $\mathcal{L}[u_a(t)f(t-a)] = e^{-as}F(s)$.

For the example above, if $g(t) = u_4(t)e^{-(t-4)/5}$, then the Laplace transform of $g(t)$ is

$$\mathcal{L}[g] = e^{-4s}\mathcal{L}[e^{-t/5}] = \frac{e^{-4s}}{s+(1/5)}.$$

Completion of the RC circuit example

Recall that we started with the initial-value problem

$$2\frac{dv_c}{dt} + v_c = 3u_2(t), \quad v_c(0) = 4.$$

We found above that the Laplace transform of the solution $v_c(t)$ was given by

$$\mathcal{L}[v_c] = \frac{4}{s+1/2} + \frac{3e^{-2s}}{s} - \frac{3e^{-2s}}{s+1/2}.$$

We already know that

$$\mathcal{L}[4e^{-t/2}] = \frac{4}{s + 1/2} \quad \text{and} \quad \mathcal{L}[3u_2] = \frac{3e^{-2s}}{s},$$

so we need only find the function whose Laplace transform is equal to the third summand of $\mathcal{L}[v_c]$. Using the fact that

$$\mathcal{L}[3e^{-t/2}] = \frac{3}{s + 1/2}$$

and the rule derived above, we have that

$$\mathcal{L}[3u_2(t)e^{-(t-2)/2}] = \frac{3e^{-2s}}{s + (1/2)}.$$

Hence

$$v_c(t) = 4e^{-t/2} + 3u_2(t) - 3u_2(t)e^{-(t-2)/2}$$

$$= 4e^{-t/2} + 3u_2(t)(1 - e^{-(t-2)/2}).$$

To interpret this formula for the solution we look at the slope field for the differential equation

$$2\frac{dv_c}{dt} + v_c = 3u_2(t)$$

(see Figure 8.6) and at the graph of the solution (see Figure 8.7). For $0 \le t \le 2$, the function $v_c(t)$ decays from its initial value of $v_c(0) = 4$ toward zero. At $t = 2$, the right-hand side of the equation jumps from 0 to 3. The value of $v_c(t)$ starts to rise at $t = 2$ toward $v_c = 3$.

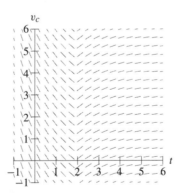

Figure 8.6

Slope field for $2\dfrac{dv_c}{dt} + v_c = 3u_2(t)$.

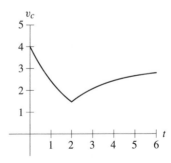

Figure 8.7

Graph of $v_c(t) =$
$4e^{-t/2} + 3u_2(t)(1 - e^{-(t-2)/2})$.

Analysis of the Laplace Transform Technique

Finding a solution of a differential equation using Laplace transforms is quite different from other procedures we have studied. First, the initial conditions enter the calculations very early instead of at the end. Second, the integrals that are necessary in the other methods are replaced by algebra. Third, several steps in the process involve using previously computed formulas for Laplace transforms.

In a sense, the integrals necessary for solving a differential equation are done when computing Laplace transforms. By compiling a list of Laplace transforms (see the table on page 594), we can replace the integration steps necessary in solving a differential equation with algebra and "table look-ups" of Laplace transforms.

Exercises for Section 8.1

1. Compute the Laplace transforms of the following functions using the definition of the transform $\mathcal{L}$.

   (a) $f(t) = 3$ the constant function,

   (b) $g(t) = t$,

   (c) $h(t) = -5t^2$,

   (d) $k(t) = t^{15}$.

2. (a) Using the results of the previous question as a guide, give a formula for the Laplace transform of a polynomial of the form $f(t) = a_0 + a_1 t + a_2 t^2 + \ldots + a_n t^n$ where the a_i are constants.

   (b) Using the definition of Laplace transform, verify that your formula is correct.

3. Compute the Laplace transform of $g(t) = 1 - u_3(t)$. (You may do this using the definition of Laplace transform or using the Laplace transform of $u_3(t)$ and the properties of Laplace transform.)

4. Find a formula for the Laplace transform of the function

$$
g_a(t) = \begin{cases} 1 & \text{if} \quad t < a \\ 0 & \text{if} \quad t \geq a. \end{cases}
$$

5. (a) Compute the Laplace transform of the function

$$
r_a(t) = \begin{cases} 0 & \text{if} \quad t < a \\ k(t - a) & \text{if} \quad t \geq a. \end{cases}
$$

   for $a \geq 0$. [Hint: Your solution will contain a and k as parameters.]

   (b) Sketch the graph of $r_a(t)$ and comment on why it is called a *ramp function*.

6. For each of the following, find a function which has the given function as its Laplace transform

(a) $e^{-2s}/(s-3)$,

(b) $e^{-3s}/((s-1)(s-2))$.

7. Find the Laplace transform of $f(t) = u_2(t)\cos(2t)$.

8. Find the general solution of the equation

$$\frac{dy}{dt} = 2y + 2e^{-3t}.$$

(This equation is linear, but please use the method of Laplace transforms.)

In Exercises 9–15,

(a) Compute the Laplace transform of both sides of the equation.

(b) Substitute the initial conditions and solve for the Laplace transform of the solution.

(c) Find a function whose Laplace transform is the same as the solution.

(d) Check that you have found the solution of the initial-value problem.

9. $\dfrac{dy}{dt} = -y + e^{-2t}; \quad y(0) = 2$ **10.** $\dfrac{dy}{dt} + 5y = e^{-t}; \quad y(0) = 2.$

11. $\dfrac{dy}{dt} + 7y = u_2(t); \quad y(0) = 3$ **12.** $\dfrac{dy}{dt} + 9y = u_5(t); \quad y(0) = -2$

13. $\dfrac{dy}{dt} = -y + t^2; \quad y(0) = 1$ **14.** $\dfrac{dy}{dt} = -y + 2u_3(t), \quad y(0) = 4$

15. $\dfrac{dy}{dt} = -y + u_2(t)e^{-2(t-2)}, \quad y(0) = 1$

16. Suppose $g(t) = \int f(t)dt$, that is, $g(t)$ is the antiderivative of $f(t)$. Express the Laplace transform of $g(t)$ in terms of the Laplace transform of $f(t)$.

17. Compute the Laplace transform of the function

$$g_a(t) = \begin{cases} t/a & \text{if} \quad t < a \\ 1 & \text{if} \quad t \geq a. \end{cases}$$

for $a \geq 0$. [Hint: Your solution will contain a as a parameter.]

18.   (a) Given a function $f(t)$ with Laplace transform $\mathcal{L}[f(t)] = F(s)$, find the Laplace transform of the function $f(at)$ where a is a nonzero constant. [Hint: Use the definition of the Laplace transform to compute $\mathcal{L}[f(at)]$ using that you already know that $\mathcal{L}[f(t)]$ is $F(s)$.]

(b) Formulate your computation above as a rule.

(c) The function $f(at)$ is called a rescaling of $f(t)$, why?

(d) Use this rule to compute the Laplace transform of $\cos(2\pi t)$.

8.2 LAPLACE TRANSFORMS AND SECOND-ORDER EQUATIONS

Laplace transforms turn a differential equation for an unknown function $y(t)$ into an algebraic equation for the Laplace transform $\mathcal{L}[y] = Y(s)$ of the unknown function. The advantage of Laplace transforms is that we can construct a table of the most commonly used functions and their Laplace transforms (and very large tables are available.) The computation of the Laplace transform integral is then replaced with a "table look-up."

Therefore, for the technique to be useful, we must have a fairly extensive table of Laplace transforms of different functions and some good tools for using the table. In this section we continue to build a short version of such a table and discuss some general rules that make the table more usable. We also apply the Laplace transform to second-order equations, particularly the forced harmonic oscillator equation. The Laplace transform is especially useful when dealing with forced harmonic oscillator equations with discontinuous forcing terms.

Laplace Transforms of Sine and Cosine

Next to the exponential, the sine and cosine seem to be the most commonly encountered functions in differential equations. We compute $\mathcal{L}[\sin(\omega t)]$ where ω is a parameter.

From the definition of Laplace transform,

$$\mathcal{L}[\sin(\omega t)] = \int_0^\infty \sin(\omega t)e^{-st}dt.$$

For this integral we use integration by parts with $u = \sin(\omega t)$ and $dv = e^{-st}dt$. This gives

$$\int_0^\infty \sin(\omega t)e^{-st}dt = -\sin(\omega t)\frac{e^{-st}}{s}\Big|_0^\infty + \int_0^\infty \omega\cos(\omega t)\frac{e^{-st}}{s}dt$$

$$= 0 + \int_0^\infty \omega\cos(\omega t)\frac{e^{-st}}{s}dt.$$

Integrating by parts again (using $u = \cos(\omega t)$ and $dv = (e^{-st}/s)dt$) yields

$$\int_0^\infty \sin(\omega t)e^{-st}dt = -\omega\cos(\omega t)\frac{e^{-st}}{s^2}\Big|_0^\infty - \int_0^\infty \omega^2\sin(\omega t)\frac{e^{-st}}{s^2}dt$$

$$= \frac{\omega}{s^2} - \frac{\omega^2}{s^2}\int_0^\infty \sin(\omega t)e^{-st}dt.$$

Combining the integrals of $\sin(\omega t)e^{-st}$, we obtain

$$\left(1 + \frac{\omega^2}{s^2}\right)\int_0^\infty \sin(\omega t)e^{-st}dt = \left(1 + \frac{\omega^2}{s^2}\right)\mathcal{L}[\sin(\omega t)] = \frac{\omega}{s^2}.$$

Solving for $\mathcal{L}[\sin(\omega t)]$ yields

$$\mathcal{L}[\sin(\omega t)] = \frac{\omega}{s^2 + \omega^2}.$$

We could compute $\mathcal{L}[\cos(\omega t)]$ in the same manner, however, it is easier to take advantage of how Laplace transforms deal with derivatives. Recall that $\mathcal{L}[df/dt] = s\mathcal{L}[f] - f(0)$. We can use this and the fact that

$$\frac{d(\sin(\omega t))}{dt} = \omega \cos(\omega t)$$

to compute

$$\mathcal{L}[\omega \cos(\omega t)] = \mathcal{L}\left[\frac{d(\sin(\omega t))}{dt}\right]$$

$$= s \cdot \frac{\omega}{s^2 + \omega^2} - \sin(\omega \cdot 0)$$

$$= \frac{s\omega}{s^2 + \omega^2}.$$

Hence,

$$\mathcal{L}[\cos(\omega t)] = \frac{s}{s^2 + \omega^2}.$$

Shifting the Origin on the s-Axis

We have frequently encountered functions of the form $e^{\alpha t}\sin(\omega t)$. Although the Laplace transform of a product is generally quite complicated, it turns out that the situation is easier when one of the multiplicands is an exponential.

Suppose we are given a function $f(t)$ and we know that its Laplace transform is $F(s)$. To compute the Laplace transform of $e^{at} f(t)$, we recall the definition of Laplace transform and write

$$\mathcal{L}[e^{at} f(t)] = \int_0^\infty e^{at} f(t) e^{-st} dt$$

$$= \int_0^\infty f(t) e^{-(s-a)t} dt$$

$$= F(s - a)$$

That is,

If $\mathcal{L}[f] = F(s)$ then $\mathcal{L}[e^{at} f(t)] = F(s - a)$.

Multiplying $f(t)$ by e^{at} corresponds to shifting the argument of the Laplace transform of f by a.

For example, the Laplace transform of $e^{-3t}\cos(2t)$ is

$$\mathcal{L}[e^{3t}\cos(2t)] = \frac{s + 3}{(s + 3)^2 + 2^2} = \frac{s + 3}{s^2 + 6s + 13}.$$

Inverse Laplace Transform

To use Laplace transforms to solve a differential equation, we first compute the Laplace transform of both sides of the equation. Next we solve for the Laplace transform of the solution. Finally, to find the solution, we find a function with the given Laplace transform. The last step is called taking the **inverse Laplace transform**. The notation is $\mathcal{L}^{-1}$, that is,

$$\mathcal{L}^{-1}[F] = f \quad \text{if and only if} \quad \mathcal{L}[f] = F.$$

There is a uniqueness property for Laplace transforms which states that for each function $f(t)$ for which $\mathcal{L}[f]$ exists, there is precisely one function $F(s)$ such that $\mathcal{L}^{-1}[F] = f$. This allows us to say "the" inverse Laplace transform instead of "an" inverse Laplace transform. Since the integral in the definition of the Laplace transform has limits zero and infinity, only nonnegative values of t are used. The inverse Laplace transform is defined only for $t \geq 0$.

Because the Laplace transform is a linear operator, that is,

$$\mathcal{L}[f + g] = \mathcal{L}[f] + \mathcal{L}[g] \text{ and } \mathcal{L}[cf] = c\mathcal{L}[f]$$

for any functions $f(t)$ and $g(t)$ and any constant c, it follows that the inverse Laplace transform is also a linear operator,

$$\mathcal{L}^{-1}[f + g] = \mathcal{L}^{-1}[f] + \mathcal{L}^{-1}[g] \text{ and } \mathcal{L}^{-1}[cf] = c\mathcal{L}^{-1}[f].$$

This is important because it allows us to compute the inverse Laplace transform of a complicated sum by computing the inverse Laplace transform of each summand.

Inverse Laplace transforms are computed by "table look-up." To find the inverse Laplace transform of a function $F(s)$, we go to our table and look down the list of Laplace transforms until we find one that matches $F(s)$. To compute the inverse Laplace transform of a complicated function $F(s)$, we first break it into a sum of functions that we can find in our table.

An example of the inverse Laplace transform

For example, to compute

$$\mathcal{L}^{-1}\left[\frac{3s + 2}{(s^2 + 4)(s - 2)}\right],$$

we use partial fractions to write

$$\frac{3s + 2}{(s^2 + 4)(s - 2)} = \frac{-s + 1}{s^2 + 4} + \frac{1}{s - 2}$$

$$= -\frac{s}{s^2 + 4} + \frac{1}{s^2 + 4} + \frac{1}{s - 2}.$$

Looking in the table, we see that the first term fits into the form

$$\mathcal{L}[\cos(\omega t)] = \frac{s}{s^2 + \omega^2}$$

with $\omega = 2$. Hence

$$\mathcal{L}^{-1}\left[-\frac{s}{s^2 + 4}\right] = -\cos(2t).$$

The middle term almost fits the form

$$\mathcal{L}[\sin(\omega t)] = \frac{\omega}{s^2 + \omega^2}.$$

If we multiply and divide the middle term by 2, then

$$\mathcal{L}^{-1}\left[\frac{1}{s^2 + 4}\right] = \frac{1}{2}\mathcal{L}^{-1}\left[\frac{2}{s^2 + 2^2}\right] = \frac{1}{2}\sin(2t).$$

Finally, the third term appears in the table, $\mathcal{L}^{-1}[1/(s-2)] = e^{2t}$. Putting these together using linearity, we obtain

$$\mathcal{L}^{-1}\left[\frac{3s + 2}{(s^2 + 4)(s - 2)}\right] = -\cos(2t) + \frac{1}{2}\sin(2t) + e^{2t}.$$

A messier example

Not all quadratics are perfect squares nor do they always have integers as roots. We must be ready to use additional algebraic techniques to massage a given expression into pieces that appear in the table of Laplace transforms. As an example that we will encounter in the next section, we compute

$$\mathcal{L}^{-1}\left[\frac{3}{s(s^2 + s + 3)} + \frac{3e^{-4s}}{s(s^2 + s + 3)}\right].$$

To find the inverse Laplace transform, we first compute that the partial fractions decomposition of $3/(s(s^2 + s + 3))$ is

$$\frac{3}{s(s^2 + s + 3)} = \frac{1}{s} - \frac{s + 1}{s^2 + s + 3}.$$

The first term appears in the table of Laplace tranforms, $\mathcal{L}^{-1}[1/s] = 1$. To put the second term into a form we can find in the table, we complete the square,

$$s^2 + s + 3 = \left(s + \frac{1}{2}\right)^2 + \frac{11}{4}.$$

This gives a term of the form $s + 1/2$ in the denominator while $s + 1$ appears in the numerator. We can remedy this by breaking this expression into two pieces

$$\frac{s + 1}{s^2 + s + 3} = \frac{s + (1/2)}{(s + (1/2))^2 + (11/4)} + \frac{1/2}{(s + (1/2))^2 + (11/4)},$$

and each of these terms appears in the table. Thus

$$\mathcal{L}^{-1}\left[\frac{s + 1}{s^2 + s + 3}\right] = e^{-t/2}\left[\cos\left(\frac{\sqrt{11}}{2}t\right) + \frac{1}{\sqrt{11}}\sin\left(\frac{\sqrt{11}}{2}t\right)\right].$$

To find $\mathcal{L}^{-1}[3e^{-4s}/(s(s^2 + s + 3))]$, we use the rule derived in Section 8.1, combined with the same computation as above. Hence

$$\mathcal{L}^{-1}\left[\frac{3}{s(s^2 + s + 3)} + \frac{3e^{-4s}}{s(s^2 + s + 3)}\right] = 1 - e^{-t/2}\left[\cos\left(\frac{\sqrt{11}}{2}t\right) + \frac{1}{\sqrt{11}}\sin\left(\frac{\sqrt{11}}{2}t\right)\right]$$

$$+ u_4(t) - u_4(t)e^{-(t-4)/2}\left[\cos\left(\frac{\sqrt{11}}{2}(t - 4)\right) + \frac{1}{\sqrt{11}}\sin\left(\frac{\sqrt{11}}{2}(t - 4)\right)\right].$$

A Forced Harmonic Oscillator with Discontinuous Forcing

Consider a harmonic oscillator, say a mass attached to a spring sliding back and forth on a table. Suppose the mass is at rest, but then, at time $t = 0$, the table is tilted, so that gravity is provides a force that stretches the spring. We expect that the mass will slide down, then tend to a new equilibrium position on the tilted table (see Figure 8.8). Finally, suppose that at time $t = 7$, the table is suddenly returned to the level position, so that the force of gravity no longer affects the spring. The mass will tend back toward its original rest position. We could treat this problem as two initial-value problems, one for $0 \leq t < 7$ and another for $t \geq 7$; however, Laplace transforms allow us to treat this as one problem.

For definiteness, we fix the parameters of the mass-spring system described above as follows; mass $m = 1$, spring constant $k_s = 5$, coefficient of friction $k_d = 2$. We also assume that the angle of the tilted table is such that gravity provides a unit force stretching the spring. The initial-value problem is given by

$$\frac{d^2y}{dt^2} + 2\frac{dy}{dt} + 5y = h(t), \quad y(0) = 0, \, y'(0) = 0,$$

where $h(t) = 1 - u_7(t)$ is given by

$$h(t) = \begin{cases} 1, & \text{if } t < 7 \\ \\ 0, & \text{if } t \geq 7 \end{cases}$$

(see Figure 8.9).

To use Laplace transforms to find the solution of this initial-value problem, we begin as before by taking Laplace transforms of both sides of the equation. Using linearity, we obtain

$$\mathcal{L}\left[\frac{d^2y}{dt^2}\right] + 2\mathcal{L}\left[\frac{dy}{dt}\right] + 5\mathcal{L}[y] = \mathcal{L}[h(t)].$$

To simplify $\mathcal{L}\left[d^2y/dt^2\right]$, we use the rule for Laplace transforms of a derivative twice; that is,

$$\mathcal{L}\left[\frac{d^2y}{dt^2}\right] = s\mathcal{L}\left[\frac{dy}{dt}\right] - y'(0)$$

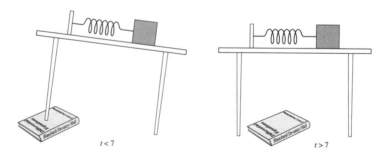

$t < 7$ $t > 7$

Figure 8.8

Schematic of mass spring on a tilted table ($t < 7$) and a level table ($t \geq 7$).

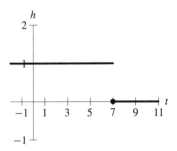

Figure 8.9
Graph of the function $h(t)$.

$$= s(s\mathcal{L}[y] - y(0)) - y'(0)$$

$$= s^2\mathcal{L}[y] - sy(0) - y'(0).$$

Returning to the differential equation, we have

$$(s^2\mathcal{L}[y] - sy(0) - y'(0)) + 2(s\mathcal{L}[y] - y(0)) + 5\mathcal{L}[y] = \mathcal{L}[h(t)].$$

Collecting terms, substituting in the initial conditions $y(0) = 0$, $y'(0) = 0$ and evaluating $\mathcal{L}[h(t)] = \mathcal{L}[1 - u_7(t)]$, we have

$$(s^2 + 2s + 5)\mathcal{L}[y] = \frac{1}{s} - \frac{e^{-7s}}{s}$$

or

$$\mathcal{L}[y] = \frac{1}{s(s^2 + 2s + 5)} - \frac{e^{-7s}}{s(s^2 + 2s + 5)};$$

that is,

$$y = \mathcal{L}^{-1}\left[\frac{1}{s(s^2 + 2s + 5)} - \frac{e^{-7s}}{s(s^2 + 2s + 5)}\right].$$

To compute the right-hand side, we first use partial fractions to write $1/(s(s^2 + 2s + 5))$ as

$$\frac{1}{s(s^2 + 2s + 5)} = \frac{1}{5s} - \frac{s + 2}{5(s^2 + 2s + 5)}.$$

Completing the square of the quadratic in the denominator gives

$$s^2 + 2s + 5 = (s + 1)^2 + 4.$$

So we can write

$$\frac{s + 2}{5(s^2 + 2s + 5)} = \frac{1}{5}\left(\frac{s + 1}{(s + 1)^2 + 2^2} + \frac{1}{2}\frac{2}{(s + 1)^2 + 2^2}\right).$$

Thus

$$\mathcal{L}^{-1}\left[\frac{1}{5s} - \frac{s + 2}{5(s^2 + 2s + 5)}\right] = \frac{1}{5} - \frac{1}{5}e^{-t}\left(\cos(2t) + \frac{1}{2}\sin(2t)\right).$$

The inverse Laplace transform of $e^{-7s}/(s(s^2+2s+5))$ involves the Heaviside function $u_7(t)$ and the same partial fraction decomposition as before. Our final result is

$$y(t) = \frac{1}{5} - \frac{e^{-t}}{5}\left(\cos(2t) + \frac{1}{2}\sin(2t)\right)$$

$$- \frac{u_7(t)}{5} + \frac{u_7(t)e^{-(t-7)}}{5}\left(\cos(2(t-7)) + \frac{1}{2}\sin(2(t-7))\right).$$

Now that we have a formula for the solution, it is interesting to graph it in the ty-, tv-, and yv-planes where $v = dy/dt$ as usual. See Figure 8.10.

Note that the solution crosses itself in the yv-plane. This is not a violation of the Uniqueness Theorem because the system is nonautonomous. For $t < 7$, the forcing term pushes the solution away from $y(0) = 0$. When the forcing is turned off at $t = 7$, the system returns to being an underdamped oscillator with no forcing and the solution spirals toward $(0,0)$.

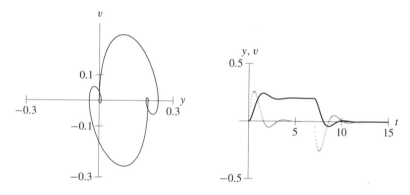

Figure 8.10

Solutions of $\frac{d^2y}{dt^2} + 2\frac{dy}{dt} + 5y = h(t)$, with $y(0) = 0$, $y'(0) = 0$ with constant forcing that "turns off" at $t = 7$.

Sinusoidal Forcing

We can imagine many different types of forcing that can suddenly appear. For example, suppose that we again have a mass attached to a spring on a table. For $t \leq 5$, we suppose the table is horizontal so there is no external force on the mass. At time $t = 5$, an earthquake begins to shake the table, rocking it back and forth, providing a periodic forcing term.

As an example, let the mass $m = 1$ and the spring constant $k_s = 3$. We assume the oscillator is undamped, so $k_d = 0$. For the forcing term, we take $2u_5(t)\sin(t-5)$, which has amplitude 2 and period 2π and begins at time $t = 5$. The model equation is

$$\frac{d^2y}{dt^2} + 3y = 2u_5(t)\sin(t-5).$$

We take initial conditions $y(0) = 1$ and $y'(0) = 0$.

Applying the Laplace transform to both sides of this equation, we have

$$\mathcal{L}\left[\frac{d^2y}{dt^2}\right] + 3\mathcal{L}[y] = \mathcal{L}[2u_5(t)\sin(t-5)]$$

or

$$s^2\mathcal{L}[y] - sy(0) - y'(0) + 3\mathcal{L}[y] = \frac{2e^{-5s}}{s^2+1}.$$

Substituting in the initial conditions and simplifying, we obtain

$$(s^2+3)\mathcal{L}[y] - s = \frac{2e^{-5s}}{s^2+1}.$$

So

$$y = \mathcal{L}^{-1}\left[\frac{2e^{-5s}}{(s^2+3)(s^2+1)}\right] + \mathcal{L}^{-1}\left[\frac{s}{s^2+3}\right].$$

Using partial fractions, we can rewrite the first term on the right as

$$\frac{2e^{-5s}}{(s^2+3)(s^2+1)} = \frac{e^{-5s}}{s^2+1} - \frac{e^{-5s}}{s^2+3}.$$

This gives

$$y = \mathcal{L}^{-1}\left[\frac{e^{-5s}}{s^2+1}\right] + \mathcal{L}^{-1}\left[-\frac{e^{-5s}}{s^2+3}\right] + \mathcal{L}^{-1}\left[\frac{s}{s^2+3}\right].$$

Using the table of Laplace transforms, we obtain

$$y(t) = u_5(t)(\sin(t-5) - \sin(\sqrt{3}(t-5))) + \cos(\sqrt{3}t).$$

Again, we take a moment to study this solution. The graph of $y(t)$ is given in Figure 8.11. The solution behaves like an unforced, undamped harmonic oscillator, oscillating with constant amplitude until time $t = 5$. For $t \geq 5$ the amplitude rises and falls periodically depending on the relationship between the natural frequency and the forcing frequency (see Chapter 5).

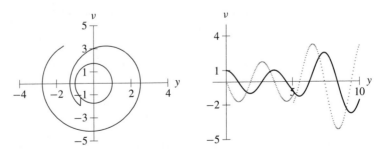

Figure 8.11

Graphs of the solution of the initial-value problem $\frac{d^2y}{dt^2} + 3y = 2u_5(t)\sin(t-5)$.

Limitations of the Laplace Transform Method

The Laplace transform method is remarkably effective when it works. The idea that, instead of doing tedious integrals we merely look up the answer in a table can be very seductive. However, there are some sobering things to remember.

First, the equations to which the Laplace transform method can be easily applied are linear. For nonlinear equations we must turn to other analytical techniques or to qualitative or numerical techniques. Second, using Laplace transforms replaces calculus with algebra, but this algebra can be complicated, particularly when computing the inverse Laplace transform. Finally, the solution comes to us in the form of a formula. We must do more work to obtain a qualitative understanding of the solution.

It is sometimes possible to obtain the information we require about the solution of an initial-value problem from its Laplace transform. That is, we analyze the solution in the *s* domain and avoid computing the inverse Laplace transform. This is frequently done in electric circuit theory. We give an example of this type of analysis in Section 8.4.

Exercises for Section 8.2

1. Compute the Laplace transform for

    (a) $e^{at} \cos(\omega t)$,

    (b) $e^{at} \sin(\omega t)$

    (that is, verify that the entries in the table on page 594 are correct).

In Exercises 2–6, compute the Laplace transform for the given function. [Hints: In Exercise 2, differentiate both sides of the formula for $\mathcal{L}[\cos(\omega t)]$ with respect to ω. Use a similar method for the other problems.]

2. $f(t) = t \sin(\omega t)$ 3. $f(t) = t \cos(\omega t)$

4. $f(t) = t e^{at}$ 5. $f(t) = t^2 e^{at}$

6. $f(t) = t^n e^{at}$

In Exercises 7–12:

   (a) Compute the Laplace transform of both sides of the equation.

   (b) Substitute in the initial conditions and simplify to obtain the Laplace transform of the solution.

   (c) Find the solution by taking the inverse Laplace transform.

7. $\dfrac{d^2y}{dt^2} + 2\dfrac{dy}{dt} + 5y = \sin(3t), \quad y(0) = 2, y'(0) = 0.$

8. $\dfrac{d^2y}{dt^2} + 3y = u_4(t) \cos(5(t-4)), \quad y(0) = 0, y'(0) = -2.$

9. $\dfrac{d^2y}{dt^2} + 4y = \cos(2t), \quad y(0) = -2, y'(0) = 0.$

10. $\dfrac{d^2y}{dt^2} + 9y = u_5 \sin(3(t-5))$, $\quad y(0) = 2$, $y'(0) = 0$.

11. $\dfrac{d^2y}{dt^2} + 3y = w(t)$, $\quad y(0) = 2$, $y'(0) = 0$ where

$$w(t) = \begin{cases} t & \text{if} \quad 0 \le t < 1 \\ 1 & \text{if} \quad t \ge 1. \end{cases}$$

12. $\dfrac{d^2y}{dt^2} + \dfrac{dy}{dt} + 3y = w(t)$, $y(0) = 2$, $y'(0) = -2$, where $w(t)$ is as in Exercise 11.

13. (a) Find the solution of the initial-value problem

$$\dfrac{d^2y}{dt^2} + 9y = 0.1\cos(3t), \qquad y(0) = 2, \ y'(0) = 0.$$

[*Hint*: It is not necessary to use Laplace transforms, but it will help in doing the next problem if you do.]

(b) In a short paragraph, describe the behavior of the solution in part (a).

14. (a) Find the solution of the initial-value problem

$$\dfrac{d^2y}{dt^2} + 9y = 0.1\left((1 - u_5(t))\cos(3t) + u_{12}(t)\cos(3t)\right), \ y(0) = 2, \ y'(0) = 0.$$

(b) In a short paragraph, describe the behavior of the soluion in part (a).

15. Suppose $f(t)$ is a periodic function with period T, that is,

$$f(t + T) = f(t) \quad \text{for all } t.$$

Show that

$$\mathcal{L}[f] = \dfrac{1}{1 - e^{-Ts}} \int_0^T f(t)e^{-st} \, dt.$$

16. Compute the Laplace transform of the square wave

$$w(t) = \begin{cases} 1 & \text{if} \quad 2n \le t < 2n + 1 \text{ for some integer } n \\ \\ -1 & \text{if} \quad 2n + 1 \le t < 2n + 2 \text{ for some integer } n \end{cases}$$

(see Figure 8.12).

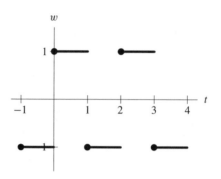

Figure 8.12
Periodic square wave with period 2.

17. Compute the Laplace transform of the saw-tooth function

$$z(t) = t - [t]$$

where $[t]$ denotes the greatest integer less than t (see Figure 8.13). [*Hint:* Use the formula developed in Exercise 15.]

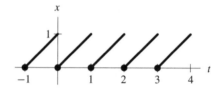

Figure 8.13
A saw-tooth wave with period 1.

18. If $w(t)$ is the square wave of Exercise 16,

(a) Compute the Laplace transform of the solution of the initial-value problem

$$\frac{d^2 y}{dt^2} + 20\frac{dy}{dt} + 200y = w(t), \qquad y(0) = 1, y'(0) = 0.$$

(b) Describe the behavior of the solution. [Hint: Knowing the Laplace transform of the solution may not help with this part.]

19. If $z(t)$ is the saw-tooth wave of Exercise 17,

(a) Compute the Laplace transform of the solution of the initial-value problem

$$\frac{d^2 y}{dt^2} + 20\frac{dy}{dt} + 200y = z(t), \qquad y(0) = 1, y'(0) = 0.$$

(b) Describe the behavior of the solution. [Hint: Knowing the Laplace transform of the solution may not help with this part.]

8.3 IMPULSE FORCING AND DELTA FUNCTIONS

We have seen that Laplace transforms make it possible to deal with forced harmonic oscillator equations that have discontinuous forcing terms. In this section we consider another type of discontinuous forcing function known as an impulse function.

Impulse Forcing

Impulse forcing is the term used to describe a very quick push or pull on a system, like a blow with a hammer or the effect of an explosion. For example, suppose we have an unforced harmonic oscillator with initial conditions

$$\frac{d^2y}{dt^2} + \frac{dy}{dt} + 3y = 0 \quad y(0) = 1, \, y'(0) = 0.$$

Now suppose we strike the mass with a hammer once at time $t = 4$. See Figure 8.14.

We can write the forced equation as

$$\frac{d^2y}{dt^2} + \frac{dy}{dt} + 3y = g(t)$$

where

$$g(t) = \begin{cases} \text{very big when } t \text{ is very near } t = 4 \\ \\ 0 \text{ for } t \text{ away from } t = 4. \end{cases}$$

To make this precise enough to have any hope of obtaining a solution for this initial-value problem, we need some sort of formula for $g(t)$. As a first attempt to derive such a formula, we might assume that the hammer delivers a large constant force over a short time interval. More precisely, we assume that the hammer is in contact with the mass for times $4 - \Delta t$ to $4 + \Delta t$ for some small time $\Delta t > 0$, and that the force on the mass during this interval is the constant k. Then the forcing function becomes

$$g_{\Delta t}(t) = \begin{cases} k & \text{if } 4 - \Delta t \leq t \leq 4 + \Delta t \\ \\ 0 & \text{otherwise.} \end{cases}$$

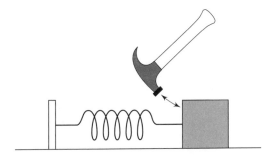

Figure 8.14
Harmonic oscillator struck by hammer.

Here we think of both Δt and k as parameters.

In order to simplify the situation somewhat, we choose $k = 1/(2\Delta t)$, so that k becomes large as Δt tends to zero. Therefore

$$g_{\Delta t}(t) = \begin{cases} 1/(2\Delta t) & \text{if } 4 - \Delta t \leq t \leq 4 + \Delta t \\ \\ 0 & \text{otherwise.} \end{cases}$$

The reason for this strange choice of k is that the area under the graph of $g_{\Delta t}(t)$ is the same as the area of a rectangle with base $2\Delta t$ and height $1/(2\Delta t)$. That is, the area is 1 no matter what we choose for Δt.

Since we are modeling a hammer strike which is delivered very quickly, we would like to take the parameter Δt as small as posible. So we consider what happens as Δt tends to zero. The value of $g_{\Delta t}$ near $t = 4$ increases as Δt decreases in such a way that the total area under the graphs remains equal to one. See Figure 8.15. Informally, as Δt tends to zero, we are compressing the same amount of force into a shorter and shorter time interval.

An "instantaneous" force would then be represented by the limit as $\Delta t \to 0$. But $\lim_{\Delta t \to 0} g_{\Delta t}(t)$ is a confusing object. For any $t \neq 4$, $\lim_{\Delta t \to 0} g_{\Delta t}(t) = 0$ because for small enough Δt, t is outside the interval on which $g_{\Delta t}(t)$ is positive. On the other hand, $\lim_{\Delta t \to 0} g_{\Delta t}(4)$ does not exist because $g_{\Delta t}(4)$ tends to infinity. The "function"

$$\delta_4(t) = \lim_{\Delta t \to 0} g_{\Delta t}(t)$$

is called the **Dirac delta function**. This "function" is zero for all values of t except $t = 4$, where it is infinity.

To a mathematician, the definition above is useless, or worse. According to this definition there is no way to tell the difference between $\delta_4(t)$ and $2\delta_4(t)$ because both

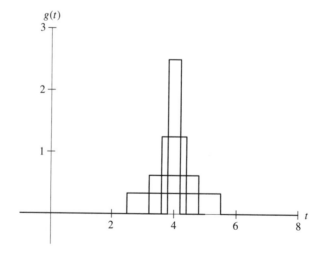

Figure 8.15
Graph of the function $g_{\Delta t}$ for several values of Δt.

of these "functions" are equal to zero for $t \neq 4$ and equal to infinity for $t = 4$. However, using Laplace transforms, we can rescue the Dirac delta function, turning it from nonsense into a useful mathematical object.

Laplace Transforms of Dirac Delta Functions

To make sense of this mess, we turn to the Laplace transform. For any $\Delta t > 0$ we can compute

$$\mathcal{L}[g_{\Delta t}] = \int_0^\infty g_{\Delta t}(t) e^{-st} dt$$

$$= \int_{4-\Delta t}^{4+\Delta t} \frac{1}{2\Delta t} e^{-st} dt$$

$$= \frac{1}{2\Delta t} \frac{e^{-st}}{(-s)} \Big|_{t=4-\Delta t}^{t=4+\Delta t}$$

$$= \frac{1}{2\Delta t} \left(-\frac{e^{-s(4+\Delta t)} - e^{-s(4-\Delta t)}}{s} \right)$$

$$= \frac{e^{-4s}}{s} \left(\frac{e^{\Delta ts} - e^{-\Delta ts}}{2\Delta t} \right).$$

Next we take the limit of this quantity as $\Delta t \to 0$,

$$\lim_{\Delta t \to 0} \mathcal{L}[g_{\Delta t}] = \lim_{\Delta t \to 0} \frac{e^{-4s}}{s} \left(\frac{e^{\Delta ts} - e^{-\Delta ts}}{2\Delta t} \right)$$

$$= \frac{e^{-4s}}{s} \lim_{\Delta t \to 0} e^{-\Delta ts} \left(\frac{e^{2\Delta ts} - 1}{2\Delta t} \right)$$

$$= e^{-4s}$$

(see Exercise 1 for hints on computing this limit).

The limit of the Laplace transform of $g_{\Delta t}$ as $\Delta t \to 0$ is a perfectly normal function. As long as we take Laplace transform *before* taking the limit, everything stays nice. Hence, we make the following definition of the Laplace transform of $\delta_4(t)$:

$$\mathcal{L}[\delta_4] = e^{-4s}.$$

Now there is nothing special about the time $t = 4$. We could just as easily put the impulse force at time $t = a$, obtaining a Dirac delta function $\delta_a(t)$ that "is zero for $t \neq a$ and ∞ for $t = a$." So we have the following definition:

Definition: The Laplace transform of $\delta_a(t)$ is $\mathcal{L}[\delta_a(t)] = e^{-as}$.

To double the strength of the hammer blow, we would like to use the forcing $2\delta_4(t)$. As we observed above, as a function, $2\delta_4(t)$ is impossible to distinguish from $\delta_4(t)$. However, in terms of Laplace transforms, we may use linearity to find

$$\mathcal{L}[2\delta_4] = 2\mathcal{L}[\delta_4] = 2e^{-4s}.$$

Impulse Forcing of an Harmonic Oscillator

We return to the harmonic oscillator equation with mass $m = 1$, spring constant $k_s = 3$, and coefficient of damping $k_d = 1$, which is struck by a hammer at $t = 4$. Using the Dirac delta function, we can write the initial-value problem

$$\frac{d^2y}{dt^2} + \frac{dy}{dt} + 3y = \delta_4(t), \quad y(0) = 1, y'(0) = 0.$$

Qualitative analysis

Before finding a formula for the solution, let's think about what we expect. For $0 \le t < 4$, we have an unforced, underdamped harmonic oscillator. The y-coordinate of the solution with initial condition $y(0) = 1$, $y'(0) = 0$ oscillates with decreasing amplitude. For the corresponding system, the origin is a spiral sink (see Figure 8.16).

At $t = 4$ the impulse force changes d^2y/dt^2 suddenly. So the velocity should have a discontinuity at $t = 4$ and the y-coordinate should abruptly change in amplitude.

After $t = 4$ we again have an underdamped oscillator. Therefore we expect the solution in the phase plane to spiral toward $(0, 0)$ for $0 < t < 4$, jump at $t = 4$, and then spiral toward $(0, 0)$ again for all $t > 4$. The $y(t)$-graph should oscillate with decreasing amplitude for $0 \le t \le 4$, then have a sudden change in amplitude followed again by oscillations with decreasing amplitude. The velocity $v(t)$ also oscillates with

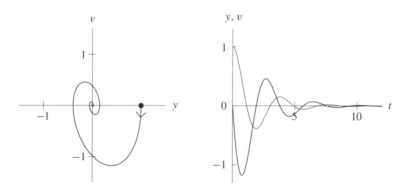

Figure 8.16
Phase plane and $y(t)$- and $v(t)$-graphs of the solution of unforced oscillator
$\frac{d^2y}{dt^2} + \frac{dy}{dt} + 3y = 0$ with initial conditions $y(0) = 1$, $y'(0) = 0$.

decreasing amplitude before and after $t = 4$, but at $t = 4$ the $v(t)$-graph should have a discontinuity.

Unfortunately, this is only a rough, qualitative description of the behavior of the solution. To see better what is going on, we must compute a formula for the solution.

Analytic solution using Laplace transform

We can find a formula for the solution to this initial-value problem by using Laplace transforms. Taking the Laplace transform of both sides of the equation, we obtain

$$\mathcal{L}\left[\frac{d^2y}{dt^2}\right] + \mathcal{L}\left[\frac{dy}{dt}\right] + 3\mathcal{L}[y] = \mathcal{L}[\delta_4(t)].$$

Using the formula for the Laplace transform of a derivative, we have

$$s^2\mathcal{L}[y] - sy(0) - \frac{dy}{dt}(0) + s\mathcal{L}[y] - y(0) + 3\mathcal{L}[y] = \mathcal{L}[\delta_4(t)].$$

Substituting the initial conditions gives

$$s^2\mathcal{L}[y] - s + s\mathcal{L}[y] - 1 + 3\mathcal{L}[y] = \mathcal{L}[\delta_4]$$

$$(s^2 + s + 3)\mathcal{L}[y] = s + 1 + \mathcal{L}[\delta_4]$$

$$\mathcal{L}[y] = \frac{s+1}{s^2 + s + 3} + \frac{\mathcal{L}[\delta_4]}{s^2 + s + 3}.$$

Now $\mathcal{L}[\delta_4] = e^{-4s}$, so

$$y = \mathcal{L}^{-1}\left[\frac{s+1}{s^2+s+3}\right] + \mathcal{L}^{-1}\left[\frac{e^{-4s}}{s^2+s+3}\right]$$

$$= \mathcal{L}^{-1}\left[\frac{s+1/2}{(s+1/2)^2 + 11/4}\right] + \mathcal{L}^{-1}\left[\frac{1/2}{(s+1/2)^2 + 11/4}\right]$$

$$+ \mathcal{L}^{-1}\left[\frac{e^{-4s}}{(s+1/2)^2 + 11/4}\right].$$

Now recall that we encountered several of these inverse transforms in that messy example we worked out in the last section (see page 571). Using these computations, we find

$$y(t) = e^{-\frac{1}{2}t}\cos\left(\frac{\sqrt{11}}{2}t\right) + \frac{1}{\sqrt{11}}e^{-\frac{1}{2}t}\sin\left(\frac{\sqrt{11}}{2}t\right)$$

$$+ u_4(t)\frac{2}{\sqrt{11}}e^{-\frac{1}{2}(t-4)}\sin\left(\frac{\sqrt{11}}{2}(t-4)\right)$$

(recall that $u_4(t)$ is zero when $t < 4$ and 1 when $t \geq 4$).

The function $y(t)$ does have the expected form. See Figure 8.17.

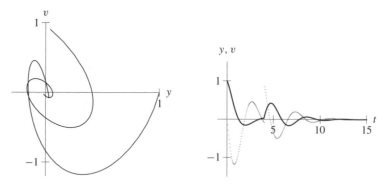

Figure 8.17

Graphs of the solution $y(t)$ of the initial-value problem $\frac{d^2y}{dt^2} + \frac{dy}{dt} + 3y = \delta_4(t)$ with $y(0) = 1$, $y'(0) = 0$.

Periodic Impulse Forcing

We note that the qualitative approach can be useful, even when complicated forcing terms with discontinuities are involved. As an example, consider the initial-value problem

$$\frac{d^2y}{dt^2} + y = \sum_{n=1}^{\infty} \delta_{na}(t), \quad y(0) = 0, \ y'(0) = 0.$$

The equation has a delta function impulse at $t = a$, $t = 2a$, $t = 3a$, ..., for example, a blow with a hammer every a time units. This is an undamped oscillator that is tapped in the positive direction every a time units.

The long-term effect of this striking will depend on what the mass is doing when the strike arrives. If the mass is moving toward the hammer when it is hit, then the blow will take energy from the system. If the mass is moving away from the hammer, then the hammer strike will speed up the mass, adding energy to the system.

This is like the situation of sinusoidal forcing — if the period of the forcing is the same as the natural period of the oscillator, then the impact of the hammer will always catch the mass in the same situation. Hence a resonance phenomenon will occur and the amplitude of the oscillation will increase without bound. If the forcing period and natural period are different, then the strikes of the hammer sometimes add energy to and sometimes take energy from the system.

Exercises for Section 8.3

1. Compute the limit

$$\lim_{\Delta t \to 0} e^{-\Delta t s} \left(\frac{e^{2\Delta t s} - 1}{2\Delta t} \right).$$

In Exercises 2–6, find the explicit solution (that is, a formula) for each of the following initial-value problems. [Hint: Use Laplace Transforms].

2. $\dfrac{d^2y}{dt^2} + 3y = 5\delta_2(t), \quad y(0) = 0, \, y'(0) = 0.$

3. $\dfrac{d^2y}{dt^2} + \dfrac{dy}{dt} + 3y = 5\delta_1(t), \quad y(0) = 0, \, y'(0) = 0.$

4. $\dfrac{d^2y}{dt^2} + 2\dfrac{dy}{dt} + 5y = \delta_3(t), \quad y(0) = 1, \, y'(0) = 1$

5. $\dfrac{d^2y}{dt^2} + 2\dfrac{dy}{dt} + 2y = -2\delta_2(t), \quad y(0) = 2, \, y'(0) = 0.$

6. $\dfrac{d^2y}{dt^2} + 2\dfrac{dy}{dt} + 3y = \delta_1(t) - 3\delta_4(t), \quad y(0) = 0, \, y'(0) = 0.$

7. There is a relationship between the Heaviside function $u_a(t)$ and the $\delta_a(t)$ which can be noticed by looking at their Laplace transforms.

  (a) Show that for $a > 0$, $\mathcal{L}[\delta_a] = s\mathcal{L}[u_a] - u_a(0)$.

  (b) What relationship does this suggest between $u_a(t)$ and $\delta_a(t)$?

  (c) Is there any way we could understand this relationship in terms of everyday calculus (without Laplace tranforms)?

8. Calculate the Laplace transform of

$$g(t) = \sum_{n=1}^{\infty} \delta_{na}(t).$$

where a is a positive parameter. [Hint: See Exercise 15 in Section 8.2.]

9. **(a)** Find the Laplace transform of the solution of the initial-value problem

$$\frac{d^2y}{dt^2} + 2y = \sum_{n=1}^{\infty} \delta_n(t), \quad y(0) = 0, \, y'(0) = 0.$$

  (b) Find the solution of the initial-value problem in part (a) (that is, take the inverse Laplace transform of your answer in part (a)).

  (c) State what assumptions you are making in parts (a) and (b) (for example, what assumptions are you making about convergence and about inverse Laplace transforms of infinite sums).

10. **(a)** Find the Laplace transform of the solution of the initial-value problem

$$\frac{d^2y}{dt^2} + y = \sum_{n=1}^{\infty} \delta_{2n\pi}(t), \quad y(0) = 0, \, y'(0) = 0.$$

  (b) Find the solution of the initial-value problem in part (a) (that is, take the inverse Laplace transform of your answer to part (a)).

  (c) What assumptions are you making in part (a) and (b) (for example, about convergence and about the inverse Laplace transform of an infinite sum).

8.4 THE QUALITATIVE THEORY OF LAPLACE TRANSFORMS

This chapter is different from the previous chapters in several ways. Probably the most obvious difference is the fact that there are fewer pictures than in the previous chapters. The method of Laplace transforms seems to focus on finding formulas for solutions of initial-value problems. However, Laplace transforms can also be used to provide considerable qualitative insight into the behavior of the solutions of nonhomogeneous second-order differential equations, sometimes with considerably less computation than is necessary to obtain the formula for the solution. We study a small part of this qualitative theory in this section.

Homogeneous Second-Order Equations

We begin our investigation of the use of Laplace transforms in the qualitative theory of differential equations by studying a familiar friend, the second-order homogeneous equation

$$\frac{d^2y}{dt^2} + p\frac{dy}{dt} + qy = 0.$$

We have several methods available for attacking this equation that are easier to use than Laplace transforms, but our goal here is to learn more about Laplace transforms.

Taking the Laplace transform of both sides of this equation, we obtain

$$\mathcal{L}\left[\frac{d^2y}{dt^2}\right] + p\mathcal{L}\left[\frac{dy}{dt}\right] + q\mathcal{L}[y] = 0.$$

Using the rules concerning Laplace transforms of derivatives, this equation becomes

$$(s^2\mathcal{L}[y] - sy(0) - y'(0)) + p(s\mathcal{L}[y] - y(0)) + q\mathcal{L}[y] = 0,$$

which can be simplified to

$$(s^2 + ps + q)\mathcal{L}[y] - (s + p)y(0) - y'(0) = 0.$$

Thus the Laplace transform of the solution with initial conditions $y(0)$ and $y'(0)$ is

$$\mathcal{L}[y] = \frac{(s + p)y(0)}{s^2 + ps + q} + \frac{y'(0)}{s^2 + ps + q}.$$

The important point to notice here is that the denominator of the Laplace transform of the solution is the polynomial

$$s^2 + ps + q.$$

This is precisely the same as the characteristic polynomial of the system

$$\frac{dy}{dt} = v$$

$$\frac{dv}{dt} = -qy - pv.$$

that corresponds to this second-order equation. This is not a coincidence. If the roots of the characteristic polynomial (which are the eigenvalues of the system) are λ and μ, then we can write

$$s^2 + ps + q = (s - \lambda)(s - \mu).$$

We first assume that λ and μ are real. Then the solution y is given by

$$y = \mathcal{L}^{-1}\left[\frac{(s+p)y(0)}{(s-\lambda)(s-\mu)}\right] + \mathcal{L}^{-1}\left[\frac{y'(0)}{(s-\lambda)(s-\mu)}\right].$$

Using partial fractions, we can break this into a sum of fractions with denominators $(s - \lambda)$ and $(s - \mu)$. Every term of the solution will have a factor involving either $e^{\lambda t}$ or $e^{\mu t}$, as we expect from our previous work. Qualitatively, we know that if λ and μ are both positive, then the origin is a source; if they are both negative, then the origin is a sink; and if one is positive and the other is negative, then the origin is a saddle.

If $\lambda = \alpha + i\beta$, $\mu = \alpha - i\beta$, then

$$(s - \lambda)(s - \mu) = (s - (\alpha + i\beta))(s - (\alpha - i\beta))$$

$$= ((s - \alpha) - i\beta)((s - \alpha) + i\beta)$$

$$= (s - \alpha)^2 + \beta^2$$

In this case we can write y as the inverse Laplace transform of functions with $(s - \alpha)^2 + \beta^2$ in the denominator and with either a constant or a constant times s in the numerator. The inverse Laplace transforms of these terms are multiples of either $e^{\alpha t}\sin(\beta t)$ or $e^{\alpha t}\cos(\beta t)$. Again, this is exactly what we expect. In this case the solutions oscillate and the long-range behavior is determined by the sign of the real part α. The amplitude increases if α is positive and decreases if α is negative. The frequency of oscillation is determined by the imaginary part β and is given by $\beta/(2\pi)$.

We can summarize the discussion above as follows: For homogeneous, second-order, constant coefficient equations, the qualitative behavior of solutions is determined by the values of s for which the denominator of the Laplace transform of the solution is zero. (These values of s are precisely the same as the eigenvalues of the corresponding linear system.) This motivates some terminology.

Definition: Suppose $F(s)$ is a rational function; that is

$$F(s) = \frac{G(s)}{H(s)}$$

where $G(s)$ and $H(s)$ are polynomials with no common factors and the degree of G is less than the degree of H. The **poles** of F are the values of s for which $H(s) = 0$. If $F(s)$ is the sum of rational functions,

$$F(s) = \frac{G_1(s)}{H_1(s)} + \frac{G_1(s)}{H_1(s)} + \ldots + \frac{G_n(s)}{H_n(s)},$$

the poles of F are found by first rewriting F as a single fraction, canceling any common terms in the numerator and denominator, and then finding the poles of the resulting rational function.

We can summarize the computation above by saying that for a homogeneous, second-order, constant coefficient, linear equation, the poles of the Laplace transform of the general solution are the same as the eigenvalues of the corresponding linear system.

Nonhomogeneous Second-Order Equations

When we consider a nonhomogeneous linear differential equation, the qualitative techniques that we used for homogeneous linear equations and systems no longer apply. A nonautonomous system is very different from an autonomous system because the vector field changes with time.

However, when using the Laplace transform, there is not much difference between a homogeneous and a nonhomogeneous equation. The arithmetic for nonhomogeneous equations is slightly more complicated, but the basic method is the same. Hence we are led to consider the poles of the Laplace transform of solutions of nonhomogeneous equations.

For example, consider the initial-value problem

$$\frac{d^2y}{dt^2} + \frac{dy}{dt} + 4y = 3e^{-t/10}, \quad y(0) = 1, \, y'(0) = 1.$$

Taking the Laplace transform of both sides of the equation and solving for $\mathcal{L}[y]$, we obtain

$$\mathcal{L}[y] = \frac{(s+1)y(0)}{s^2 + s + 4} + \frac{y'(0)}{s^2 + s + 4} + \frac{3}{(s^2 + s + 4)(s + 1/10)},$$

and substituting the initial conditions gives

$$\mathcal{L}[y] = \frac{s+1}{s^2 + s + 4} + \frac{1}{s^2 + s + 4} + \frac{3}{(s^2 + s + 4)(s + 1/10)}.$$

To find the poles, we note that the least common denominator of the sum above is

$$(s^2 + s + 4)(s + 1/10).$$

The roots of $s^2 + s + 4$ are $-(1/2) \pm i\sqrt{15}/2$, and so the poles are $s = -(1/2) \pm i\sqrt{15}/2$ and $s = -1/10$. Using partial fractions, we can rewrite the Laplace transform of solutions as a sum of terms with denominators

$$s + 1/10 \text{ and } s^2 + s + 4 = (s + (1/2))^2 + 15/4.$$

From the table of Laplace transforms we see that the only terms that can appear in the solution are $e^{-t/10}$, $e^{-t/2}\sin(\sqrt{15}t/2)$, and $e^{-t/2}\cos(\sqrt{15}t/2)$.

Just as for the homogeneous case, the poles of the Laplace transform, $s = -1/10$ and $s = -1/2 \pm i\sqrt{15}/2$, tell us the qualitative behavior of the solution. The solution decreases toward zero because all of the poles are negative or have negative real parts. The solution is not monotonic, but oscillates with a natural frequency of $\sqrt{15}/(4\pi)$ because of the complex poles. Finally, the rate at which solution approaches zero is determined by the exponential term with the exponent whose real part is closest to zero. Hence this solution approaches zero at the same rate as $e^{-t/10}$

It is important to note that, although this solution tends to zero, the origin is not an equilibrium point for this nonhomogeneous equation. Indeed, the constant function $y(t) = 0$ is never a solution of a nonhomogeneous linear equation, as may be checked by substituting zero into the equation.

This qualitative information agrees with the graph of the solution given in Figure 8.18. What is remarkable is that the description we obtain using the poles of the Laplace transform is very similar to the description of solutions of linear, homogeneous equations that we derive using the eigenvalues. We informally think of the poles of the Laplace transform of solutions as an extension of the idea of eigenvalues to nonhomogeneous equations.

It turns out that the poles of the Laplace transform $F(s)$ along with the **zeroes** (the values of s where $F(s) = 0$) can be used to reconstruct the inverse Laplace Transform of F. This is a sort of automation of the partial fractions technique which is used extensively in electric circuit and systems theories.

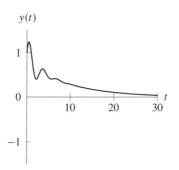

Figure 8.18

The solution of $\frac{d^2 y}{dt^2} + \frac{dy}{dt} + 4y = 3e^{-t/10}$ with initial conditions $y(0) = 1$, $y'(0) = 1$.

Classification Using Poles

Collecting the ideas above, we can state some rules of thumb concerning the relation between the poles of the Laplace transform of a solution and the qualitative behavior of that solution.

If all of the poles are either negative or have negative real parts, then the solution tends toward the origin as t increases. The rate of approach is exponential with exponent equal to the real part of the pole closest to zero. If one or more of the poles is positive or has positive real part, then the solution is unbounded. It tends to infinity exponentially if the pole with largest real part is real. If the pole with the largest real part is complex, then the solution oscillates and the amplitude of the oscillation grows exponentially without bound. We emphasize that so far we have only talked about solutions which tend to infinity at an exponential rate.

The part of the complex plane to the left of the imaginary axis is called the left half-plane. The left half-plane contains the negative real numbers and complex numbers that have negative real part. Those with positive real part are in the right half-plane. The condition above is informally summarized by the sayings:

> If all the poles are in the left half-plane,
> then the solution tends to zero exponentially

and

> If there is a pole in the right half-plane,
> then the solution tends to infinity.

We summarize this information qualitatively in Figure 8.19. In the special case in which all of the poles lie on the imaginary axis, the situation is more complicated. The solution may oscillate or we may encounter resonance. We will not deal further with this special but important case. However, see Exercises 7–9.

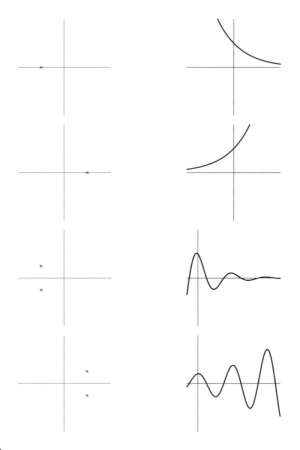

Figure 8.19
Schematic representation of the information contained in the poles of the Laplace transform. The picture on the left gives the location of the poles in the complex s-plane, the picture on the right is the $y(t)$-graph of the corresponding solution.

Another Example with a Moral

Consider the differential equation

$$\frac{d^2y}{dt^2} + \frac{dy}{dt} + 3y = u_2(t)\sin(t-2).$$

This is an underdamped harmonic oscillator with a sinusoidal forcing term that is turned on at time $t = 2$. Taking the Laplace transforms, we obtain

$$s^2\mathcal{L}[y] - sy(0) - y'(0) + s\mathcal{L}[y] - y(0) + 3\mathcal{L}[y] = \mathcal{L}[u_2(t)\sin(t-2)].$$

Solving for $\mathcal{L}[y]$ gives

$$\mathcal{L}[y] = \frac{(s+1)y(0)}{s^2+s+3} + \frac{y'(0)}{s^2+s+3} + \frac{e^{-2s}}{(s^2+1)(s^2+s+3)}.$$

The poles are $-(1/2) \pm i\sqrt{11}/2$ and $\pm i$. The first two poles are the eigenvalues of the unforced equation and represent the natural response of the system. The poles $\pm i$ represent the forced response of the system. From this we see that the natural response decays exponentially to zero (like $e^{-t/2}$), and the solution approaches a steady-state oscillation with bounded amplitude and period 2π (see Figure 8.20).

However, we must be careful. If we change the equation to

$$\frac{d^2 y}{dt^2} + \frac{dy}{dt} + 3y = (1 - u_2(t)) \sin(t - 2)$$

then we can compute that the poles of the Laplace transform are exactly the same as before (see exercise 5). However, the second equation has a forcing term that turns off at time $t = 2$, so the long-term behavior of solutions is the same as for the unforced, underdamped harmonic oscillator; that is, it tends to zero. See Figure 8.21.

The moral is that one should always expect the unexpected. Relying blindly on a technique, without examining the equation or the underlying question (or physical system, if there is one) invites disaster.

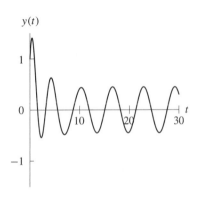

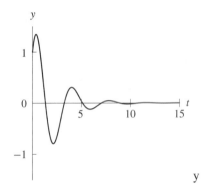

Figure 8.20
Solution of
$\frac{d^2 y}{dt^2} + \frac{dy}{dt} + 3y = u_2(t) \sin(t - 2)$ with
initial conditions $y(0) = 1$, $y'(0) = 2$.

Figure 8.21
Solution of
$\frac{d^2 y}{dt^2} + \frac{dy}{dt} + 3y = (1 - u_2(t)) \sin(t - 2)$
with initial conditions $y(0) = 1$,
$y'(0) = 2$.

A Final Comment

There's an old saying that goes

If the only tool you have is a hammer, then every problem looks like a nail.

This book is about many different types of tools that can be used to study differential equations. There is a temptation to become attached to a particular type of tool or technique. This leads to despair if the technique doesn't work, or it leads to embarrassment if the technique works, but another technique is much easier to use. We must always remember that the problem, be it theoretical or physical in nature, comes first, and the technique of solution of the problem is a means to an end and not, in itself, the end.

Exercises for Section 8.4

1. Suppose

$$\mathcal{L}[y] = \frac{1}{(s^2 + 3)(s^2 + 4)}.$$

(a) What are the poles?

(b) Rewrite $\mathcal{L}[y]$ using partial fractions.

(c) Compute $y(t)$.

2. (a) Compute the Laplace transform of the solution of

$$\frac{d^2y}{dt^2} + 16y = 0, \quad y(0) = 1, y'(0) = 1.$$

(b) What are the poles the result of part (a).

In Exercises 3–6:

(a) Compute the Laplace transform of the solution.

(b) Find the poles of the Laplace transform.

(c) Discuss the behavior of the solution.

3. $\dfrac{d^2y}{dt^2} + 2\dfrac{dy}{dt} + 2y = e^{-2t}\sin(4t), \quad y(0) = 2, y'(0) = -2.$

4. $\dfrac{d^2y}{dt^2} + \dfrac{dy}{dt} + 5y = u_2(t)\sin(4(t - 2)), \quad y(0) = -2, y'(0) = 0.$

5. $\dfrac{d^2y}{dt^2} + \dfrac{dy}{dt} + 3y = (1 - u_2(t))\sin(t - 2), \quad y(0) = 1, y'(0) = 4.$ [Hint: Recall $\sin(t - 2) = \cos(2)\sin(t) - \sin(2)\cos(t)$.]

6. $\dfrac{d^2y}{dt^2} + \dfrac{dy}{dt} + 8y = (1 - u_4(t))\cos(t - 4), \quad y(0) = 0, y'(0) = 0.$

7. Suppose the poles of a rational function $F(s)$ are all on the imaginary axis. What can you conclude about the function $\mathcal{L}^{-1}[F]$? Justify your answer.

8. (a) Compute the Laplace transform for the solution of the initial-value problem

$$\frac{d^2y}{dt^2} + 2\frac{dy}{dt} + y = 0, \quad y(0) = 1, \frac{dy}{dt}(0) = 2.$$

(b) What are the poles of the Laplace transform?

(c) Use this to formulate a conjecture concerning what the occurence of a "double pole" in the Laplace transform means for the qualitative behavior of solutions.

9. (a) Compute the Laplace transform of the solution of the initial-value problem

$$\frac{d^2y}{dt^2} + 4y = \sin(2t), \quad y(0) = 0, y'(0) = 0.$$

(b) Compute the poles of the Laplace transform of the solution.

(c) Use your results to make a conjecture for what sort of poles are present for a harmonic oscillator in resonance.

10. In Exercise 18 of Section 8.2, we considered the equation

$$\frac{d^2y}{dt^2} + 20\frac{dy}{dt} + 200y = w(t) \quad y(0) = 1, \, y'(0) = 0$$

where $w(t)$ is the square wave.

(a) Compute the Laplace transform of the solution of this initial-value problem.

(b) What are the poles of the solution?

(c) Describe the long-term behavior of the solution.

11. In Exercise 19 of Section 8.2, we considered the equation

$$\frac{d^2y}{dt^2} + 20\frac{dy}{dt} + 200y = z(t) \quad y(0) = 1, \, y'(0) = 0$$

where $z(t)$ is the saw tooth wave.

(a) Compute the Laplace transform of the solution of this initial-value problem.

(b) What are the poles of the solution?

(c) Describe the long-term behavior of the solution.

Table 8.1   Frequently Encountered Laplace Transforms.

$y(t) = \mathcal{L}^{-1}[Y]$	$Y(s) = \mathcal{L}[y]$	$y(t) = \mathcal{L}^{-1}[Y]$	$Y(s) = \mathcal{L}[y]$
$y(t) = e^{at}$	$Y(s) = \dfrac{1}{s-a} \quad (s > a)$	$y(t) = t^n$	$Y(s) = \dfrac{n!}{s^{n+1}} \quad (s > 0)$
$y(t) = \sin(\omega t)$	$Y(s) = \dfrac{\omega}{s^2 + \omega^2}$	$y(t) = \cos(\omega t)$	$Y(s) = \dfrac{s}{s^2 + \omega^2}$
$y(t) = e^{at}\sin(\omega t)$	$Y(s) = \dfrac{\omega}{(s-a)^2 + \omega^2}$	$y(t) = e^{at}\cos(\omega t)$	$Y(s) = \dfrac{s-a}{(s-a)^2 + \omega^2}$
$y(t) = t\sin(\omega t)$	$Y(s) = \dfrac{2\omega s}{(s^2 + \omega^2)^2}$	$y(t) = t\cos(\omega t)$	$Y(s) = \dfrac{s^2 - \omega^2}{(s^2 + \omega^2)^2}$
$y(t) = u_a(t)$	$Y(s) = \dfrac{e^{-as}}{s} \quad (s > 0)$	$y(t) = \delta_a(t)$	$Y(s) = e^{-as}$

Table 8.2   Rules for Laplace Transforms:
Given functions $y(t)$ and $w(t)$ with $\mathcal{L}[y] = Y(s)$ and $\mathcal{L}[w] = W(s)$ and constants α and a.

Rule for Laplace Transform	Rule for Inverse Laplace Transform
$\mathcal{L}\left[\dfrac{dy}{dt}\right] = s\mathcal{L}[y] - y(0) = sY(s) - y(0)$	
$\mathcal{L}[y + w] = \mathcal{L}[y] + \mathcal{L}[w] = Y(s) + W(s)$	$\mathcal{L}^{-1}[Y + W] = \mathcal{L}^{-1}[Y] + \mathcal{L}^{-1}[W] = y(t) + w(t)$
$\mathcal{L}[\alpha y] = \alpha\mathcal{L}[y] = \alpha Y(s)$	$\mathcal{L}^{-1}[\alpha Y] = \alpha\mathcal{L}^{-1}[Y] = \alpha y(t)$
$\mathcal{L}[u_a(t)y(t - a)] = e^{-as}\mathcal{L}[y] = e^{-as}Y(s)$	$\mathcal{L}^{-1}[e^{-as}Y] = u_a(t)y(t - a)$
$\mathcal{L}[e^{at}y(t)] = Y(s - a)$	$\mathcal{L}^{-1}[Y(s - a)] = e^{at}\mathcal{L}^{-1}[Y] = e^{at}y(t)$

Lab 8.1   A Bouncing Ball

A ball dropped from a few feet above a hard floor will bounce several times before coming to rest. The forces acting on the ball are gravity (a constant force pulling the ball down), a slight amount of air resistance, and whatever forces are imparted in the impact of the ball with the ground. For some balls it is clear that interaction with the ground is the most significant factor in the change in the motion of the ball. For example, a lump of clay will hit the ground and not bounce at all. On the other hand, some hard rubber balls seem to bounce back almost to the height from which they were dropped. In this lab we model different types of bouncing balls.

In your report you should address the following items:

1. Write a differential equation model for the vertical motion of a ball. Assume the force of gravity is constant and that there is a small amount of air resistance giving some damping force. Model the effect of the impact of the ball with the ground as an impulse force with the size of the impulse force depending on the velocity of the ball and the times of the delivery of the impulse forces equal to the times of impacts of the ball with the ground. (You do not know the exact times of the impacts with the ground, so these must be parameters in your model. These are not "free" parameters, they are determined by the other parameters.)

2. Determine the size of the impulse force necessary to obtain a perfectly elastic bounce, that is, a bounce where the velocity just before the impact going down is equal to the velocity just after the impact going up.

3. Modify your model so that the impulse force is a fraction of that required to give a perfect elastic bounce where the fraction is another parameter.

4. Modify your model so that the impulse force is a constant amount less than that required to give a perfect elastic bounce where the amount of decrease is a parameter.

5. By observing a bouncing ball and comparing its behavior with the predictions of the model, determine which of the models in parts 3 and 4 is more accurate.

Your report: In your report you should address the questions above. Be sure to define all variables and parameters and justify all terms in your model equations. Include any computations and graphs of solutions as appropriate. You may need to enlist the aid of an assistant when gathering the data for part 5.

Lab 8.2   Paddle Ball

A popular toy, called a paddle ball, consists of an elastic string attached to a small rubber ball at one end and to a wooden paddle at the other. The operator of this device hits the ball with the paddle causing the ball to fly away. The elastic string stretches, eventually causing the ball to return to vicinity of the operator. If the paddle is stationary when the ball hits it, then it bounces with a small loss of velocity (decrease of energy). The operator can increase the velocity (energy) of the ball by moving the paddle towards the ball at the moment of impact. Air resistance provides a small amount of damping to the motion of the ball. The goal is to repeat the process as many times as possible. In this lab you will model the motion of the ball.

In your report you should address the following questions:

1. Ignoring the effect of gravity, write a differential equation for the motion of the ball. Include the effect of the elastic string and of the strike with the paddle. [Hint: Your

equation will resemble a harmonic oscillator. Use a delta function to model the effect of the paddle and assume every paddle strike has the same force. Your model should contain several parameters, including the spring constant of the elastic string, the mass of the ball and the strength of the strikes with the paddle.]

2. For most of these devices, the elastic string is "slack" (that is, exerts no force) when the ball is closer to the paddle than the rest length of the string. If you did not account for this in your model above, how would you modify your model to include this observation.

3. For your models, what relationship must hold between the parameters in order for the ball to continue bouncing. Describe the long term behavior of the ball for various values of the strength of the forcing (paddle strikes).

Your report: In your report you should address the problems above. Be sure to justify carefully your model equations in parts 1 and 2. Include sketches of graphs of solutions as appropriate.

Hints and Answers for Section 1.1

1.(a) For $0 < P < 230$.

(b) For $P > 230$ or the non-physical values $P < 0$.

(c) $P = 0$ and $P = 230$.

3.(a) For $-3 < y < 0$ or $y > 4$.

(b) For $y < -3$ or $0 < y < 4$.

(c) For $y = -3$, $y = 0$ and $y = 4$.

5. $L = 0$.

7.(a) Beth.

(b) Jillian.

(c) They have the same rate.

9.(a) $\lambda = \ln(2)/5230 \approx 0.000132533$.

(b) $\lambda = \ln(2)/8 \approx 0.0866434$.

(c) The units are per year, per day.

(d) Yes.

11. Count the number of C-14 atoms that decay in a given time period. Divide this number by the time elapsed to obtain an approximation for dr/dt. Then use $dr/dt = -\lambda r$ to find r.

13.(a) $dP/dt = k(1 - P/N)P - 100$.

(b) $dP/dt = k(1 - P/N)P - P/3$.

(c) $dP/dt = k(1 - P/N)P - a\sqrt{P}$, where a is a positive parameter.

15. $dm/dt = k_1 + k_2 m(1 - m)$ where k_1, k_2 are constants.

17.(a) Logistic model.

(b) Carrying capacity 64; growth-rate ≈ 0.38.

(c) Prediction for population today is 64.

19.(a) System (i).

(b) System (ii).

(c) System (i).

21.(a) The species cooperate.

(b) The species compete.

Hints and Answers for Section 1.2

1.(a) Bob and Glen.

(b) The equilibrium solution $y = -1$.

3. $dy/dt = 3t^2 y$.

5. $y = ke^{t^2/2}$, where k is any real number.

7. $y = ke^{2t} - 1/2$, where k is any real number.

9. $y(t) = \ln(t + C)$, where C is any real number.

11. $y(t) = \pm\sqrt{\ln\left(k(t^2 + 1)\right)}$, where k is any real number. The sign is determined by the initial condition.

13. $y = (-1 \pm \sqrt{4t + C})/2$ where the sign is determined by the initial condition.

15. $y(t) = ke^t/(ke^t + 1)$, where k is any real number, and the equilibrium solution $y = 1$. **Remark:** This is a special case of the logistic equation with growth-rate parameter 1 and carrying capacity 1.

17. $y = -1 + ke^{t+t^3/3}$, where k is any real number.

19. $y^2/2 + \ln|y| = e^t + C$, where C is any real number, and the equilibrium solution $y = 0$.

21. $w = kt$, where k is any real number, for $t > 0$ or $t < 0$. (The differential equation is not defined at $t = 0$.)

23. $y^5/5 + 3y^2/2 = t^3/3 + t + C$, where C is any real number.

25. $y = 7e^{2t}/2 - 1/2$.

27. $y(t) = 1/(t + 2)$.

29. $y(t) = 0$.

31. $y(t) = -\sqrt{4 + (2/3)\ln(t^3 + 1)}$.

33. $y(t) = \tan(t^2/2 + \pi/4)$.

35.(a) Amount of salt ≈ 0.238 lbs.

(b) Amount of salt ≈ 1.58 lbs.

(c) Amount of salt ≈ 2.49 lbs.

(d) Amount of salt ≈ 2.50 lbs.

(e) Amount of salt ≈ 2.50 lbs.

37.(a) The initial-value problem is:
$$dT/dt = -0.2(T - 70), \quad T(0) = 170.$$
  (b) $t = \ln 2.5/0.2 \approx 4.6$.

39. C(t)=$20/3 + ke^{-3t/100}$ where k is any real number. $C(t) = 20/3$ is an equilibrium solution.

41.(a) At 7%, $278,735; at 6.85% with points, $280,009.
  (b) Choose the 7% option.
  (c) Choose the 6.85% option as the interest of $7,732 more than makes up for the difference of $1,274.

Hints and Answers for Section 1.3

1.

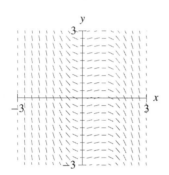

7.(a)

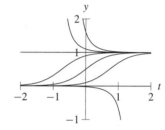

  (b) The solution approaches the equilibrium value 1 from below.

9.(a)

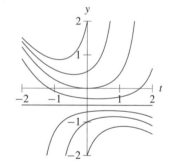

  (b) the solution escapes to infinity as t increases

11. (a) iv. (b) v. (c) viii. (d) iii.

13.(a) On the line $y = 3$ in the ty plane, the slope field is a segment with slope -1.
  (b) No. Solutions with $y(0) < 3$ will have $y(t) < 3$ for all $t > 0$.

3.

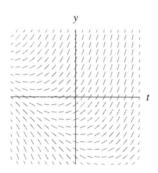

5.

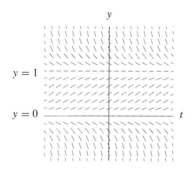

15.

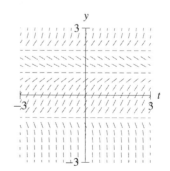

17.(a) This is enough information to give the slope field for all points (t, y) with $y > 0$.

(b)

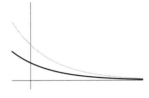

The solution with $y(0) = 2$ will be a translation "left" of the given solution.

19. $v(t) = K + ke^{-t/RC}$

21.(a)

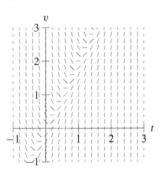

(b)

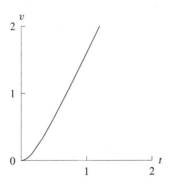

(c)

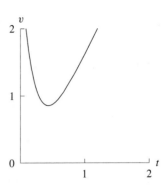

23.(a)

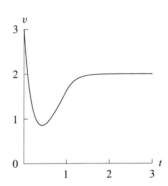

(b)

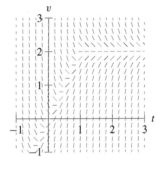

(c)

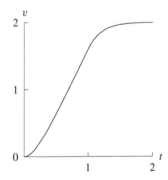

(d) For $t < 1$ look at Exercise 21. For $t \geq 1$ look at Exercise 22

Hints and Answers for Section 1.4

1. When $t = 2$ the Euler approximation is $y \approx 55.5$.

3. When $t = 2$ the Euler approximation is $y \approx 25.49$.

5. When $t = 5$ the Euler approximation is $w \approx -1$.

7. When $t = 2$ the Euler approximation is $y \approx 5.81$.

9. The answer of Exercise 8 is a translate right of the answer of Exercise 7.

11. Hint: What are the constant solutions? Remember Existence and Uniqueness Theorem.

13. Hint: What is the concavity of the solution? Are the approximations from Euler's method going to above or below the real solution?

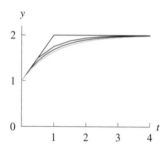

15.

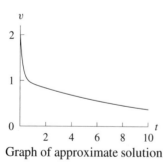

Graph of approximate solution

17.

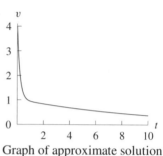

Graph of approximate solution

19.(a)

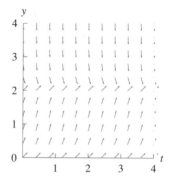

(b)

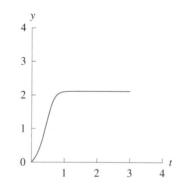

(c) $y \approx 0.2.115$, $y \approx -1.861$ and $y \approx -0.254$

Hints and Answers for Section 1.5

1. $y(t) < 3$ for all t in the domain of definition of $y(t)$

3. $-t^2 < y(t) < t + 2$ for all t

5. $y(t) > 3$ for all t in the domain of definition of $y(t)$, $y(t)$ tends to 3 as t decreases and $y(t)$ increases without bound as t increases

7. $0 < y(t) < 2$ for all t, $y(t)$ tends to 2 as t increases and to 0 as t decreases

9.(a) Plug $y_1(t)$ into both sides of the differential equation and compute.

(b) Use the uniqueness theorem.

11. Hints:

(a) Differentiate $y_1(t)$ at t_0 and remember that $y_1(t)$ is a solution.

(b) Remember that the equation is autonomous.

(c) Look at the slope field–but check by plugging $y_2(t)$ into both sides of the differential equation.

(d) Uniqueness theorem.

(e) Do the same four steps again.

13.(a) If $y_1(t) = 0$ then $dy_1/dt = 0 = y_1(t)/t^2$.

(b) For any real number, c, let

$$y_c(t) = \begin{cases} 0, & \text{for } t \le 0 \\ ce^{-1/t}, & \text{for } t > 0. \end{cases}$$

$y_c(t)$ is a solution which is 0 for $t < 0$ and non–zero for $t > 0$.

(c) y/t^2 is not continuously differentiable with respect to t (or even defined) at $t = 0$

15.(a) $y(t) = -1 + \sqrt{1 + \ln((1 - t/2)^2)}$

(b) Domain of solution is $t \le 2(1 - 1/\sqrt{e})$.

(c) $\displaystyle\lim_{t \to 2(1 - 1/\sqrt{e})} y(t) = -1$ where the differential equation does not exist,

$$\lim_{t \to -\infty} y(t) = \infty$$

17.(a) $y(t) = 2 - \sqrt{t^2 + 3}$

(b) Domain of definition is all real numbers.

(c) $\displaystyle\lim_{t \to \pm\infty} y(t) = -\infty$

Hints and Answers for Section 1.6

1. $y = 0$ is a source and $y = 1$ is a sink.

3. $\ldots -\frac{3\pi}{2}, \frac{\pi}{2}, \frac{5\pi}{2}, \ldots$ are sinks and $\ldots -\frac{\pi}{2}, \frac{3\pi}{2}, \frac{7\pi}{2}, \ldots$ are sources.

5. The sinks are $w = 0, -2\pi, -4\pi\ldots$ and $\pi, 3\pi, 5\pi, \ldots$, the sources are $w = 2$, $w = -\pi, -3\pi, \ldots$ and $w = 2\pi, 4\pi, 6\pi, \ldots$

7. There are no equilibrium points.

9.

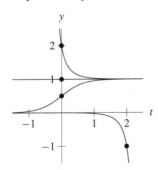

11.

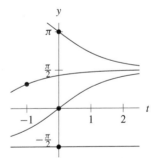

13.

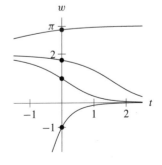

15.

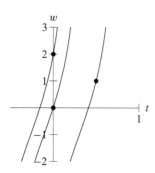

17. Since the initial condition, 1, is between the roots of the function, $y^2 - 4y + 2$, the solution to the initial-value problem, $y(t)$, is always between $2 \pm \sqrt{2}$. The limit of $y(t)$ as $t \to \infty$ is $2 - \sqrt{2}$ and the limit of $y(t)$ as $t \to -\infty$ is $2 + \sqrt{2}$.

19. The solution remains below the equilibrium point $2 - \sqrt{2}$ and is increasing for all t for which it is defined. The limit of $y(t)$ as $t \to \infty$ is $2 - \sqrt{2}$.

21. Since the initial condition, 1, is between the roots of the function, $y^2 - 4y + 2$, the solution to the initial-value problem, $y(t)$, is always between $2 \pm \sqrt{2}$. The limit of $y(t)$ as $t \to \infty$ is $2 - \sqrt{2}$ and the limit of $y(t)$ as $t \to -\infty$ is $2 + \sqrt{2}$.

23.

25.

27.

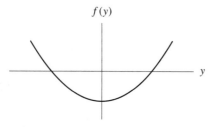

29.

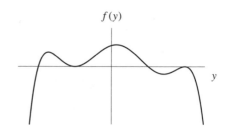

31.(a)

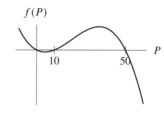

(b)

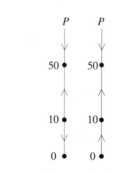

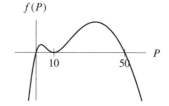

(c) $f(P) = P(P - 10)(50 - P)$ and
$f(P) = P(P - 10)^2(50 - P)$

33.(a) *Hint*: Use the Intermediate Value Theorem.

(b) *Hint*: At a source f crosses the y axis from negative to positive. Use the Intermediate Value Theorem twice.

35.(a) Source.

(b) Sink.

(c) Node.

37. The term βx is the effect of the passengers and $-\alpha$ is the term giving the rate of decrease of the time between trains when no passengers are present. Both α and β should be positive.

39. Because the only equilibrium point, $x = \alpha/\beta$, is a source, if the initial gap between trains is too large then x will increase without bound and if it is too small x will decrease to zero. It is very unlikely that the time between trains will stay exactly at the source for long.

Hints and Answers for Section 1.7

1.

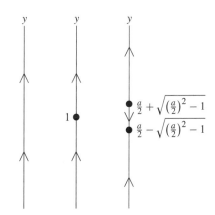

Phase lines for $a < 0$, $a = 0$, and $a > 0$.

3.

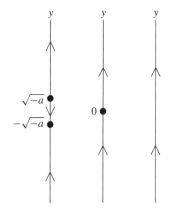

Phase lines for $a < 2$, $a = 2$ and $a > 2$.

5. Only one bifurcation at $\alpha = 1$. A sink–source pair of equilibria for $\alpha < 1$ collide and form a node when $\alpha = 1$. No equilibria for $\alpha > 1$.

7.(a) $L = 16$

(b) If the population is above 40 when the licenses are issued, the population will head toward 60. If the population is below 40 when fishing begins, the fish will become extinct.

(c) As long as the fish population is well over 40 when 16 licenses are issued, unexpected perturbations in the system will not drastically effect the fish and the population will tend toward the equilibrium point near 60.

9.(a)

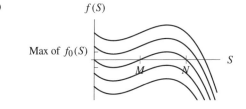

(b) Assuming that the fox squirrel population begins above the threshold, the population will decrease slowly, hovering at the right-most fixed point. After the bifurcation value, the population will plummet toward zero.

(c) $E = (k)(27MN)\big((2M - N)(M + N)(2N - M) - 2(M^2 - MN + N^2)^{3/2}\big)$

11. The discriminant, $D = \alpha^2 - 4\beta$. When D is positive, there are two equilibria, when D is zero, one equilibrium and when D is negative, no equilibria. In the figure shaded region corresponds to phase lines with two equilibria. On the parabola, one. Above the parabola, none.

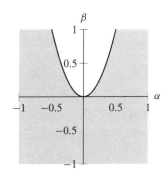

13. Hint: What are the signs of $f_0(0)$ and $f_0(1)$ and when (if ever) can they change as α is changed?

15. Since the sum of the indices is zero, the number of sinks and sources are same. Moreover, because $f_\alpha(0) > 0$ and $f_\alpha(1) \neq 0$ for all α and f_α is continuous, there is negative minimum value for f_α in $0 \leq y \leq 1$. Adding the value bigger than that minimum to f_α, one obtains $f_\alpha = m_\alpha > 0$. Hint: Sketch several functions which satisfy the conditions of $f_0(y)$ on the interval from 0 to 1.

17. 1 root, 3 roots.

19. $dy/dt = y(y + \alpha)(y^2 + \alpha)(y^4 + \alpha)$ is one of an infinite number of examples.

Hints and Answers for Section 1.8

1. $y(t) = t + c/t$

3. $y(t) = \dfrac{3e^{-t}}{2} + ce^t$

5. $y(t) = t + (1 + t^2)(\arctan(t) + c)$

7. $y(t) = t^2 - 2t + 2 + ce^{-t}$

9. $y(t) = \dfrac{t^2 + 2t + 3}{1 + t}$

11. $y(t) = t + \dfrac{2}{t}$

13. $y(t) = 2t^3 + 5t^2$

15. $y(t) = 4e^{-\cos t} \displaystyle\int e^{\cos t}\, dt$

17. $y(t) = 4e^{-1/t} \displaystyle\int e^{1/t} \cos t\, dt$

19. $a = -2$

21.(a) $dP/dt = .04P + 520$
  (b) about \$3,488.94

23. Approximately 5.08 years (61 months)

25. 1,700 parts per billion

27.(a) $dy/dt = 1/2 - y/(V_0 + t)$
  (b) Notice that if $V_0 = 0$, then the differential equation is undefined at $t = 0$. The amount of salt in the tank at time t is $t/4$.

Hints and Answers for Section 1.9

1. $du/dt = u^2 + u$

3. $du/dt = u/t + t(u + u^2 + \cos u)$

5. For $u = y - t$ the equation becomes $du/dt = u^2 - u - 2$.

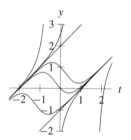

7. For $u = ty$ the equation becomes $du/dt = u \cos u$.

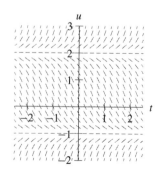

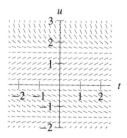

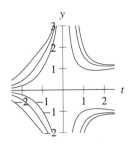

9. $y(t) = t^3(1+t)/4 + k(1 + (1/t))$

11. $dy/dt = -y^2 + 6ty + y - 9t^2 - 3t + 3$

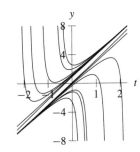

13. $dy/dt = 2y(1 - \sqrt{y})$

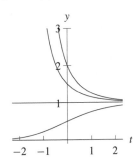

15.(a) $dS/dt = 3 - 3S/(10 + 3t)$

(b) $dC/dt = 3(1 - 2C)/(10 + 3t)$

(c) $C = 1/2$

(d) $C(5) = 0.42$

17.(a) $y = 10^{-1/3}$

(b) With $u = y - 10^{-1/3}$, the linear approximation is $du/dy = 3 \cdot 10^{\frac{1}{3}} u$. The origin is a source.

(c) The solution grows like $e^{3 \cdot 10^{\frac{1}{3}} t}$. The time necessary to double the distance from the equilibrium point is $t = \ln 2/(3 \cdot 10^{\frac{1}{3}})$.

19.(a) $y = -1, 3$

(b) With $u = y + 1$, the linearization is $du/dt = 4u$. $u = 0$ is a source. With $v = y - 3$, the linearization is $dv/dt = -4v$. $v = 0$ is a sink.

(c) For $y = -1$, the solution grows like e^{4t}. The time necessary to double the distance from the equilibrium point is $t = \ln 2/4$. For $y = 3$, the solution grows like e^{-4t}. The time necessary to halve the distance from the equilibrium point is $t = \ln 2/4$.

21.(a) $y - y^3/3!$

(b) $dy/dt = -3y$

(c) $y(2) = 0.04e^{-6}$

23.(a) With $u = P - 100$,
$du/dt = -0.05u(u/100 + 1) - 0.02(1 + \sin t)$.

(b) $du/dt = -0.05u - 0.02(1 + \sin t)$

(c) $u(t) = -0.4 - 2\sin t/2005 + 8\cos t/401 + ke^{-0.05t}$

(d) For large t, $ke^{-0.05t}$ becomes negligible and, also, $u(t)$ fluctuates between -0.38 and -0.42. P fluctuates around 99.6.

25.(a) $du/dt = f(u + y_0)$

(b) Using the Taylor expansion at $u = 0$, one obtains $f(u + y_0) = f(y_0) + f'(y_0)u + \dots$. The linearization is $du/dt = f'(y_0)u$.

Hints and Answers for Section 2.1

1. System (i) corresponds to large predators and small prey. System (ii) corresponds to small predators and large prey.

3. *Hint*: Compute dy/dt for $y = 0$.

5. *Hint*: Compute dx/dt for $x = 0$.

7. The populations oscillate with decreasing amplitude about an equilibrium point with both populations nonzero.

9. Change only the dR/dt equation,

   (i) $dR/dt = 2R - 1.2RF - \alpha$

   (ii) $dR/dt = R(2 - R) - 1.2RF - \alpha$

11. In both systems, make $dF/dt = kF + 0.9RF$, where $k > 0$ is the growth rate parameter for the predator population (ignoring any effects of overcrowding).

13. In both systems, make $dF/dt = -F + 0.9FR + k(5R - F)$, where k is an immigration rate parameter.

15. System (i) corresponds to boa constrictors and system (ii) to small cats. *Hint*: Suppose there were one predator ($y = 1$), how many units of prey are required to keep $dy/dt = 0$?

17. (a) *Hint*: Follow the behavior of a solution with

initial condition near $(0, 0)$ (after application of the pesticide kills almost all of both predator and prey).

   (b) After applying the pest control, you may see the an explosion of the pest population due to the absence of the predator.

19. $da/dt = -\alpha ab$, $db/dt = -\alpha ab$, where α is the reaction rate parameter.

21. $da/dt = k_1 - \alpha ab$, $db/dt = k_2 - \alpha ab$, where k_1 and k_2 are the rates A and B are added, respectively.

23. $da/dt = k_1 - \alpha ab + \gamma b^2/2$, $db/dt = k_2 - \alpha ab - \gamma b^2$ where γ is the reaction rate parameter for the reaction turning 2 B's into an A.

25. (a) Equilibria for (i) are: whenever $x = y$; or $x = 60$, $y = 55$. For (ii) it is $x = 58.333...$, $y = 56.666...$. If (x, y) is an equilibrium point and the cars start with speeds of x and y, then they stay at those speeds for all time.

   (b) The better model is (ii) as (i) has unreasonable solutions — such as, if they start their cars at the same time, they never go anywhere ($x = y = 0$). (Maybe that's not so unrealistic for the cars that Tom and Ray drive.)

Hints and Answers for Section 2.2

1. The dimension is 2, t is the independent variable, x and y are the dependent variables and the system is autonomous and nonlinear.

3. The dimension is 2, t is the independent variable, w and z are dependent variables, β and ϵ are parameters and the system is nonlinear and nonautonomous

5. The dimension is 2, t is the independent variable, x and y are dependent variables, ϵ is a parameter, and the system is nonautonomous and nonlinear.

7. The dimension is 3, s is the independent variable, x, y, and t are the dependent variables, ϵ is the parameter, and the system is autonomous and nonlinear.

9. $(x(t), y(t))$ is a solution.

11. $(x(t), y(t))$ is not a solution.

13. $(x(t), y(t))$ is not a solution.

15. $(x(t), y(t))$ is not a solution.

17. System ii.

19. System viii.

21. $(x(t), y(t))$ is a solution.

23. $(x(t), y(t))$ is not a solution.

25. $(x(t), y(t))$ is not a solution.

27. $(x(t), y(t))$ is not a solution.

29.(a) $(R, F) = (2, 1)$

(b)

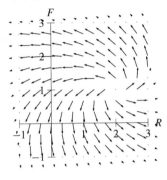

31.(a) Equilibria: $(\frac{\pi}{2} + k\pi, \frac{\pi}{2} + k\pi)$, k any integer.

(b)

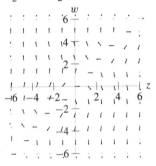

33.(a) Equilibria: $(\frac{\pi}{2} + k\pi, 0)$, k any integer.

(b)

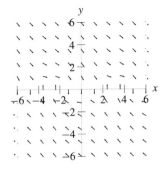

35.(a) Equilibrium is $(-1/9, 2/9)$.

(b)

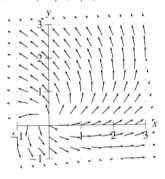

37. If $(x(t), y(t))$ is a solution to the system in Exercise 6, then $(x(t), y(t), \alpha_0)$ is a solution to the system in Exercise 8. If $(x(t), y(t), \alpha_0)$ is a solution to the system in Exercise 8, then $(x(t), y(t))$ is a solution to the system in Exercise 6 with $\alpha = \alpha_0$.

Hints and Answers for Section 2.3

1.(a) Equilibria: $(0, 0)$, $(1, 0)$, $(-1, 0)$.

(b)

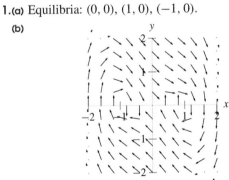

(c)

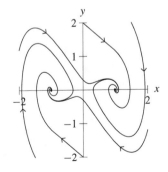

3.(a) Equilibria: $(\frac{\pi}{2} + k\pi, \frac{\pi}{2} + k\pi)$, k any integer.

(b)

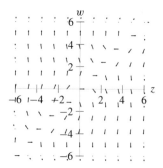

(c)

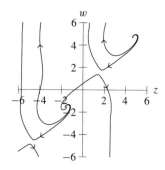

5.(a) Equilibria $(0, 0)$ and every point on the circle with radius one.

(b)

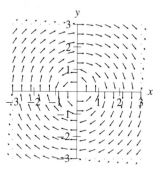

(Every point on the circle with radius one is an equilibrium point. Note direction of arrows close to origin and far from the origin.)

(c)

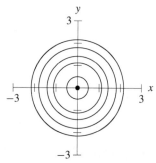

(Every point on the circle with radius one is an equilibrium point. The orbits go clockwise outside the unit circle and counter-clockwise inside the unit circle, see direction field.)

7.(a) Equilibria:
$(\pi/4 + k\pi/2, \pi/4 + k\pi/2)$, k any integer.

(b)

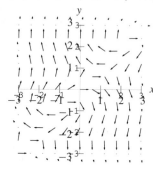

(c)

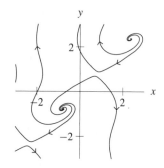

9.(a) $\mathbf{Y}(t) = \left(C_2 e^{C_0 e^{2t}/2}, C_0 e^{2t} \right)$

(b) $\mathbf{Y}(t) = (e^{(e^{2t} - 1)/2}, e^{2t})$

(c)

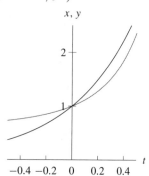

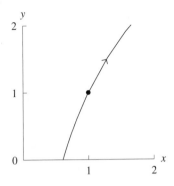

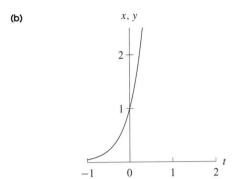

The $x(t)$-graph is on the x-axis

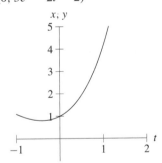

(d)

11.(a) $\left(C_2 \exp\left(C_0^2 e^{2t}/2 - 4C_0 t e^t + 4(t+1)^3/3\right),\right.$
$\left. C_0 e^t - 2t - 2\right)$

(b) $\mathbf{Y}(t) = (0, 3e^t - 2t - 2)$

(c)

The solution in the xy-phase plane lies on the y-axis.

(c)

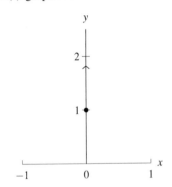

(The solution has $x(t) = 0$ for all t, hence the $x(t)$-graph is on the x-axis in the tx plane. In the xy-phase plane, the solution lies on the y-axis (since $x = 0$).

(d) Direction field not applicable since the system is nonautonomous.

13.(a) $\mathbf{Y}(t) = (0, e^{3t})$

15.(a) $\mathbf{Y}(t) = \left(1 - \sqrt{2}\tan(\sqrt{2}t + \arctan(1/\sqrt{2})), 2\right)$

(b)

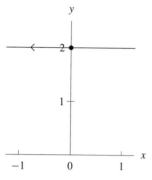

(c)

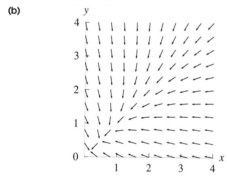

17.(a) System is: $dy/dt = (z - y) - y^2$,
$dz/dt = (y - z) - z^2$.

(b)

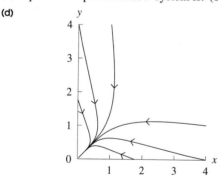

(c) Equilibrium point for new system is: $(0, 0)$.

(d)

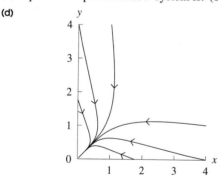

(e) In each case, both countries will eventually disarm.

19.(a) $dy/dt = (z - y/2) - 2y^2$, $dz/dt = (y - z/2) - z^2$

(b)

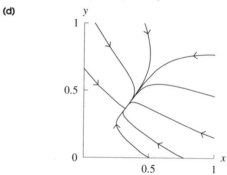

(c) Equilibria for the new system are:
$(0, 0)$ and $\approx (0.324, 0.371)$.

(d)

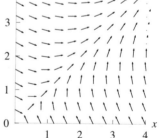

(e) In each case, both countries settle at the equilibrium point $\approx (0.324, 0.371)$.

21.(a) System is: $dy/dt = z - y/2$, $dz/dt = y - z/2$.

(b)

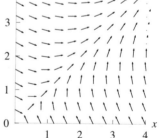

(c) Equilibrium point for the new system: $(0, 0)$.

(d)

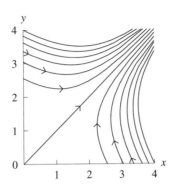

(e) Both countries increase the amount of arms without the limit.

23.(a)

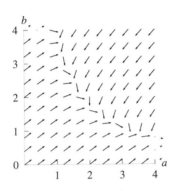

(b) No equilibria.

(c)

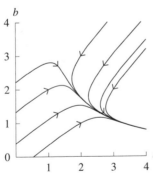

(d) All solutions eventually are asymptotic to zero in b and unbounded in a.

25.(a)

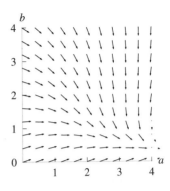

(b) No equilibria.

(c)

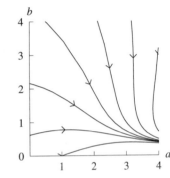

(d) All solutions eventually are asymptotic to zero in b and unbounded in a.

27. First we deal with $x(0)$, $y(0)$ positive.

(a) First check that $(0, 0)$ is an equilibrium point and that $\mathbf{Y}_1(t) = (2e^{-2t}, e^{-2t})$ and $\mathbf{Y}_2(t) = (e^{-t}, -2e^{-t})$ are solutions.

(b) Graph $\mathbf{Y}_1(t)$ and $\mathbf{Y}_2(t)$ in the xy-plane.

(c) Now apply the Existence and Uniqueness of Theorem (the uniqueness part).

The case $x(0)$, and $y(0)$ negative is similar, using $-\mathbf{Y}_1(t)$ and $-\mathbf{Y}_2(t)$.

29. *Hint*: Substitute $\mathbf{Y}_1(t) = (e^{-t}\sin(3t), e^{-t}\cos(3t))$ into the differential equation and check that left hand side equals right hand side.

31. *Hint*: They both give the same spiral. This doesn't contradict the Uniqueness Theorem because they are never in the same place at the same time.

Hints and Answers for Section 2.4

1. The system is

$$\frac{dy}{dt} = v$$

$$\frac{dv}{dt} = -4v - 3y.$$

This is a two-dimensional, autonomous, linear system with independent variable t, and dependent variables y and v.

3. The system is:

$$\frac{dy}{dt} = v \qquad \frac{dv}{dt} = w \qquad \frac{dw}{dt} = w - 2tv + 3y - 6.$$

This is a three-dimensional, nonautonomous, linear system with independent variable t and dependent variables y, v and w.

5. Letting the origin be the rest position of the mass in the absence of gravity, the differential equation is

$$\frac{d^2y}{dt^2} + \frac{k_d}{m}\frac{dy}{dt} + \frac{k_s}{m}y - g = 0.$$

7. The mass will oscillate in our system, but not around the origin. Moving the origin to the rest position of the mass in the presence of gravity (vertical spring) amounts to making the change of variables $z = y - mg/k_s$. This transforms our system into the one for the standard harmonic oscillator.

9.(a)
$$\frac{dy}{dt} = v$$
$$\frac{dv}{dt} = -v - 3y$$

(b)

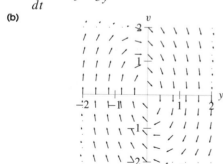

(c)

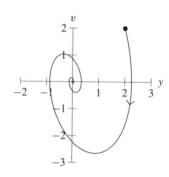

(d)

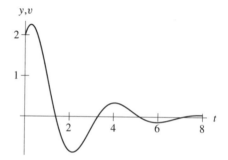

(e) The mass starts when the spring is stretched 2 units with an initial velocity of 2 away from the rest position. The mass will oscillate around the rest position with an amplitude which tends to zero very quickly. Our model correctly predicts that the mass gets closer and closer to the rest position as time increases.

11.(a)
$$\frac{dy}{dt} = v$$
$$\frac{dv}{dt} = -0.05v - 9y$$

(b)

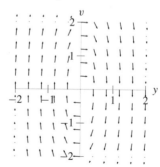

(c)

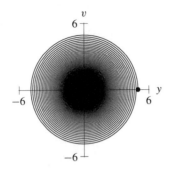

(d)

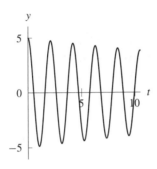

(e) The mass starts when the spring is stretched 5 units with no initial velocity. The mass will oscillate around the rest position with an amplitude which tends to zero slowly. Our model correctly predicts that the mass gets closer and closer to the rest position as time increases.

13. $dy/dt = v,$
$dv/dt = ((k_2 - k_1)/m)y - (k_d/m)v + (k_1 l_1 - k_2 L_2)/m + (k_2 - k_1)/(2m).$ Here $y = 0$ is the center, $y < 0$ is to the left and $y > 0$ is to the right.

15. Model (a) is better. The parameter α controls the response for y near zero and β controls the nonlinear effect as y moves away from zero.

17.

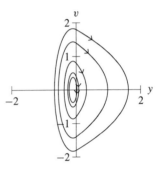

19. Unique continuous solutions exist for any initial condition. Hint: Unique continuous solutions exist on both sides of $y = 0$ and for each solution hitting $y = 0$ from the left there is exactly one solution starting at $y = 0$ on the right (and vice versa).

Hints and Answers for Section 2.5

1.(a) $(x(1.25), y(1.25)) \approx (.65, -.59)$

(b)

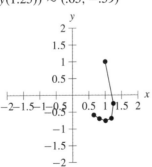

(c)

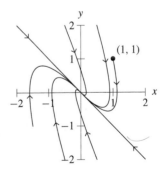

(d) It is consistent.

3.(a) $(x(1.25), y(1.25)) \approx (1.94, -.72)$

(b)

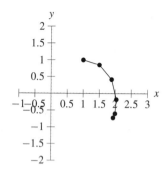

(c)

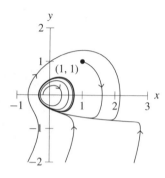

(d) It is consistent.

5.(a) It is a solution.

(b) At $t = 10$ the Euler approximation is approximately $(-9.21, 1.41)$ and the actual solution is approximately $(-0.84, -0.54)$.

(c) At $t = 10$ the Euler approximation is approximately $(-1.41, -0.85)$ and the actual solution is approximately $(-0.84, -0.54)$.

(d) The approximation will always spiral away from the origin.

7. System (ii).

9. System (i).

11. Observe the period of oscillation as a function of amplitude.

Hints and Answers for Section 2.6

1. Equilibria: $(1, 1)$ and $(-2, 4)$. All three solutions will go down and to the right without bound.

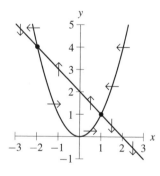

3. Equilibria: $(0, 0)$ and $(1, 1)$. Solutions passing through (a) and (b) will move toward the equilibrium point $(0, 0)$, while solution through (c) will go up and to the right without bound.

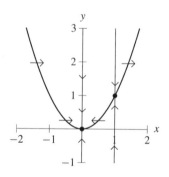

5.(a) Equilibria: $(0, 0)$, $(0, 30)$, $(10, 0)$ (and $(20, -10)$, but this is outside the region of interest).

(b) Solutions which start on an axis stay on that axis. On the positive x-axis, solutions move toward the equilibrium point at $x = 10$. On the positive y-axis, solutions move toward the equilibrium point at $y = 30$.

(c)

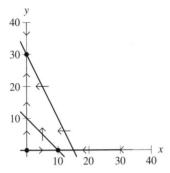

(d) Solutions with an initial condition in the interior of the first quadrant approach the equilibrium point at $(0, 30)$.

7. Note that the y component of the vector field is equal to twice the x component at every point. This implies that the vector field has slope 2 at every point and hence, straight lines with slope 2 are solution curves. The line $y = 1 - x$ is a line of equilibrium points. Above the line, solutions move down and left, below the line solutions move up and right.

9.(a) Change the signs in front of B and D from minuses to pluses.

(b) The nullclines are pairs of straight lines. The points $(0, 0)$, $(0, F/E)$ and $(C/A, 0)$ are equilibrium points. We need to have $AE - BD > 0$ in order for there to be an equilibrium point with both x and y positive.

11. Equilibrium points (in the first quadrant): $(0, 0)$, $(0, 50)$, $(40, 0)$ Solutions with an initial condition in the interior of the first quadrant approach the equilibrium at $(0, 50)$.

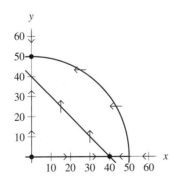

13. Equilibria: $(0, 0), (0, 50), (60, 0), (30, 40), (234/5, 88/5)$

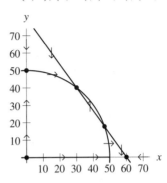

Solutions with an initial condition in the interior of the first quadrant approach either the equilibrium at $(60, 0)$ or the one at $(30, 40)$.

15.(a)

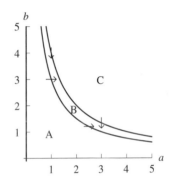

(b) In region A, the vector field points up and to the right. In region B, the vector field points down and to the right. In region C, the vector field points down and to the left.

(c) Once in region B, a solution can not leave. As time increases, these solutions are asymptotic to the positive x–axis from above.

17.(a)

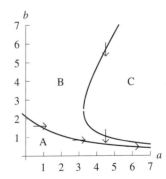

(b) In region A, the vector field points up and to the right. In region B, the vector field points down and to the right. In region C, the vector field points down and to the left.

(c) Once in region B, a solution can not leave. As time increases, these solutions are asymptotic to the positive x–axis from above.

19.

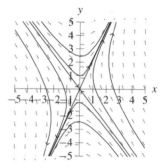

21. Solutions circle around the equilibrium point $(1, 0)$ in a clockwise direction.

Hints and Answers for Section 2.7

1.(a) *Hint*: Substitute $(0, 0, 0)$ and $(\pm 6\sqrt{2}, \pm 6\sqrt{2}, 27)$ to see if all derivatives are zero.

(b) *Hint*: Set $dx/dt = dy/dt = dz/dt = 0$ and solve three simultaneous equations.

3.(a) *Hint*: The plane $y = x$.

(b) *Hint*: The surface $y = 28x - xz$.

(c) *Hint*: The surface $z = 3xy/8$.

5.(a)

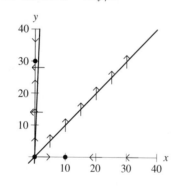

(b)

(c) *Hint*: This is a phase line with a sink at zero and no other equilibrium points.

(d) *Hint*: Use the picture in part (b) in the horizontal direction and the picture in part (c) in the vertical direction.

Hints and Answers for Section 3.1

1.(a) $a = 1$: If Paul's store is making a profit ($x > 0$) then Paul's profit increases more quickly ($ax > 0$).

(b) $b = -1$: If Bob's store is making a profit ($y > 0$) then Paul's profit increases more slowly ($by < 0$).

(c) $c = 1$: If Paul's store is making a profit ($x > 0$) then Bob's profit increases more quickly ($cx > 0$).

(d) $d = -1$: If Bob's store is making a profit ($y > 0$) then Bob's profit increases more slowly ($dy < 0$).

3.(a) $a = 1$: If Paul is making a profit ($x > 0$), then Paul's profit increases more quickly ($ax > 0$).

(b) $b = 0$: Bob's store profit doesn't affect Paul's store.

(c) $c = 2$: If Paul's store is making a profit ($x > 0$) then Bob's profit increases more quickly ($cx > 0$).

(d) $d = 1$: If Bob's store is making a profit ($y > 0$) then Bob's profit increases more quickly ($dy > 0$).

5. $\mathbf{Y} = \begin{pmatrix} x \\ y \end{pmatrix}, \dfrac{d\mathbf{Y}}{dt} = \begin{pmatrix} 2 & 1 \\ 1 & 1 \end{pmatrix} \mathbf{Y}$

7. $\mathbf{Y} = \begin{pmatrix} p \\ q \\ r \end{pmatrix}, \dfrac{d\mathbf{Y}}{dt} = \begin{pmatrix} 3 & -2 & -7 \\ -2 & 0 & 6 \\ 0 & 7.3 & 2 \end{pmatrix} \mathbf{Y}$

9.
$$\dfrac{dx}{dt} = \beta y$$
$$\dfrac{dy}{dt} = \gamma x - y$$

11.(a)

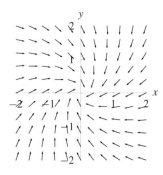

(b)

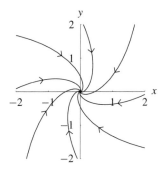

(c)

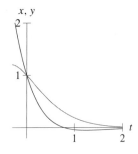

13.(a)

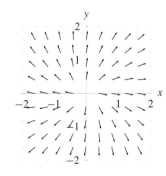

(b)

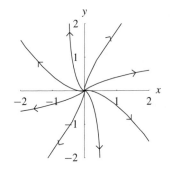

(c)

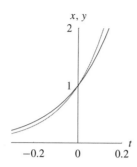

(d)

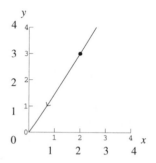

(e)

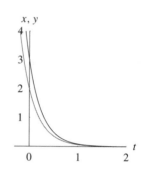

15.(a) If $a \neq 0$ then any point on the line $ax + by = 0$ (where the x-component of the vector field is zero) is of the form $(-by/a, y)$. Using this in the y-component of the vector field gives zero (using that the determinant is zero).

(b) If $a = 0$ then $bc = 0$, and if one of them is nonzero proceed as above. If both are zero, then all points with coordinates $(x, 0)$ are equilibrium points.

17.(a)
$$\frac{dy}{dt} = v$$
$$\frac{dv}{dt} = -pv$$
Equilibrium line $v = 0$.

(b)
$$\frac{dy}{dt} = v$$
$$\frac{dv}{dt} = 0$$
Same equilibrium line.

19.(a) Example ii): $d^2x/dt^2 + 3dx/dt + 2x = 0$

(b) Example iii): $d^2y/dt^2 - 2dy/dt - y = 0$

(c) Example v): $d^2x/dt^2 - x = 0$, or $d^2y/dt^2 - y = 0$

21. $\gamma > 0$

23. $\delta < 0$

25.(a) Substitute $\mathbf{Y}(t)$ into the differential equation and check that the left-hand side equals the right-hand side.

(b) The solution with this initial condition is $(-2te^{2t}, (2t + 2)e^{2t})$.

27.(a) They are solutions.

(b) They are linearly independent.

(c) $\mathbf{Y}(t) = (e^{-3t} + e^{-4t}, e^{-3t} + 2e^{-4t})$

29. $\mathbf{Y}_1$ is a solution but $\mathbf{Y}_2$ is not.

31. Want to show that one of the vectors is a multiple of the other.

(a) (x_2, y_2) and $(x_1, y_1) = (0, 0)$ are on the same line through the origin.

(b) This is the same as saying that (x_2, y_2) is a multiple of (x_1, y_1) which implies that they are on the same line through the origin.

(c) If $x_1 \neq 0$ have: $(x_2, y_2) = (x_2/x_1)(x_1, y_1)$ and use part (b). If $x_1 = 0$ then if $y_1 \neq 0$ proceed as before, and if $y_1 = 0$ you are in the situation of part (a).

33. Can give explicit solutions for (a), (c) and (d).

35.(a) $dW/dt = y_2(dx_1/dt) + x_1(dy_2/dt) - y_1(dx_2/dt) - x_2(dy_1/dt)$

(b) Replace dx_1/dt by $ax_1 + by_1$, and so on, in the above formula.

(c) $W(t) = Ce^{(a+d)t}$, where C is a constant. Hence if $C = W(0) = 0$ then $W(t) = 0$ for all t, and if $C = W(0) \neq 0$ then $W(t) \neq 0$ for any t.

(d) See Exercise 32. If $\mathbf{Y}_1(0)$ and $\mathbf{Y}_2(0)$ are linearly independent then $W(0) \neq 0$. But then $W(t) \neq 0$ for any t, hence $\mathbf{Y}_1(t)$ and $\mathbf{Y}_2(t)$ are linearly independent.

Hints and Answers for Section 3.2

1.(a)

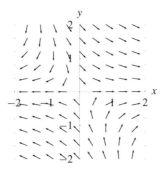

(b)

There are two straight-line solutions. On the *x*-axis, solutions tend away from the origin. On the line through the second and fourth quadrants, solutions tend toward the origin.

3.(a)

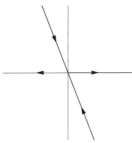

(b)

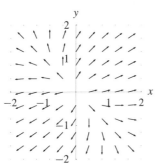

There are two straight-line solutions. Solutions along both lines tend away from the origin.

5.(a)

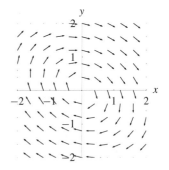

(b) There are no straight-line solutions.

7.(a)

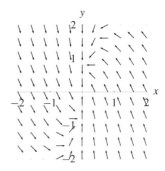

(b)

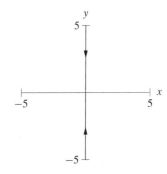

The *y*-axis is the only straight-line solution and solutions along it tend toward the origin.

9.(a) Eigenvalues: $\lambda_1 = -2, \lambda_2 = 3$;
Eigenvectors: $v_1 = (-2, 5), v_2 = (1, 0)$

(b) $\mathbf{Y}(t) = k_1 e^{-2t} \begin{pmatrix} -2 \\ 5 \end{pmatrix} + k_2 e^{3t} \begin{pmatrix} 1 \\ 0 \end{pmatrix}$

11.(a) Eigenvalues: $\lambda_1 = 2, \lambda_2 = 5$;
Eigenvectors: $v_1 = (-1, 1), v_2 = (2, 1)$

(b) $\mathbf{Y}(t) = k_1 e^{2t} \begin{pmatrix} -1 \\ 1 \end{pmatrix} + k_2 e^{5t} \begin{pmatrix} 2 \\ 1 \end{pmatrix}$

13. The eigenvalues are complex.

15. The eigenvalues are equal.

17.(a) $\lambda_1 = -3, \lambda_2 = 2, v_1 = (5, 1), v_2 = (0, 1)$

(b) $\mathbf{Y}(t) = k_1 e^{-3t} \begin{pmatrix} 5 \\ 1 \end{pmatrix} + k_2 e^{2t} \begin{pmatrix} 0 \\ 1 \end{pmatrix}$

19. $\lambda_1 = -2, \lambda_2 = -5, v_1 = (1, 2), v_2 = (1, -1)$

$\mathbf{Y}(t) = k_1 e^{-2t} \begin{pmatrix} 1 \\ 2 \end{pmatrix} + k_2 e^{-5t} \begin{pmatrix} 1 \\ -1 \end{pmatrix}$

21.(a) Eigenvalues: $\lambda_1 = 0, \lambda_2 = -3$;
Eigenvectors: $v_1 = (1, 1), v_2 = (-2, 1)$

(b) $\mathbf{Y}(t) = k_1 \begin{pmatrix} 1 \\ 1 \end{pmatrix} + k_2 e^{-3t} \begin{pmatrix} -2 \\ 1 \end{pmatrix}$

23.(a) $\mathbf{Y}(t) = \frac{1}{5} e^{3t} \begin{pmatrix} 5 \\ 1 \end{pmatrix} - \frac{1}{5} e^{-2t} \begin{pmatrix} 0 \\ 1 \end{pmatrix}$

(b) $\mathbf{Y}(t) = e^{-2t} \begin{pmatrix} 0 \\ 1 \end{pmatrix}$

(c) $\mathbf{Y}(t) = \frac{2}{5} e^{3t} \begin{pmatrix} 5 \\ 1 \end{pmatrix} + \frac{8}{5} e^{-2t} \begin{pmatrix} 0 \\ 1 \end{pmatrix}$

(d)

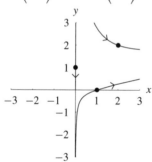

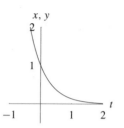

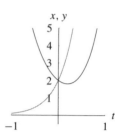

(e) A lot more work is required in order to get the formula.

25.(a) $\mathbf{Y}(t) = \begin{pmatrix} \frac{1}{3} \\ \frac{1}{3} \end{pmatrix} - \frac{1}{3} e^{-3t} \begin{pmatrix} -2 \\ 1 \end{pmatrix}$

(b) $\mathbf{Y}(t) = \begin{pmatrix} 2 \\ 2 \end{pmatrix}$

(c) $\mathbf{Y}(t) = e^{-3t} \begin{pmatrix} -2 \\ 1 \end{pmatrix}$

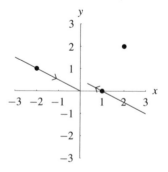

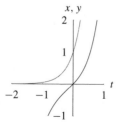

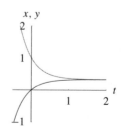

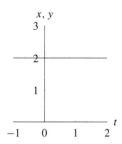

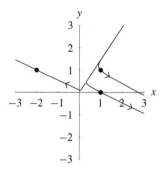

(d)

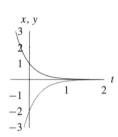

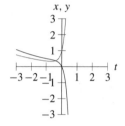

(d) A lot more work is required in order to get a formula.

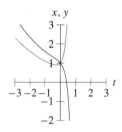

27.(a) $\mathbf{Y}(t) = (2/\sqrt{33})e^{(5-\sqrt{33})t/2} \begin{pmatrix} (-3+\sqrt{33})/4 \\ 1 \end{pmatrix} -$

$(2/\sqrt{33})e^{(5+\sqrt{33})t/2} \begin{pmatrix} (-3-\sqrt{33})/4 \\ 1 \end{pmatrix}$

(b) $\mathbf{Y}(t) = \left(1/2 + 7/(2\sqrt{33})\right) e^{(5-\sqrt{33})t/2} \cdot$

$\begin{pmatrix} (-3+\sqrt{33})/4 \\ 1 \end{pmatrix} +$

$\left(1/2 - 7/(2\sqrt{33})\right) e^{(5+\sqrt{33})t/2} \cdot$

$\begin{pmatrix} (-3-\sqrt{33})/4 \\ 1 \end{pmatrix}$

(c) $\mathbf{Y}(t) = \left(1/2 - 5/(2\sqrt{33})\right) e^{(5-\sqrt{33})t/2} \cdot$

$\begin{pmatrix} (-3+\sqrt{33})/4 \\ 1 \end{pmatrix} +$

$\left(1/2 + 5/(2+\sqrt{33})\right) e^{(5+\sqrt{33})t/2} \cdot$

$\begin{pmatrix} (-3-\sqrt{33})/4 \\ 1 \end{pmatrix}$

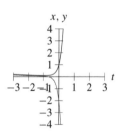

(e) A lot more work is required in order to get a formula.

29. The characteristic polynomial, $(\lambda - a)^2$, has a as its only root. Thus, $\mathbf{A}$ has a as its only eigenvalue. To find the eigenvectors, we solve the system

$$\begin{cases} ax_0 = ax_0 \\ ay_0 = ay_0. \end{cases}$$

Since all vectors (x_0, y_0) solve this system, every vector is an eigenvector.

31. The characteristic polynomial is $\lambda^2 - (a + d)\lambda + ad - b^2$. This quadratic has the discriminant $D = (a - d)^2 + 4b^2$. Since $D \geq 0$, the characteristic polynomial always has real roots, i.e. the matrix always has real eigenvalues. If $b \neq 0$, then $D > 0$ so that the characteristic polynomial always has real distinct roots. Hence **B** has two distinct eigenvalues.

33.(a) $\dfrac{d^2y}{dt^2} + 3\dfrac{dy}{dt} + 2y = 0,\quad \dfrac{dy}{dt} = v,\quad \dfrac{dv}{dt} = -2y - 3v$

(b) $\mathbf{Y}(t) = k_1 e^{-2t}\begin{pmatrix} -1 \\ 2 \end{pmatrix} + k_2 e^{-t}\begin{pmatrix} -1 \\ 1 \end{pmatrix}$

(c) $\mathbf{Y}(t) = e^{-2t}\begin{pmatrix} -1 \\ 2 \end{pmatrix} - 2e^{-t}\begin{pmatrix} -1 \\ 1 \end{pmatrix}$

35.(a) $\dfrac{d^2y}{dt^2} + 4\dfrac{dy}{dt} + y = 0,\quad \dfrac{dy}{dt} = v,\quad \dfrac{dv}{dt} = -y - 4v$

(b) $\mathbf{Y}(t) = k_1 e^{(-2-\sqrt{3})t}\begin{pmatrix} -2 + \sqrt{3} \\ 1 \end{pmatrix} +$
$k_2 e^{(-2+\sqrt{3})t}\begin{pmatrix} -2 - \sqrt{3} \\ 1 \end{pmatrix}$

(c) $\mathbf{Y}(t) = -\dfrac{1}{\sqrt{3}} e^{(-2-\sqrt{3})t}\begin{pmatrix} -2 + \sqrt{3} \\ 1 \end{pmatrix} +$
$\dfrac{1}{\sqrt{3}} e^{(-2+\sqrt{3})t}\begin{pmatrix} -2 - \sqrt{3} \\ 1 \end{pmatrix}$

37. The characteristic polynomial is $\lambda^2 + \lambda + 4$ which has complex roots: $-\frac{1}{2}(1 \pm \sqrt{15}i)$. Therefore the eigenvalues are not real, i.e. there are no straight-line solutions.

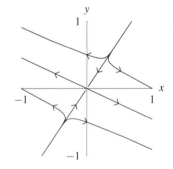

Hints and Answers for Section 3.3

1.(a) Eigenvalues: $\lambda_1 = 3$, $\lambda_2 = -2$.

(b) Eigenvectors: $v_1 = (5, 1)$, $v_2 = (0, 1)$.

(c) See (d).

(d)

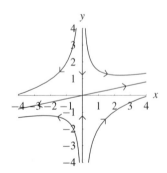

3.(a) Eigenvalues: $\lambda_1 = (5 - \sqrt{33})/2$, $\lambda_2 = (5 + \sqrt{33})/2$.

(b) Eigenvectors: $v_1 = (2, 1 - \sqrt{33}/3)$, $v_2 = (2, 1 + \sqrt{33}/3)$.

(c) See (d).

(d)

5.

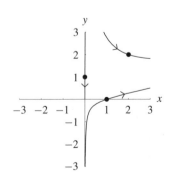

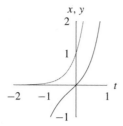

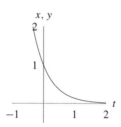

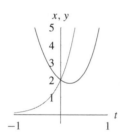

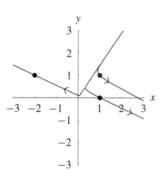

7.

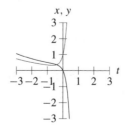

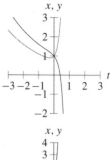

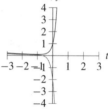

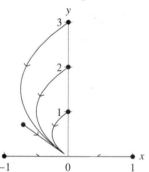

9.(a) The presence of the new species will decrease the population of the native fish. The native fish do not effect the population of the new fish.

(b) The model agrees with the premises of the system.

(c)

The straight line solutions are the *x*-axis and the line $y = -x$. Solutions along these lines tend toward the origin.

(d) If a small number of new fish are added to the lake, the native population drops from equilibrium. The new fish die out as the native fish return to equilibrium.

11.(a) The native fish, *x*, have no affect on the population of the new fish. The introduction of new fish increases the population of the native fish.

(b) The model agrees with the system described in the problem.

(c) *Hint*: The straight-line solutions are the *x*-axis and $y = x$. The solutions along these lines tend toward the origin. Hence the origin is a sink.

(d) The population of the native fish decreases to the equilibrium and the new fish will die out.

13. For any initial condition, y settles back to 0. For an initial condition (x_0, y_0) not at the origin, if $y_0 + 3x_0 > 0$ then x's profit grow unboundedly. If $y_0 + 3x_0 < 0$ then x's debts grow unboundedly. If $y_0 + 3x_0 = 0$ then x settles back to 0.

15.(a) $d^2y/dt^2 + 3dy/dt + y = 0$ $dy/dt = v$ $dv/dt = -y - 3v$

(b) Eigenvalues: $-\frac{1}{2}(3 \pm \sqrt{5})$;
Eigenvectors: $(3 + \sqrt{5}, -2)$ and $(3 - \sqrt{5}, -2)$

(c)

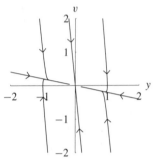

(d)

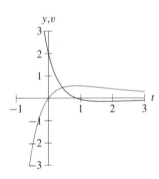

17.(a) $d^2y/dt^2 + 10dy/dt + 2y = 0$;
$dy/dt = v$, $dv/dt = -2y - 10v$

(b) Eigenvalues: $-5 \pm \sqrt{23}$; Eigenvectors:
$(5 + \sqrt{23}, -2)$ and $(5 - \sqrt{23}, -2)$

(c)

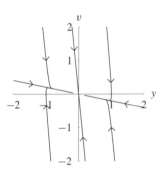

(d)

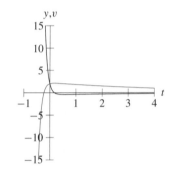

19. The mass had passed though zero at $t = -\ln\frac{5}{3}$. Also, as $t \to \infty$, $y(t) \to 0^-$.

21.(a) Sink.

(b) x-axis and $y = 2x$.

(c) *Hint:* : $x(t)$ and $y(t)$ tend to zero in all cases, watch for where the solution curves on the phase planes cross the x and y axes and where they cross the $x = y$ line.

23.(a) Calculate the discriminant of the characteristic equation.

(b) $\lambda_1 = 2$, $v_1 = (1, 4)$, $\lambda_2 = -2$, $v_2 = (1, 0)$ so the phase plane is a saddle

(c) Hint: one goes "up" along $y = 4x$, the other goes "down" along $y = 4x$

(d) $t = 1.9$

Hints and Answers for Section 3.4

1. $\text{Re}\mathbf{Y}(t) = e^t \begin{pmatrix} 2\cos(3t) - \sin(3t) \\ \cos(3t) \end{pmatrix}$,

$\text{Im}\mathbf{Y}(t) = e^t \begin{pmatrix} \cos(3t) + 2\sin(3t) \\ \sin(3t) \end{pmatrix}$.

3.(a) $\lambda_1 = 2i$, $\lambda_2 = -2i$

(b) Center.

(c) Natural period π; Natural frequency $\frac{1}{\pi}$.

(d) Clockwise.

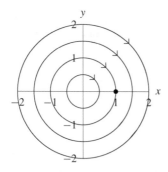

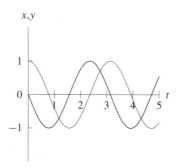

5.(a) Eigenvalues: $\lambda_1 = 4 - 2i$, $\lambda_2 = 4 + 2i$

(b) $(0, 0)$ is a spiral source.

(c) Natural period: π, natural frequency: $\frac{1}{\pi}$.

(d) Clockwise.

(e)

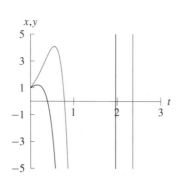

7.(a) $\lambda_1 = (3 + \sqrt{47}i)/2$; $\lambda_2 = (3 - \sqrt{47}i)/2$

(b) Spiral source

(c) Natural period $\frac{4\pi}{\sqrt{47}}$; Natural frequency $\frac{\sqrt{47}}{4\pi}$.

(d) Counterclockwise.

(e)

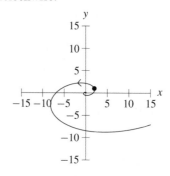

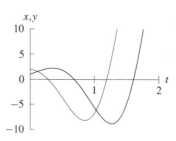

9.(a) $\mathbf{Y}(t) = k_1 \begin{pmatrix} \cos(2t) \\ -\sin(2t) \end{pmatrix} + k_2 \begin{pmatrix} \sin(2t) \\ \cos(2t) \end{pmatrix}$

(b) $\mathbf{Y}(t) = \begin{pmatrix} \cos(t) \\ -\sin(2t) \end{pmatrix}$

(c)

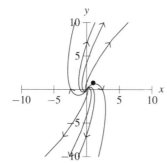

11.(a) General solution: $\mathbf{Y}(t) =$

$$k_1 e^{4t} \begin{pmatrix} \cos 2t \\ \cos 2t - \sin 2t \end{pmatrix} + k_2 \begin{pmatrix} \sin 2t \\ \cos 2t + \sin 2t \end{pmatrix}$$

(b) Particular solution: $\mathbf{Y}(t) = e^{4t} \begin{pmatrix} \cos 2t \\ \cos 2t - \sin 2t \end{pmatrix}$

(c)

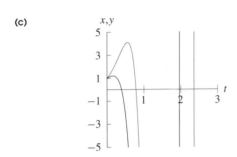

13.(a) $Y(t) = k_1 e^{3t/2} \begin{pmatrix} 12\cos(\frac{\sqrt{47}}{2}t) \\ \cos(\frac{\sqrt{47}}{2}t) + \sqrt{47}\sin(\frac{\sqrt{47}}{2}t) \end{pmatrix}$

$+$

$k_2 e^{3t/2} \begin{pmatrix} 12\sin(\frac{\sqrt{47}}{2}t) \\ -\sqrt{47}\cos(\frac{\sqrt{47}}{2}t) + \sin(\frac{\sqrt{47}}{2}t) \end{pmatrix}$

(b) $Y(t) = e^{3t/2} \begin{pmatrix} 2\cos(\frac{\sqrt{47}}{2}t) - \frac{10}{\sqrt{47}}\sin(\frac{\sqrt{47}}{2}t) \\ \cos(\frac{\sqrt{47}}{2}t) + \frac{7}{\sqrt{47}}\sin(\frac{\sqrt{47}}{2}t) \end{pmatrix}$

(c)

15.(a) For complex eigenvalues, $x(t)$ oscillates around $x(t) = 0$ with constant period. Also, by the real part of eigenvalues, the amplitude increases or decreases monotonically. The pictures satisfying these characters are (ii) and (v).

(b) For(ii), the natural period is 1 and, since the amplitude tends to zero as t increases, the origin is a sink. For (v), the natural period is 1 and, since the amplitude tends to increase as t increases, the origin is a source.

(c) (i) the natural period is not constant over time.
(iii) the amplitude is not monotonically decreasing or increasing.
(iv) Oscillation stops at some t. Therefore, the period is not constant over t.
(vi) Oscillation starts at some t, and no oscillation before theat. Therefore, the period is not constant over t.

17. If $\lambda_1 = \alpha + i\beta$ satisfies the equation then $\lambda_2 = \alpha - i\beta$ does the same. But a quadratic equation has at most two solutions, so they must be λ_1 and λ_2.

19. Suppose $Y_2 = kY_1$ for some constant k. Then, $Y_0 = (1 + ik)Y_1$. Since $AY_0 = \lambda Y_0$, substitution yields

$$(1 + ik)AY_1 = \lambda(1 + ik)Y_1,$$

or $AY_1 = \lambda Y_1$. Since AY_1 is real vector and λ is complex, λY_1 is complex and this leads to contradiction. Therefore, Y_1 and Y_2 are linearly independent.

21.(a) π/β

(b) π/β

(c) $2\pi/\beta$

(d) $(\arctan(\beta/\alpha))/\beta$

23.(a) The second-order equation is $\frac{d^2 y}{dt^2} = -2y - 2\frac{dy}{dt}$ and the first-order system is $dY/dt = AY$,

where $A = \begin{pmatrix} 0 & 1 \\ -2 & -2 \end{pmatrix}$.

(b) Eigenvalues are: $\lambda_1 = -1 - i$ and $\lambda_2 = -1 + i$

(c) Underdamped oscillator.

(d)

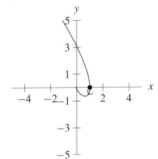

(e) General solution: $Y(t) =$

$k_1 e^{-t} \begin{pmatrix} -\cos t + \sin t \\ 2\cos t \end{pmatrix} + k_2 e^{-t} \begin{pmatrix} \cos t + \sin t \\ -2\sin t \end{pmatrix}$

25.(a) The second-order equation is: $\frac{d^2 y}{dt^2} = -4y - \frac{dy}{dt}$
The first-order system is: $dY/dt = AY$, where

$A = \begin{pmatrix} 0 & 1 \\ -4 & -1 \end{pmatrix}$

(b) The eigenvalues are: $\lambda_1 = -1/2 + i\sqrt{15}/2$ and $\lambda_2 = -1/2 - i\sqrt{15}/2$.

(c)

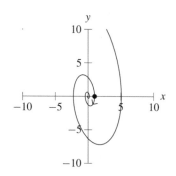

(d) General solution is: $\mathbf{Y}(t) =$

$$k_1 e^{-t/2} \left(\begin{array}{c} -\cos \frac{\sqrt{15}}{2}t + \sqrt{15} \sin \frac{\sqrt{15}}{2}t \\ 8\cos \frac{\sqrt{15}}{2}t \end{array} \right) +$$

$$k_2 e^{-t/2} \left(\begin{array}{c} \sqrt{15}\cos \frac{\sqrt{15}}{2}t + \sin \frac{\sqrt{15}}{2}t \\ -8\sin \frac{\sqrt{15}}{2}t \end{array} \right)$$

27. The mass is: $m = 1/(2\pi^2)$

29.(a) The clock will run slow.

  (b) The clock will run slow.

  (c) *Hint*: : Compute the derivative of the natural period with respect to m.

  (d) *Hint*: : Can the effects of changing different parameters cancel each other?

31.(a) Plug into the equation.

  (b) This is Euler's formula.

  (c) The hint tells you everything.

33. *Hint*: Try to draw the phase plane.

Hints and Answers for Section 3.5

1.(a) Eigenvalue: -3

  (b) Eigenvector: $(0, 1)$

  (c)

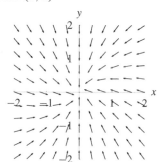

  (d)

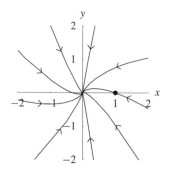

(e)

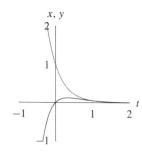

3.(a) Eigenvalue: -3

  (b) Eigenvector: $(1, 1)$

  (c)

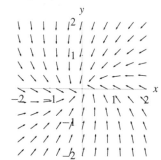

(d)

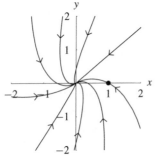

(e)

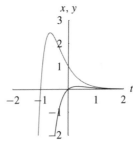

5.(a) $\mathbf{Y}(t) = k_1 e^{-3t} \begin{pmatrix} 0 \\ 1 \end{pmatrix} + k_2 e^{-3t} \left(t \begin{pmatrix} 0 \\ 1 \end{pmatrix} + \begin{pmatrix} 1 \\ 0 \end{pmatrix} \right)$

(b) $\mathbf{Y}(t) = e^{-3t} \left(t \begin{pmatrix} 0 \\ 1 \end{pmatrix} + \begin{pmatrix} 1 \\ 0 \end{pmatrix} \right)$

(c)

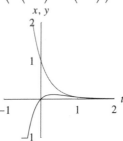

7.(a) $\mathbf{Y}(t) = k_1 e^{-3t} \begin{pmatrix} 1 \\ 1 \end{pmatrix} + k_2 e^{-3t} \left(t \begin{pmatrix} 1 \\ 1 \end{pmatrix} + \begin{pmatrix} 1 \\ 0 \end{pmatrix} \right)$

(b) $\mathbf{Y}(t) = e^{-3t} \left(t \begin{pmatrix} 1 \\ 1 \end{pmatrix} + \begin{pmatrix} 1 \\ 0 \end{pmatrix} \right)$

(c)

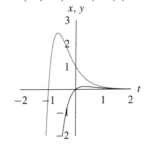

9.(a) $\alpha^2 = 4\beta$

  (b) $\beta = 0$

11.(a) $p^2 + 4q > 0$

  (b) $p^2 + 4q < 0$

  (c) $p^2 + 4q = 0$

13. $\mathbf{A} \begin{pmatrix} 1 \\ 0 \end{pmatrix} = \begin{pmatrix} a \\ c \end{pmatrix} = \begin{pmatrix} \lambda \\ 0 \end{pmatrix}$ so we have that
$a = \lambda$ and $c = 0$.
Likewise, $\mathbf{A} \begin{pmatrix} 0 \\ 1 \end{pmatrix} = \begin{pmatrix} b \\ d \end{pmatrix} = \begin{pmatrix} 0 \\ \lambda \end{pmatrix}$ so we have
that $d = \lambda$ and $b = 0$.

15. $e^{-\sqrt{3}t} \begin{pmatrix} -1 + (3 - \sqrt{3})t \\ 3 + (3 - 3\sqrt{3})t \end{pmatrix}$

17.(a) $k_s = 9/4$

  (b) Eigenvalue: $\lambda = -3/2$
    Eigenvector: $(-2, 3)$

  (c)

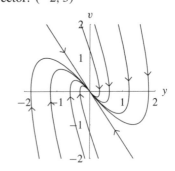

  (d) $\mathbf{Y}(t) = k_1 e^{-3t/2} \begin{pmatrix} -2 \\ 3 \end{pmatrix}$
    $+ k_2 e^{-3t/2} \left(t \begin{pmatrix} -2 \\ 3 \end{pmatrix} + \begin{pmatrix} 2 \\ -5 \end{pmatrix} \right)$

  (e) $\mathbf{Y}(t) = \frac{-5}{2} e^{-3t/2} \begin{pmatrix} -2 \\ 3 \end{pmatrix}$
    $+ \frac{-3}{2} e^{-3t/2} \left(t \begin{pmatrix} -2 \\ 3 \end{pmatrix} + \begin{pmatrix} 2 \\ -5 \end{pmatrix} \right)$

19.(a) $p^2 = 4q$

  (b) If λ is the only root, then the quadratic must have
    the form $(s - \lambda)^2 = s^2 - 2\lambda s + \lambda^2$. The second-
    order differential equation must be $\frac{d^2 y}{dt^2} - 2\lambda \frac{dy}{dt} + \lambda^2 y$. Plugging in $y_2(t) = t e^{st}$ yields: $e^{st}(\lambda - s)((\lambda - s)t - 2)$. If $s = \lambda$ then $y_2(t)$ is a solution.

21.(a) Eigenvalues: 0,8

(b) Eigenvectors: $(-2, 1)$, $(2, 3)$

(c)

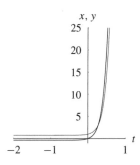

(d)

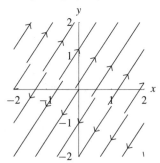

(e) $\mathbf{Y}(t) = k_1 \begin{pmatrix} 6 + 2e^{8t} \\ -3 + 3e^{8t} \end{pmatrix} + k_2 \begin{pmatrix} -2 + 2e^{8t} \\ 1 + 3e^{8t} \end{pmatrix}$

(f) $\mathbf{Y}(t) = \dfrac{1}{8} \left(\begin{pmatrix} 6 \\ -3 \end{pmatrix} + e^{8t} \begin{pmatrix} 2 \\ 3 \end{pmatrix} \right)$

23.(a) The characteristic polynomial is $\lambda^2 - \mathrm{Tr}\mathbf{A}\lambda + \det\mathbf{A}$. If 0 is an eigenvalue of $\mathbf{A}$, then 0 is a root of the characteristic polynomial and we must have $\det\mathbf{A} = 0$.

(b) The characteristic polynomial is $\lambda^2 - \mathrm{Tr}\mathbf{A}\lambda + \det\mathbf{A}$. If the $\det\mathbf{A} = 0$, then 0 is a root of the characteristic polynomial — that is, 0 is an eigenvalue of $\mathbf{A}$.

25.(a) $\mathbf{Y}(t) = k_1 \begin{pmatrix} 1 \\ 0 \end{pmatrix} + k_2 \begin{pmatrix} 2t \\ 1 \end{pmatrix}$

(b) $\mathbf{Y}(t) = k_1 \begin{pmatrix} 1 \\ 0 \end{pmatrix} + k_2 \begin{pmatrix} -2t \\ 1 \end{pmatrix}$

27.(a)

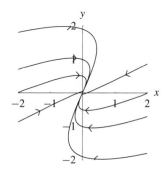

(b) The origin is a sink.

(c) Below is solution curve and a blow up near the origin

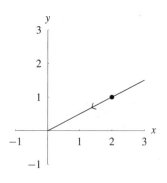

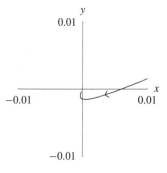

Hints and Answers for Section 3.6

1. *Hint*: There are 3 types in Sections 3.2 and 3.3, 3 types in Section 3.4 (these are the major types), while the rest are discussed in Section 3.5.

3.(a) $\lambda_1 = (-3 + \sqrt{65})/2$, $\lambda_2 = (-3 - \sqrt{65})/2$
$v_1 = (4, 7 + \sqrt{65})$, $v_2 = (4, 7 - \sqrt{65})$

(b) saddle

(c)

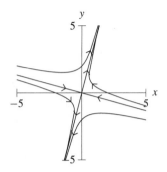

(d)

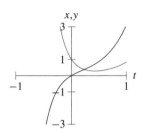

5.(a) $\lambda_1 = (-1 + \sqrt{53})/2$, $\lambda_2 = (-1 - \sqrt{53})/2$
$v_1 = (2, 7 + \sqrt{53})$, $v_2 = (2, 7 - \sqrt{53})$

(b) saddle

(c)

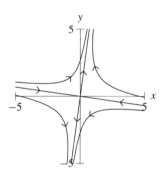

(d)

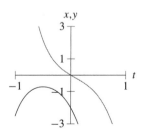

7.(a) $\lambda_1 = (-1 + i\sqrt{15})/2$, $\lambda_2 = (-1 - i\sqrt{15})/2$
$v_1 = (2, -1 + i\sqrt{15})$, $v_2 = (2, -1 - i\sqrt{15})$

(b) spiral sink

(c)

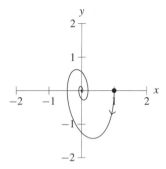

(d)

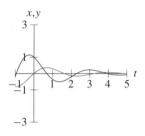

9.(a) $\mathbf{Y}(t) =$
$$k_1 e^{-3t/2} \begin{pmatrix} 8\cos(\sqrt{79}t/2) \\ \cos(\sqrt{79}t/2) + \sqrt{79}\sin(\sqrt{79}t/2) \end{pmatrix} +$$
$$k_2 e^{-3t/2} \begin{pmatrix} 8\sin(\sqrt{79}t/2) \\ \sin(\sqrt{79}t/2) - \sqrt{79}\cos(\sqrt{79}t/2) \end{pmatrix}$$

(b) $\mathbf{Y}(t) = e^{-3t/2}$
$$\begin{pmatrix} 2\cos(\sqrt{79}t/2) - (6\sqrt{79}/79)\sin(\sqrt{79}t/2) \\ \cos(\sqrt{79}t/2) + (19\sqrt{79}/79)\sin(\sqrt{79}t/2) \end{pmatrix}$$

(c)

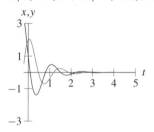

11.(a) $\mathbf{Y}(t) = k_1 e^{(-2+\sqrt{3})t} \begin{pmatrix} 1 \\ -2 + \sqrt{3} \end{pmatrix}$
$$+ k_2 e^{(-2-\sqrt{3})t} \begin{pmatrix} 1 \\ -2 - \sqrt{3} \end{pmatrix}$$

(b) $\mathbf{Y}(t) = e^{(-2+\sqrt{3})t} \begin{pmatrix} 1 + 2\sqrt{3}/3 \\ -\sqrt{3}/3 \end{pmatrix}$
$$+ e^{(-2-\sqrt{3})t} \begin{pmatrix} 1 - 2\sqrt{3}/3 \\ \sqrt{3}/3 \end{pmatrix}$$

(c)

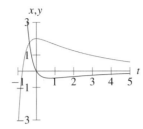

(d)

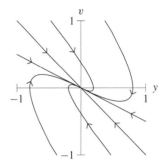

13.(a) $y'' + y' + 2y = 0$, $dy/dt = v$, $dv/dt = -2y - v$.

(b) $\lambda_1 = (-1 + i\sqrt{7})/2$, $\lambda_2 = (-1 - i\sqrt{7})/2$, $v_1 = (2, -1 + i\sqrt{7})$, $v_2 = (2, -1 - i\sqrt{7})$.

(c) Underdamped

(d) *Hint*: : Spirals clockwise.

(e) *Hint*: What are the periods of the oscillation of $y(t)$ and $v(t)$?

15.(a) $y'' + 0.4y' + 0.1y = 0$,
$dy/dt = v$, $dv/dt = -0.1y - 0.4v$.

(b) $\lambda_1 = -1/5 + i\sqrt{3/50}$, $\lambda_2 = -1/5 - i\sqrt{3/50}$, $v_1 = (1, -1/5 + i\sqrt{3/50})$, $v_2 = (1, -1/5 - i\sqrt{3/50})$.

(c) Underdamped

(d)

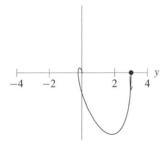

(e)

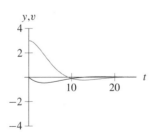

17.(a) $y'' + 3y'/2 + y/2 = 0$,
$dy/dt = v$, $dv/dt = -y/2 - 3v/2$.

(b) $\lambda_1 = -1$, $\lambda_2 = -1/2$,
$v_1 = (1, -1)$, $v_2 = (2, -1)$.

(c) overdamped

(e)

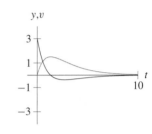

19.(a) $y'' + 3y/2 = 0$,
$dy/dt = v$, $dv/dt = -3y/2$.

(b) $\lambda_1 = i\sqrt{3/2}$, $\lambda_2 = -i\sqrt{3/2}$,
$v_1 = (1, i\sqrt{3/2})$, $v_2 = (1, -i\sqrt{3/2})$.

(c) Undamped

(d)

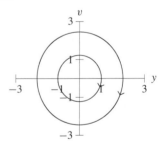

(e)

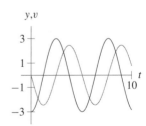

21.(a) $\mathbf{Y}(t) =$
$$k_1 e^{-t/2} \begin{pmatrix} 2\cos(\sqrt{7}t/2) \\ -\cos(\sqrt{7}t/2) - \sqrt{7}\sin(\sqrt{7}t/2) \end{pmatrix} +$$
$$k_2 e^{-t/2} \begin{pmatrix} 2\sin(\sqrt{7}t/2) \\ \sqrt{7}\cos(\sqrt{7}t/2) - \sin(\sqrt{7}t/2) \end{pmatrix}$$

(b) $\mathbf{Y}(t) = e^{-t/2} \begin{pmatrix} 2\cos(\sqrt{7}t/2) + \frac{2}{7}\sqrt{7}\sin(\sqrt{7}t/2) \\ -\frac{8}{7}\sqrt{7}\sin(\sqrt{7}t/2) \end{pmatrix}$

(c)

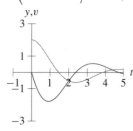

(b) $\mathbf{Y}(t) = \begin{pmatrix} 2\cos(3t/2) - 2\sin(3t/2) \\ -3\sin(3t/2) - 3\cos(3t/2) \end{pmatrix}$

(c)

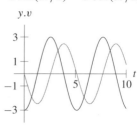

23. (a) $\mathbf{Y}(t) = k_1 e^{-t/5}$
$$\begin{pmatrix} \cos(\sqrt{3/50}\,t) \\ -(1/5)\cos(\sqrt{3/50}\,t) - \sqrt{3/50}\sin(\sqrt{3/50}\,t) \end{pmatrix} +$$
$$k_2 e^{-t/5}$$
$$\begin{pmatrix} \sin(\sqrt{3/50}\,t) \\ \sqrt{3/50}\cos(\sqrt{3/50}\,t) - (1/5)\sin(\sqrt{3/50}\,t) \end{pmatrix}$$

(b) $\mathbf{Y}(t) = e^{-t/5} \begin{pmatrix} 3\cos(\sqrt{3/50}\,t) + \sqrt{6}\sin(\sqrt{3/50}\,t) \\ -(4/5)\sqrt{6}\sin(\sqrt{3/50}\,t) \end{pmatrix}$

(c)

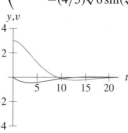

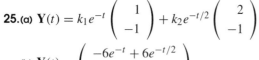

25. (a) $\mathbf{Y}(t) = k_1 e^{-t} \begin{pmatrix} 1 \\ -1 \end{pmatrix} + k_2 e^{-t/2} \begin{pmatrix} 2 \\ -1 \end{pmatrix}$

(b) $\mathbf{Y}(t) = \begin{pmatrix} -6e^{-t} + 6e^{-t/2} \\ 6e^{-t} - 3e^{-t/2} \end{pmatrix}$

(c)

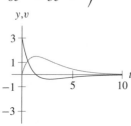

27. (a) $\mathbf{Y}(t) = k_1 \begin{pmatrix} \cos(3t/2) \\ -(3/2)\sin(3t/2) \end{pmatrix}$
$$+ k_2 \begin{pmatrix} \sin(3t/2) \\ (3/2)\cos(3t/2) \end{pmatrix}$$

29. (a) $\lambda_1 = (a + \sqrt{a^2 - 36})/2$, $\lambda_2 = (a - \sqrt{a^2 - 36})/2$.

(b) $a = 0, \pm 6$

(c)

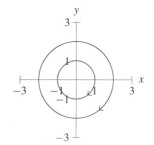

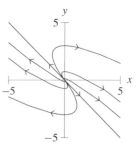

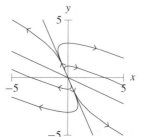

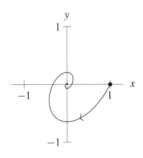

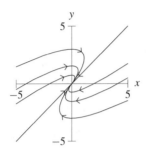

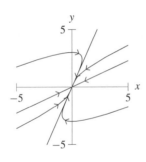

(d) See series of pictures and describe the transitions.

31.(a) The types of phase plane are saddle for $a < -1/2$, line of equilibria for $a = -1/2$, source for $-1/2 < a < 0$ and for $a > 4$, degenerate source for $a = 0, 4$, and spiral source for $0 < a < 4$.

(b) For $a < -1/2$, $x(t)$ and $y(t)$ tend to infinity both as t goes to infinity and as t goes to negative infinity. For $a = -1/2$, $x(t)$ and $y(t)$ are constant and equal 2 and 1 respectively. For $-1/2 > a >$

0, both $x(t)$ and $y(t)$ go to minus infinity as t increases and to 0 when t decreases. For $0 < a < 4$, $x(t)$ and $y(t)$ oscillate as their amplitudes become large as t increases. For $a = 4$, $x(t)$ goes to infinity and $y(t)$ goes to negative infinity as t goes to infinity. For $a > 4$, $x(t)$ decreases to negative infinity and $y(t)$ increases to infinity as t increases to infinity.

33.(a) $\lambda = \pm i$

(b) Calculate the eigenvalues and see when they satisfy the required conditions.

(c) Calculate the eigenvalues and see when they satisfy the required conditions.

35.(a) The second-order differential equation is $y'' + (k_{mf}/m)y' + (k_s/m)y = 0$, and the first-order system is $dy/dt = v$ and $dv/dt = -(k_s/m)y - (k_{mf}/m)v$.

(b) Modify the paragraph of classification of harmonic oscillator.

37.(a) Solution with the initial condition $(200, 300)$ approaches to $(300, 600)$.

(b) x and y tend to 1400 and 2800 respectively as t increases.

(c) Both populations explode as t increases.

(d) Both populations explode as t increases.

39. $\begin{pmatrix} I \\ r \end{pmatrix} = \frac{1}{s\beta + k\alpha} \begin{pmatrix} -\beta((1-s)T - G) + \alpha M \\ -k((1-s)T - G) - sM \end{pmatrix}$

41.(a) I and r can be adjusted in either direction by adjusting these parameters appropriately

(b) I and r can be adjusted in either direction by adjusting these parameters appropriately

(c) Changing either variable does not accomplish the decrease in r and the increase in I. Do nothing.

(d) Increase in M does the job.

Hints and Answers for Section 3.7

1.(a) Differentiate $\mathbf{Y}_2$ coordinate-wise to see if the system of differential equations is satisfied.

(b) Differentiate $\mathbf{Y}_3$ coordinate-wise to see if the system of differential equations is satisfied.

3.(a) Not linearly independent.

(b) Linearly independent.

(c) Linearly independent.

(d) Linearly independent.

5.(a) $\lambda = \pm 1, -5$

(b) The system decouples into a two-dimensional system in the xy-plane and a one-dimensional system in z.

(c) Hint: sink on the z-axis, saddle in the xy-plane

(d) Combine the pictures from part c.

7.(a) $\lambda = 1, 3$

(b) The system decouples into a two-dimensional system in the yz-plane and a one-dimensional system in x.

(c) *Hint*: Source on x-axis, source on yz-plane

(d) Combine the pictures from part c.

9. From $p(a + ib) = 0$, show that the real and negative parts of $p(a - ib)$ are both zeros.

11.(a) $\lambda = -2, -1$

(b) Hint: source on the z-axis, repeated eigenvalue sink on the xy-plane

(c) Combine the two pictures in part b.

13.(a) $\lambda = 0, -5$

(b) *Hint*: z-axis is all equilibrium points, in the xy-plane there is a line of equilibrium points and all solutions approach this line

(c) Combine the pictures in part b (three-dimensional phase space contains a plane of equilibrium points).

15.(a) $\lambda = 0$

(b) $(1, 0, 0)$

(c) z is constant over time, for some constant z_0. For $z_0 = 0$, y is constant and solution curves are the straight lines parallel to x-axis. For $y_0 > 0$, x is increasing and, for $y_0 < 0$, x is decreasing over time. For $z_0 \neq 0$, x is quadratic in y. Therefore, solution curves with $z = z_0$ initially, stay on the plane $z = z_0$ and trace a parabola.

17.(a) The eigenvalue is -1.

(b) The other eigenvalues are $-2 \pm i$.

(c) Spiral sink.

(d) Hint: find the plane containing the real and imaginary parts of an eigenvector for the complex eigenvalue, this is the plane where solutions spiral, on the line through the origin and $(1, 1, 0)$ solutions tend directly toward the origin

19.(a) Glen's profit helps Paul and Bob's profits.

(b) Whether Paul and Bob are making profits or not is irrelevant to Glen's profit.

21.(a) Hint: when $z = 0$, $dz/dt = 0$ so this is a two-dimensional problem

(b) Same hint as part a.

(c) Glen's profit stays constant at 0. Paul and Bob tend toward the break-even point (always staying equal)

Hints and Answers for Section 4.1

1. Systems (i) and (iii) have the same "local picture" near $(0, 0)$.

3.(a) The linearized system is
$dx/dt = -2x + y, dy/dt = -y.$

(b) Sink.

(c)

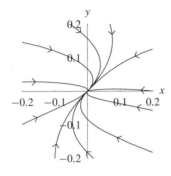

5.(a) Solve $-x = 0, -4x^3 + y = 0.$

(b) $dx/dt = -x, dy/dt = y.$

(c) Saddle.

7.(a) Equilibria $(0, 0), (0, 100), (150, 0), (30, 40)$ and they are a source, sink, sink, and saddle respectively.

(b)

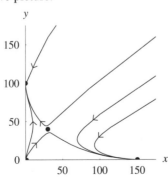

(c) See above picture.

(d)

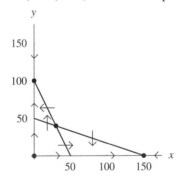

9.(a) Equilibria $(0, 0), (0, 25), (100, 0), (75, 12.5)$ and they are in order source, saddle, saddle, sink.

(b)

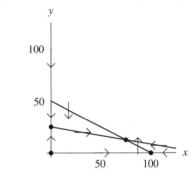

(c) See above picture.

(d)

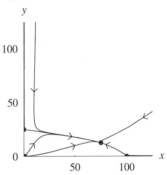

11.(a) Equilibria $(0, 0), (0, 50), (40, 0)$ and they are in order source, sink, saddle.

(b)

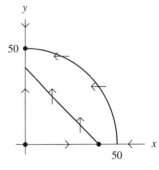

(c) See above picture.

(d)

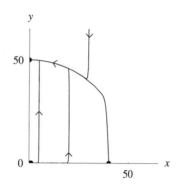

(b)

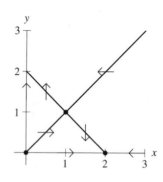

(c) See above picture.

(d)

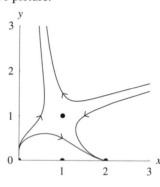

13.(a) Equilibria $(0, 0)$, $(0, 50)$, $(60, 0)$, $(30, 40)$, $(234/5, 88/5)$ and they are in order source, saddle, sink, sink, saddle.

(b)

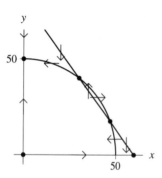

(c) See above picture.

(d)

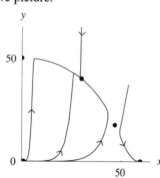

15.(a) Equilibria $(0, 0)$, $(1, 1)$, $(2, 0)$, $(0, 0)$ has an eigenvalue equal to zero (so see phase plane to determine local behavior), $(1, 1)$ is a saddle and $(2, 0)$ is a sink.

17.(a) Equilibria $(0, 0)$, $(0, 1)$, $(-1, 2)$, $(2, 2)$ and they are in order spiral source, saddle, source, sink.

(b)

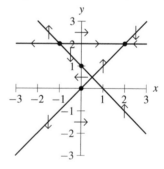

(c) See above picture.

(d)

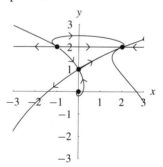

19.(a) Equilibria $(0, 0)$, $(0, 1)$, $(2, -2)$, $(3, -2)$ and they are in order saddle, spiral sink, source, saddle.

(b)

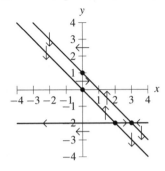

(c) See above picture.

(d)

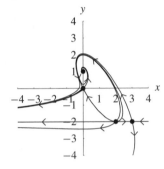

21.(a) $dx/dt = 0$, $dy/dt = -y$

(b) Eigenvalues 0, -1 with eigenvectors $(1, 0)$, $(0, 1)$.

(c) $dx/dt = 0$, $dy/dt = y$

(d) Eigenvalues 0, 1 with eigenvectors $(0, 1)$, $(1, 0)$.

(e) The equilibrium point $(0, 0)$ is a sink and the equilibrium point $(0, 1)$ looks like a saddle.

(f) The equation for dx/dt is all "higher order."

23.(a) *Hint*: When $a = -1$ the dx/dt term is always positive so all solutions move to the right.

(b) *Hint*: When $a = 0$ first draw the phase line for $dx/dt = x^2$.

(c) *Hint*: When $a = 1$ first draw the phase line for $dx/dt = x^2 - 1$.

25.(a) *Hint*: When $a \neq 0$ there are no equilibrium points, when $a = 0$ the curve $y = x^2$ is all equilibrium points.

(b) $a = 0$.

(c) *Hint*: When $a < 0$ all solutions move up, when $a > 0$ all solutions move down.

27.(a) *Hint*: $(0, 0)$ for all values of a, $(\pm 1/\sqrt{a}, \pm 1/\sqrt{a})$ when $a > 0$.

(b) $a = 0$

(c) *Hint*: y nullcline is always $y = x$, examine how the x nullcline changes as a is changed.

29.(a) *Hint*: Equilibrium points $x = \pm\sqrt{a/2}$, $y = -a/2$ when $a > 0$.

(b) $a = 0$.

(c) *Hint*: The y nullcline is $y = -x^2$ for all values of a, examine how the x nullcline changes as x is varied.

31.(a) If X and Y are not present on the island, then neither immigrate to the island.

(b) They are large positive.

(c) They are negative.

(d) Source or saddle.

(e) *Hint*: See Chapter 3.

33.(a) $\partial f/\partial x(0, 0)$ is big, $\partial g/\partial y(0, 0)$ is small.

(b) $\partial f/\partial y(0, 0)$ is large negative and $\partial g/\partial x(0, 0) = 0$.

(c) Source.

(d) *Hint*: See Chapter 3.

Hints and Answers for Section 4.2

1. Hints:

(a) compute $\partial H/\partial x$, $\partial H/\partial y$ and compare

(b) $(0, 0)$ is a local minimum, $(\pm 1, 0)$ are saddles

(c) $(0, 0)$ is a center, $(\pm 1, 0)$ are saddles that are connected by separatrix orbits

3. *Hints:*

(a) Compute $\partial H/\partial x$ and $\partial H/\partial y$ and compare.

(b) The equilibrium points are on the lines $y = \pm x/2$ and y must equal $\pm\pi/2$, $\pm 3\pi/2$, ...

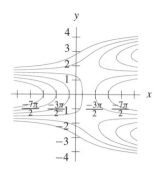

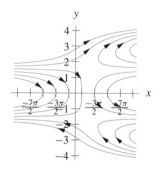

(c)

5. No, longer.

7.(a) Multiply by $\sqrt{2}$.

  (b) $\pi/\sqrt{gl}$.

9. Yes, $H(x, y) = xy - y^3$.

11. Not Hamiltonian.

13. (b) and (c) can not be Hamiltonian.

15.(a) It is not Hamiltonian.

  (b)

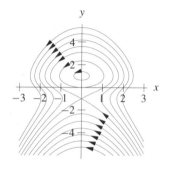

  (c) Yes.

  (d) $H(x, y) = -y + 2\arctan y - x^2/2$

  (e) Changing the length of the vectors doesn't change the direction field, so the phase planes of the two systems are the same.

17. If $y > 2$, $H(x, y) =$
$-x^2/2 + 2y + y^2/2 + 3\ln(y - 2)$; if $y < 2$,
$H(x, y) = -x^2/2 + 2y + y^2/2 + 3\ln(2 - y)$.

19. *Hint*: For $a \neq 0$ the system has two equilibrium points, but for $a = 0$, it has only one.

Hints and Answers for Section 4.3

1.(a) Let $(x(t), y(t))$ be a solution and differentiate $L(x(t), y(t))$ with respect to t.

  (b) Hint: The level sets are circles.

  (c) All solutions tend toward the origin as t increases.

3.(a) The eigenvalues are $-0.05 \pm i\sqrt{15.99}/2$.

  (b) Let $(x(t), y(t))$ be the solution with $x(0) = -1$, $y(0) = 1$ and compute the derivative with respect to t of $L(x(t), y(t))$ at $t = 0$.

  (c) Let $(x(t), y(t))$ be any solution and compute the derivative with respect to t of $K(x(t), y(t))$.

5. It runs fast.

7. $m > k_d\sqrt{l}/(2\sqrt{g})$.

9. *Hint*: The amplitude of the solutions near $(0, 0)$ decreases like $e^{-tk_d/(2m)}$.

11.(a) $(0, 0)$ is a spiral source, $(\pi, 0)$ is a saddle.

  (b) *Hint*: : It looks like the damped pendulum except solutions spiral outward.

  (c) Oscillate with increasing amplitude.

13.(a) $dx/dt = 2x$, $dy/dt = -2y$.

  (b) Saddle.

  (c) *Hint*: It is a saddle surface.

  (d) *Hint*: The axes are the lines of eigenvectors.

15.(a) The eigenvalues are 1 and -1.

  (b) The eigenvalues are -2 and -1.

  (c) *Hint*: $(0, 1)$ is an eigenvector for eigenvalue -1.

  (d) The Jacobian matrices are the same.

17.(a)
$$\frac{dx}{dt} = 2x - x^3 - 6xy^2$$
$$\frac{dy}{dt} = 2y - y^3 - 6x^2y.$$

(b)

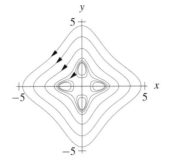

(c) Four dead fish.

(d)

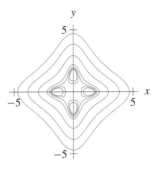

(e) S is negative if x or y is too big.

19.(a) *Hint*: Mixed partials are equal.

(b) $\partial f/\partial y \neq \partial g/\partial y$.

21. Let $(y(t), v(t))$ be a solution of the second system and compute $d(H(y(t), v(t))/dt$.

Hints and Answers for Section 4.4

1.(a)
$$\begin{pmatrix} -1 & -1 & 0 \\ 0 & 2 & 0 \\ 0 & 0 & 1 \end{pmatrix}$$

(b) Eigenvalues 1, 2, -1 and eigenvectors $(0, 0, 1)$, $(1, -3, 0)$ and $(1, 0, 0)$ respectively.

(c) Saddle.

(d) *Hint*: The linearized system decouples.

(e) Solutions with $y = z = 0$ move toward the equilibrium point, all others move away.

3.(a)
$$\begin{pmatrix} 1 & 0 & 0 \\ 0 & 0 & 0 \\ 0 & 1 & -1 \end{pmatrix}$$

(b) The eigenvalues are 1, 0 and -1 and eigenvectors are $(1, 0, 0)$, $(0, 1, 1)$ and $(0, 0, 1)$ respectively.

(c) The linearized system has a line of equilibrium points so it does not have a name.

(d) *Hint*: The x and y, z equations decouple.

(e) Solutions with $x = 0$ move toward an equilibrium point, solutions with $x \neq 0$ move in the direction of increasing x.

5.(a) *Hint*: Set $y = 0$ and look at the system of equations, then try $x = 0$ and $z = 0$.

(b) *Hint*: What is the connection between wolves and trees if moose are extinct?

7.(a) $(0, 0, 0)$, $(1, 0, 0)$, $(0, 1, 0)$, $(0, 0, 1)$, $(1, 0, 1)$, and $((\gamma + 1)(2\gamma + 1), \gamma/(2\gamma + 1), (2\gamma + 2)/(2\gamma + 1))$.

(b) As γ decreases, x_0 increases, y_0 decreases and z_0 increases.

(c) *Hint*: compare how γ comes into the equation for dz/dt with the system in the section.

Hints and Answers for Section 5.1

1.(a) *Hint*: Substitute $y_1(t) + y_p(t)$ into the nonhomogeneous equation.

(b) *Hint*: You need to show that $ky_1(t) + y_p(t)$ is a solution for every k and that every initial value can be obtained by adjusting k.

3. General solution $y(t) = ke^{-5t} + 2e^{-2t}/3$
Particular solution $y(t) = 4e^{-5t}/3 + 2e^{-2t}/3$.

5. General solution $y(t) = ke^{2t} - e^{-2t}/2$
Particular solution $y(t) = -3e^{2t}/2 - e^{-2t}/2$

7. General solution:
$y(t) = k_1 \sin(\sqrt{2}t) + k_2 \cos(\sqrt{2}t) + e^{-2t}/6$
Particular solution:
$y(t) = -\sqrt{2}\sin(\sqrt{2}t)/3 + 11\cos(\sqrt{2}t)/6 + e^{-2t}/6$.

9. General solution $y(t) = k_1 e^{-t}(-\cos 2t + 2\sin 2t) + k_2 e^{-t}(2\cos 2t + \sin 2t) + e^t/8$
Particular solution $y(t) = -e^{-t}\cos(2t)/40 - 2e^{-t}\sin(2t)/40 + 37(e^{-t}(2\cos 2t + \sin 2t)/40 + e^t/8$

11. General solution $y(t) = k_1 + k_2 e^t + 2te^t$
Particular solution $y(t) = -1 - e^t + 2te^t$

13. General solution $y(t) = k_1 e^{-3t} + k_2 e^{-2t} + e^{-t}$

15. General solution $y(t) = k_1 e^{-t} + k_2 te^{-t} + 3t^2 e^{-t}/2$

17. General solution is $y(t) = k_1 \sin(2t) + k_2 \cos(2t) + 3t/4 + 1/2$

19. Substitute $y_1 + y_2$ into $d^2y/dt^2 + p\,dy/dt + qy$ and use the facts that y_1 and y_2 are solutions of $y'' + py' + qy = g(t)$ and $y'' + py' + qy = h(t)$ respectively.

21. General solution $y(t) = k_1 \sin 2t + k_2 \cos 2t + e^{-2t}/8 + t^2 - 1/2$

23. General solution $y(t) = k_1 \sin(\sqrt{2}t) + k_2 \cos(\sqrt{2}t) + t + 5e^{-3t}/11$

25. Solution is $y(t) = 83\sin(2t)/80 + 9\cos(2t)/5 + e^{-t}/5 + 5t^3/4 - 15t/8$

27.(a) $v(t) = k_1 e^{-t}(-\cos\sqrt{5}t + \sqrt{5}\sin\sqrt{5}t) + k_2 e^{-t}(\sqrt{5}\cos\sqrt{5}t + \sin\sqrt{5}t) + 32e^{-t/4}/89$

(b) $v(t) = k_1 e^{-t}(-\cos\sqrt{5}t + \sqrt{5}\sin\sqrt{5}t) + k_2 e^{-t}(\sqrt{5}\cos\sqrt{5}t + \sin\sqrt{5}t) + 32e^{-t/4}/89 + 2/3$

(c) For (a), as t increases, the solution oscillates and approaches zero. For (b), as t increases, the solution oscillates and approaches $2/3$.

Hints and Answers for Section 5.2

1.(a) $y(t) = 6\cos(2t)/5 - \cos(3t)/5$

(b) $|y(t)| \le 7/5$

3.(a) $y(t) = 15e^{-2t}/16 + e^{2t}/16 - \sin(2t)/8$

(b) $|y(t)|$ is unbounded as t goes to plus or minus infinity.

5. Particular solution $y(t) = 4\sqrt{3}e^{-t}\sin(\sqrt{3}t)/15 + e^{-t}\cos(\sqrt{3}t) + \sin(2t)/10$. As t goes to infinity, the maximum of $|y(t)|$ become slightly larger than $1/10$. As t goes to minus infinity, $|y(t)|$ blows up.

7. *Hint*: Look at the form of the general solution of the forced equation and use the fact that damping is present to show that the terms coming from the unforced equation tend to zero as t increases.

9. The beat of Hey Jude is the same (or very close to) the natural frequency of the swaying motion of the stadium.

11. $\phi = \arctan(-b/a)$ and $k = a/\cos(\phi) = -b/\sin(\phi)$

Hints and Answers for Section 5.3

1.(a) The singer has to sing a lower note to break the glass.

 (b) It will become harder to break the glass half-filled with water.

3. 1/180 miles apart

5.(a) When the velocity is negative, H is decreasing and when the velocity is positive, H is increasing.

 (b) After the long time, the system becomes almost harmonic oscillator with the frequency $\sqrt{q}/2\pi$.

 (c) The solution of the forced system tends to the solution of the unforced system.

7.(a) For the first system,

$$\frac{dH}{dt} = -2yv + 0.2v \cos 2t,$$

and for the second system,

$$\frac{dH}{dt} = -2yv + 0.2vt.$$

 (b) Both grow at the same rate.

Hints and Answers for Section 5.4

1.(a) β is increased for this modification.

 (b) The system looks more like a linear system.

3.(a) γ is increased.

 (b) The system looks less like a linear system.

5. $\dfrac{d^2 y}{dt^2} + \dfrac{\epsilon}{m}\dfrac{dy}{dt} - \dfrac{pa^2}{m}y = -g$

7. $m\dfrac{d^2 y}{dt^2} + \epsilon\dfrac{dy}{dt} = -mg + \begin{cases} pa^2 y, & \text{if } y \geq 0; \\ 0, & \text{if } y < 0. \end{cases}$

Hints and Answers for Section 5.5

1. Graph (ii).

3. Graph (i).

5. Graph (iii).

7. Graph (iv).

9.(a) $y(t) = -\sqrt{3}\sin(\sqrt{3}t)/30 + \cos(\sqrt{3}t) + \sin t/10$

 (b) Only the terms corresponding to the homogeneous solution affect the Poincaré map.

 (c) The same only larger.

 (d) *Hint*: There are no saddles for the (unforced) harmonic oscillator.

Hints and Answers for Section 6.1

1. The orbit is eventually fixed.

3. The orbit tends to infinity.

5. The orbit is periodic of period 2.

7. The orbit is eventually fixed.

9. $x = 1$ is fixed, all other points are periodic with period 2

11. There are no fixed points and no periodic point of period 2.

13. $x = 0$ is the only fixed point and there are no periodic points of period 2.

15. For fixed points, $x = 0, -3$. For periodic points of period 2, $x = (-1 \pm \sqrt{5})/2$.

17. There is one point which is fixed and no periodic points of period 2.

19. There is one fixed point $x = 1$ and every point is a periodic point of period two.

21. There is one fixed point $x = 2$ and no other periodic points.

23. The orbit of any real number is a periodic orbit of period 2, except $x = 2$, which is fixed.

25. It is eventually fixed.

27. It is periodic with period 3.

29. It is eventually fixed.

31. It is periodic with period 5.

33. It is a fixed point.

35. It is a prefixed point.

37. For $c = 1/4$, there is one fixed point. For $c < 1/4$, there are two fixed points. For $c > 1/4$, there are no fixed points.

39. For $c \geq -3/4$, there are no periodic points of period 2, and for $c < -3/4$, there are two periodic points of period 2.

Hints and Answers for Section 6.2

1. The fixed points $x = 0, 3$ are both repelling.

3. $x = 0$ is attracting.

5. $x = 0$ is attracting.

7. $x = 0$ is repelling, and $x = \pm\pi/2$ are attracting.

9. $x = \pm 1$ are neutral.

11. There are no fixed points.

13. The period is 2 and the orbit is attracting.

15. The period is 5 and the cycle is repelling.

17. The period is 2 and the cycle is neutral.

19. The fixed points are $x = \pm 1$ and they are neutral.

21. The fixed point is $x = 0$ and it is repelling.

23. The fixed point is $x = 0$ and it is repelling.

25. The fixed point is attracting.

27. There are infinitely many fixed points and they are all repelling.

29. All cycles are repelling. Note that $|(T^n)'(x_0)| = 4^n$.

Hints and Answers for Section 6.3

1. The bifurcation is a tangent bifurcation. For small enough $-\alpha$, the fixed point $x = \sqrt{-\alpha}$ is repelling and the other fixed point $x = -\sqrt{-\alpha}$ is attracting. For $\alpha = 0$, $x = 0$ is neutral.

3. This is a pitchfork bifurcation. For α slightly smaller than or equal to 1, the fixed point is attracting. For $\alpha > 1$, two more fixed points appear and they are attracting. The origin becomes a repelling fixed point.

5. This is a tangent bifurcation. There is one neutral fixed point for $\alpha = -1/4$, and there are two fixed points for α slightly larger than $-1/4$. One is repelling and the other is attracting.

7. The bifurcation is none of the above. If $\alpha = 0$, everything is maps to zero, which is a fixed point. If $0 < |\alpha| < 1$, then the origin is an attracting fixed point and $x = (1 - \alpha)/\alpha$ is a repelling fixed point.

9. If $\mu < 1$, the origin is the only fixed point. At $\mu = 1$ every point satisfying $x \leq 1/2$ is a fixed point. If $\mu > 1$ there are two fixed points.

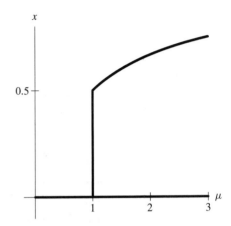

11. Hint: The function $F_c(x)$ is a parabola with minimum at $x = 0$ and $F_c(0) = c$. The function $F_c^2(x)$ is a quartic with local maximum at $x = 0$. Try looking at $F_c^2(x)$ upside down near $x = 0$.

13. The orbit tend to infinity.

15. Hint: L_k^3 undergoes a tangent bifurcation at this value of k.

Hints and Answers for Section 6.4

1. Hint: The histograms look the same but after the first few iterates the time series look very different.

3. $T^n(p/2^n) = 0$

5. Each of the intervals $[i/2^n, (i + 1)/2^{n+1})$ for $i = 0, 1, 2, \ldots, 2^n - 2$ contains a fixed point of T^n.

7. The eventually fixed points have the form $p/2^n$ for any integer p between zero and $2^n - 1$.

9. T^n has $2^n - 1$ fixed points

11. List all possible finite strings of zeros and ones, then concatinate them to form the binary expansion of x_0.

13. Hint: T^n has a fixed point in the interval $[i/2^n, (i + 1)/2^n]$.

Hints and Answers for Section 7.1

1.(a) $y(t) = \sqrt[3]{1 - 3t^2}$

 (b) $y_{20} \approx 0.19334189$

 (c) $e_{20} \approx 0.0066581009$

 (d) $y_{1000} \approx 0.19987415, \ldots, y_{6000} \approx 0.19997904$

 (e) $e_{100} \approx 0.0012709461$, $e_{200} \approx 0.00063198471$, $e_{300} \approx 0.00042055770, \ldots, e_{6000} \approx .000020955836$

 (f) $k \approx 0.126731$

3. To make sure you are on the right track, a few values are $e_{100} \approx 0.044489615$, $e_{200} \approx 0.022446456$, $e_{300} \approx 0.015009699, \ldots, e_{6000} \approx 0.00075483852$. The best fit for the error is approximately $4.47018/n$.

5. To make sure you are on the right track, a few values are $e_{100} \approx 1.0595505$, $e_{200} \approx 0.54084346$, $e_{300} \approx 0.36308636, \ldots, e_{6000} \approx 0.018398709$. The best fit for the error is approximately $106.854/n$.

7.(a) $M_1 = 8t^2 y^3 - 2y^2$, $M_2 = -4ty$

 (b) The estimated error is 0.00039984000 while the real error is approximately 0.00039984006.

 (c) The estimated error is 0.00079744230 while the real error is approximately 0.00079744409.

 (d) The third and fourth estimated errors are 0.0011896358 and 0.0015733200. The third and fourth real errors are 0.0011896455 and 0.0015733534.

 (e)

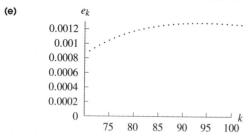

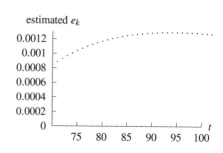

9. We cannot compute the actual error e_k because we do not have a closed-form solution for the differential equation.

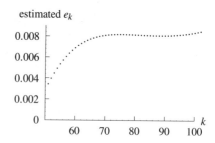

11. *Hint:* For parts (a)–(c), refer to Figure 7.4. For (d) and (e), use the inequality in (c). For (f), recall geometric sums from calculus. Parts (g)–(i) are just algebra. The expression $M_1(e^{M_2(t_n - t_0)} - 1)/(2M_2)$ is a constant in part (i) once the function $f(t, y)$ and the interval of time are given; so $C = M_1/(2M_2)(e^{M_2(t_n - t_0)} - 1)$. The bounds do not depend on n.

Hints and Answers for Section 7.2

1. You should get the points: $(0, 3)$, $(0.5, 8.25)$, $(1, 21.375)$, $(1.5, 54.1875)$, $(2, 136.219)$.

3. You should get the points: $(0, 1.4)$, $(0.5, 1.4976)$, $(1, 1.65604)$, $(1.5, 1.95339)$, $(2, 2.67616)$.

5. You should get the points: $(0, 0)$, $(1, 1.5)$, $(2, -3.65625)$, $(3, -259.965)$, $(4, -2.33657 \times 10^9)$, $(5, -1.49035 \times 10^{37})$

7. You should get the points: $(0, 2)$, $(0.5, 3.13301)$, $(1, 4.01452)$, $(1.5, 4.80396)$, $(2, 5.54124)$

9.(a) The analytic solution is $y(t) = 1 - e^{-t}$.

 (b) $y_4 \approx 0.627471$, $e_4 \approx .00465012$

 (c) We know that $e_n = K/n^2$ for some constant K.

Since we calculated $e_4 \approx .00465012$, we can estimate $K = 0.0744019$. We can now set e_n in $e_n = K/n^2$ to be 0.0001 and solve for $n = 27.27$. Since n must be an integer, we expect $e_{28} < .0001$. In fact, $e_{28} = 0.0000808718$.

11.(a) The analytic solution is $y(t) = 1/(t+2)$.

(b) $y_4 \approx 0.252281$, $e_4 \approx .00228017$

(c) We know that $e_n = K/n^2$ for some constant K. Since we calculated $e_4 \approx .00228017$, we can estimate $K = 0.0744019$. We can now set e_n in $e_n = K/n^2$ to be 0.0001 and solve for $n = 19.10$. Since n must be an integer, we expect $e_{20} < .0001$. In fact $e_{20} = 0.0000803174$.

13.(a) We know that $e_n = K/n^2$ for some constant K. Since we calculated $e_{20} \approx .000695$, we can estimate $K = 0.278$. We can now set e_n in $e_n = K/n^2$ to be 0.0001 and solve for $n = 52.72$. Since n must be an integer, we expect $e_{53} < .0001$.

(b) $y_{53} = 0.200095$

(c) $e_{53} = 0.000095$

15.

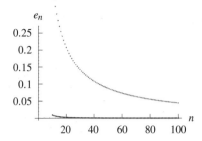

17.

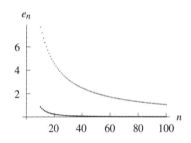

Hints and Answers for Section 7.3

1. You should get the points: $(0, 3)$, $(0.5, 8.97917)$, $(1, 25.1727)$, $(1.5, 69.0303)$, $(2, 187.811)$.

3. You should get the points: $(0, 0)$, $(0.5, 1.8229)$, $(1, 2.70058)$, $(1.5, 2.90368)$, $(2, 2.96803)$, $(2.5, 2.98935)$, $(3, 2.99645)$, $(3.5, 2.99882)$, $(4, 2.99961)$, $(4.5, 2.99987)$, $(5, 2.99996)$.

5. You should get the points: $(1, 2)$, $(1.5, 3.10456)$, $(2, 3.98546)$, $(2.5, 4.77554)$, $(3, 5.51352)$.

7.(a) In pairs of the form (R, F), you should get the points (in order): $(1, 1)$, $(1.50412, 1.91806)$, $(0.641301, 2.48192)$, $(0.416812, 1.62154)$, $(0.636774, 1.08434)$, $(1.27734, 1.25344)$

(b)

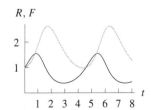

(c)

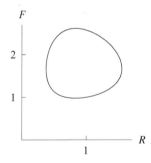

The $R(t)$-graph is black; the $F(t)$-graph is gray.

Hints and Answers for Section 7.4

1. To make sure you are on the right track:
   $y_2 \approx 0.941885$, $y_4 \approx 0.924849$, $y_8 \approx 0.94101$.

3. To make sure you are on the right track:
   $y_2 \approx 0.262027$, $y_4 \approx 0.261532$, $y_8 \approx 0.261506$.

Hints and Answers for Section 8.1

1.(a) $\mathcal{L}[f] = 3/s$
  (b) $\mathcal{L}[g] = 1/s^2$
  (c) $\mathcal{L}[h] = -10/s^3$
  (d) $\mathcal{L}[k] = 15!/s^{16}$

3. $\mathcal{L}[1 - u_3(t)] = (1 - e^{-3s})/s$

5.(a) $\mathcal{L}[r_a(t)] = ke^{-as}/s^2$

  (b)

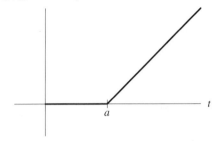

7. $\mathcal{L}[f] = e^{-2s}(s \cos 4 - 2 \sin 4)/(s^2 + 4)$

9.(a) $\mathcal{L}[dy/dt] = s\mathcal{L}[y] - y(0)$ and
    $\mathcal{L}[-y + e^{-2t}] = -\mathcal{L}[y] + 1/(s + 2)$

(b) $\mathcal{L}[y] = (2s + 5)/((s + 1)(s + 2))$
  (c) $y(t) = 3e^{-t} - e^{-2t}$

11.(a) $\mathcal{L}[dy/dt + 7y] = s\mathcal{L}[y] - y(0) + 7\mathcal{L}[y]$
    and $\mathcal{L}[u_2(t)] = e^{-2s}/s$
  (b) $\mathcal{L}[y] = 3/(7 + s) + e^{-2s}/s(s + 7)$
  (c) $y(t) = 3e^{-7t} + u_2(t)(1 - e^{-7(t-2)})/7$

13.(a) $\mathcal{L}[dy/dt] = s\mathcal{L}[y] - y(0)$ and
    $\mathcal{L}[-y + t^2] = -\mathcal{L}[y] + 2/s^3$
  (b) $\mathcal{L}[y] = (2 + s^3)/(s^3(s + 1))$
  (c) $y(t) = t^2 - 2t + 2 - e^{-t}$

15.(a) $\mathcal{L}[dy/dt] = s\mathcal{L}[y] - y(0)$ and
    $\mathcal{L}[-y + u_2(t)e^{-2(t-2)}] = -\mathcal{L}[y] + e^{-2s}/(s + 2)$
  (b) $\mathcal{L}[y] = (s + 2 + e^{-2s})/((s + 1)(s + 2))$
  (c) $y(t) = e^{-t} + u_2(t)\left(e^{-(t-2)} - e^{-2(t-2)}\right)$

17. $(1 - e^{-as})/as^2$

Hints and Answers for Section 8.2

1.(a) $\mathcal{L}[e^{at} \cos(\omega t)] = (s - a)/(\omega^2 + (s - a)^2)$
  (b) $\mathcal{L}[e^{at} \sin(\omega t)] = \omega/(\omega^2 + (s - a)^2)$

3. $\mathcal{L}[f] = (s^2 - \omega^2)/(\omega^2 + s^2)^2$

5. $\mathcal{L}[f] = 2/(s - a)^3$

7.(a) $\mathcal{L}[d^2y/st^2 + 2dy/dt + 5y] =$
    $(s^2 + 2s + 5)\mathcal{L}[y] - 2(s + 2)$
    $\mathcal{L}[\sin(3t)] = 3/(s^2 + 9)$
  (b) $\mathcal{L}[y] =$
    $2(s+2)/(s^2+2s+5) + 3/((s^2+9)(s^2+2s+5))$
  (c) $y(t) = (55/26)e^{-t} \cos(2t) - (3/26) \cos(3t) + (61/52)e^{-t} \sin(2t) - (1/13) \sin(3t)$

9.(a) $\mathcal{L}[d^2y/dt^2 + 4y] = (s^2 + 4)\mathcal{L}[y] + 2s$
    $\mathcal{L}[\cos(2t)] = s/(s^2 + 4)$
  (b) $\mathcal{L}[y] = (-2s^3 - 7s)/(s^4 + 8s^2 + 16)$
  (c) $y(t) = -2 \cos(2t) + (t/4) \sin(2t)$

11.(a) $\mathcal{L}[d^2y/dt^2 + 3y] = (s^2 + 3)\mathcal{L}[y] - 2s$
    $\mathcal{L}[w] = (1 - e^{-s})/s^2$
  (b) $\mathcal{L}[y] = 2s/(s^2 + 3) + (1 - e^{-s})/(s^2(s^2 + 3))$

(c) $y(t) = 2 \cos(\sqrt{3}t) + (t/3) - \sin(\sqrt{3}t)/(3\sqrt{3}) + u_1(t)((1/3) - (t/3) + \sin(\sqrt{3}(t - 1))/(3\sqrt{3}))$

13.(a) $y(t) = 2 \cos(3t) + t \sin(3t)/60$
  (b) *Hint*: The forcing frequency is the natural frequency and there is no damping.

15. *Hint*: Write $\mathcal{L}[f] =$

$$\int_0^\infty f(t)e^{-st}\,dt = \int_0^T f(t)e^{-st}\,dt + \int_T^\infty f(t)e^{-st}\,dt,$$

then do the *u*-substitution $u = t - T$ on the second integral on the right.

17. $\mathcal{L}[z] = (1/s^2) - e^{-s}/(s(1 - s^2))$

19.(a) $\mathcal{L}[y] = (s + 20)/(s^2 + 20s + 200) + 1/(s^2(s^2 + 20s + 200)) - e^{-s}/(s(1 - e^{-s})(s^2 + 20s + 200))$
  (b) *Hint*: Compute the eigenvalues of the unforced oscillator and note the speed at which oscillations die out.

Hints and Answers for Section 8.3

1. *Hint*: Remember L'Hôpital's rule, differentiate numerator and denominator with respect to Δt.

3. $y(t) = u_1(t)(10/\sqrt{11})e^{-(t-1)/2} \sin(\sqrt{11}(t-1)/2)$

5. $y(t) = 2e^{-t}(\cos t + \sin t) - 2e^{-(t-2)}u_2(t)\sin(t-2)$

7.(a) $\mathcal{L}[\delta_a] = e^{-as}$
   $s\mathcal{L}[u_a] - u_a(0) = s(e^{-as}/s) - 0 = e^{-as}$

(b) This relationship suggests that $du_a/dt = \delta_a$.

(c) Approximate $u_a(t)$ with piecewise linear continuous functions and take their derivatives.

9.(a) $\mathcal{L}[y] = 1/(s^2 + 2) \sum_{n=1}^{\infty} e^{-ns}$

(b) $y(t) = (1/\sqrt{2}) \sum_{n=1}^{\infty} u_n(t) \sin(\sqrt{2}(t - n))$

(c) Whenever dealing with infinite sums we have to be nervous about convergence. Why does the infinite series for $y(t)$ converge?

Hints and Answers for Section 8.4

1.(a) $s = \pm\sqrt{3}i$ and $s = \pm 2i$

(b) $\mathcal{L}[y] = 1/(s^2 + 3) - 1/(s^2 + 4)$

(c) $y(t) = \sin(\sqrt{3}t)/\sqrt{3} - \sin(2t)/2$

3.(a) $\mathcal{L}[y] = (2s - 2)/(s^2 + 2s + 2) + 4/((s^2 + 4s + 20)(s^2 + 2s + 2))$

(b) Poles are at $s = -2 \pm 4i$ and at $s = -1 \pm i$.

(c) Since the poles have negative real part, solutions decrease to zero (at a rate of e^{-t}). Since the poles are complex, the solutions oscillate (the oscillations have periods 2π and $\pi/2$).

5.(a) $\mathcal{L}[y] = (s + 5)/(s^2 + s + 3) + \cos(2)/((s^2 + 1)(s^2 + s + 3)) - \sin(2)s/((s^2 + 1)(s^2 + s + 1))$

(b) $s = \pm i$ and $s = -1/2 \pm \sqrt{11}i/2$

(c) Solution decreases to zero (see text).

7. We cannot say that the solution remains bounded. In case of resonance, the solution may tend to infinity (note that the Laplace transforms of t and t^2 have poles at zero).

9.(a) $\mathcal{L}[y] = 2/(s^2 + 4)^2$

(b) Double roots at $s = \pm 2i$.

(c) *Hint*: Remember that undamped resonant harmonic oscillators with natural period $2\pi/2$ have terms of the form $t \sin(2t)$ and $t \cos(2t)$ in their solutions.

11.(a) $\mathcal{L}[y] = (s + 20)/(s^2 + 20s + 200) + 1/(s^2(s^2 + 20s + 200)) - e^{-s}/(s(1 - e^{-s})(s^2 + 20s + 200))$

(b) $s = -10 \pm 10i$ and $s = 0$

(c) Each push of the forcing pushes the solution up a little, but this decays like e^{-10t}, hence the solution remains bounded.

INDEX